Modeling of Physiological Flows

MS&A

Davide Ambrosi
Alfio Quarteroni
Gianluigi Rozza (Eds.)

Modeling of Physiological Flows

 Springer

Davide Ambrosi
MOX – Department of Mathematics
Politecnico di Milano, Italy

Gianluigi Rozza
Mathematics Institute of Computational
Science and Engineering
École Polytechnique Fédérale de
Lausanne, Switzerland

Alfio Quarteroni
MOX – Department of Mathematics
Politecnico di Milano, Italy
and
Mathematics Institute of Computational
Science and Engineering
École Polytechnique Fédérale de
Lausanne, Switzerland

MS&A – Modeling, Simulation & Applications
ISSN print version: 2037-5255   ISSN electronic version: 2037-5263

ISBN 978-88-470-1934-8   ISBN 978-88-470-1935-5 (eBook)
DOI 10.1007/978-88-470-1935-5

Library of Congress Control Number: 2011920260

Springer Milan Dordrecht Heidelberg London New York

© Springer-Verlag Italia 2012

Cover design: Beatrice Ꙡ, Milan
Cover image: Internal view of blood flow in the aortic arch. Reproduced with permission from
J. Bonnemain and P. Crosetto, EPFL-CMCS

Typesetting with LaTeX: PTP-Berlin, Protago TeX-Production GmbH, Germany (www.ptp-berlin.eu)

Springer-Verlag Italia S.r.l., Via Decembrio 28, I-20137 Milano
Springer is a part of Springer Science+Business Media (www.springer.com)

Preface

Cardiovascular diseases have a major impact in several countries and a widespread awareness exists that mathematical models and numerical simulations can help to better understand the physiological and pathological processes and their correlations. This book is addressed to graduate students and researchers in the field of life sciences, bioengineering, applied and numerical mathematics and medicine wishing to engage themselves in the challenging field of modeling physiological flows, with a special emphasis on cardiovascular flows.

The expertise required to a researcher wishing to work in this field is vast and multidisciplinary. This book, written by world recognized experts in the field, offers a sound and up-to-date state of the art of the development of mathematical models, numerical simulation codes, pre and post-processing tools.

The present volume is a natural continuation of the book "Cardiovascular Mathematics", MS&A series, published in 2009, and it contains selected invited papers of the Fourth International Symposium on Modeling of Physiological Flows, held in Chia Laguna, Sardinia, Italy on June 2–5, 2010 (http://www.mathcard.eu/mpf2010), sponsored by the European Research Council within the project Mathcard "Mathematical Modeling for the cardiovascular system" (ERC-2008-AdG-227058).

The first chapters provide a broad overview on the field, the focus then narrows on fluid and structural models, and heart electrophysiology. The complexity of the cardiovascular system calls for the integration of different models, operating at different levels of complexity and different spatial and temporal scales, and that can flexibly adapt to the need of the research at hand. Then, the last chapters address specific methodological contributions.

Chapter 1 critically reviews some of the common assumptions about the constitutive properties of blood flow and arterial walls, and their potential impact on the computed haemodynamics. Chapter 2 is devoted to the formulation of a simplified one-dimensional time-dependent non-linear mathematical model for blood flow in vessels with discontinuous material properties. Chapter 3 deals with blood flow components and their mutual interactions: in particular it illustrates the mathematical models for the formation and dissolution of blood clots known as hemostasis process. Chapter 4 presents a concise overview of various mathematical and numerical problems raised by the simulation of electrocardiograms (ECGs). A model for the propagation of the electrical activation in the heart and in the torso is proposed. A review of current mathematical and numerical models of the electrical activity in the ventricular myocardium is introduced in Chapter 5. The degenerate reaction-diffusion system called Bidomain model is introduced and interpreted as macroscopic averaging of a cellular model on a periodic assembling of myocytes. This model is coupled with an extracardiac medium and extracardiac potentials, computed from given cardiac sources in order to obtain body surface maps and electrograms. The focus of Chapter 6 is on the damage in the arterial wall model, in terms of experimental studies and computational applications. Structurally motivated multi-mechanism models, which explicitly fibre orientation as well as isotropic damage, are illustrated. A recently developed experimental system, which enables quantitative assessment of the microstructure simultaneous with mechanical loading experiments, is presented with application to the arterial wall, namely cerebral angioplasty. Chapter 7 recalls how experimental observations highlight the importance of altered haemodynamics on arterial function and adaptation. It contains a discussion of mechano-biological models for growth and remodeling of the arterial wall, by describing the intimate interaction between haemodynamics, cell activity, and wall mechanics. When the artery is described as a thin walled structure, and basic adaptations to perturbed pressure and flow, cerebral aneurysms, and vasospasms ca be successfully modelled treating the vascular wall as a membrane.

A general overview of the VPH/Physiome project developing multiscale tools and model databases for computational physiology is given in Chapter 8. Here the aim is to introduce the reader to the most important guidelines concepts in multiscale physiological simulations.

Chapter 9 addresses several phenomena bound to different time, constitutive and geometrical scales in the cardiovascular modeling. More specifically, the problem of integrating various geometrical scales is considered from a kinematical point of view, which amounts to integrate models with different kinematics, and in particular different dimensionality by making use of heterogeneous representations. Chapter 10 deals with the modeling of hematologic disorders associated with major changes in the shape and viscoelastic properties of red blood cells. Such changes can disrupt blood flow and even brain perfusion. A seamless multiscale approach is described, where blood cells and blood flow in the entire arterial tree are represented accurately using physiologically consistent parameters. A computational methodology based on dissipative particle dynamics which models red blood cells as well as whole blood in health and disease is illustrated.

Since degradable materials have found a wide variety of applications in the biomedical field ranging from sutures, local drug delivery, tissue engineering scaffolds, and endovascular stents. Chapter 11 introduces a bottom-up multiscale analysis applied to model the degradation mechanism which takes place in polymers matrices used in stents. The macroscale model is based on diffusion-reaction equations for hydrolytic polymer degradation and erosion, while the microscale model is based on atomistic simulations.

The development of new technologies for acquiring measures and images in order to investigate cardiovascular diseases raises new challenges in scientific computing. In fact, these data can be merged with the numerical simulations for improving the accuracy and reliability of the computational tools. Assimilation of measured data and numerical models is relatively new field in computational haemodynamics. Different approaches are possible for the mathematical setting of this problem. With this aim, Chapter 12 considers a variational approach, based on the minimization of the mismatch between data and numerical results by acting on a suitable set of control variables: in this way a mathematically sound (variational) assimilation of data can significantly improve the reliability of the numerical models, but also provide important features in view of the progressive adoption of numerical tools in medicine.

Finally, Chapter 13 describes efficient algorithms for generating high quality computational tetrahedral meshes for cardiovascular blood flow simulations starting from low quality triangulations, usually obtained from the segmentation of patient specific medical images.

All the contributions of this book have been accurately revised by anonymous reviewers, who are deeply acknowledged. Finally, we would like to gratefully acknowledge Francesca Bonadei and Pierpaolo Riva from Springer Milan for their kind help and precious collaboration in the realization of this book.

Milano and Lausanne, June 2011   *Davide Ambrosi*
Alfio Quarteroni
Gianluigi Rozza

Contents

1

Assumptions in modelling of large artery hemodynamics

David A. Steinman

Abstract. The last decade has seen tremendous growth in the use of computational methods for simulating large artery hemodynamics. As computational models become more sophisticated and their applications more varied, it is worth (re)considering the simplifying assumptions that are traditionally, and often implicitly, made. This chapter reviews some of the common assumptions about the constitutive properties of the arteries and the blood within, and their potential impact on the computed hemodynamics. It will be seen, for example, that the assumption of rigid walls, while reasonable and expedient, may be questionable for extensive domains and/or heterogeneities in the arterial wall structure and properties, and that this has implications for the way in which prevailing flow conditions are imposed. Simplifying assumptions about the properties of blood are undoubtedly necessary, but the Newtonian/non-Newtonian dichotomy may prove too simplistic, especially as simulations move from laminar flows to unstable and turbulent flows. Rather than dwelling upon the potential limitations arising from these assumptions, this chapter attempts to highlight some of the potentially interesting research opportunities that may arise in investigating and overcoming them.

1.1 Overview

The vascular system is a complex network of branching vessels spanning length scales from meters to microns. Key features of the normal arterial system include compliant artery walls, complex branching or tortuous artery lumens, and non-Newtonian blood rheology. As Fig. 1.1 illustrates, and as outlined below, the relative importance of these features depends upon the hemodynamic scales of interest.

David A. Steinman (✉)
Biomedical Simulation Laboratory, Department of Mechanical & Industrial Engineering, University of Toronto, Toronto, ON, Canada
email: steinman@mie.utoronto.ca

Ambrosi D., Quarteroni A., Rozza G. (Eds.): Modeling of Physiological Flows.
DOI 10.1007/978-88-470-1935-5_1, © Springer-Verlag Italia 2012

	Macrocirculation	Individual Arteries	Microcirculation
	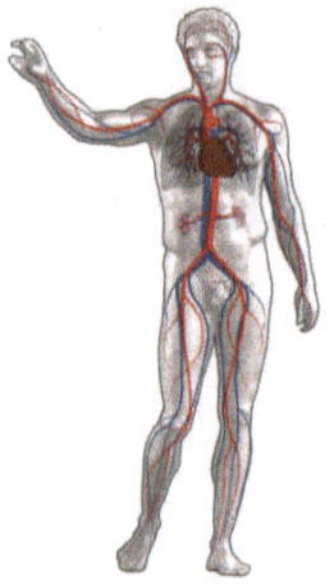		
Compliance	++	??	--
Geometry	--	++	--
Rheology	--	??	++

Fig. 1.1. Representation of the relative importance of various factors in models of the different hemodynamic scales: ++ indicates primary importance; – indicates secondary or negligible importance; ?? indicates potential or unclear importance (taken from Creative Commons)

At the largest scales, it has been shown reasonable to model the vascular network analogously to an AC electrical circuit [1], where inductance and resistance (i.e., impedance) represent inertia and frictional (viscous) losses of the pulsating blood. Capacitance represents the compliance of the blood vessel walls, which act analogously via the alternating local storage and release of blood down the vessel. This latter feature largely determines the wave propagation phenomena – attenuation, dispersion, reflection, etc. – that can be used to infer the functional state of the vasculature from macroscopic pressure and flow measurements. For estimating vascular impedance, vessel diameters and lengths are required, but otherwise the specific shapes and connections of the vessel are largely immaterial, except perhaps when its effects are modelled via the imposition of correction factors or impedance mismatches.[1] Blood may be modelled as a Newtonian fluid, perhaps with adjustments to the apparent viscosity based on the vessel sizes included in the circuit.

At the smallest scales, in the arterioles and capillaries where vessel diameters approach within one or two orders of magnitude of red blood cell (RBC) diameters, the blood, normally comprising about 40–50 % RBC by volume, may no longer be treated as a homogeneous, constant-viscosity fluid. Well-known phenomena such as the development of cell-free layers near the wall (Fahreaus-Lindqvist effect) or deformation of RBC through capillaries alter not only the apparent viscosity of the fluid, but arterial transport phenomena in general. Here again the specific shape of a vessel is relatively unimportant: simple tubes mimicking the lengths, connections and vessel density per tissue-volume are often adequate domains. As will be discussed later, at these scales it is no longer appropriate to think of blood as a contin-

---

[1] Even then, reflections arising from the microcirculation tend to dominate over the effects of such local impedance mismatches.

uum; however, simple models have been developed to incorporate the effect of this discreteness on the apparent viscosity [2].

Between these extremes, and the focus of this chapter, are individual arteries, usually branches, bends, sacs and constrictions, which are the cause or consequence of focal vascular diseases like atherosclerosis and aneurysms. Owing to cycle-averaged Reynolds numbers[2] on the order of a few hundreds, inviscid or creeping flow approximations are not possible. Owing to the heartbeat, pulsatile (unsteady) flow effects can often not be neglected, expect perhaps for cases having low Womersley numbers[3] or low-amplitude flow rate dynamics.

Decades of research have highlighted the important role of vascular geometry in giving rise to distinct and surprisingly complicated fluid mechanics in these vessels. As a result, the three-dimensional nature of the artery shape is critical to capture. Such models are usually considered in isolation from the rest of the vascular network, or at least those effects are incorporated into the inlet and outlet boundary conditions. Rigid walls and Newtonian rheology are almost always assumed, although the appropriateness of these assumptions remains the subject of ongoing debate.[4]

This chapter will briefly review the rationale and evidence behind these two key assumptions, particularly as they pertain to studies of large artery hemodynamics that have been the subject of the author's investigations over the past two decades. Along the way, important links between these assumptions and the prescription of flow boundary conditions or the nature of turbulent blood flow will also be considered. It must be stated from the outset that this chapter does not aim for comprehensiveness, but rather seeks to inspire the reader to think critically about assumptions in hemodynamic modelling for their own applications, especially those employing image-based or "patient-specific" modelling. It also aims to highlight the challenges, many ongoing, about when and how to relax these assumptions, and some of the surprising opportunities that arise, or have arisen, from their consideration.

1.2 Rigid vs. compliant walls

As anyone who has taken his or her pulse can plainly feel, arteries are distensible or compliant.[5] In other words, they deform in response to the pressure pulse produced by the heart. For a typical large artery, pulse wave velocities (PWV) are on the order of several metres per second. For a typical resting heart rate, this implies a

---

[2] Re is defined as VD/v, where V is the lumen-averaged velocity, D is the lumen diameter, and v is the blood viscosity, usually assumed to be 0.03–0.04 cm²/s.

[3] Wo is defined as $R(\omega/v)^{1/2}$, where R is the lumen radius and ω is the frequency of the heart beat in rad/s. For a typically blood viscosity of 0.035 cm²/s and a heart rate of 60 bpm, large artery Womersley numbers range from around 2 (coronaries) to 20 (aorta).

[4] The questions this author is most often asked after presentations are: "How can you assume the wall is rigid?" and "How can you assume blood is a Newtonian fluid". The former tends to be asked by clinicians, the latter by engineers; I'll not hazard a guess why.

[5] These descriptive terms, as well as "stiffness" are often used interchangeably, but have specific meanings for quantifying arterial wall properties [3].

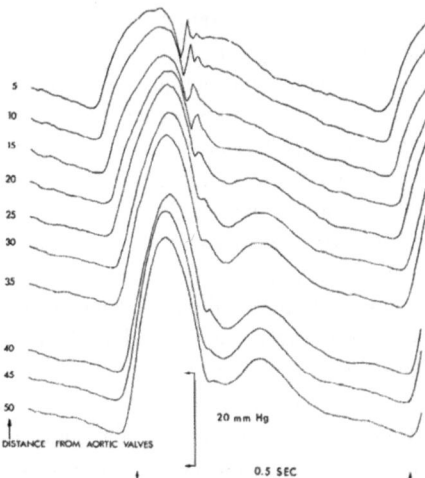

Fig. 1.2. Changes in pressure waveform shape and timing along the aorta [5]. Reproduced with permission from the American Physiological Society

wavelength, λ, of several metres. When looked at "from a distance", any long segment of the vascular tree may harbor an appreciable fraction of a pulse cycle. This is manifested as a temporal shift in the peak of the propagating pulse along the vessel length,[6] and changes in amplitude and shape due to attenuation, dispersion and wave reflection (Fig. 1.2).

For studies of local hemodynamics and their relationship to focal vascular disease, vessel segments of interest are typically of lengths, L, on the order of cm. It then follows that the segment harbors only small fraction of a pressure pulse (i.e., $L/\lambda \ll 1$, the long wavelength approximation), meaning that the entire segment essentially pulses in synchrony. While there may be some wave propagation – as noted above this is the principle of measuring PWV – the effects of any differential compliance are assumed to be relatively minor.

What remains then is the question of whether the "uniform" pulsation is large enough to warrant consideration of the flow effects. Considering that wall shear stress (WSS, often denoted τ_w), the target of many of these studies, is thought to scale with the cube of the lumen radius, at least according to Poiseuille's eponymous law, one can deduce via Taylor expansion that a diameter pulsation of $\pm\Delta D/D$ will result in a WSS variation of approximately $\pm 3\Delta\tau_w/\tau_w$ relative to the nominal rigid-wall values. In other words, for a $\pm 5\%$ deformation, rigid wall simulations should overestimate peak systolic WSS by roughly 15 %; however, when averaged over the cardiac cycle these differences are muted or cancelled out depending on the nature of the presumed pressure pulse and inertial flow (i.e., Womersley number) effects.

---

[6] In fact, PWV is often measured in vivo by measuring this shift at two or more locations spaced a known distance apart [4].

Of course this simplistic perturbation analysis ignores important non-linear effects of wall motion for large artery flows having Reynolds numbers on the order of hundreds. Nevertheless, early studies seemed to confirm these simplistic analyses. For example, the author's early CFD studies of flow in distensible vs. rigid 2D bypass graft models reported a 15–20 % reduction in WSS variations over the cardiac cycle for a ± 4 % distension [6]. Perktold's pioneering CFD studies of a compliant, idealised compliant 3D carotid bifurcation [7] reported 25 % WSS reductions compared to rigid wall simulations, but pointed out that "global structure of the flow and stress patterns remains unchanged", findings later echoed by Zhao et al [8] in their studies of anatomically realistic carotid bifurcations. Those authors also nicely summarized the time-varying nature of these effects: "the largest difference occurs at peak systole with less influence of wall distensibility at diastole. This may be attributed partly to the wall motion which transiently enlarged the cross-sectional area, *causing the instantaneous velocities everywhere in the cross section to decrease in comparison to the rigid model*[7], in accordance with the conservation of mass."

Despite the conclusions of the above and other similar studies, the rigid wall assumption is not necessarily reasonable for all large artery flows. Remembering that it presumes a modest and relatively uniform radial wall motion, the situation is less clear for vessels that undergo large and/or non-uniform motions. Examples of this are the ascending aorta and coronary arteries, for which the bulk motion rather than the compliance alone may have an appreciable impact on the predicted flow patterns [9, 10]. Between these extremes are cases of strong compliance mismatch, such as artificial grafts or implanted vascular devices such as stents. Plaques or stenoses also introduce compliance mismatches, although it should be remembered that most vessels harboring a stenosis, or those requiring some kind of intervention, are probably stiffer to begin with owing to hypertension or other vascular risk factors, thus serving in practice to decrease the actual mismatch.

For cases where compliance and/or vessel motion are shown or believed to have a non-negligible quantitative or qualitative effect, a popular modelling approach is fluid-structure interaction (FSI), whereby the wall motion must also be solved. In this case one must confront the practicalities of obtaining the necessary wall structure and properties, and the pressure boundary conditions, particularly for patient-specific studies. For example, wall properties and thicknesses are often assumed to be uniform and based on literature values, which tends to overlook inter- and intraindividual variations. Measurement of individual wall properties may be possible through imaging and inverse modelling approaches, but such non-linear analyses often introduce appreciable uncertainties. Similarly, although we have shown that wall *thickness* can be measured and mapped three dimensionally [11], we have also shown how imaging can distort wall thickness measurements in a way that can be difficult to detect or correct [12, 13]. Pressure dynamics, also essential as boundary conditions for FSI, are typically measured at the brachial artery, whereas the pressures may well be different at other vascular sites of interest [14].

---

[7] The italics are mine, and I will return to this point in the next section, as it has implications for the measurement and prescription of prevailing flow conditions, and for the validation of so-called "patient-specific" CFD models against in vivo measurements.

In short, uncertainties or inaccuracies in the data needed to model the effects of compliance may mask any perceived benefit of doing so, especially in light of the extra conceptual and computational effort required for including the effects of compliance. This is especially true for patient-specific studies, where these uncertainties may well be subsumed by the uncertainties associated with the reconstruction of the (rigid) geometry itself [15]. If the *effects* of compliance on flow are to be included, one way to do this without resorting to full-blown FSI simulations would be to impose the wall motion directly (e.g., based on time-resolved images, as is often done for coronary arteries). Alternatively, for cases with relatively small motions, one may employ transpiration (normal wall velocity) boundary conditions [16], based on the premise that the effects of compliance are mainly due to the storage and release of fluid at the wall, rather than the wall deformations themselves.[8]

In summary, for predicting flow and WSS patterns in finite segments of large arteries, rigid wall CFD models are probably a reasonable approximation to that which occurs in a compliant vessel. This is not to imply, by the way, that structural (wall) stresses are unworthy of investigation, for undoubtedly they play a central role in regulating normal and pathological vascular responses, alone or in concert with WSS [17]. Rather, the reader is encouraged to contextualize the effects of compliance with the effect of other sources of error/imprecision. As will be seen in the next section, however, the effects of compliance must be confronted when imposing *in vivo* flow boundary conditions, necessarily measured from compliant vessels, onto rigid CFD models.

1.3 Compliance and flow boundary conditions

Notwithstanding Moore's law and the development of increasingly clever computational modelling techniques, 3D CFD models of large arteries must still be isolated from the entirety of the circulatory system. This means that assumptions must be made about the nature of the velocity profile at the truncated inlet(s) and outlet(s), irrespective of whether these profiles are derived form measured or assumed flow rates, or indeed multiscale models. The most common assumption, especially for inlets, is a fully developed (Womersley) velocity profile, which implicitly assumes that the upstream vasculature is straight enough for the prevailing flow to be approximately fully-developed, or that the CFD model has been truncated sufficiently far upstream so that the effect of the assumed velocity profile shape is negligible. At the carotid bifurcation, for example, we have shown that it is not always clear what constitutes "straight enough" when it comes to fully-developed flow (or lack thereof) in the upstream common carotid artery [18]. On the other hand, we have demonstrated

---

[8] This was the principle used by Womersley to derive his famous analytic solution for flow in compliant tubes. It was later exploited by the author for his so-called "hybrid" implicit-explicit 2D FSI approach [6], which solved for the transpiration velocities simultaneously with the rest of the velocity field, and then iteratively updated the wall position based on those wall velocities.

that this has little impact if the inlet is truncated at least three diameters upstream of bifurcation [19].[9]

Setting aside the technicalities of what kind of velocity profile to impose onto such 3D CFD models, the section turns its attention to subtler, but potentially more deleterious effects of the rigid-wall assumption on large artery CFD models. As noted earlier, compliance results in the alternating storage and release of flow caused by the periodic expansion and contraction of the artery walls in response to the pulse pressure. Although periodicity of the cardiac cycle presumes that inlet and outlet flow rates must match in a time-averaged sense, there is no such requirement of instantaneous conservation of flow rates. On the other hand, for rigid-wall CFD simulations, the continuity equation requires that mass/flow be conserved at each instant of time. This sets up the potential for a fundamental incompatibility between inlet and outlet flow rates derived from in vivo measurements on compliant vessels, which must be reconciled with the use of rigid-wall CFD models at the potential cost of accuracy.

One obvious example of this arises from CFD simulation of long arterial domains, where the local effects of compliance might be relatively small, but over long distances may result in appreciable discrepancies between the inlet and outlet flow rates. This is an issue we were first forced to confront in our PIV vs. CFD studies of aneurysm flows, which employed silicone-based flow models for the PIV measurements [20]. As shown in Fig. 1.3, there was a significant delay between the outlet vs. inlet flow rates, owing to the small but non-negligible compliance of the silicone, coupled with the placement of the flow meters necessarily far upstream and downstream. Although the (rigid-walled) CFD domain included these long inlet and outlet segments, we were focused on flow within the aneurysm sac, and thus we could simply align the inlet and outlet flow waveforms temporally (i.e., reflecting their actual relative timings close to the aneurysm sac) without penalty, as suggested by the good agreement between the CFD simulations and PIV measurements [20].

Less clear cut is our recent study of retrograde flow in the mouse aorta and its role in accelerating plaque formation [21]. Doppler ultrasound measurements obtained previously suggested that flow rate dynamics might be appreciably attenuated along the length of the descending aorta [22], which constituted our (rigid-walled) CFD domain. In this case we chose to impose a flow waveform appropriate to the proximal descending aorta inlet, and accepted the fact that flow at the (traction-free) distal descending aorta outlet would not match the actual flow rate dynamics as measured by Doppler ultrasound. This precluded any robust quantitative analyses, but we considered it a reasonable first order approximation in light of our goal to qualitatively associate patterns of hemodynamic extrema with patterns of plaque distribution.

Finally, and to demonstrate that the effects of compliance on flow waveforms are not restricted to long CFD domains, consider the left panel of Fig. 1.4, which illustrates the observation that flow rate waveforms at the internal and external (ICA and ECA) outlets of the carotid bifurcation do not necessarily align temporally. As part of a magnetic resonance imaging (MRI) study of carotid artery flow rates in older

---

[9] Different conclusions may be drawn for other vascular territories depending on, say, the curvature of the inlet or the Reynolds number (i.e., entrance length), so the interested reader is encouraged to review the literature and/or perform their own tests specific to their application of interest.

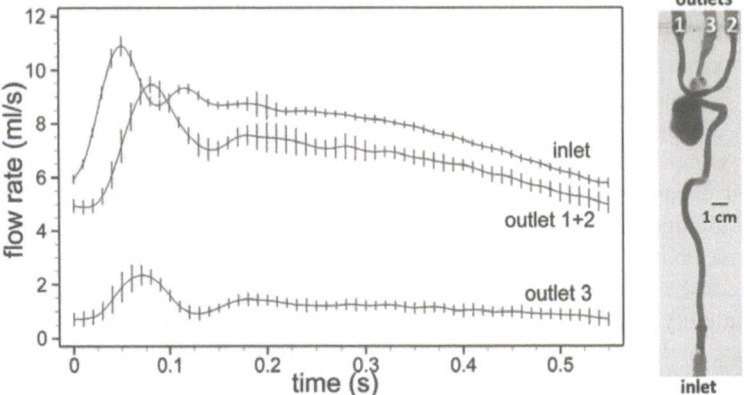

Fig. 1.3. Compliance-induced mismatch in the timing of inlet vs. outlet flow rates (left) measured simultaneously from a life-sized silicone-based aneurysm model (right). Adapted from [20]

adults, we noted a small but significant mean delay of 13 msec between ICA and ECA peak flow rates [23], despite the fact that these were measured from the same MRI slice location. This observation was consistent with an earlier MRI-based study by Marshall et al. [24], which showed what appeared to be a $\sim$ 50 msec delay for younger adults. This may be explained by the fact that the ICA, usually having thin-

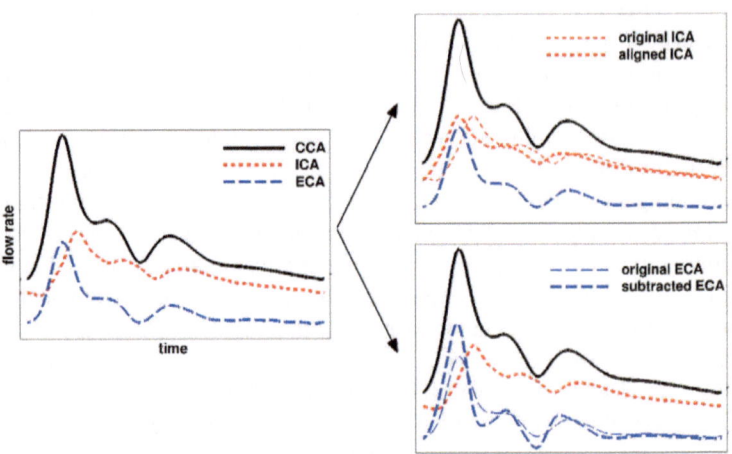

Fig. 1.4. Two options for reconciling measured flow rate waveforms with a rigid CFD model. Left: Starting point is a set of typical normal carotid artery inlet (CCA) and outlet (ICA, ECA) waveforms, exhibiting a 50 msec delay due to the more compliant ICA. Top right: First option aligns the ICA waveform. Bottom right: Second option maintains the ICA delay, but then overestimates ECA flow dynamics to maintain instantaneous mass conservation

ner walls, is more compliant, especially in younger adults.[10] As a result, there can be non-negligible peak outflow deficits of 10–20 % during systole, counterbalanced by smaller instantaneous outflow surpluses spread out over the longer diastolic phase.

Since the continuity equation requires instantaneous conservation of mass/flow, rigid wall CFD simulations must somehow correct for these instantaneous outflow:inflow mismatches. As illustrated by the top right panel of Fig. 1.4, and as noted previously, a common approach is to align the inflow and outflow waveforms. Even then, errors or uncertainties in the measured flow rates may results in mismatches, which must be corrected via ad hoc methods [11]. Alternatively, as illustrated by the bottom right panel of Fig. 1.4, the ICA delay may be maintained, purposely or inadvertently, by subtracting the inherently-aligned CCA and ICA waveforms, but with a resulting overestimation of flow rate dynamics at the ECA. This approach is especially common when flow rates are explicitly prescribed for all but one of the outlets, which effectively assigns any missing flow to the single, traction-free outlet (e.g., [25, 26, 27]).

Although it would seem that aligning the flow waveforms is more correct, in the sense that it properly simulates what should happen if the artery was actually rigid, it is not clear whether it is a more faithful representation of the biologically-relevant hemodynamic parameters such as wall shear stresses in the compliant artery. As the author has seen in his own research, temporal alignment of the ICA and ECA makes it difficult to validate CFD predictions against velocity measurements since, at a given time point (e.g., MRI image), one branch of the CFD model is bound to disagree by virtue of it being out of synch with the measured velocities.

This raises a broader issue that we have only recently begun to investigate as part of our efforts to validate our CFD predictions against in vivo measurements. Specifically, by imposing measured flow rates on a rigid CFD model, the velocity levels in the lumen will only match those in vivo when the in vivo lumen area matches that of the (rigid) CFD model. Consider, for example, a simulation for which the peak systolic diameter is 10 % greater than the nominal (mean) diameter. In this case, if the actual flow rate is imposed, the mean velocities in the rigid model will necessarily overestimate the actual velocities by 20 %. One way around this while still maintaining a rigid-walled CFD model is to match the mean velocities rather than the flow rates. In that case, the rigid CFD model will experience higher flow rates, but velocity levels should more closely match those in vivo. One could also aim to match instantaneous Reynolds numbers as a compromise between these two extremes, although in either inlet:outlet flow rates must still be somehow constrained to ensure instantaneous continuity of mass. Alternatively, one could apply transpiration boundary conditions to the CFD model, as originally suggested by Cebral ct al. [16] for the carotid bifurcation, to allow for the prescription of measured flow waveforms, irrespective of any instantaneous flow/velocity mismatches.

In summary, the use of rigid wall models, while undoubtedly convenient and probably justified in many cases, should not overlook the fact that boundary con-

---

[10] This differential compliance may serve to act as a kind of low pass filter to smooth the transients of the cardiac cycle and thus ensure a more quasi-steady supply of blood to the brain vs. other cerebral vascular territories.

ditions invariably come from real, compliant vessels, and that care must be taken when attempting to impose them onto rigid models. Out of these considerations, it is worthwhile to consider incorporating the effects of compliance in an approximate way that avoids the need for expensive and cumbersome fluid structure analyses, which anyway may be inaccurate or imprecise.

1.4 Newtonian vs. non-Newtonian rheology

Blood is a complex fluid, comprised mainly of fluid-filled vesicles in a water-like carrier. This carrier (plasma) is mostly water with some dissolved proteins, and is widely considered to behave as a constant viscosity (Newtonian) fluid. The fluid filled vesicles (erythrocytes or, literally, red blood cells) normally comprise 40–45 % of the blood by volume. RBC are normally shaped like biconcave disks having an 8-μm diameter and 2-μm height [28], although they can take on a variety of shapes depending on the prevailing flow conditions. It is this concentrated suspension of microscopic, flexible cells that gives blood its most well known non-Newtonian property: shear-thinning.[11]

As shown in Fig. 1.5, at shear rates above about $100\ \text{s}^{-1}$ blood viscosity asymptotes to a constant value, variously reported to be 3–4 cPoise.[12] For shear rates below $100\ \text{s}^{-1}$ blood viscosity starts to increase exponentially, as the stagnating RBC assemble into stacks or rouleaux. For most large arteries under normal, fully-developed

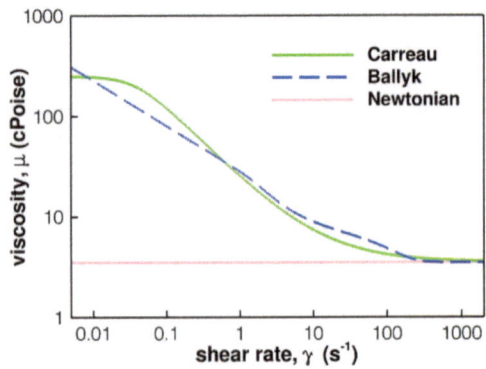

Fig. 1.5. Shear-thinning behaviour of blood, approximated by the Carreau and generalized power law (Ballyk) curve fits. Adapted from [30]

---

[11] Although blood exhibits a variety of non-Newtonian properties, as briefly discussed later, in the large artery hemodynamics literature the term "non-Newtonian" often explicitly or implicitly refers to shear-thinning properties alone. Many shear-thinning models have been proposed, as recently reviewed by [29].

[12] Poise is the cgs unit of dynamic viscosity, named for Poiseuille. 1 Poise = 1 dyne-s/cm^2 = 0.1 Pa-s. The cgs unit for kinematic viscosity is the Stokes, equivalent to cm^2/s, and is often calculated from the dynamic viscosity assuming a blood density of 1.06 g/cm^3.

flow conditions, shear rates are expected to be on order of $100\,\mathrm{s}^{-1}$, and for this reason the Newtonian approximation is widely justified. In regions of slow, recirculating flow, however, shear rates can fall below $10\,\mathrm{s}^{-1}$, raising questions about the Newtonian approximation for such complex flow conditions.

An early attempt to make sense of the effects of non-Newtonian rheology under physiological flow conditions was that of Ballyk et al. [31]. Employing the same CFD techniques and 2D models used in author's investigations of wall distensibility effects, we showed that the Newtonian approximation was reasonable, insofar as non-Newtonian effects on the wall shear stress patterns were minor, and indeed comparable to the effects of wall distensibility. Perhaps more importantly, our study explained why earlier investigations of steady flows were likely to have overestimated non-Newtonian effects: under pulsatile flow conditions, inertial (Womersley) flow effects dominate throughout much of the velocity profile, with viscous effects confined to the thin viscous (Stokes) layer near the wall, where shear rates tend to be above $100\,\mathrm{s}^{-1}$ anyway. In fact, we argued that the importance of non-Newtonian effects could be anticipated by consideration of the Stokes layer thickness relative to the lumen radius (i.e., the Womersley number) rather than the lumen radius itself, ideas that were later demonstrated for more realistic, 3D coronary artery flows by Johnston et al. [32, 33].

Another important and related insight was provided by Gijsen et al. [34], who demonstrated that the effects of non-Newtonian rheology could be approximated by considering a Newtonian viscosity based not on its value at the high shear limit but rather its value at the characteristic or effective average shear rate for that particular flow. This recognizes that, and as Fig. 1.5 shows, even around $100\,\mathrm{s}^{-1}$ small differences in shear rate can produce potentially non-negligible changes in viscosity (i.e., effective Reynolds number). In other words, care should be taken to choose an appropriate reference against which non-Newtonian effects are compared.

This issue of how to contextualize the effects of non-Newtonian vs. Newtonian models is at the heart of any such attempts to evaluate the impact of relaxing assumptions in hemodynamic modelling. For example, there have been many studies on the effects of shear-thinning in large artery flows, yet they have often drawn opposite conclusions despite observing similar levels of effect. One reason for this may be that, without some objective reference as to what constitutes a significant or important effect, authors can only draw subjective conclusions.

With this in mind, in 2007 we published a study that attempted to contextualize the impact of non-Newtonian effects on carotid bifurcation hemodynamics [30]. As Fig. 1.6 summarizes, we found that use of a shear-thinning model has only a minor effect on the computed wall shear stress patterns, especially when compared to the effects of image-based CFD (i.e., geometric) uncertainty that we had previously demonstrated on the same models [15]. Moreover, we showed that the already modest non-Newtonian effects could be captured by using a characteristic rather than shear-limit viscosity, as Gijsen et al. had suggested. Finally, by varying the Reynolds number in a way that roughly mimicked the effect of RBC concentration (i.e., hematocrit) we demonstrated that the choice of hematocrit could have an effect comparable to shear-thinning. In other words, efforts to improve the "accuracy"

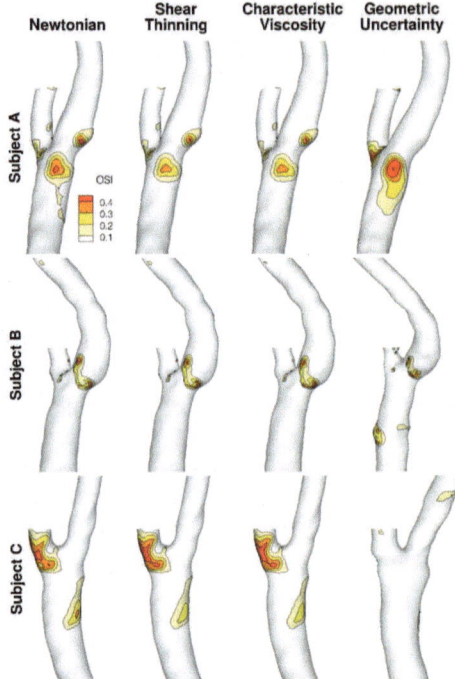

Fig. 1.6. Demonstration that shearthinning has a minor effect on carotid bifurcation shear patterns relative to the effects of geometric uncertainty, and can be approximated well by choosing an appropriate characteristic (Newtonian) viscosity. Adapted from [30]

of patient-specific CFD models by incorporating non-Newtonian rheologies may be wasted if interindividual differences in hematocrit or plasma viscosity are not similarly accounted for [35].

For aneurysms the story may be different, because there is a greater likelihood of large and persistent recirculation zones, and hence persistent low shear rates, within the aneurysm sac. Although our own unpublished studies of shear-thinning effects for the giant saccular aneurysm CFD model of [36] suggested only a minor influence, Rayz et al. [37] have convincingly demonstrated that regions of stagnant flow predicted by a non-Newtonian CFD model of a fusiform basilar aneurysm matched the subsequently observed distribution of thrombus better than that predicted by a Newtonian CFD model. The author has also been told by neurosurgeons that differently-shaded streams of blood are often visible through the thin, near-transparent walls of cerebral aneurysms during surgery, suggesting spatiotemporal variations in the properties of blood that have also been observed by ultrasound imaging of the carotid arteries [38].

As noted earlier, implicit in most studies on the effect of rheology for large artery flows is the assumption that shear thinning dominates over other possible non-Newtonian properties of blood. For example, all shear-thinning models presume that blood viscosity changes instantaneously with shear rate. While disaggregation of RBC rouleaux with increasing shear is thought to occur more or less instantaneously, the kinetics of RBC aggregation are slower and more complex. The end result is that simple shear-thinning rheology models may be reasonable for flow fields that have persistent recirculation regions (e.g., aneurysms), but may *overestimate* non-Newtonian effects when recirculation zones are transient (e.g., carotid bulb), such that rouleaux may not have time to form within a periodic cardiac cycle. As we pointed out in [30], the implication of this is the truth of non-Newtonian effects may fall somewhere between the already-narrow space between constant viscosity and instantaneously shear-thinning models.

One final assumption that is inevitably made for large artery CFD studies is that blood may be treated as if it were a homogeneous fluid, which is usually justified by the orders of magnitude difference in the length scales of large arteries (e.g., lumen diameters) vs. RBC.[13] This presumes that the length scales of the *flow features* are similarly large, an implicit assumption that may well be violated under conditions of pathological flow, as shall be discussed in the next and final section.

1.5 Rheology and turbulence

It is generally accepted that blood flow in large arteries is laminar under normal physiological conditions, except perhaps distal to heart valves and in the ascending aorta [39]. It is also widely considered that turbulence occurs only under severe pathological conditions, notably downstream of severe constrictions or stenoses,[14] although there is some evidence that mild stenoses may induce turbulence [41].

Until relatively recently, CFD studies have focused almost exclusively on nominally turbulent flows through idealized stenoses, for which detailed experimental data are available (e.g., [42]). We have found, for example, that so-called two-equation turbulence models may be satisfactory for modelling steady turbulent flows, but tend to be overly dissipative for the transitional, relaminarizing flows experienced under physiological conditions [43]. More recent investigations by others have focused on large-eddy simulation (LES) and even direct numerical simulation (DNS) methods [44], the latter essentially a brute force solution of the Navier-Stokes equations resolved down to the anticipated length scales of the smallest (i.e., Kolmogorov) eddies at which viscous dissipation occurs. Irrespective of CFD methodology, there is ongoing debate about the nature of physiological "turbulence", namely, whether it is turbulence in the strict fluid mechanical sense of random and uncorrelated flow exhibiting a classical cascade of energy, or indeed whether in some cases it

---

[13] For example, the $\sim$3-μm RBC-free plasma layer adjacent to the arterial wall has an appreciable effect on the apparent viscosity of blood (i.e., the Fahraeus-Lindqvist effect) for arteries below ~ 0.25 mm radius [28], namely a two-order-of-magnitude scale difference.

[14] Turbulence caused by the external compression of an artery is thought to give rise to the so-called Korotkoff sounds used for blood pressure measurement. Noises ("bruits") can also be detected by a stethoscope placed over a stenosed superficial vessel like the carotid artery, or intracranial aneurysms, which may also harbor turbulent flow [40].

is simply unstable flow, characterized by vortex-shedding phenomena at frequencies substantially higher than the heart rate (e.g., [20, 45, 46, 47]).

Setting aside the question of whether these pathological flow conditions are turbulent, transitional or merely unstable, an important question arises regarding the physical scale of the flow structures. Remember that the assumption that blood can be modelled as homogeneous fluid rests on the assumption that the length scales of the arteries ($\sim$ mm) are orders of magnitude larger than the length scales of the RBC ($\sim$ μm). Now consider that both numerical and experimental studies of turbulent blood flow observe, or at least infer, length scales of viscous eddies on the order of tens of microns (e.g., [41, 48]). This raises the question of whether the homogeneity assumption – upon which, ironically, these estimates of turbulent eddy length scales are invariably based – is valid under pathological conditions of flowing *blood*.

The assumption that blood can be treated as a homogeneous fluid in the presence of such small-scale flow features seems to have been only recently questioned by Antiga, who in collaboration with the author described what might occur physically if this assumption is indeed violated [49]. Specifically, we hypothesized that turbulent kinetic energy must at some point be transferred to individual RBC, which, *being closely packed*, must dissipate their energy through laminar cell-cell interactions mediated by the plasma. An order of magnitude analysis revealed that these laminar shear stresses could approach the level of the Reynolds stresses.[15] This is a crucial observation, for it potentially resolves a ongoing paradox in turbulent blood flow research, namely, that "fictional" Reynolds stresses, which are not actual stresses on the RBC, are excellent predictors of RBC damage (hemolysis), whereas the "actual" viscous stresses experienced by RBC in turbulent flow are found to be one to two orders of magnitude less (e.g., [50, 51]). Such findings, however, tend to be based on CFD models of homogeneous fluids upon which an individual RBC is superimposed, which presumes that the RBC are far enough apart so as to not affect the flow field or each other. In our thought experiment, on the other hand, the close packing of RBC is a central and important feature.

That the homogeneity assumption for blood has not been questioned before in this way (or at least not as overtly) might be attributed to it being all too easy to forget that RBC are so closely packed together, even at a "half-full" concentration of 45 %. Consider Fig. 1.7, which presents a common picture of close packing, namely the $\sim$ 90 % optimal packing of circles in two dimensions. Next to this, 50 % packing, representative of a normal blood hematocrit, looks positively sparse, with inter-cell spacing on the order of a cell diameter. In three dimensions, where intuition tends to fail us, optimal packing of ellipsoids, representing RBC under high shear conditions preceding hemolysis, is on the order of 70 %. It can then be shown that the spacing of ellipsoidal RBC at 50 % packing implies an inter-cell spacing on the order of one micron [49], which is broadly consistent with the simulation of 50 % deformable droplets shown in Fig. 1.7.

---

[15] Reynolds stresses are terms that arise from time averaging of the Navier-Stokes equation after decomposition of the velocity into mean and fluctuating components. They are not stresses in a physical sense, but rather embody the effects of the fluctuating velocities on the mean flow.

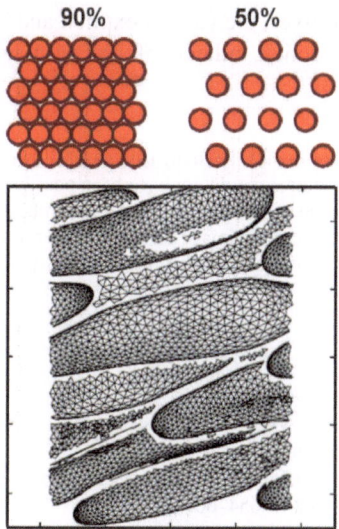

Fig. 1.7. Top: maximum (90 %) and 50 % packing densities for circles in 2D. Bottom: 50 % concentration of deformable droplets (i.e., blood at a normal hemotocrit) in 3D [52]

Only time, and state-of-the-art physical and numerical experiments, will tell whether our suppositions have merit. For now, they reinforce how the simple questioning of a widespread assumption may open up fundamental and interesting lines of inquiry, in this case about the nature of turbulence in blood. More prosaically, they question the role of DNS for *validating* our predictions about turbulent blood flow vs. merely *verifying* the accuracy of our CFD solvers ("solving the right equations" vs. merely "solving the equations right" [53]).

1.6 Conclusions

CFD modelling of large arteries has advanced to the stage where it is now almost routine to reconstruct anatomically realistic models from medical images, or relax traditional assumptions of rigid walls, imposed flow boundary conditions, Newtonian rheology, and laminar flow [54]. Nevertheless, meaningful relaxation of these assumptions invariably requires physiological measurements that may not be available; may be difficult or impossible to obtain; or may come with sufficiently large uncertainties as to render any perceived improvement in the predictive capability of a hemodynamic model purely academic.

In presenting this largely personal journey of discovery regarding the limitations and opportunities of hemodynamic assumptions, admittedly circumscribed by the author's own experiences in modelling flow in carotid arteries and aneurysms, it is hoped that the lessons learned and questions still-to-be-answered will inspire investigators tackling these and other vascular territories to at least consider, and then dismiss or exploit for themselves, the various explicit and implicit assumptions that, depending upon one's perspective, either confound or invigorate the study of arterial hemodynamics.

Acknowledgements. The author thanks the many students, fellows, colleagues and study participants, without whom these adventures would not have been possible. Numerous funding agencies have supported this research, none more so than Heart and Stroke Foundation of Canada, whose early and ongoing support for the author's image-based CFD investigations has allowed him to ask questions that are sometime uncomfortable but ultimately rewarding.

References

[1] Westerhof N., Bosman F., De Vries C.J., Noordergraaf A.: Analog studies of the human systemic arterial tree. J. Biomech. **2**(2): 121–143, 1969.

[2] Pries A.R., Secomb T.W., Gaehtgens P.: Biophysical aspects of blood flow in the microvasculature. Cardiovasc. Res. **32**(4): 654–667, 1996.

[3] O'Rourke M.F., Staessen J.A., Vlachopoulos C., Duprez D., Plante G.E.: Clinical applications of arterial stiffness; definitions and reference values. Am. J. Hypertens. **15**(5): 426–444, 2002.

[4] Davies J.I., Struthers A.D.: Pulse wave analysis and pulse wave velocity: a critical review of their strengths and weaknesses. J. Hypertens. **21**(3): 463–472, 2003.

[5] O'Rourke M.F.: Pressure and flow waves in systemic arteries and the anatomical design of the arterial system. J. Appl. Physiol. **23**(2), 139–149, 1967.

[6] Steinman D.A., Ethier C.R.: The effect of wall distensibility on flow in a two-dimensional end-to-side anastomosis. J. Biomech. Eng. **116**(3): 294–301 , 1994.

[7] Perktold K., Rappitsch G.: Computer simulation of local blood flow and vessel mechanics in a compliant carotid artery bifurcation model. J. Biomech. **28**(7): 845–856, 1995.

[8] Zhao S.Z., Xu X.Y., Hughes A.D., Thom S.A., Stanton A.V., Ariff B., Lon, Q.: Blood flow and vessel mechanics in a physiologically realistic model of a human carotid arterial bifurcation. J. Biomech. **33**(8): 975–984, 2000.

[9] Jin S., Oshinski J., Giddens D.P.: Effects of wall motion and compliance on flow patterns in the ascending aorta. J. Biomech. Eng. **125**(3), 347–354, 2003.

[10] Torii R., Keegan J., Wood N.B., Dowsey A.W., Hughes A.D., Yang G.Z., Firmin D.N., Thom S.A., Xu X.Y.: MR image-based geometric and hemodynamic investigation of the right coronary artery with dynamic vessel motion. Ann. Biomed. Eng. **38**(8), 2606–2620, 2010.

[11] Steinman D.A., Thomas J.B., Ladak H.M., Milner J.S., Rutt B.K., Spence J.D.: Reconstruction of carotid bifurcation hemodynamics and wall thickness using computational fluid dynamics and MRI. Magn. Reson. Med. **47**(1), 149–159, 2002.

[12] Antiga L., Wasserman B.A., Steinman D.A.: On the overestimation of early wall thickening at the carotid bulb by black blood MRI, with implications for coronary and vulnerable plaque imaging. Magn. Reson. Med. **60**(5), 1020–1028, 2008.

[13] Steinman D.A., Antiga L., Wasserman B.A.: Overestimation of cerebral aneurysm wall thickness by black blood MRI? J. Magn. Reson. Imaging **31**(3), 766, 2010.

[14] Kips J., Vanmolkot F., Mahieu D., Vermeersch S., Fabry I., de Hoon J., Van Bortel L., Segers P.: The use of diameter distension waveforms as an alternative for tonometric pressure to assess carotid blood pressure. Physiol. Meas. **31**(4), 543–553, 2010.

[15] Thomas J.B., Milner J.S., Rutt B.K., Steinman D.A.: Reproducibility of image-based computational fluid dynamics models of the human carotid bifurcation. Ann. Biomed. Eng. **31**(2), 132–141, 2003.

[16] Cebral J.R., Putman C.M., Pergolesi R., Burgess J., Yim P.: Multi-modality image-based models of carotid artery hemodynamics. Proc. SPIE Medical Imaging **5369**, 529–538, 2004.

[17] Qiu Y., Tarbell J.M.: Numerical simulation of pulsatile flow in a compliant curved tube model of a coronary artery. J. Biomech. Eng. **122**(1), 77–85, 2000.

[18] Ford M.D., Xie J., Wasserman B.A., Steinman D.A.: Is flow in the common carotid artery fully developed? Physiol. Meas. **29**(11), 1335–1349, 2008.

[19] Hoi Y., Wasserman B.A., Lakatta E.G., Steinman D.A.: Effect of common carotid artery inlet length on normal carotid bifurcation hemodynamics. J. Biomech. Eng. **132**(12), 121008, 2010.

[20] Ford M.D., Nikolov H.N., Milner J.S., Lownie S.P., Demont E.M., Kalata W., Loth, F., Holdsworth, D.W., Steinman, D.A.: PIV-measured versus CFD-predicted flow dynamics in anatomically realistic cerebral aneurysm models. J. Biomech. Eng. **130**(2), 021015, 2008.

[21] Hoi Y., Zhou Y.Q., Zhang X., Henkelman R.M., Steinman D.A.: Correlation between local hemodynamics and lesion distribution in a novel aortic regurgitation murine model of atherosclerosis. Ann. Biomed. Eng. **39**(5), 1414–1422, 2011.

[22] Zhou Y.Q., Zhu S.N., Foster F.S., Cybulsky M.I., Henkelman R.M.: Aortic regurgitation dramatically alters the distribution of atherosclerotic lesions and enhances atherogenesis in mice. Arterioscler. Thromb. Vasc. Biol. **30**(6), 1181–1188, 2010.

[23] Hoi Y., Wasserman B.A., Xie Y.J., Najjar S.S., Ferruci L., Lakatta E.G., Gerstenblith G., Steinman D.A.: Characterization of volumetric flow rate waveforms at the carotid bifurcations of older adults. Physiol. Meas. **31**(3), 291–302, 2010.

[24] Marshall I., Papathanasopoulou P., Wartolowska K.: Carotid flow rates and flow division at the bifurcation in healthy volunteers. Physiol. Meas. **25**(3), 691–697, 2004.

[25] Milner J.S., Moore J.A., Rutt B.K., Steinman D.A.: Hemodynamics of human carotid artery bifurcations: computational studies with models reconstructed from magnetic resonance imaging of normal subjects. J. Vasc. Surg. **28**(1), 143–156, 1998.

[26] Cebral J.R., Yim P.J., Lohner R., Soto O., Choyke P.L.: Blood flow modeling in carotid arteries with computational fluid dynamics and MR imaging. Acad. Radiol. **9**(11), 1286–1299, 2002.

[27] Younis H.F., Kaazempur-Mofrad M.R., Chan R.C., Isasi A.G., Hinton D.P., Chau A.H., Kim L.A., Kamm R.D.: Hemodynamics and wall mechanics in human carotid bifurcation and its consequences for atherogenesis: investigation of inter-individual variation. Biomech. Model. Mechanobiol. **3**(1), 17–32, 2004.

[28] Ethier C.R., Simmons C.A.: Introductory Biomechanics: From Cells to Organisms. Cambridge University Press, Cambridge, 2007.

[29] Yilmaz F., Gundogdu M.Y.: A critical review on blood flow in large arteries; relevance to blood rheology, viscosity models, and physiologic conditions. Korea Australia Rheol. J. **20**(4), 197–211, 2008.

[30] Lee S.W., Steinman D.A.: On the relative importance of rheology for image-based CFD models of the carotid bifurcation. J. Biomech. Eng. **129**(2), 273–278, 2007.

[31] Ballyk P.D., Steinman D.A., Ethier C.R.: Simulation of non-Newtonian blood flow in an end-to-side anastomosis. Biorheology **31**(5), 565–586, 1994.

[32] Johnston B.M., Johnston P.R., Corney S., Kilpatrick D.: Non-Newtonian blood flow in human right coronary arteries: steady state simulations. J. Biomech. **37**(5), 709–720, 2004.

[33] Johnston, B.M., Johnston, P.R., Corney, S., Kilpatrick, D.: Non-Newtonian blood flow in human right coronary arteries: transient simulations. J Biomech **39**(6), 1116–1128 (2006).

[34] Gijsen F.J., Allanic E., van de Vosse F.N., Janssen J.D.: The influence of the non-Newtonian properties of blood on the flow in large arteries: unsteady flow in a 90 degrees curved tube. J. Biomech. **32**(7), 705–713, 1999.

[35] Box F.M., van der Geest R.J., Rutten M.C., Reiber J.H.: The influence of flow, vessel diameter, and non-newtonian blood viscosity on the wall shear stress in a carotid bifurcation model for unsteady flow. Invest. Radiol. **40**(5), 277–294, 2005.

[36] Steinman D.A., Milner J.S., Norley C.J., Lownie S.P., Holdsworth D.W.: Image-based computational simulation of flow dynamics in a giant intracranial aneurysm. AJNR Am. J. Neuroradiol. **24**(4), 559–566, 2003.

[37] Rayz V.L., Boussel L., Lawton M.T., Acevedo-Bolton G., Ge L., Young W.L., Higashida R.T., Saloner D.: Numerical modeling of the flow in intracranial aneurysms: prediction of regions prone to thrombus formation. Ann. Biomed. Eng. **36**(11), 1793–1804, 2008.

[38] Paeng D.G., Nam K.H., Shung K.K.: Cyclic and radial variation of the echogenicity of blood in human carotid arteries observed by harmonic imaging. Ultrasound. Med. Biol. **36**(7), 1118–1124, 2010.

[39] Nerem R.M., Seed W.A.: An in vivo study of aortic flow disturbances. Cardiovasc. Res. **6**(1), 1–14, 1972.

[40] Ferguson G.G.: Turbulence in human intracranial saccular aneurysms. J. Neurosurg. **33**(5), 485–497, 1970.

[41] Lee S.E., Lee S.W., Fischer P.F., Bassiouny H.S., Loth F.: Direct numerical simulation of transitional flow in a stenosed carotid bifurcation. J. Biomech. **41**(11), 2551–2561, 2008.

[42] Ahmed S.A., Giddens D.P.: Pulsatile poststenotic flow studies with laser Doppler anemometry. J. Biomech. **17**(9), 695–705, 1984.

[43] Ryval J., Straatman A.G., Steinman D.A.: Two-equation turbulence modeling of pulsatile flow in a stenosed tube. J. Biomech. Eng. **126**(5), 625–635, 2004.

[44] Varghese S., Frankel S., Fischer P.: Direct numerical simulation of stenotic flows. Part 2. Pulsatile flow. Journal of Fluid Mechanics **582**, 281, 2007.

[45] Baek H., Jayaraman M.V., Richardson P.D., Karniadakis G.E.: Flow instability and wall shear stress variation in intracranial aneurysms. J. R. Soc. Interface **7**(47), 967–988, 2009.

[46] Les A.S., Shadden S.C., Figueroa C.A., Park J.M., Tedesco M.M., Herfkens R.J., Dalman R.L., Taylor C.A.: Quantification of hemodynamics in abdominal aortic aneurysms during rest and exercise using magnetic resonance imaging and computational fluid dynamics. Ann. Biomed. Eng. **38**(4), 1288–1313.

[47] Wang C., Pekkan K., de Zelicourt D., Horner M., Parihar A., Kulkarni A., Yoganathan A.P.: Progress in the CFD modeling of flow instabilities in anatomical total cavopulmonary connections. Ann. Biomed. Eng. **35**(11), 1840–1856, 2007.

[48] Liu J.S., Lu P.C., Chu S.H.: Turbulence characteristics downstream of bileaflet aortic valve prostheses. J. Biomech. Eng. **122**(2), 118–124 (2000).

[49] Antiga L., Steinman D.A.: Rethinking turbulence in blood. Biorheology **46**(2), 77–81, 2009.

[50] Ge L., Dasi L.P., Sotiropoulos F., Yoganathan A.P.: Characterization of hemodynamic forces induced by mechanical heart valves: Reynolds vs. viscous stresses. Ann. Biomed. Eng. **36**(2), 276–297 (2008).

[51] Quinlan N.J., Dooley P.N.: Models of flow-induced loading on blood cells in laminar and turbulent flow, with application to cardiovascular device flow. Ann. Biomed. Eng. **35**(8), 1347–1356, 2007.

[52] Cristini V., Kassab G.S.: Computer modeling of red blood cell rheology in the microcirculation: a brief overview. Ann. Biomed. Eng. **33**(12), 1724–1727, 2005.

[53] Roache P.J.: Quantification of uncertainty in computational fluid dynamics. Annu. Rev. Fluid Mech. **29**, 123–160, 1997.

[54] Taylor C.A., Steinman D.A.: Image-based modeling of blood flow and vessel wall dynamics: applications, methods and future directions: Sixth International Bio-Fluid Mechanics Symposium and Workshop, March 28–30, 2008 Pasadena, California. Ann Biomed Eng **38**(3), 1188–1203, 2010.

2

Simplified blood flow model with discontinuous vessel properties: analysis and exact solutions

Eleuterio F. Toro, and Annunziato Siviglia

Abstract. We formulate a simplified one-dimensional time-dependent non-linear mathematical model for blood flow in vessels with discontinuous material properties. The resulting 3×3 hyperbolic system is analysed and the associated Riemann problem is solved exactly, including tube collapse. Our exact solutions constitute useful reference solutions for assessing the performance of numerical methods intended for simulating more general situations. In addition the presented model may be a useful starting point for numerical calculations involving rapid and discontinuous material properties variations.

2.1 Introduction

The theoretical study of blood flow phenomena in humans through mathematical models is closely related to the study of flow of an incompressible liquid in thin-walled collapsible tubes. In fact the applicability of theoretical models for thin-walled collapsible tubes covers a wider variety of physiological phenomena as well the design of clinical devises for practical medical applications. In this paper we are interested in theoretical models for blood flow in medium to large arteries and veins regarded as thin-walled collapsible tubes. We centre our attention on one-dimensional, time-dependent non-linear models. Classical works on this subject are, for example, [12, 17] and the many references therein. For more recent works see [1, 3, 4, 7, 8, 18, 20] to name but a few.

This paper is motivated by physical situations of medical interest in which certain properties that characterize blood vessels, external pressures and body forces

Eleuterio F. Toro (✉)
Laboratory of Applied Mathematics, Faculty of Engineering, University of Trento, Italy
e-mail: toro@ing.unitn.it

Annunziato Siviglia
Laboratory of Applied Mathematics, Faculty of Engineering, University of Trento, Italy
e-mail: nunzio.siviglia@ing.unitn.it

Ambrosi D., Quarteroni A., Rozza G. (Eds.): Modeling of Physiological Flows.
DOI 10.1007/978-88-470-1935-5_2, © Springer-Verlag Italia 2012

change rapidly, or even discontinuously. Physical quantities of interest are vessel wall thickness, equilibrium cross sectional area and Young's modulus. A prominent example arises in the surgical treatment of Abdominal Aortic Aneurysms (AAA) [23] that includes the insertion of stents. These devises do not always match the compliance properties of natural vessels and discontinuous jumps of physical properties may arise, influencing significantly the wave propagation phenomena associated with the hemodynamics. External pressures and body forces are another source of potentially rapid or even discontinuous variations, which again will influence the wave phenomenon [12]. Here we formulate a mathematical model that allows for discontinuous variation of certain vessel properties, all in the context of simplified one-dimensional flow. In spite of the very strong assumptions, we still expect the one-dimensional model to provide by itself useful information for practical purposes. Moreover, one-dimensional models are an integral part of large models in multiscale approaches [18] and thus the present work may also be useful in the construction of more realistic models.

In current models used for numerical simulation of blood flow phenomena the effect of the variation of the above mentioned quantities enters the equations in the form of source terms; see [20], for example. In particular, for external forces such as muscle forces, the corresponding source term involves a pressure gradient source term, analogous to the geometric source term given by bottom variation in shallow water models [21]. In the numerical analysis literature it is well known that such source terms are likely to cause serious numerical difficulties. An important issue of that of constructing *well balanced schemes* that achieve equilibrium between advective and source terms in the equations near the steady state [9, 13, 16]. The severity of the numerical difficulties increases as spatial gradients of the physical quantities of interest increase.

In this paper we formulate and study a simplified model in which discontinuities of two parameters are permitted, namely wall thickness and Youngs modulus. We study the mathematical properties of the resulting 3×3 hyperbolic system and obtain the exact solution for the Riemann problem. Exact solutions include the case of tube collapse and constitute reference solutions for assessing the performance of numerical methods intended for general use. Potentially, the proposed formulation would facilitate the numerical treatment of source terms due to spatial variation of material properties and external forces. There are, however, two major difficulties with our formulation. One is the potential occurrence of *resonance* [10] and [14], and the possibility of non-uniqueness. The second problem is the non-conservative character of the model, with all the attendant mathematical and numerical implications [5]. The issue of resonance is currently the subject of further studies by the authors and results will be published elsewhere.

The rest of the paper is structured as follows. In Sect. 1 we review the governing equations and the tube law to be used. In Sect. 2 we introduce and study a 3×3 hyperbolic model with discontinous property variations. In Sect. 4 we formulate and solve exactly the Riemann problem. In Sect. 5 we show sample exact solutions. In Sect. 6 we derive exact solutions for tube collapse and give some examples. Conclusions are drawn in Sect. 7.

2.2 Mathematical models

Consider the geometric situation described in Fig. 2.1, which depicts a model for a blood vessel. The mathematical model will assume one-dimensional flow in the axial direction x.

2.2.1 Review of the basic equations

The basic equations for the flow of blood in medium-size to large arteries and veins are obtained from the principles of conservation of mass

$$\partial_t A + \partial_x (uA) = 0 \tag{2.1}$$

and momentum

$$\partial_t (uA) + \partial_x (\hat{\alpha} A u^2) + \frac{A}{\rho} \partial_x p = -Ru . \tag{2.2}$$

$A(x,t)$ is the cross-sectional area of the vessel or tube at position x and time t, $u(x,t)$ is the averaged velocity of blood at a cross section, $p(x,t)$ is pressure, ρ is the density of blood, assumed constant, and $R > 0$ is the viscous resistance of the flow per unit length of the tube, assumed to be known. We assume $\hat{\alpha} = 1$ in the momentum equation (2.2). There are two governing partial differential equations, (2.1), (2.2), and three unknowns, namely $A(x,t)$, $u(x,t)$ and $p(x,t)$. An extra relation is required to close the system. This is provided by the *tube law*, which relates the pressure $p(x,t)$ to the wall displacement via the cross-sectional area $A(x,t)$. The tube law couples the elastic properties of the vessel to the fluid dynamics and is analogous to the *equation of state* in gas dynamics [22].

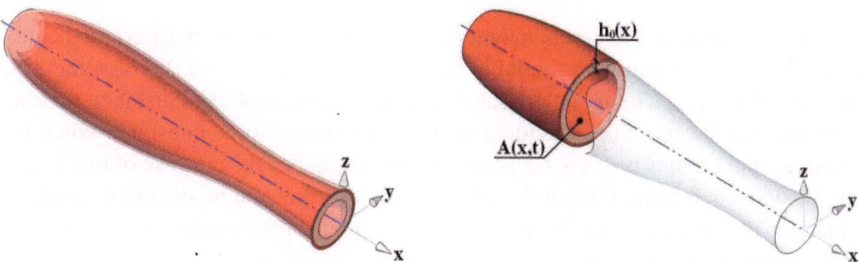

Fig. 2.1. Assumed axially symmetric vessel configuration in three space dimensions at time t. Cross sectional area $A(x,t)$ and wall thickness $h_0(x)$ are illustrated

2.2.2 Tube law

Here we adopt a very simple tube law of the form

$$p = p_e(x) + \psi(A;K) , \qquad (2.3)$$

where

$$\psi(A;K) = p - p_e \equiv p_{trans} \qquad (2.4)$$

is the *transmural pressure*, the difference between the pressure in the vessel, the *internal pressure*, and the external pressure. Here we choose

$$\psi(A;K) = K(x)\text{sign}(m)\left[\left(\frac{A}{A_0}\right)^m - 1\right] , \qquad (2.5)$$

with

$$K(x) = \frac{\sqrt{\pi}}{(1 - v^2)} \frac{E(x)h_0(x)}{\sqrt{A_0(x)}} . \qquad (2.6)$$

Here $h_0(x)$ is the vessel thickness; $A_0(x)$ is the cross-sectional area of the vessel at equilibrium, $p_{trans} = 0$; $E(x)$ is the Young's modulus, v is the Poisson ratio and m is a real number different from zero.

The external pressure, assumed to be known, may be decomposed as follows

$$p_e(x) = p_{atm} + p_{musc}(x) , \qquad (2.7)$$

where p_{atm} is the atmospheric pressure, assumed constant here, and $p_{musc}(x)$ is the pressure exerted by the surrounding tissue. It is reasonable that $p_{musc}(x)$ be a function of time as well. See [24] for a discussion on $p_{musc}(x)$ in the context of chronic venous disease. For a fuller discussion on tube laws see, for example, [3, 4, 8, 17, 18].

2.3 Model for discontinuous properties

In this section we reformulate the mathematical model (2.1)–(2.6) so as to accommodate discontinuous variations of material properties such as Young's modulus and wall thickness.

2.3.1 Equations

We consider a simple mathematical model consisting of the partial differential equations (2.1)–(2.2), along with the tube law (2.3)–(2.5) with $m > 0$. We assume wall thickness $h_0(x)$ and Young's modulus $E(x)$ to be functions of axial distance x. However, the equilibrium cross-sectional area A_0 is assumed to be constant, which for practical applications is of course a serious limitation. For the purpose of this paper, we consider such assumption to be adequate. A more general model comprising six equations is currently being studied, which will include the variation of A_0. Then the pressure gradient in (2.2) is

$$\partial_x p = \psi_A \partial_x A + \psi_K \partial_x K + \partial_x p_e(x) , \qquad (2.8)$$

with

$$\psi_A = \frac{\partial \psi}{\partial A} = \frac{mK}{A}\left(\frac{A}{A_0}\right)^m, \quad \psi_K = \frac{\partial \psi}{\partial K} = \left(\frac{A}{A_0}\right)^m - 1. \tag{2.9}$$

The complete system reads

$$\left.\begin{array}{l} \partial_t A + \partial_x(uA) = 0, \\[2mm] \partial_t(uA) + \partial_x(Au^2) + \frac{A}{\rho}\psi_A \partial_x A = -\frac{A}{\rho}\psi_K \partial_x K - \partial_x p_e(x) - Ru. \end{array}\right\} \tag{2.10}$$

We note that the principal part of the equations (left-hand side) does not have conservation-law form. Note also that there are source terms on the right hand side which depend on gradients of the vessel properties $E(x)$ and $h_0(x)$ and the external pressure. The external pressure $p_e(x)$ is analogous to bottom variation in shallow water flows [21], both giving rise to a source term involving a spatial gradient. In the rest of this paper we assume $p_e(x)$ to be constant. In the numerical literature it is well known that the treatment of such source terms, sometimes known as *geometric source terms*, is notoriously difficult. Common difficulties include the generation of spurious oscillations and the lack of balance between convective terms and source terms in the steady state. For a discussion on these issues see, for example, [9, 13, 16] and references therein. In principle, for slowly-varying vessel properties one can still proceed with formulation (2.10). However, for significant vessel property variations, or even in the case of discontinuous properties, formulation (2.10) is not suitable.

In this paper we present an alternative formulation of the model by considering the variable vessel properties $h_0(x)$ and $E(x)$ as additional unknowns of the problem. As a matter of fact it is sufficient to consider the coefficient $K(x)$ as the new unknown, as this includes the combined variations of $h_0(x)$ and $E(x)$. We then add the following obvious partial differential equation

$$\partial_t K(x) = 0. \tag{2.11}$$

The enlarged system from (2.10) and (2.11) in quasi-linear form reads

$$\partial_t \mathbf{Q} + \mathbf{A}(\mathbf{Q})\partial_x \mathbf{Q} = \mathbf{S}(\mathbf{Q}), \tag{2.12}$$

where

$$\mathbf{Q} = \begin{bmatrix} q_1 \\ q_2 \\ q_3 \end{bmatrix} \equiv \begin{bmatrix} A \\ Au \\ K \end{bmatrix}, \quad \mathbf{S}(\mathbf{Q}) = \begin{bmatrix} s_1 \\ s_2 \\ s_3 \end{bmatrix} \equiv \begin{bmatrix} 0 \\ -Ru \\ 0 \end{bmatrix},$$

$$\mathbf{A}(\mathbf{Q}) = \begin{bmatrix} 0 & 1 & 0 \\ \frac{A}{\rho}\psi_A - u^2 & 2u & \frac{A}{\rho}\psi_K \\ 0 & 0 & 0 \end{bmatrix}. \tag{2.13}$$

Next we study some mathematical properties of the equations.

2.3.2 Eigenstructure and characteristic fields

The eigenstructure of the first-order system (2.12), (2.13) is that of the principal part of the system and is given by the eigenvalues and corresponding eigenvectors.

Proposition 3.1. The eigenvalues of (2.12) are all real and given by

$$\lambda_1 = u - c, \quad \lambda_2 = 0, \quad \lambda_3 = u + c, \tag{2.14}$$

where

$$c = \sqrt{\frac{A}{\rho} \psi_A} \tag{2.15}$$

is the *wave speed*, analogous to the *sound speed in gas dynamics* [22].

Proof. By definition the eigenvalues of system (2.12), (2.13) are the eigenvalues of the matrix **A**, which in turn are the roots of the characteristic polynomial

$$P(\lambda) = Det(\mathbf{A} - \lambda \mathbf{I}) = 0, \tag{2.16}$$

where **I** is the identity matrix and λ is a parameter. Simple calculations give

$$P(\lambda) = \lambda \left(\lambda^2 - 2u\lambda + u^2 - c^2 \right) = 0,$$

from which the result (2.14) follows.   □

Proposition 3.2. The right eigenvectors of **A** corresponding to the eigenvalues (2.14) are

$$\mathbf{R}_1 = \gamma_1 \begin{bmatrix} 1 \\ u - c \\ 0 \end{bmatrix}, \quad \mathbf{R}_2 = \gamma_2 \begin{bmatrix} \frac{c^2}{u^2 - c^2} \\ 0 \\ 1 \end{bmatrix}, \quad \mathbf{R}_3 = \gamma_3 \begin{bmatrix} 1 \\ u + c \\ 0 \end{bmatrix}, \tag{2.17}$$

where γ_1, γ_2 and γ_3 are arbitrary scaling factors.

Proof. For an arbitrary right eigenvector $\mathbf{R} = [r_1, r_2, r_3]^T$ we have

$$\mathbf{AR} = \lambda \mathbf{R}, \tag{2.18}$$

which gives the algebraic system

$$\left. \begin{array}{r} r_2 = \lambda r_1, \\ (u^2 - c^2)r_1 + 2ur_2 + \frac{A}{\rho}\psi_K r_3 = \lambda r_2, \\ 0 = \lambda r_3. \end{array} \right\} \tag{2.19}$$

By substituting λ in (2.19) by the appropriate eigenvalues in (2.14) in turn we arrive at the sought result.   □

Proposition 3.3. The λ_1- and λ_3- characteristic fields are genuinely non-linear if the tube law exponent $m \neq -2$ and the λ_2- characteistic field is linearly degenerate.

Proof. Since $\lambda_2 = 0$ it follows that $\nabla \lambda_2 = \mathbf{0}$ and thus $\nabla \lambda_2 \cdot \mathbf{R}_2 = 0$. Therefore the λ_2-characteristic field is linearly degenerate as claimed. For the other two characteristic fields, some algebraic manipulations give

$$\nabla \lambda_1 \cdot \mathbf{R}_1 = \frac{m(m+2)K\left(\frac{A}{A_0}\right)^m}{2A\sqrt{m\rho K\left(\frac{A}{A_0}\right)^m}} \, , \quad \nabla \lambda_3 \cdot \mathbf{R}_3 = -\frac{m(m+2)K\left(\frac{A}{A_0}\right)^m}{2A\sqrt{m\rho K\left(\frac{A}{A_0}\right)^m}} \, .$$

Therefore the $\lambda_1(\mathbf{Q})$- and $\lambda_3(\mathbf{Q})$- characteristic fields are genuinely non-linear provided $m \neq -2$, and the proof is complete. $\square$

2.3.3 Generalized Riemann invariants

The generalized Riemann invariants are relations that are valid across simple waves. These are most conveniently expressed as a set of ordinary differential equations in phase space, see [11] for details.

Proposition 3.4. For a given hyperbolic system of M unknowns $[w_1, w_2, \ldots, w_M]^T$, for any λ_i-characteristic field with right eigenvector $\mathbf{R}_i = [r_{1i}, r_{2i}, \ldots, r_{Mi}]^T$ the generalized Riemann invariants are solutions of the following $M - 1$ ordinary differential equations in phase space

$$\frac{dw_1}{r_{1i}} = \frac{dw_2}{r_{2i}} = \ldots \frac{dw_n}{r_{Mi}} \, . \tag{2.20}$$

Proof. (omitted). See [11]. $\square$

Proposition 3.5. The generalized Riemann invariants for $\lambda_1 = u - c$ are

$$K = constant \, , \quad \frac{2}{m}c + u = constant \, . \tag{2.21}$$

Proof. Application of (2.20) from Proposition 3.4 to $\lambda = \lambda_1 = u - c$ with $\mathbf{R}_1 = \gamma_1[1, u - c, 0]^T$, with $\gamma_1 = 1$ gives

$$\frac{dA}{1} = \frac{d(Au)}{u - c} = \frac{dK}{0} \, . \tag{2.22}$$

The last member of the two equalities gives the first sought result $K = constant$, as desired, while the second result is obtained by manipulating the first equality, leading to

$$\frac{c(A)}{A}dA + du = 0 \, .$$

Integration in phase space gives

$$\int \frac{c(A)}{A} dA + u = constant .$$

After expressing $c(A)$ explicitly as a function of A and performing exact integration leads to the second result in (2.21).   □

Proposition 3.6. The generalized Riemann invariants for $\lambda_3 = u + c$ are

$$K = constant , \quad \frac{2}{m}c - u = constant . \tag{2.23}$$

Proof. The proof is entirely analogous to the previous case and is thus omitted.   □

2.4 The Riemann problem

Here we pose and solve exactly the Riemann problem for system (2.12), that is the special Cauchy problem with piece-wise constant initial condition, namely

$$\left. \begin{array}{l} \partial_t \mathbf{Q} + \mathbf{A}(\mathbf{Q})\partial_x \mathbf{Q} = \mathbf{0} , \quad x \in \mathcal{R} , \quad t > 0 , \\[2ex] \mathbf{Q}(x,0) = \left\{ \begin{array}{l} \mathbf{Q}_L \ \ \text{if} \ \ x < 0 , \\[2ex] \mathbf{Q}_R \ \ \text{if} \ \ x > 0 . \end{array} \right. \end{array} \right\} \tag{2.24}$$

The structure of the similarity solution of the problem is shown in Fig. 2.2 in the entire x-t half plane. There are three wave families. The left family is associated with the eigenvalue λ_1, the middle family is associated with λ_2 and the right wave family is associated with λ_3. Waves associated with the *genuinely non-linear* characteristic fields λ_1 and λ_3 are either shocks (discontinuous solutions) or rarefactions (smooth solutions), while the wave associated with the *linearly degenerate* characteristic field λ_2 is a contact discontinuity. The entire solution consists of four constant states, namely $\mathbf{Q}_L$ (data), $\mathbf{Q}_{*L}$, $\mathbf{Q}_{*R}$ and $\mathbf{Q}_R$ (data), separated by three waves. The unknown states to be found are $\mathbf{Q}_{*L}$ (left of the contact) and $\mathbf{Q}_{*R}$ (right of the contact). If any of the λ_1 and λ_3 waves is a rarefaction then there will be a smooth transition between two adjacent constant states. In order to solve exactly the entire initial-value problem we need to establish appropriate jump conditions across each characteristic field to connect the unknown states $\mathbf{Q}_{*L}$ and $\mathbf{Q}_{*R}$ to the initial conditions $\mathbf{Q}_L$ (left) and $\mathbf{Q}_R$ (right) respectively. In what follows we establish such jump conditions across each characteristic field.

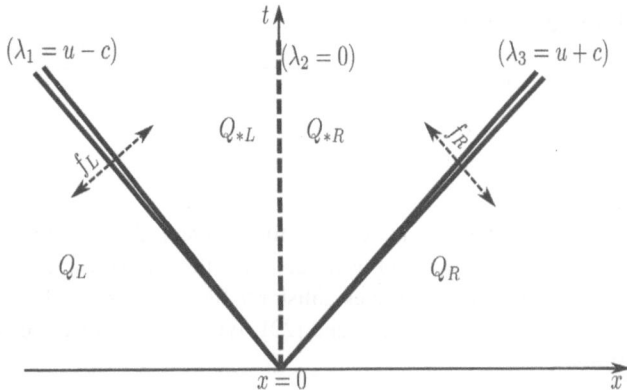

Fig. 2.2. Structure of the solution of the Riemann problem (2.24) for the simplified 3×3 blood flow model of this paper

2.4.1 Jump across shocks: Rankine-Hugoniot conditions

An important feature of the proposed model for variable vessel properties is that the system cannot be expressed in conservation-law form. However, for the case of the Riemann problem the vessel properties h_0 and E, and thus K, are constant across the non-linear waves. Hence, across the genuinely non-linear characteristic fields (rarefactions and shocks) it is possible to express the equations in conservation-law form. In fact it is sufficient to consider the reduced 2×2 conservative system, excluding the equation for K in (2.12). The homogeneous part of the equations in conservation-law form are

$$\partial_t \mathbf{Q} + \partial_x \mathbf{F}(\mathbf{Q}) = \mathbf{0} , \tag{2.25}$$

in terms of the redefined vector of conserved variables

$$\mathbf{Q} = \begin{bmatrix} q_1 \\ q_2 \end{bmatrix} \equiv \begin{bmatrix} A \\ Au \end{bmatrix} \tag{2.26}$$

and flux vector

$$\mathbf{F}(\mathbf{Q}) = \begin{bmatrix} f_1 \\ f_2 \end{bmatrix} \equiv \begin{bmatrix} Au \\ Au^2 + BA^{m+1} \end{bmatrix} , \tag{2.27}$$

with

$$B = \frac{mK}{\rho(m+1)A_0^m} : constant . \tag{2.28}$$

Proposition 4.1 (left shock). If the left λ_1-wave is a left-facing shock wave of speed S_L then

$$u_{*L} = u_L - f_L , \quad f_L = \sqrt{\frac{B_L(A_{*L} - A_L)(A_{*L}^{m+1} - A_L^{m+1})}{A_L A_{*L}}} \tag{2.29}$$

and the shock speed is given as

$$S_L = u_L - \frac{M_L}{A_L} , \quad M_L = \sqrt{\frac{B_L A_L A_{*L}(A_{*L}^{m+1} - A_L^{m+1})}{A_{*L} - A_L}} . \tag{2.30}$$

Proof. In Fig. 2.2 we illustrate the function f_L that connects the velocity u_{*L} to the left data state and the unknown A_{*L}. Let us assume that the left λ_1-wave is a left-facing shock wave of speed S_L. We need to establish relations across the shock, for which one uses standard techniques, see [21] and [22]. We first transform the equations to a stationary frame via

$$\hat{u}_L = u_L - S_L , \quad \hat{u}_{*L} = u_{*L} - S_L . \tag{2.31}$$

Then the jump conditions become

$$\left.\begin{array}{r} A_{*L}\hat{u}_{*L} = A_L\hat{u}_L , \\ A_{*L}\hat{u}_{*L}^2 + B_L A_{*L}^{m+1} = A_L\hat{u}_L^2 + B_L A_L^{m+1} . \end{array}\right\} \tag{2.32}$$

Now from the first equation in (2.32) define $M_L = -A_{*L}\hat{u}_{*L} = -A_L\hat{u}_L$. In fact this is the mass flux through the wave, which is constant. Use of M_L into the second equation in (2.32) followed by suitable manipulations leads to the sought relations (2.29). Details of the calculation of the shock speed S_L are omitted. □

Proposition 4.2 (right shock). If the right λ_3-wave is a right-facing shock wave of speed S_R then

$$u_{*R} = u_R + f_R , \quad f_R = \sqrt{\frac{B_R(A_{*R} - A_R)(A_{*R}^{m+1} - A_R^{m+1})}{A_R A_{*R}}} \tag{2.33}$$

and the shock speed is given as

$$S_R = u_R + \frac{M_R}{A_R} , \quad M_R = \sqrt{\frac{B_R A_R A_{*R}(A_{*R}^{m+1} - A_R^{m+1})}{A_{*R} - A_R}} . \tag{2.34}$$

Proof. The proof follows the same methodology as for a left shock and details are thus omitted. □

2.4.2 Jump conditions across rarefactions

It is possible to establish jump relations across rarefactions waves by means of generalized Riemann invariants introduced in Sect. 3.3.

Proposition 4.3 (left rarefaction wave). Across a left rarefaction wave associated with the characteristic field $\lambda_1 = u - c$ the following relations hold

$$u_{*L} = u_L - f_L , \quad f_L = \frac{2}{m}(c_{*L} - c_L) . \tag{2.35}$$

Proof. From the left generalized Riemann invariants (2.21) we can write

$$\frac{2}{m}c_{*L} + u_{*L} = \frac{2}{m}c_L + u_L ,$$

from which we have

$$u_{*L} = u_L - (\frac{2}{m}c_{*L} - \frac{2}{m}c_L)$$

and the result follows.   □

Proposition 4.4 (right rarefaction wave). Across a right rarefaction wave associated with the characteristic field $\lambda_3 = u + c$ the following relations hold

$$u_{*R} = u_R + f_R , \quad f_R = \frac{2}{m}(c_{*R} - c_R) . \tag{2.36}$$

Proof. The proof uses the right generalized Riemann invariants (2.23) and is entirely analogous to the previous case.   □

2.4.3 Jump conditions across the stationary contact

We wish to establish jump conditions across the stationary contact discontinuity associated with the eigenvalue $\lambda_2 = 0$. As stated earlier, for variable material properties it is not possible to express the equations in conservation-law form and therefore it is not possible to apply the classical Rankine-Hugoniot conditions. Thus to establish jump conditions we follow two alternative approaches, leading to identical results.

Proposition 4.5. Across the contact discontinuity the following relations hold

$$Au = constant , \quad \frac{1}{2}\rho u^2 + \psi = constant , \tag{2.37}$$

leading to

$$A_{*L}u_{*L} = A_{*R}u_{*R} , \quad \frac{1}{2}\rho u_{*L}^2 + \psi_{*L} = \frac{1}{2}\rho u_{*R}^2 + \psi_{*R} . \tag{2.38}$$

Proof. We first apply generalized Riemann invariants across the linearly degenerate field. This approach was advocated by Embid and Baer [6] to analyse the Baer-Nunziato equation for two-phase compressible flow, a well-known non-conservative system. We obtain

$$\frac{dA}{\frac{A}{\rho}\frac{\psi_K}{u^2-c^2}} = \frac{d(Au)}{0} = \frac{dK}{1} . \tag{2.39}$$

The second member of the above equation states immediately that across the contact discontinuity $d(Au) = 0$ and thus $Au = constant$, proving the first result in (2.37). Relations (2.39) also state that across the contact wave, see Fig. 2, A and K do change. Equating the first and third members gives

$$\frac{A}{\rho}\left(\frac{\psi_K}{u^2 - c^2}\right) = dA .$$
(2.40)

Using $d(Au) = 0$ just proved and $c^2 = \frac{A}{\rho}\psi_A$ from the definition of wave speed we may write

$$\psi_K dK = -\rho u du - \psi_A dA ,$$
(2.41)

or

$$\rho u du + \psi_A dA + \psi_K dK .$$
(2.42)

But $\psi = \psi(A; K)$ and thus $d\psi = \psi_A dA + \psi_K dK$. Therefore $\rho u du + d\psi = 0$, which after integration gives the second sought result in (2.37).   □

Now we adopt the thin-layer approach advocated by Schwendemann et al. [19], also used to analyse the Baer-Nunziato equations; see also [2]. It is assumed that the transition layer containing the contact discontinuity is vanishingly thin and that the solution is smooth within the layer. Assuming the layer travels with constant speed S we define the independent variable

$$\xi = x - St , \quad S = constant ,$$
(2.43)

which measures distance across the layer. We now study the governing equations locally. For any function $G(x,t)$ we have

$$\frac{\partial G}{\partial x} = \frac{\partial G}{\partial \xi}\frac{\partial \xi}{\partial x} , \quad \frac{\partial G}{\partial t} = \frac{\partial G}{\partial \xi}\frac{\partial \xi}{\partial t} .$$
(2.44)

Then the continuity equation (2.1) gives

$$\frac{\partial A}{\partial \xi}\frac{\partial \xi}{\partial t} + \frac{\partial Au}{\partial \xi}\frac{\partial \xi}{\partial x} = 0 ,$$
(2.45)

or

$$d((u - S)A) = 0$$
(2.46)

and thus with $S = 0$ we obtain $d(Au) = 0$, which is the first sought result in (2.37). Analogous manipulations for the momentum equation give $\frac{1}{2}\rho u^2 + \psi = constant$, which is the second sought result in (2.37) and the result is thus proved.

It is worth noting that conditions (2.37) are identical to those proposed by [7] in the context of treating the discontinuous coefficients case by a domain decomposition approach. They derived the same conditions adopting an energetic approach.

2.4.4 Solution of Riemann problem

In previous sections we have put in place all the necessary relations to obtain the solution of the Riemann problem in the Star Region, which is the region in physical space located between the waves associated with the outer characteristic fields. The procedure to find the solution is embodied in the following proposition.

Proposition 4.6 (solution of Riemann problem). The solution of the Riemann problem in the *Star Region* is given by the solution of the following non-linear system

$$
\left.
\begin{aligned}
f_1(x_1,x_2) &= x_2 - u_L + f_L(x_1) &&= 0, \\
f_2(x_1,x_2,x_3,x_4) &= x_2 x_1 - x_4 x_3 &&= 0, \\
f_3(x_1,x_2,x_3,x_4) &= \tfrac{1}{2}\rho(x_2^2 - x_4^2) + K_L\left[\left(\tfrac{x_1}{A_{0L}}\right)^m - 1\right] - K_R\left[\left(\tfrac{x_3}{A_{0R}}\right)^m - 1\right] &&= 0, \\
f_4(x_3,x_4) &= x_4 - u_R - f_R(x_3) &&= 0,
\end{aligned}
\right\}
\tag{2.47}
$$

where the unknowns of the problem are

$$
\mathbf{X} = [x_1, x_2, x_3, x_4] \equiv [A_{*L}, u_{*L}, A_{*R}, u_{*R}],
\tag{2.48}
$$

with

$$
f_L(x_1) =
\begin{cases}
\sqrt{\dfrac{B_L(x_1 - A_L)(x_1^{m+1} - A_L^{m+1})}{A_L x_1}} & \text{if} \quad A_{*L} > A_L, \\[2ex]
\dfrac{2}{m}\left(D_L x_1^{m/2} - c_L\right) & \text{if} \quad A_{*L} \leq A_L,
\end{cases}
\tag{2.49}
$$

and

$$
f_R(x_3) =
\begin{cases}
\sqrt{\dfrac{B_R(x_3 - A_R)(x_3^{m+1} - A_R^{m+1})}{A_R x_3}} & \text{if} \quad A_{*R} > A_R, \\[2ex]
\dfrac{2}{m}\left(D_R x_3^{m/2} - c_R\right) & \text{if} \quad A_{*R} \leq A_R,
\end{cases}
\tag{2.50}
$$

$$
D_L = \sqrt{\dfrac{m K_L}{\rho A_{0L}^m}}, \quad D_R = \sqrt{\dfrac{m K_R}{\rho A_{0R}^m}}.
\tag{2.51}
$$

The wave speeds c_L and c_R are evaluated on the data according to (2.15). The constants K_L and K_R are evaluated on the data from to (2.6) and B is given by (2.28).

Proof. The proof involves putting together the results stated previously. Details are omitted. □

Remarks:

- **Complete solution.** The numerical solution of the non-linear system (2.47) gives the four unknowns in the *Star Region*. The rest of the solution follows by applying the wave relations studied in the previous sections. Part of the process of finding the complete solution involves a procedure for the solution at a point inside a rarefaction fan. Details of this are given in Sect. 6.

- **Resonance.** A serious disadvantage of our formulation for discontinous material properties is the potential occurrence of *resonance*. This happens when either of the eigenvalues $\lambda_1 = u - c$ or $\lambda_3 = u + c$ vanish, creating locally critical flow with a non-linear wave overlapping the stationary wave associated with the eigenvalue $\lambda_2 = 0$. Theoretical issues regarding resonance are found in [10, 14], and references there in. A serious difficulty is the loss of uniqueness, as shown by our preliminary results to be reported elsewhere.

2.5 Sample solutions

We consider two test problems for a long, straight tube of length $1\ m$, of constant equilibrium cross sectional area $A_0 = 2.1124 \times 10^{-4}\ m^2$, of wall thickness $h_0 = 8.2 \times 10^{-4}\ m$, Young's modulus $E = 3.0 \times 10^5\ N/m^2$, exponent in tube law $m = \frac{1}{2}$ and Poisson ratio $v = \frac{1}{2}$. The coefficient K is taken as

$$K_0 = \frac{4\sqrt{\pi}}{3} \frac{E(x)h_0}{\sqrt{A_0}}\ ,$$

while blood density is $\rho = 1050\ kg/m^3$. Initial data for two Riemann problems are given in Table 2.1. Table 2.2 shows the values of the exact solution in the Star Region for areas A_{*L}, A_{*R} and velocities u_{*L}, u_{*R}. These numbers can be useful to test numerical methods.

Test 1: left rarefaction and right shock. Fig. 2.3 shows profiles of vessel diameter, velocity and Froude number at the output time $T_{out} = 0.012\ s$. Also shown is the resulting wave configuration in the x-t plane. The solution consists of a left facing rarefaction, a stationary contact and right facing shock.

Test 2: two rarefactions. Fig. 2.4 shows solution profiles of vessel diameter, velocity and Froude number at $T_{out} = 0.0075\ s$. See also wave configuration in the x-t plane. The solution consists of a left rarefaction, a stationary contact and a right rarefaction.

Table 2.1. Initial conditions for two test Riemann problems

Test	$A_L\ [m^2]$	$u_L\ [m/s]$	K_L	$A_R\ [m^2]$	$u_R\ [m/s]$	K_R
1	3.000×10^{-4}	-2.6575×10^{-5}	$50 \times K_0$	$3.000E-04$	6.1230×10^{-6}	K_0
2	3.100×10^{-4}	-3.9967	K_0	3.100×10^{-4}	11.9943	$40 \times K_0$

Table 2.2. Exact solution in the Star Region for two test Riemann problems

Test	$A_{*L}\ [m^2]$	$u_{*L}\ [m/s]$	K_{*L}	$A_{*R}\ [m^2]$	$u_{*R}\ [m/s]$	K_{*R}
1	2.0119×10^{-4}	12.8103	K_L	6.3403×10^{-4}	4.0648	K_R
2	9.3917×10^{-5}	0.9625	K_L	2.0784×10^{-4}	0.4349	K_R

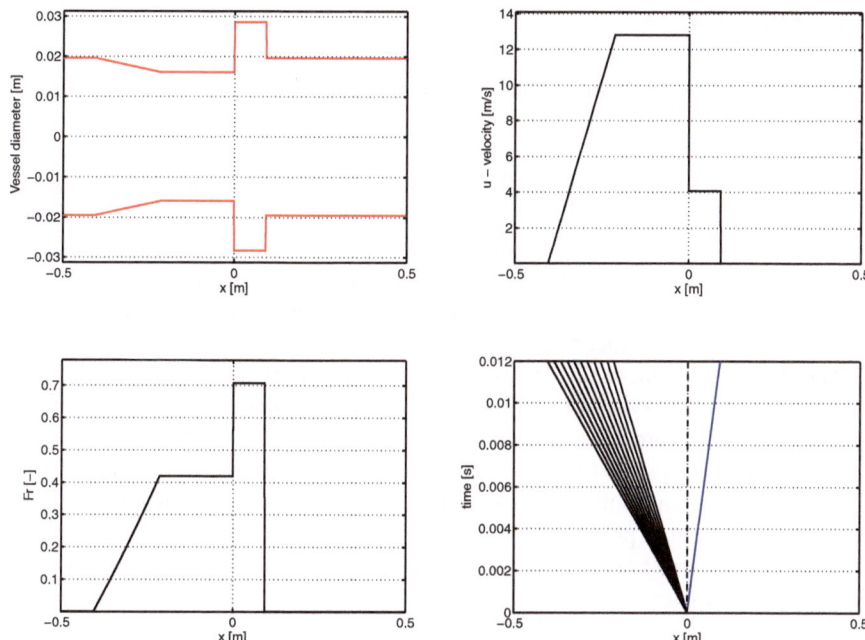

Fig. 2.3. Exact solution for Test 1 at the output time $T_{out} = 0.012$ s

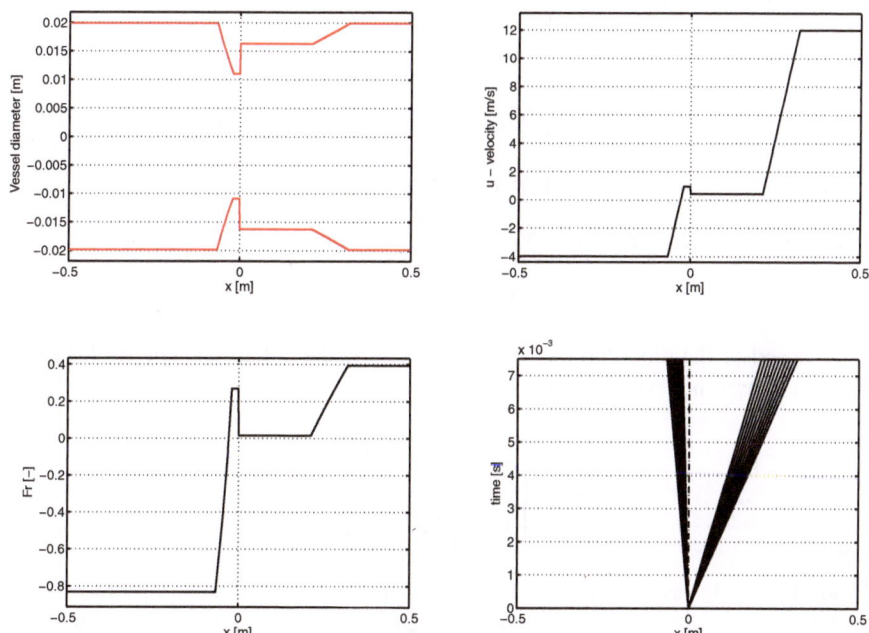

Fig. 2.4. Exact solution for Test 2 at the output time $T_{out} = 0.0075$ s

2.6 Collapse of vessels

Veins collapse under normal physiological conditions. For example, in normal individuals venous blood is drained from the brain mainly through the jugular veins in the supine position, whereas in the upright position these veins collapse and most of the venous flow takes place through the vertebral veins [25]. Arteries do collapse, but rarely do so under normal physiological conditions. In this section we construct exact solutions for tube collapse that can be used as reference solutions for validating numerical methods. To be consistent with the main body of the paper we continue the analysis of the equations for a tube law intended for arteries, and thus the solution have a limited physical meaning.

The Riemann problem (2.24) is considered, neglecting variations of K, leading to the reduced 2×2 system (2.25) with $m = 1/2$. Consider the physical variables $\mathbf{W} = [A, u]^T$. There are three cases of interest. Case 1 is that in which the left hand portion of the collapsible tube $(-\infty, 0]$ is not collapsed but the right hand portion $[0, \infty)$ is collapsed and is therefore *dry*. Case 2 is the opposite case, the left portion is collapsed but the right portion is not. In case 3 the initial condition does not include collapse but this arises as time evolves. Below we provide a complete solution to all three possible cases.

Proposition 6.1. The solution of the Riemann problem (2.24) with data $A_L > 0$, u_L arbitrary; $A_R = 0$ and $u_R = 0$ has exact solution consisting of a single rarefaction wave associated with the eigenvalue $\lambda_1 = u - c$ and is given as

$$
\mathbf{W}_{LO}(x,t) = \begin{cases} \mathbf{W}_L & \text{if } x/t \leq u_L - c_L , \\ \mathbf{W}_{Lfan} & \text{if } u_L - c_L \leq x/t \leq S_{*L} , \\ \mathbf{W}_0 & \text{if } S_{*L} \leq x/t , \end{cases}
\tag{2.52}
$$

where $\mathbf{W}_L$ is the left data state, $\mathbf{W}_R = \mathbf{W}_0$ is the collapsed right data state and

$$
S_{*L} = u_L + 4c_L
\tag{2.53}
$$

is the speed of the blood/no blood front separating the collapsed state (right) from the uncollapsed new state (left) given as

$$
\mathbf{W}_{Lfan} \equiv \begin{cases} u = \dfrac{1}{5}\left(u_L + 4c_L + 4\dfrac{x}{t}\right) , \\ c = \dfrac{1}{5}\left(u_L + 4c_L - \dfrac{x}{t}\right) . \end{cases}
\tag{2.54}
$$

Proof. First it is easy to show that a shock solution of the Riemann problem with initial condition as for case 1 is not possible [21]. Therefore the only possible way for connecting the left state $\mathbf{W}_L$ and the collapsed state $\mathbf{W}_0$ is through a rarefaction wave associated with the left eigenvalue $\lambda_1 = u - c$. This wave, denoted by $\mathbf{W}_{Lfan}$, has head given by the characteristic $\frac{x}{t} = S_{*L} = \hat{u} - \hat{c}$ which coalesces with the blood/no blood front. To determine S_{*L} we select a point P on the characteristic line $\frac{x}{t} = S_{*L} = \hat{u} - \hat{c}$ on which the particle velocity and wave speed take on respectively the values $\hat{u}$ and $\hat{c}$. Then we connect this unknown point to the left data state via the left generalized Riemnan invariant $u + 4c = constant$ to obtain $u_L + 4c_L = \hat{u} + 4\hat{c}$. But since $A = 0$ right at the blood/no blood front, $\hat{c} = 0$ and therefore $S_{*L} = u_L + 4c_L$ follows.

To find the solution inside the fan $\mathbf{W}_{Lfan}$ it is sufficient to select a point $P = (x,t)$ inside $\mathbf{W}_{Lfan}$. A characteristic through the origin and P has slope $\frac{x}{t} = u - c$, with u and c two unknowns. Then, connecting the left data state to the unknown point P via the left generalized Riemann invariant gives $u_L + 4c_L = u + 4c$. Solving these two linear equations for the two unknowns u and c leads to the claimed solution (2.54) and the proposition is thus proved.   □

Now we consider the mirror-image problem case.

Proposition 6.2. The solution of the Riemann problem (2.24) with data $A_L = 0$, $u_L = 0$; $A_R > 0$ and u_R arbitrary has exact solution consisting of a single rarefaction wave associated with the right eigenvalue $\lambda_3 = u + c$ given as

$$\mathbf{W}_{RO}(x,t) = \begin{cases} \mathbf{W}_0 & \text{if } x/t \leq S_{*R}, \\ \mathbf{W}_{Rfan} & \text{if } \leq S_{*R} \leq x/tu_R + c_R, \\ \mathbf{W}_R & \text{if } u_R + c_R \leq x/t, \end{cases} \tag{2.55}$$

where $\mathbf{W}_R$ is the right data state, $\mathbf{W}_L = \mathbf{W}_0$ is the collapsed left data state and

$$S_{*R} = u_R - 4c_R \tag{2.56}$$

is the speed of the blood/no blood front separating the collapsed state (left) from the uncollapsed new state (right) $\mathbf{W}_{Rfan}$, with

$$\mathbf{W}_{Lfan} \equiv \begin{cases} u = \dfrac{1}{5}\left(u_R - 4c_R + 4\dfrac{x}{t}\right), \\ c = \dfrac{1}{5}\left(-u_R + 4c_R + \dfrac{x}{t}\right). \end{cases} \tag{2.57}$$

Proof. The proof follows analogous steps to those of Proposition 6.1 and details are therefore omitted.   □

Now we consider the most interesting case in which no collapsed state is present at the initial time $t = 0$ but a collapsed state arises as the result of the interaction of the data states via the differential equations.

Proposition 6.3. The solution of the Riemann problem (2.24) with data A_L, u_L; A_R, u_R arbitrary but subject to the condition

$$4(c_L + c_R) \leq u_R - u_L \equiv \Delta u_{crit} \qquad (2.58)$$

has the form

$$\mathbf{W}(x,t) = \begin{cases} \mathbf{W}_{L0}(x,t) & \text{if } x/t \leq S_{*L}, \\[2mm] \mathbf{W}_0 \text{ (collapsed state)} & \text{if } S_{*L} \leq x/t \leq S_{*R}, \\[2mm] \mathbf{W}_{R0}(x,t) & \text{if } S_{*R} \leq x/t, \end{cases} \qquad (2.59)$$

where $\mathbf{W}_{L0}$ and $\mathbf{W}_{R0}$ are given by (2.52) and (2.55) respectively.

Proof. For this case to apply the condition $S_{*L} \leq S_{*R}$ must hold, from which (2.58) follows. The rest of the proof follows from Propositions 6.1 and 6.2.   □

We now show some examples of tube collapse with uniform data: Young's modulus $E = 3.0 \times 10^5$ N/m^2, exponent in tube law $m = 1/2$, wall thickness $h_0 = 5.0 \times 10^{-4}$ m, radius at zero transmural pressure $R_0 = 6.5 \times 10^{-3}$ m. Fig. 2.5 illustrates a solution of the type constructed in Proposition 6.1. The solution consists of a single left rarefaction wave associated with the eigenvalue $\lambda_1 = u - c$, in which

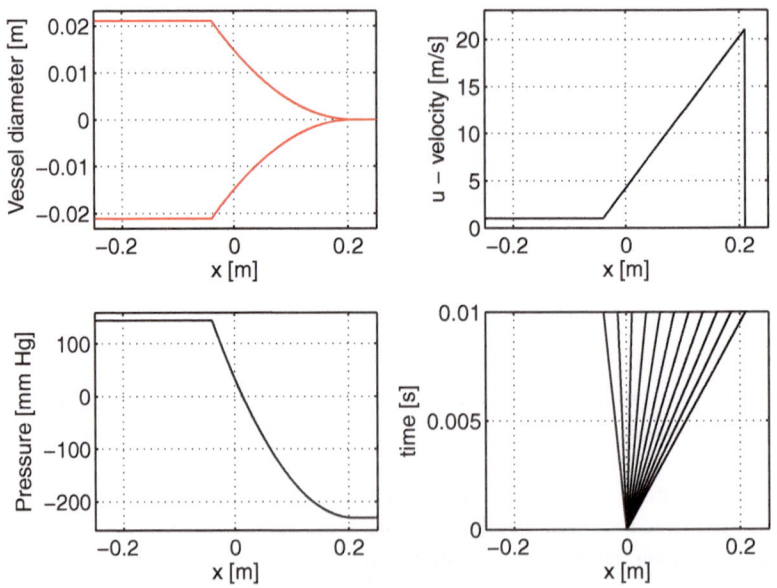

Fig. 2.5. Exact solution dry-bed Riemann problem at time 0.01 s. Initial discontinuity is positioned at $x_0 = 0.0$ m. Initial conditions are: $A_L = 3.5 \times 10^{-4}$ m^2, $u_L = 1.0$ m/s, $A_R = 0.0$ m^2, $u_R = 0.0$ m/s

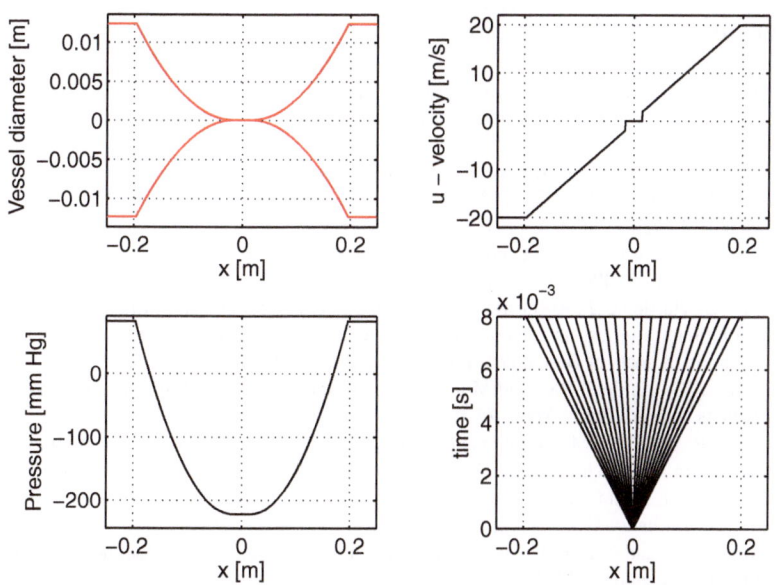

Fig. 2.6. Exact solution of tube-collapse Riemann problem at time $0.008s$ with initial discontinuity is positioned at $x_0 = 0.0$ m. Initial conditions are: $A_L = 1.2 \times 10^{-4}$ m^2, $u_L = -20.0$ m/s, $A_R = 1.2 \times 10^{-4}$ m^2, $u_R = 20.0$ m/s

the tail of the rarefaction coalesces with the blood/no blood front. Fig. 2.6 shows a typical solution constructed in Proposition 6.3 in which the middle portion of the vessel becomes dry as time evolves.

2.7 Conclusions

A simple model for blood flow in arteries with discontinuous material properties has been presented. The equations have been analysed and exact solutions have been derived, including tube collapse. These solutions can be useful in assessing the performance of numerical methods for solving more realistic problems. The formulation presented also holds promise as the basis for numerical methods for treating the more general initial-boundary value problem. Current work includes the formulation of models with more physical parameters with discontinuous variations, a detailed study of the resonance phenomenon and the implementation of numerical methods.

Acknowledgements. This research has been partially funded by the Italian Ministry of University and Research (MIUR) under the project PRIN 2007.

References

[1] Alastruey J., Parker K.H., Peiro J., Byrd S.M. and Sherwin S.J.: Modelling the circle of Willis to assess the effects of anatomical variations and occlusions on cerebral flows. Journal of Biomechanics **40**: 1794–1805, 2007.

[2] Asay B.W., Son S.F. and Bdzil J.B.: The role of gas permeation in convective burning. Int. J. Multiphase Flow **23**(5): 923–952, 1996.

[3] Brook B.S., Falle S.A.E.G. and Pedley T.J.: Numerical solutions for unsteady gravity-driven flows in collapsible tubes: evolution and roll-wave instability of a steady state. Journal of Fluid Mechanics **396**: 223–256, 1999.

[4] Canic S., Hartley C.J., Rosenstrauch D., Tamba J., Guidoboni G. and Mikelic A.: Blood flow in compliant arteries: an effective viscoelastic reduced model, numerics, and experimental validation. Annals of Biomedical Engineering. **34**(4): 575–592, 2006.

[5] Dalmaso G., LeFloch P.L. and Murat F.: Definition and weak stability of non-conservative products. J. Math Pures Appl. **74**: 483–548, 1995.

[6] Embid P. and Baer M.: Mathematical analysis of a two-phase continuum mixture theory. Continuum Mech. Thermodyn. **4**: 279–312, 1992.

[7] Formaggia L., Lamponi D. and Quarteroni A.: One-dimensional models for blood flow in arteries. Journal of Engineering Mathematics **47**: 251–276, 2003.

[8] Fullana M.J. and Zaleski S.: A branched one-dimensional model of vessel networks. Journal of Fluid Mechanics **621**: 183–204, 2009.

[9] Greenberg J.M., LeRoux A.Y., Barailles R. and Noussair A.: Analysis and approximation of conservation laws with source terms. SIAM J. Numerical Analysis **34**: 1980–2007, 1997.

[10] Isaacson E. and Temple B.: Nonlinear resonance in systems of conservation-laws, SIAM J Appl. Math. **52**(5): 1260–1278, 1992.

[11] Jeffrey A.: Quasilinear Hyperbolic Systems and Waves. Pitman, London, 1976.

[12] Kamm R.D. and Shapiro A.H.: Unsteady flow in a collapsible tube subjected to external pressure or body forces. Journal of Fluid Mechanics **95**: 1–78, 1979.

[13] LeVeque R.J.: Balancing source terms and flux gradients in high-resolution Godunov methods. Journal of Computational Physics **146**: 346–365, 1998.

[14] Liu T.P.: Nonlinear resonance for quasi-linear hyperbolic equation. J. Math. Phys. **28**(11): 2593–2602, 1987.

[15] Marchandise E. and Flaud P.: Accurate modelling of unsteady flows in collapsible tubes. Computer Methods in Biomechanics and Biomedical Engineering **13**(2): 279–290, 2010.

[16] Parés C.: Numerical methods for nonconservative hyperbolic systems: a theoretical framework. SIAM Journal on Numerical Analysis **44**: 300–321, 2006.

[17] Pedley T.J.: The fluid dynamics of large blood vessels. Cambridge University Press, 1980.

[18] Quarteroni A., Tuveri M. and Veneziani A.: Computational vascular fluid dynamics: problems, models and methods. Survey article. Comput. Visual Sci. **2**: 163–197, 2000.

[19] Schwendeman D.W., Wahle C.W. and Kapila A.K.: The Riemann problem and a high-resolution Godunov method for a model of compressible two-phase flow. J. Comput. Phys. **212**: 490–526, 2006.

[20] Sherwin S.J., Formaggia L., Peiro J. and Franke V.: Computational modelling of 1D blood flow with variable mechanical properties and its application to the simulation of wave propagation in the human arterial system. International Journal for Numerical Methods in Fluids **43**: 673–700, 2003.

[21] Toro E.F.: Shock capturing methods for free surface shallow water flows. Wiley and Sons, 2001.

[22] Toro E.F.: Riemann solvers and numerical methods for fluid dynamics. Third edition, Springer-Verlag, 2009.

[23] Umscheid T. and Stelter W.J.: Time-related alterations in shape, position, and structure of self-expanding modular aortic stent grafts: a 4-year single-centre follow up. J. Endovasc. Surg. **6**: 17–32, 1999.

[24] Zamboni P., Lanzara S., Mascoli F., Caggiati A., Liboni A.: Inflammation in venous disease. International Angiology, ProQuest Science Journals **27**(5): 361–369, 2008.

[25] Zamboni P., Galeotti R., Menegatti E., Malagoni A.M., Tacconi G., Dall'Ara S., Bartolomei I., Salvi F.: Chronic cerebrospinal venous insufficiency in patients with multiple sclerosis. J. Neurol. Neurosurg. Psychiatry **80**: 392–399, 2009.

3

Blood coagulation: a puzzle for biologists, a maze for mathematicians

Antonio Fasano, Rafael F. Santos, and Adélia Sequeira

Abstract. We present a concise summary of mathematical models for the formation and dissolution of blood clots (in other words for the process of hemostasis). For lack of space we restrict our attention to very few models, selected from a very large literature, trying to emphasize the variety of methods and viewpoints. A peculiar aspect concerning hemostasis is the fact that a new interpretation of its extremely complex biological mechanism has been found rather recently, so that most of the mathematical models should be revisited. Also in view of this fact we believed that it was absolutely necessary to write an extensive introduction to the various aspects of hemostasis, including some history, and not disregarding a description of bleeding disorders (another large field of investigation for mathematical modelling), from which much has been learned about the role and importance of each of the numerous elements intervening in hemostasis. We realize that our work is necessarily incomplete. Indeed, our conclusion is that mathematicians are still in front of the huge task

Antonio Fasano
University of Firenze, Department of Mathematics "U. Dini", Viale Morgagni 67/a, 50134 Firenze, Italy
and
IASI-CNR, Viale Manzoni 30, 00185 Roma, Italy
e-mail: fasano@math.unifi.it

Rafael F. Santos
University of Algarve, Department of Mathematics, Campus de Gambelas, 2005-139 Faro, Portugal
and
CEMAT/IST, Av. Rovisco Pais, 1049-001, Lisboa, Portugal
e-mail: rsantos@ualg.pt

Adélia Sequeira (✉)
Instituto Superior Técnico, Technical University of Lisbon, Department of Mathematics
and
CEMAT/IST, Av. Rovisco Pais, 1049-001 Lisboa, Portugal
e-mail: adelia.sequeira@math.ist.utl.pt

Ambrosi D., Quarteroni A., Rozza G. (Eds.): Modeling of Physiological Flows.
DOI 10.1007/978-88-470-1935-5_3, © Springer-Verlag Italia 2012

of keeping up with the developments of the medical theory and of the therapeutical practice of this multifaceted subject.

3.1 Introduction

Blood coagulation is a well familiar process following a wound and leading to bleeding termination (*hemostasis*). Such a goal is achieved by sealing the lesion with a clot (or *thrombus*). It is in fact much more important than that, since it intervenes in repairing small internal injuries that occur spontaneously in blood vessels. It is a physiological process which keeps alive any being provided with blood, but it can also be very harmful if it derails from its carefully planned path. Arteries or veins occlusion[1] (*thrombosis*) can be the consequence of excessive coagulation, while e.g. spontaneous bleeding is related to defective parts of the coagulation machinery. The aim of this paper is to present a concise review of the literature concerning mathematical modelling of blood coagulation. Before we come to the mathematical models, it will be necessary to synthesize the present view of this incredibly complex biological process (Sect. 2), as well as the bleeding disorders that arise as a consequence of deficiency or dysfunctions of some of the many elements playing a role in it (Sect. 3.3.1). Previous physiological models will be shortly mentioned (Sect. 3.3.2, 3.3.4).

First of all, it can be useful to go through some history. Of course bleeding and blood coagulation have been observed since ancient times. A good review article on written historical documents is [4]. We quote also the book [60]. To Huang Ti (or Di), the legendary Yellow Emperor who reigned sometime between 2698 BC and 2599 BC, and credited to have founded the Traditional Chinese Medicine[2], it is attributed the description of symptoms that clearly suggest arterial thrombosis. Hippocrates (ca. 460 BC – 370 BC) used the term *leucophlegmatia* for classes of symptoms including limb swelling, which might well include thrombotic episodes. Aristotle (384 BC – 322 BC) attributed blood coagulation to the presence in blood of a fibrous material (we know that Fibrin is a basic constituent of clots), whose removal would prevent the phenomenon. Only at the beginning of the XVIII century, F. Ruysch in Amsterdam did report experiments of that kind [77]. Aristotle, however, said that clotting was due to the loss of the heat generated in blood (kind of congealing), an idea which influenced later scientists (as other wrong statements by him), due to his enormous reputation. The term thrombosis (from the Greek thrombos = clot) is due to Galen (ca. 130 AD – 200 AD), who proposed in particular a scheme (though far from reality) of the circulatory system. However he still worked

---

[1] Artery thrombosis is painful, while venous thrombosis is not, since veins are not equipped with nerves. Venous thrombosis presents the risk of pulmonary embolism, which occurs when a thrombus fragment gets into the bloodstream reaching a lung.

[2] Many discoveries and inventions (the compass, the Chinese calendar) have been attributed to the Yellow Emperor in a legendary form, but his actual existence is a matter of debate.

in a framework (the Hippocratic *humoral theory*[3]) quite far from the modern scientific approach and that was going to influence the studies of medicine for long time. We must bear in mind that the discovery of cells dates back to 1655 (Robert Hooke) and that Red Blood Cells (RBCs) were first described in a frog by the Dutch biologist Jan Swammerdam in 1658 and later (1674) by the initiator of microbiology, Antoni Ph. van Leeuwenhoek (1632–1723) (a Dutch too). Therefore, before those dates, only a qualitative approach to the blood coagulation problem can be found. Many physicians described pathological conditions in women during pregnancy and post partum that were actually related to thrombosis, but without identifying the real cause, though the idea that oedema of a limb can be caused by the obstruction of a blood vessel was occasionally put forward. The celebrated Italian physician Marcello Malpighi (1628–1694) isolated a fibrous component in the clot. The French Jean-Louis Petit (1674–1750) related the formation of clots with the phenomenon of hemostasis in a time in which medicine was already gradually adopting a rigorous scientific attitude[4]. Nevertheless, a real progress in the study of blood coagulation was still quite far.

Thrombi formation and pulmonary embolism were described in 1846 by Rudolf Virchow [93], still today remembered for the "Virchow triad", schematically expressing that vessel wall, altered flow conditions and coagulation are mutually interacting entities [49] (his contributions to medicine were far more important than the formulation of such a triad). However, a substantial approach to understanding coagulation had to wait until the discovery of *platelets* (see [11, 29, 70]), due to Max Schultze (in 1865) and to Giulio Bizzozero (1846–1901), who understood their role in blood coagulation (1881) as Fibrin producers and coined the Italian name *piastrine* and the German *Blutplächtten*. Fibrin and its precursor *Fibrinogen* were discovered by Alexander Schmidt (1831–1894), who also conjectured the existence of an enzyme responsible for the corresponding transition, to which he gave the name presently used *(Thrombin)* and of its precursor *(Prothrombin)* [81]. The way was opened to a real scientific investigation of blood clotting. We will return to the early attempts of modelling the blood coagulation process in Sect. 3.3.4. Here we stress that all branches of medicine dealing with blood are intimately related to molecular biology ([67]), and blood clotting can only be explained through a cascade of chemical reactions. In conclusion, we are dealing with a substantially young and still evolving discipline. Moreover, another frightening step is awaiting the daring mathematicians: blood rheology. Indeed, from the analysis which follows it will be clear that no model for blood clotting can be formulated without specific reference to the fluid dynamical regime in which it takes place. Thus, while the biological puzzle has gradually taken shape (through laborious assembling and disassembling), mathematicians (in the authors' opinion) are still wandering in a maze…

---

[3] According to Hippocrates four *humours* (blood, phlegm, black bile and yellow bile) had to be in a proper balance in healthy individuals.

[4] A cornerstone in this process was the publication of the book The *Philosophical Principles of Medicine* (1725) by Thomas Morgan, philosophically inspired to Newton's *Principia* (1687).

3.2 The modern view of blood coagulation

The aim of this section is to explain the most recent model of the blood coagulation process in a form accessible to people not having a specific knowledge in the field. Therefore we shall avoid the intricacy of the chemistry of the many proteins participating in the formation of a blood clot (as well as the description of their highly complicated structure), and we will concentrate on illustrating their action, presenting a general sketch of the phenomenon and trying to use a plain language. The circulatory system is equipped with an incredibly complex chemical and mechanical machinery ready to repair lesions which may occur to blood vessels by sealing them with a *clot* (or *thrombus*). A clot is a gel like structure consisting of a polymer (*Fibrin*) network entrapping various blood components. The ingredients necessary to lead to the clot formation (and that will be described in detail very soon) are either present in the blood, or reside in the endothelium of the blood vessels, namely in the external membrane (*adventitia*), in the main body of the vessel (*tunica media*), and in the innermost, thin membrane (*intima*). Actually, the ones contained in the blood vessel walls become available immediately after a lesion of the intima, so even in the presence of a tiny internal lesion. The important thing to notice is that in normal conditions they do not come in contact with blood.

Clearly, it is absolutely crucial that the clotting mechanism is set in motion only when it is really necessary, remaining silent in normal conditions, and at the same time that the coagulation process is terminated before it occludes the vessel, allowing blood to flow normally. Indeed, after the process of *hemostasis* [5] is completed, the thrombus will be gradually removed by means of another process known as *fibrinolysis*. To be more precise, it is not correct to look at these processes in a sequential way. The hemostasis machine and the fibrinolysis machine are both active at the same time and whether a thrombus is growing or retreating is the result of an unbalance between the two processes. Besides the physiological process of hemostasis following an injury, it is well known that unwanted coagulation can take place due to temporary reduced blood flow rate or stasis (or sometimes in stagnation points of the flow generally accompanying abnormal vortices), leading to *deep vein thrombosis* (DVT), a disease also induced by bone traumatization e.g. during prosthetic implants. We will briefly deal with that phenomenon too in the next section. In order to explain at least the basic steps of hemostasis it is advisable to describe the various elements contributing to the final formation of a clot. It is important to get some knowledge of the functions performed by these substances, and the reader can use the following list as a guideline through the core of the model.

---

[5] In all words with the prefix *hemo-* the spelling *haemo-* can also be used, though not frequent in the medical literature.

3.2.1 The protagonists of hemostasis

Platelets (thrombocytes)

Platelets are cells with no nucleus, produced in the bone marrow and specialized to perform a number of functions in the clotting process. They have a diameter of 2–4 µm, a life span of 5–9 days, and a discoid shape. However, during the clotting process they undergo deep morphological changes, emitting protuberances (*pseudopods*) which favor mutual aggregation, as well as adhesion to other elements constituting the clot. This is one of the many phenomenon following *platelets activation*. Their normal concentration in blood is 1.5–$4 \times 10^5/\text{mm}^3$. [6]

Their membrane participates actively in the coagulation process. It is provided with receptors (we will explain the functions of the receptors GPIb and GPIIb-IIIa), intervening in the binding process with other elements. They possess corpuscles, called α *granules* and *dense (or δ) granules*. The first class is able to secrete some among the substances taking part in the coagulation process (namely *Fibrinogen* and vWF, see below), while the dense granules contain ADP (AdenosineDiPhosphate), *serotonin* (5-hydroxytryptamine) and *Calcium* (contributing to platelets activation when released), and synthesize *Thromboxane A2* and other platelet activators, which are inhibited by aspirin (explaining the well known anticoagulant action of that drug). Already from this preliminary illustration one can infer that platelets intervene in the clotting process boosting a rapid positive feedback. The list of substances stimulating platelets is much longer (see [19]).

The numbered Factor pairs

Many proteins (generally called *Factors*) which intervene in various ways in the clotting process come in two forms: nonactivated and activated. They are numbered (usually with Roman numbers) in the chronological order of their discovery. So for instance we have the pair FVII, FVIIa, denoting the two respective forms of Factor seven. The nonactivated form is referred to as *zymogen* and once activated it becomes an *enzyme* (protease), namely a molecule able to modify other complex molecules by cleaving some strategic chemical bond. In normal conditions only the nonactivated factors are present in the blood, with the notable exception of FVIIa, which circulates along with FVII, though in minimal concentration. However, it must be said that its activity manifests itself only in the presence of the *Tissue Factor* (see below), which becomes available as a consequence of an injury. The role of each factor in

---

[6] The data referring to blood and its components are reported here just as an order of magnitude. They vary according to sex, size and health state of the individual and in the same individual even during the day. Some more information about other cells in the blood can be useful. There are approximately 5 liters of blood in the human body. 45 % of the volume is made of cells, the rest (plasma) is 92 % water: Red Blood Cells (RBCs) or erythrocytes: diam. 7 µm, concentration $5 \times 10^6/\text{mm}^3$, lifespan 120 days (approximate data), no nucleus; White Blood Cells (WBCs) or leukocytes have a nucleus and come in several different types (among which the monocytes have some importance in the coagulation process), concentration $6 \times 10^3/\text{mm}^3$, their size is much larger than the one of RBCs [12].

the clotting process has been determined experimentally by suppressing it in mice and observing the correspondingly induced bleeding disorder.

Among the many factors that will be mentioned later, we want to point out immediately the importance of FII (*Prothrombin*), FIIa (*Thrombin*), FI (*Fibrinogen*), FIa (*Fibrin*). Thrombin performs various different actions of fundamental importance that will be discussed in detail. In particular, in the most active stage of the process, it acts on Fibrinogen, changing it to a form which polymerizes rapidly, eventually giving rise to the Fibrin network which makes the skeleton of the gel trapping the various blood component (i.e. the clot).

von Willebrand Factor (vWF)

This factor deserves a special attention because of its multiple roles. It bears the name of the Finnish physician Erik von Willebrand, who first described (1926) a bleeding disorder (today called von Willebrand disease), distinguishing it from the ones known at that time. The disease was later associated with the deficiency of the factor now bearing his name. vWF is a massive multimer (a chain of identical elements, not as numerous as in a polymer, it can be an oligomer) which is present in different lengths and is sensitive to shear stress, in the sense that stretched molecules are more active. The largest variety resides in the endothelium of blood vessels wall. vWF also circulates in the blood (one of the largest circulating proteins), associated in a *complex* (see below) with FVIII (the vWF bound in the complex is not active). It intervenes in the so called *primary hemostasis* (Sect. 3.2.2) and during the main phase of clotting, because it binds with platelets in the presence of a sufficiently high stress (hence it brings a coupling of the biochemistry with the mechanics).

The vast majority of vWF is secreted in the endothelial cells, and a small percentage is stored in cytoplasmic granules (Weibel-Palade bodies) of many cells. We have seen that it is also contained in the granules of platelets and that it is secreted after their activation. The circulating FVIII-vWF complex is dissociated during the fastest phase of the coagulation process, thus providing an intense source of both factors in a free state.

A very detailed analysis of the structure and of the functions of vWF can be found in the excellent paper [78]. Another more concise and equally recommendable reading is [76].

Tissue Factor (TF)

Also known as *Thromboplastin* (or FIII), TF is expressed by cells which are normally not in contact with blood. It is known (see e.g. [19])[7] that it is also produced by some WBCs (more precisely by *monocytes*) for instance as a consequence of inflammatory processes. Today there is a general consensus on the fact that clot formation is initiated by the exposure of TF to blood at the injury site. See also the section on Deep Vein Thrombosis (Sect. 3.3.3).

---

[7] TF production by monocytes is not a recent discovery. See e.g. [34].

Complexes in the blood coagulation process

Factors may combine in complexes which have a specific and crucial task in blood clotting. We have already mentioned the FVIII-vWF complex as the natural source of FVIII in blood. Let us list other important complexes having an active part in clotting.

- FVIIa-TF: converts more FVII to FVIIa, thus providing more of the same complex; activates FIX, FX.
- FVa-FXa (*Prothrombinase*): in the presence of ions $Ca2^+$ it stimulates the *key transition FII → FIIa (Thrombin)*.
- FVIIIa-FIXa-Ca^{++} (*Tenase*): generates FXa very effectively on platelets surface.

Inhibitors of blood coagulation

Clotting is terminated (and prevented when it is not necessary) by the combined action of some inhibitors:

- TFPI (*Tissue Factor Pathway Inhibitor*), secreted by endothelium, blocks the complex FVIIa-TF and inactivates FXa (by forming a complex with it).
- AT-III (*Antithrombin-III*), present in plasma, quenches Thrombin production operating on most of the activated factors, and neutralizes existing Thrombin. Heparin[8] strongly enhances the action of AT-III, thus preventing coagulation.
- *Protein C* (PC): when Thrombin combines with the vascular *Thrombomoduline* (TM) it activates PC to APC, which complexes with *Protein S* and binds to FVa and FVIIIa, switching off the production of Thrombin. *Thus Thrombin is at the same time the engine and the brake of blood coagulation*. Protein Z is also an inhibitor of FXa.

Remark 1. A very special role in blood coagulation is played by *Vitamin K* (actually a group of vitamins) which intervenes in the activation of FII, FVII, FIX, FX and in the regulation of Proteins C, S, Z (which are said to be vitamin K dependent). The widely used anticoagulant *warfarin*[9] inhibits Vitamin K action.[10]

3.2.2 Primary hemostasis

Primary hemostasis precedes the main coagulation process and provides a rapid sealing of the wound via platelets adhesion to collagen and to vWF immobilized in the endothelial tissue exposed to blood. Such a platelet aggregate is sometimes called a *white thrombus* (to distinguish it from the blood clot still to be formed, the *red*

---

[8] Heparin was discovered in 1918 [39]. It also has the side effect of reducing the platelets count (*Heparin Induced Thrombocytopenia*, HIT), see e.g. [43], accompanied by an increase of Fibrinogen level.

[9] Better known in Europe with the trade name *Coumadin*. Patented in 1948 as a rat poison and used as anticoagulant for humans since 1954.

[10] The very first action of warfarin is to favor clotting, via the inhibition of PC, PS, PZ, while the antithrombotic effect takes place with some delay. For this reason warfarin is administered in combination with heparin.

thrombus). Platelets-vWF adhesion is by no means a simple phenomenon. It is mediated by mechanical stress and occurs in two steps. We have mentioned the two platelets membrane receptors GPIb and GPIIb-IIIa, which both have a complex internal structure that will not be described here. In the high stress conditions present at the vessel wall[11] the receptors of the first class have a large adhesion rate to vWF (at some specific site: we omit such details), but they also have a large dissociation rate. This results in a relatively slow motion of platelets over the wound site. The receptors of the second class have a much lower combination rate, but their binding to vWF is irreversible. The persistence time of platelets over the wound site is dilated by the GPIb-vWF interaction and this permits an efficient intervention of the GPIIb-IIIa receptors. It must be remarked that the role of shear stress is of crucial importance in activating the interaction between platelets and vWF. This is true also in a later phase in which the same kind of bonds are established and have a key role in the formation of the clot. Platelets cross binds are also provided by fibrinogen.

The above description is extremely concise. A basic reference is [40], which focuses on the role of receptors not only in the binding action, but also on the regulation of the platelets cytoskeleton and on their mechanical behaviour. A particularly interesting aspect is platelets "rolling" along the blood vessel wall. While the rolling of WBCs (which possess a spheroidal shape) was described already in 1839 [94], only much later [46] the mechanism of leukocyte tethering to walls was explained. That platelets also exhibit rolling followed by tethering "when required" was only established in 1995, [27]. Platelets are naturally more concentrated in the periphery of blood vessels in a regime of laminar flow [91], however their discoidal shape is not best suited for rolling. Indeed, as soon as platelets become exposed to fixed vWF they experience a rapid morphological conversion, assuming the shape of "spiny spheres", which greatly favors rolling. This and the subsequent morphological modifications require cytoskeleton remodelling in which both the receptors-vWF interaction and the shear stress play a role. As a matter of fact, it has been observed that such modifications occur very slowly under low stress conditions.

3.2.3 The cell-based model for the blood clot formation (secondary hemostasis)

A blood coagulation model that was widely accepted for almost forty years is the one that became known as the *Cascade model*, because it recognized for the first time that clotting was the result of cascades of chemical reactions. Its origin can be traced in two papers appeared independently in 1964 ([18, 48]). The model was characterized by an *intrinsic pathway* (originated in the blood stream) and an *extrinsic pathway* (initiated at the wound site), both merging into a *common pathway*. It was

---

[11] Roughly speaking, a strain rate of $1000\,s^{-1}$ is considered to fully activate the adhesion process. In arterioles in the range of 10–$50\,\mu m$ there can be strain rates in the range 500–$5000\,s^{-1}$ [88]. Stenosis (lumen reduction) due to atherosclerotic plaques can raise strain rate beyond $10000\,s^{-1}$. Of course translating strain rate into shear stress requires a rheological model for blood. This is a delicate and controversial matter to which we will return later on. Here we just point out that, in view of the relevance of stress, the size of the vessel generally plays an important role.

a fundamental step in the comprehension of the process, but as early as 2001 [37] some evidence started emerging that the theory could need a deep revision. Experimental facts that could not be explained in the framework of the cascade theory were brought to attention in 2005 [38]. We will return to the 3-pathway cascade model and its difficulties after we have exposed the currently accepted coagulation theory (the so-called cell-based model). The extreme synthesis presented below is mostly summarized from the papers already quoted about the vWF and from the following articles: [15, 19, 36, 50, 51, 71, 85, 86].

The cell-based model identifies four distinct steps.

Initiation

When blood comes in contact with TF the tiny portion of FVIIa in the blood rapidly produces a small quantity of the FVIIa-TF complex on the surface of the TF bearing cells, with a twofold consequence: more FVII is activated, thus increasing the amount of the same complex, and small amounts of FIXa and FXa are generated. FXa is an activator of FV (available in blood) to FVa (though with not a high rate constant) and then it binds to it on the TF bearing cell, forming the complex FVa-FXa *(Prothrombinase)*. The latter acts on Prothrombin (FII), so that a small amount of Thrombin (FIIa) becomes available. It is precisely this very small quantity of Thrombin which triggers the main process (Fig. 3.1).

Some more considerations about the initiation stage are in order. There is evidence that some coagulation factors can reach TF in nonvascular tissues even in the absence of a lesion [99], thus making FIXa and FXa accidentally available. However coagulation is not started because other indispensable ingredients (platelets and the complex FVIII-vWF) stay only in blood. Moreover, a legitimate question is why coagulation initiated at a wound site proceeds to the clot formation at that site, without spreading along the vessel. The answer is that FXa that happens to leak from the initiation site is rapidly inhibited by TFPI and AT-III (having a much milder action on the FVIIa-TF complex), while whatever Thrombin that reaches healthy endothelial cells is neutralized by binding with TM (see *Inhibitors*).

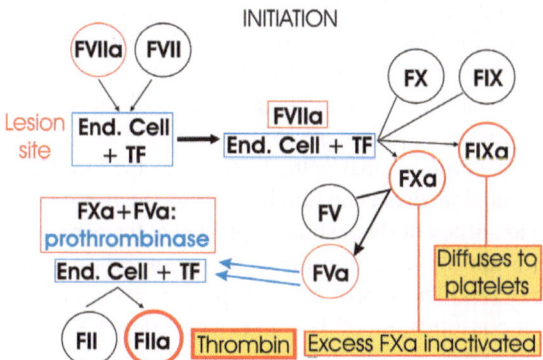

Fig. 3.1. Cell-based model: *Initiation phase* of blood coagulation

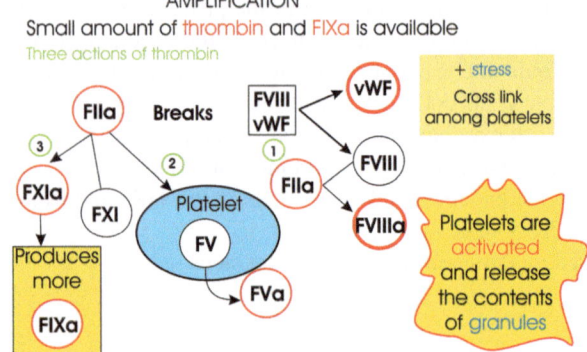

Fig. 3.2. Cell-based model: *Amplification phase* of blood coagulation

Amplification

This is a transition stage towards the final burst of Thrombin (Fig. 3.2). FIXa produced in the previous step can dissociate and diffuse to platelets that keep accumulating at the wound site. It is not affected by TFPI and (differently from FXa) only very slowly by AT-III, so it becomes available to platelets. The small amount of Thrombin produced in the previous stage activates FV to FVa on platelets surfaces and breaks the complex FVIII-vWF, at the same time activating FVIII to FVIIIa. The free vWF acts as a cross link for activated platelets in the presence of stress. A similar action is performed by fibrinogen. One more activation performed by Thrombin is the one of FXI[12] to FXIa, stimulating further production of FIXa. These events trigger the main phase of coagulation. Platelets are now activated.

Propagation

The now available FVIIIa and FIXa readily produce the *Tenase* complex FVIIIa-FIXa, whose main task is to provide FXa on the platelets surface. We recall that FXa (having the key role of producing the *Prothrombinase* complex together with FVa), present in the early stage of the process, cannot diffuse and therefore it has to be produced effectively within the growing thrombus. Factor Va, produced by Thrombin in the amplification stage, binds to FXa (as it did in the initiation stage), so that Prothrombinase produces more Thrombin, which in turn creates more FVa. Thus all the key ingredients enter a highly positive feedback process (about 95 % of Thrombin is produced during this stage [51]), see Fig. 3.3.

The abundance of Thrombin makes the final step of the Fibrin network formation practically immediate. Platelets keep binding among themselves thanks to the vWF liberated (as long as sufficient shear stress is present at the periphery of the clot) and they bind to Fibrin too (the resulting forces acting on Fibrin lead to a mechanical contraction of the clot). The Fibrin network progressively entraps all blood constituents. A further consolidation of the clot is induced by FXIIIa, also produced by

---

[12] A large debate is going on the origin and the functions of FXI [56, 83].

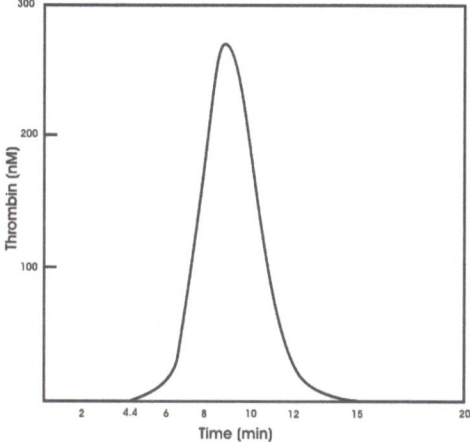

Fig. 3.3. Typical curve illustrating the time behaviour of Thrombin concentration during the clotting process

Thrombin from its precursor FXIII (also known as *Laki-Lorand* factor), which acts on polymerized Fibrin, stimulating the formation of cross links in the network (see Fig. 3.4). The importance of the latter stage is confirmed by the bleeding disorder associated to FXIII deficiency.

Termination

Stopping the self exciting mechanism of clotting is as important as initiating it (Fig. 3.5). Here too, Thrombin has a key role. We have already seen (see initiation phase) that the complex *Thrombin-TM* helps confining the clot growth at the injury site. This complex has two opposite effects on the clot. One is to contrast fibrinolysis (see the next subsection), the other is to give rise to the *APC-ProS* complex, possessing an inhibiting action on FVa and FVIIIa, so that no more Prothrombinase can be produced, interrupting the coagulation process. In the paper [86] the author questions the real ability of APC-ProS to neutralize FVa far from endothelial cells, leaving some doubt on the actual termination mechanism.

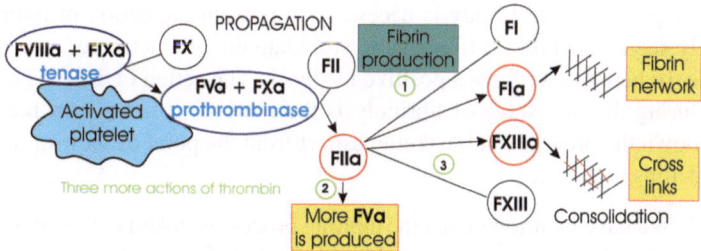

Fig. 3.4. Cell-based model: *Propagation phase* of blood coagulation

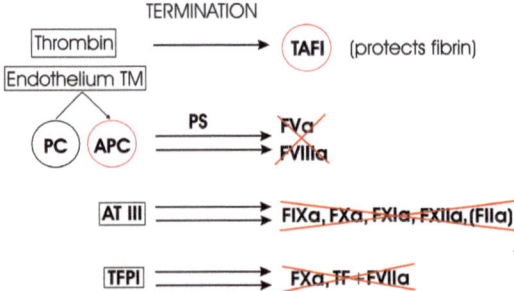

Fig. 3.5. Cell-based model: *Termination phase* of blood coagulation

3.2.4 Fibrinolysis

Clots respond to an immediate need and persist for the time needed to healing the wound, i.e. repairing the injured tissues, which is another complicated process that will not be dealt with here. However they cause an alteration of the natural blood flow and they have to be removed. Their elimination must be gradual, without delivering in the blood stream fragments that may cause embolism. Thus demolishing a blood clot is a delicate and important operation that has its own chemical machinery. A clot must incorporate some fibrinolytic element already during the growth process. Very much like coagulation, fibrinolysis proceeds through activation of (not numbered) factors, which have in turn their own inhibitors. Here is a list of the intervening substances:

- *Plasminogen* (secreted by the liver);
- *Plasmin* (activated plasminogen): it is responsible for the dissolution of Fibrin;
- *tissue Plasminogen Activator (tPA)*;
- *urokinase(uPA)*;
- *Plasmin Activator Inhibitor 1 and 2 (PAI-1, PAI-2)*;
- *antiplasmins* (plasmin inhibitors): α *2antiplasmin* and α *2macroglobulin*;
- *Thrombin Activatable Fibrinolysis Inhibitor (TAFI)*.

Plasminogen is converted to plasmin by tPA and urokinase. It binds to Fibrin during the clotting process, so its activation can immediately prompt fibrinolysis in situ. However, activators are slowly emitted at the injury site, so that plasmin is produced in a progressive way and the clot regression takes place at a low speed. Embedding of plasminogen in the clot body is necessary to prevent the action of inhibitors at a too early stage. Plasmin itself is able to stimulate the production of more tPa and urokinase, so here too we have a positive feedback. The action of TAFI is to protect Fibrin. During the first stage of fibrinolysis the Fibrin fragments may recombine, slowing down the process (an intriguing aspect from the point of view of mathematical modelling) (Fig. 3.6).

Remark 2. We have mentioned that the clotting process is linked to the fluid dynamical problem via the shear stress exerted by blood. In [85] it has been suggested that one more coupling of the chemistry with the flow can be created by convection,

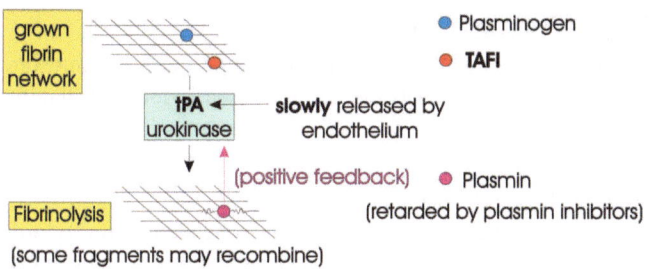

Fig. 3.6. The pathway to fibrinolysis

possibly removing activated factors. The authors illustrate this concept by means of numerical simulations. They refer mainly to FXa (entering the Prothrombinase complex on the surface of activated platelets). This aspect can be rather interesting in formulating a mathematical model.

3.3 Congenital bleeding disorders and the inadequacy of the 3-pathway Cascade model

We have placed the analysis of the drawbacks of the Cascade model in this section, because the illustration of bleeding disorders is a valuable help in understanding the reasons that led to its rejection. The shift to the cell-based model is too recent to simply ignore the existence of the Cascade model, that has been used in mathematical models till very recently. Diseases due to dysfunctions of the system regulating blood coagulation have particular interest from the point of view of mathematical modelling since a correct model for coagulation should also predict how specific anomalies produce bleeding disorders. Despite the obvious interest of this subject, our exposition will be extremely synthetic, because bleeding disorders are not the main objective of the present review, though they can be a remarkable subject for mathematical modelling.

3.3.1 Congenital bleeding disorders

Congenital bleeding disorders can be associated with the lack or deficiency of some of the factors involved in blood coagulation. This fact provides a criterion to review them. This is an impressively large chapter that we cannot treat extensively, nor exhaustively. We will just go trough a synthetic and partial list.

Hemophilia[13] A (FVIII deficiency)

The disease is caused by a defective gene located on the X chromosome. Females have two copies of the X chromosome and the disease does not manifest if one of the chromosomes is normal. It is the most common coagulation disorder.

Hemophilia B (FIX deficiency)

It is also known as Christmas disease, because it was identified (Oxford, 1952) and distinguished from Hemophilia A in a boy whose name was Stephen Christmas (he died in 1993, a victim of HIV contracted by blood transfusions). The defective gene is also located on the X chromosome.

Hemophilia C (FXI deficiency)

Also known as Rosenthal's Disease. The fourth most common coagulation disorder, after the preceding two forms and vWD (see below). It is a much milder disease, also discovered in the 1950's.

Remark 3. FVIII, FIX, FXI are also called *Antihemophilic Factors A, B, C,* respectively. FIX is sometimes called *Christmas Factor.* For a detailed description of Hemophilia A and B see [30]. According to the cell based model, FXIa is the only activator of FIX, thus one may wonder why hemophilia C is milder than hemophilia B. An explanation can be the ability of platelets to produce FXI even in individuals defective of plasma born FXI [28, 66].

Parahemophilia (FV deficiency)

A rare disease. Here we also mention ([8]) that there are mutations of FV that make FVa resistant to APC, thus prolonging Thrombin release: the so called *Leiden* mutation of FV is believed to be the most common genetic cause of thrombosis.

von Willebrand Disease (vWD: vWF deficiency)

In a small village in the Åland Islands von Willebrand studied an inherited and severe bleeding disorder hitting consanguineous families. It was 1924. He recognized that the disease was not one among the known congenital bleeding diseases, though he could not identify the cause. vWF deficiency affects both primary and secondary hemostasis. vWD comes in different types (1: partial deficiency, 2: partially defective vWF, 3: total deficiency). It may induce a deficiency of FVIII, of which vWF is a carrier in blood, as we have seen. A form of vWD-2 is hyperactivity of vWF leading to excessive platelet adhesion. Hyperactivity is caused by deficiency of the enzymes that shorten the vWF multimer, leaving only the larger, more active species. A study performed on pigs affected by vWD ([69]) showed that platelets failed to undergo significant morphological changes, thus proving that vWF has a key role

---

[13] Abbreviation of the term haemorrhaphilia, coined by the German physician Friedrich Hopff (1828).

in stimulating the emission of filopodia. Many more details on vWD can be found in [5].

Thrombocytopenia (scarcity of platelets)

It can be genetic or acquired, or induced by other illnesses or medical treatments (see e.g. [43] about Heparin Induced Thrombocytopenia (HIT)). Symptoms can be light or severe, depending on the platelets count. The adjective *purpura* often associated to this illness refers to the appearance of bruises on the skin. Two forms of *Thrombocytopenic purpura* are known: *idiopathic* (denoting that the cause is unknown) and *immune* (the patient's antibodies attacking platelets). The acronym ITP is good for both. A mathematical model aimed at predicting the efficacy of splenectomy (spleen removal) in the cure of ITP has been developed in [12].

Dysfunctions of platelets receptors

They prevent binding of platelets to vWF and other elements. In this category we find *Glanzmann's thrombastenia* [31], and *Bernard-Soulier syndrome*, [7]. It has been observed that the latter disease affects the cytoskeletal architecture (producing the so-called giant platelets), thus emphasizing that normal functioning of receptors is important in platelets morphology. There are many more disorders related to platelets dysfunctions: dysfunctions of α granules (*Grey Platelets Syndrome* and *Quebec Platelets Disorder*), of δ granules (*Hermanski-Pudlak, Chédiak-Higashi*, and *Griscelli syndromes*), defective platelets (*Wiskott-Aldrich syndrome*: affects aggregation, *Scott syndrome* and *Stormorken syndrome*: failure of the procoagulant activity of platelet surface, or non-stimulated activation, respectively). For more information on platelets related bleeding disorders see [14].

Other rare disorders

Afibrinogenemia (Fibrinogen (FI) deficiency or malfunction), Hypoprothrombinemia (FII deficiency), FVII deficiency (first recognized in 1951, may produce severe bleeding symptoms), *FX deficiency, FXIII deficiency* (may produce bleeding and can be associated with dysfunctions of the WBCs generating FXIII), *Tissue Factor deficiency, Combined FV+FVIII deficiency* [87].

Thrombophilia

Besides disorders caused by the deficiency or dysfunction of elements intervening in the coagulation process there are also disorders related to deficiency or dysfunction of elements involved in fibrinolysis. A typical example is plasminogen deficiency. The consequence is a higher risk of thrombosis, but there can also be other effects (defective wound healing, ligneous conjunctivitis, etc.). The risk of thrombosis is increased also by deficiency (type I) or dysfunction (type II) of the coagulation inhibitors AT, PC and PS. We recall that the Leiden mutation of FV (making it resistant to APC) is also a cause of thrombophilia.

Disseminated Intravascular Coagulation (DIC) is a serious pathological condition leading to the formation of clouds of microthrombi in the bloodstream. A mathematical model of DIC will be illustrated in Sect. 3.4.2.

FXII deficiency

It produces no significant symptoms and is usually discovered incidentally during clinical tests, since it causes some retardation of coagulation. In this way it was first diagnosed in 1955 by O. Rotter in a man called John Hageman (FXII, so far not mentioned in our review, is also called *Hageman Factor*). The paper [73] provides an interesting historical review of the early works on this subject. We list this disorder in the last place precisely because it provides a link with the next section. Indeed, FXII was believed to be the initiator of the so called intrinsic pathway of the Cascade model. FXII has arisen new interest in different contexts. We will return to it in Sect. 3.3.3 and 3.4.2.

3.3.2 The limitations of the 3-pathway Cascade model

It is now time to consider the 3-pathway Cascade model, based on the intrinsic, extrinsic and common pathways.

Intrinsic pathway

So called because it begins within the blood. It starts with the (slow) conversion of FXII to FXIIa, attributed to a platelet product (HMWK, *High Molecular Weight Kininogen*). FXII activation is then accelerated, since FXIIa converts *Prekallikrein* to *Kallikrein* (proteins which missed to receive a Factor number), which is a fast activator of FXII. The target of FXIIa is FXI, activated to FXIa, which we already know to produce the transition of FIX to FIXa. The next step is the activation of FVIII and the consequent production of tenase, in turn generating FXa with the help of Calcium ion. From now on the intrinsic pathway merges with the extrinsic pathway. See also Sect. 3.3.3.

Extrinsic pathway

This simply consists in the combination of FVIIa with TF, which eventually leads to FXa production.

Common pathway

It basically follows the scheme of the Propagation Phase described for the cell-based model.

Remark 4. In the Cascade model the intrinsic and the extrinsic pathways can be initiated independently from each other.

Despite the strict analogies, the Cascade and the cell-based models have two fundamental differences.

- The cell-based model ignores FXII, which in the cascade model can trigger coagulation.
- The extrinsic pathway bypasses FVIII, making it not strictly necessary. Now two important experimental features must be taken into account:
  a. FXII deficiency (described above) has little effect on coagulation;
  b. FVIII deficiency is the cause of hemophilia A, which on the contrary is an important coagulation related illness.

Differently from the 3-pathway cascade model, the cell-based model is perfectly compatible with a, b. The above argument does not exclude that some coagulation can be triggered by FXII activation (we will return to this point in the next section). The influence of the 3-pathway Cascade model has been so strong that it survives even in recent books [16].

3.3.3 Deep Vein Thrombosis (DVT) and coagulation on artificial surfaces

Having established the concept that blood coagulation is initiated by FVIIa coming in contact with TF exposed by an injury and following a scenario of shear stress mediated platelets-vWF adhesion, it is legitimate to ask how spontaneous thrombi formation can occur in a regime of reduced (or altered) blood flow or blood stagnation. We are talking about the ill famed "economy class syndrome", affecting passengers who stay seated for too long time during overseas flights[14]. In that case there is no injury and there is no mechanical stress, therefore such a pathogenic thrombosis must be initiated within the blood (i.e. has to be *intrinsic*). Atrial fibrillation can also trigger coagulation increasing the risk of strokes, thus requiring a preventive anticoagulant treatment. Coagulation on artificial surfaces poses a very similar question, since endothelium TF is not available. In both cases, if the role of FXII is excluded, a source of TF must be identified to initiate the process. The explanation of DVT onset comes from the ability of monocytes to express TF in abnormal conditions. Apparently, slow flow or stagnation (which in turn produce an abnormal environment, for instance low oxygen concentration levels) stimulate monocytes to produce TF. TF production by monocytes is a rather complex process in which other cells are involved (including platelets). For an explanation see [19]. We also remember that DVT is greatly facilitated by the tendency of RBCs to aggregate in low shear stress conditions [80]. In the paper [50] researches on the possible "pathological" delivery of Tissue Factor from the blood vessel wall are discussed. Coagulation has been observed to occur on artificial surfaces implanted in patients (e.g. blood pumps). Essential features of thrombus growth are Thrombin production (without which no Fibrin would be available) and surface adhesion. Concerning Thrombin, it can be conjectured that (like in the case of DVT) monocytes are stimulated to produce TF as a reaction of the immune system to the exposure to artificial material. More dif-

---

[14] DVT normally affects legs. It occurs much less frequently in arms. DVT of axillary or subclavian vein is known as *Paget-Schrötter disease*.

ficult is to explain the adhesion process, since platelets receptors are specifically designed to bind e. g. to vWF or to Fibrin. A reasonable explanation is that Fibrin itself, once produced, can adhere to the surface, providing a suitable substrate to platelets. In this scenario, the shear stress induced by mechanical devices activates platelets (even a modest hemolysis (distruction of RBCs), releasing ADP, can help activating platelets). Let us add that recently in [79] it has been proposed that FXII can contribute to Thrombin formation, by *autoactivation* to FXIIa in the presence of an injury or of *artificial surfaces*, somehow reviving its role in intrinsic coagulation. Indeed it was already observed by O. Rotter (the discoverer of Hageman Factor) that FXII can become activated in contact with glass [68]. Still in [79] it is suggested that FXII may have a more important function in angiogenesis.

3.3.4 Some historical remarks on coagulation models

In the Introduction we have reported some historical notes about observations of blood coagulation. Modelling attempts came much later. Whereas Thrombin formation and embolism were already known by 1860 ([93]), a first scheme of coagulation was proposed in 1905 by Paul Morawitz ([54]). It was based on just four factors, namely the last four in the larger modern scheme, FII, FIIa, FI, FIa, and assumed that thromboplastin (TF) and calcium were responsible for the conversion of Prothrombin into Thrombin. As we said, the theory of the 3-pathway Cascade, that came in 1964, dominated for about forty years. The reader may wonder why in the course of the exposition of the blood coagulation process we have mentioned Factors from I to XIII, but we omitted FIV and FVI (we also pointed out that FIII, i.e TF, has an abnormal position in the list, since it actually comes only in the active form). FIV and FVI are out of the current terminology. FIV is nothing but the ion Ca2+, and FVI used to designate an intermediate molecule between FV and FVa, but it was later identified with FVa. The existence of FVI was suggested by the discoverer of FV (see P. Owren [62]).

3.4 Mathematical models and numerical simulations

Comprehensive mathematical models of the blood coagulation and fibrinolysis processes, describing either the classical concept of *coagulation cascade* or the recently adopted *cell-based model*, and taking into account physiologic and mechanical factors, have the potential to identify regions of the arterial tree susceptible to the formation of blood clots and could contribute to a better understanding of these complex phenomena. Kinetic models generally consist of a nonlinear system of partial differential equations of advection-diffusion-reaction type or a set of ordinary differential equations (ODEs), along with appropriate initial and boundary conditions. Each equation in the system describes the evolution of the concentration of each species entering the process that leads to clot formation and dissolution. Boundary conditions simulate the beginning of the whole process following a vessel wall injury. Due to the size of the system and to its nonlinearity, a solution in a closed form is out of sought and numerical simulations are necessary. We should note that numerical simulations are useful, not only from a mathematical point of view but also from a biological perspective. In fact, a specific perturbation in the hemostatic system, which usually cannot be assessed through laboratory tests, can easily be introduced in a phenomenologic mathematical model and predictions on the effects of that specific perturbation can be evaluated through numerical simulations and their corresponding results.

In addition to the biochemical reactions, the rheology of blood also plays a crucial role in the regulation of hemostasis. While plasma can be considered as a Newtonian fluid, whole blood exhibits non-Newtonian characteristics, particularly at low shear rates. These are mainly due to erythrocytes ability to aggregate and to form a three-dimensional (3D) microstructure (*rouleaux*) at low shear rates, their deformability and their tendency to align with the flow field at high shear rates [80]. The formation and breakup of this 3D microstructure, as well as the elongation and recovery of erythrocytes, contribute to the non-Newtonian behaviour of blood, like shear–thinning viscosity, thixotropy, viscoelasticity and possibly a yield stress. In particular, the shear-thinning viscosity of blood is manifested by the fact that at rest or at low shear rates (below $1s^{-1}$) blood seems to have a high apparent viscosity, while at high shear rates there is a reduction in its viscosity. Generalized Newtonian models like the power-law, Cross, Carreau, Carreau-Yasuda, or modifications of these models have been obtained by fitting experimental data in one dimensional flows and are used to capture the shear dependence of blood viscosity [75]. Moreover, blood cells are essentially elastic membranes filled with a fluid and it seems reasonable, at least under certain flow conditions, to expect blood to behave like a viscoelastic fluid, due to its ability to store and release elastic energy from its branched 3D microstructures. Thurston [89] was among the earliest to recognize the viscoelastic nature of blood at low shear rates. Some viscoelastic constitutive models for describing blood rheology have been proposed in the recent literature, in particular the empirical three constant generalized Oldroyd-B model (Yeleswarapu model) studied in [97] and the model developed by Anand and Rajagopal [1] which includes relaxation times depending on the shear rate and gives good agreement with experimental data in steady Poiseuille and oscillatory flows. We also refer to the microstructure based model developed in [20, 61].

In most part of the arterial system of healthy individuals blood can be modelled as a Newtonian fluid [26]. However, there are flow regimes and clinical situations where non-Newtonian effects of blood can probably be observed. These include, for normal blood, regions of stable recirculation like in the venous system and parts of the arterial vasculature where geometry has been altered and RBC aggregates become more stable, like downstream a stenosis, inside a saccular aneurysm or in some cerebral anastomoses. In addition, several pathologies are accompanied by significant changes in the mechanical properties of blood and this results in alterations in blood viscosity and viscoelastic properties. For a detailed discussion of the rheological properties of blood and corresponding continuum mathematical models, see for instance [74, 75] and references therein.

As previously mentioned, blood coagulation is a very complex process. It is therefore not reasonable to expect to have a mathematical model that "fits all" the features of clot formation and lysis. Indeed, the important influence of shear stress (both on platelets activation and on the mechanical interaction of the blood flow with the clot) should be captured by the model which is therefore strongly dependent on the environmental conditions in which coagulation takes place. The path for an overall understanding of the whole process seems to be constructing mathematical models which look upon a particular phase or aspects of the process, along with numerical simulations. For example, the role of Tissue Factor in the initiation of blood coagulation, or that of other activators, or inhibitors, from a threshold point of view; or, the main step leading to clot formation, and how much Thrombin is formed; existence and control of the several feedbacks in the chain of chemical reactions (positive and negative); existence of different time scales that, in some cases, allow for a reduction in the size or in the nature of the differential system; or the important role of platelets in the whole process, how they aggregate and their behaviour under shear flow conditions. Bleeding disorders or deficiencies leading to diseases can also be taken into account in a mathematical model. Moreover, blood rheology and the specific flow regime for blood clotting cannot be neglected. These are the main issues that will now be considered, while giving a short overview of some relevant mathematical models of blood coagulation proposed in the literature. It is important to notice that there remains a pressing need for further experimental data to validate these models and to develop new ones in order to better understand the interplay between biochemical and mechanical factors under different flow conditions found in the human vasculature.

3.4.1 Early models of blood coagulation

On the wake of the celebrated *cascade* [48], or *waterfall* [18] models, already in 1966 Levine [47] proposed one of the first mathematical models for blood coagulation based on a series of enzymatic reactions leading to the formation of Fibrin, but where no feedbacks were considered. The process was modelled by a linear system or first order ODEs taking the form

$$\begin{cases} \frac{dy_{1_a}}{dt} = k_1 y_1 \left[U(t) - U(t-a) \right] - K_1 y_{1_a} \\[2mm] \frac{dy_{2_a}}{dt} = k_2 y_2 y_{1_a} - K_2 y_{2_a} \\[2mm] \quad \vdots \\[2mm] \frac{dy_{N_a}}{dt} = -k_N y_N y_{(N-1)_a} - K_N y_{N_a} \end{cases} \qquad (3.1)$$

where y_{i_a} denotes the concentration of the activated form of y_i, and k_i and K_i are rate reaction constants. It was assumed that the first reactant y_1 was activated over a time interval of amplitude a, after which the stimulus was terminated. The delay was modelled through the unit step function $U(t)$ so that $U(t) - U(t-a)$ was considered as a unit pulse of duration a, the activation time. The system was linearized by setting all y_i's equal to their initial values, and a gain function for the enzyme amplifier was defined. Linearization makes sense if the amount of y_{i_a} is supposed to be small in comparison with that initially present or if the precursor (in the chain reactions) is replaced at the same rate as it is converted.

A generalized version of this model, taking into account negative feedbacks, was studied in [52]. Fibrinolysis was modelled by considering the rate of change of Plasminogen to Plasmin transition. A similar model was also studied in [82]. A probabilistic generalization of Levine's model, where the activation time was supposed to be a random variable, was given in [55].

3.4.2 Biochemical models

Mathematical models of higher complexity, involving the kinetics of various sets of reactions, with feedback loops and inhibitors, emerged later in the eighties. Khanin and Semenov [44] proposed the first mathematical model of blood coagulation that takes into account nonlinear effects, considering the extrinsic pathway and only some enzymatic reactions important for the activation. The main purpose was to study the initiation of the process, that is the activation of FVII, along with the Thrombin generation in plasma. Included is the positive feedback essential to the acceleration of Thrombin formation.

This model consists of the following nonlinear system of ODEs:

$$
\begin{cases}
\dfrac{d\,[FVII_a]}{dt} = \alpha K_1 - H_1\,[FVII_a] \\[2mm]
\dfrac{d\,[FX_a]}{dt} = K_2\,[FVII_a] - H_2\,[FX_a] \\[2mm]
\dfrac{d\,[FV_a]}{dt} = K_3\,[FII_a] - H_3\,[FV_a] \\[2mm]
\dfrac{d\,[FII_a]}{dt} = -\dfrac{K_4\,[FX_a][FV_a]}{K_a+[FV_a]} - H_4\,[FII_a]
\end{cases}
\tag{3.2}
$$

where K_n is the n-th reaction rate, K_a is a Michaelis-Menten constant ([53]), H_n the n-th reaction product breakdown constant and α, the stimulation intensity, is the concentration of a substance released from injured vessel wall to activate FVII. Here and hereafter the concentration of a chemical C is denoted by $[C]$.

As it is well known, during hemostasis not all reactions have the same rate. Taking into account the characteristic times required to reach steady states, or by analyzing the reaction rates, it is possible to determine which reactions are slower. This allows to say that, in the initial period, the concentration of FV is relatively small and therefore the reaction leading to the formation of Thrombin is the slowest. After reducing the system by means of the Tikhonov reduction theorem, a qualitative linear stability analysis shows that if

$$
\alpha > \alpha_0 \equiv K_a\,\frac{H_1 H_2 H_3 H_4}{K_1 K_2 K_3 K_4},
$$

then system (3.2) has two steady state solutions. It also shows that if the stimulation intensity does not exceed its threshold, then there is only one zero stationary point which is stable. In this case coagulation does not take place. On the contrary, if $\alpha > \alpha_0$, there exists a stable non-zero stationary point, while the zero point becomes unstable. In this case there is Fibrin production. Note that a decrease of α_0 favors hypercoagulation, while hypocoagulation might be possible in the opposite case.

Other mathematical models are concerned with certain important issues related to the coagulation response. It is for instance the case of [6] were the threshold response of a hierarchy of enzyme-catalyzed positive feedback loops and inhibition, acting in a cascade, has been investigated and supported by numerical simulations. More recently, we can refer to [42], where the authors have characterized the threshold response of initiation of blood clotting to the size of a patch of stimulus, based on experiments and numerical simulations. The threshold response follows a scaling relationship based on the Damköler number, which describes the competition between reaction and diffusion of molecules. The influence of the flow on the threshold patch size has not been considered in this work.

As already mentioned, blood flow has a major role in the regulation of hemostasis and thrombosis, being one of the components of the famous Virchow's triad [93]. Ataullakhanov and co-workers developed in [85] a detailed quantitative mechanism-driven mathematical model of (TF)-initiated thrombus formation in flowing plasma (adapted from a previous model without flow presented in [64]). The model includes 28 partial differential equations describing biochemical reactions, diffusion and convection of the reactants, with all initial concentrations and kinetic constants obtained from experiments. It was demonstrated that blood flow can regulate clotting onset in the model in a threshold-like manner, in agreement with existing experimental evidence. This is due to a combination of the positive feedback of FVII activation with chemical inhibition of extrinsic Tenase, and effective removal of FXa by flow from the activating patch, depriving the feedback of "ignition". This mechanism is controlled by the activity of Tissue Factor Pathway Inhibitor (TFPI) (see also *Remark 2*, Sect. 3.2.4).

The role of activated platelets

Models characterized by a large number of reaction-diffusion equations with the inclusion of a greater number of factors have been developed by several authors. A typical example is the biochemical model proposed by Kuharsky and Fogelson [45] that also incorporates the effects of hemodynamic forces on the transport of reactants, catalysts and products. More precisely, it includes plasma-phase and surface-bound enzymes and zymogens, coagulation inhibitors and activated and non activated platelets, as well as membrane-phase reactions, and in a simplified way it accounts for chemical and cellular transport by flow and diffusion. The model assumes that FVII and FVIIa compete for the TF binding sites on the subendothelium, that FIX and FX compete for the FVIIa-TF complex on the subendothelium, and that each pair FII/FIIa, FV/FVa, FVIII/FVIIIa and FX/FXa has distinct binding sites on activated platelets for which each zymogen and enzyme compete. The kinetic reactions occur in a thin boundary layer above the injured surface where all species are well mixed and are described by a large system of nonlinear ODEs (more than fifty!), involving kinetic constants taken from the literature. The authors observe that even these may be not sufficient to describe the specific processes in which they are interested. The non dimensionalized system is solved by appropriate methods for stiff ODEs. Numerical simulations have shown that the increase or decrease of Thrombin concentration depends on the flow shear rate. They also have shown that an increase in TF binding sites, expected in case of a vascular injury, is responsible for a significant change in the production of Thrombin. Therefore, the availability of particular surface binding can be a useful threshold. As for the complex FVIIa-TF activity, it was suggested that platelets adhering to and covering the endothelium might play a dominant role, instead of being chemical inhibitors. From here one could say that the role of the Tenase pathway, for activating FX is to continue its production, after the covering of FVIIa-TF by platelets. Finally, regarding bleeding disorders, the model was able to give a kinetic explanation of the reduced Thrombin production in the cases of Hemophilia A and B (see also Sect. 3.3.1 and Sect. 3.4.2).

Extensions of this model to incorporate the chemistry of the Activated Protein C (APC) pathway on endothelial cells adjacent to the injury, the transport of Thrombin and APC between the injury and these endothelial cells, as well as to explore the effectiveness of coagulation inhibitors like TFPI and APC at a site of injury-induced thrombosis, have been considered in [25]. See also [41] that presented an early large scale model for thrombin generation via the extrinsic pathway, involving 18 ODEs; this model has been extended in [35] to include the role of inhibitors, namely *stoichiometric anticoagulants*; it consists of 34 ODEs with 42 rate constants, and in particular shows that for concentrations above the TF threshold the amount of Thrombin produced is quantitatively equivalent.

Another dynamic model including the role of activated platelets in the coagulation cascade and based on biochemical experiments, was introduced by Xu, Zeng and Gregersen in [96]. It relies on the concept of the biological cell-based models and provides a generalization of Khamin and Semenov's model [44]. It consists of a system of nonlinear ODEs which relates the rate of change in the concentrations of FXa, FIXa, FVa, FVIIIa and Thrombin at any time, with its rate of formation, inhibition or breakdown. Thrombin is formed in the initiation stage, in small amounts, thereby activating platelets. It is assumed that the fraction c of activated platelets at a given time depend on its initial value c_0 and on the Thrombin concentration, as follows:

$$c = c_0 + \frac{f([FII_a])}{1 + f([FII_a])}$$

where f is at least twice differentiable. The system of ODEs is given by

$$
\begin{cases}
\frac{d[FVII_a - TF]}{dt} = [FVII_a - TF]_0 - h'_1 [FVII_a - TF][TFPI][FX_a]k_{14} \\[2mm]
\frac{d[FX_a]}{dt} = k_2 [FVII_a - TF] + k_{21}[FIX_a] \left(\frac{[FVIII_a]}{d_2 + [FVIII_a]} \right) \left(c_0 + \frac{f([FII_a])}{1 + f([FII_a])} \right) \\[1mm]
\quad\quad\quad - h'_{21}[TFPI][FX_a] - h'_{22}[ATIII][FX_a] \\[2mm]
\frac{d[FIX_a]}{dt} = k_3 [FVII_a - TF] - h'_3 [ATIII][FIX_a] \\[2mm]
\frac{d[FII_a]}{dt} = k_4 [FX_a] \left(\frac{[FV_a]}{d_1 + [FV_a]} \right) \left(c_0 + \frac{f([FII_a])}{1 + f([FII_a])} \right) - h'_4 [ATIII][FII_a] \\[2mm]
\frac{d[FV_a]}{dt} = k'_5 [FII_a][FV] - h_5 [FV_a] \\[2mm]
\frac{d[FVIII_a]}{dt} = -k'_6 [FII_a][FVIII] - h_6 + [FVIII_a].
\end{cases}
\tag{3.3}
$$

The model shows that the formation of FXa is the most complex. In fact, during the initiation stage, FX is activated by the FVIIa-TF complex, and by the complex FIXa-FVIIIa during the propagation stage. Moreover, FXa is inhibited both by Antithrombin III (ATIII) and by Tissue Factor Inhibitor (TFPI). By inhibiting FXa-TF complex, FXa-TFPI can inhibit the formation of more FXa. Tissue Factor (TF) is not formed but assumed to be exposed initially. The amount of TF is supposed to be limited and forming complex FVII-TF, which quickly activates FVIIa. That is why the equation for FVIIa-TF only shows the decrease of FVIIa-TF and is inhibited by FXa-TFPI.

A stability linear analysis, similar to the one in [44] is considered. Asymptotic stability for a zero equilibrium point is shown. Zero is the only equilibrium point when a certain quantity, "the strength of pro-coagulant stimulus", β, is equal to zero. Otherwise, there exists also a non zero equilibrium point, which is acceptable (positive) if $\beta > \beta_0$, a threshold value.

The asymptotic stability leads to the conclusion that β must exceed the threshold value for the coagulation cascade to be initiated. This value is related to the rate of activation of platelets c and the binding constant d_2. For $\beta < \beta_0$ numerical calculations show concentrations of FVa and FIIa approaching zero in time, leading to the conclusion that, in this case, the system does not amplify and therefore coagulation does not take place.

For the non-zero equilibrium point numerical simulations are performed. They suggest that FVIII and FIX can contribute to the amplification stage of the cascade only when β reaches the threshold level. Simulations also show that if FVIII is lacking, blood coagulation does not occur since only small amounts of Thrombin are produced. On the other hand, when the level of FIX is below 1 % of normal levels, the rate of Thrombin generation reduces to a point that severe bleeding tendency may occur, which confirms the great dependency of coagulation on FVIII and FIX.

Protein deficiencies and diseases

The hemostatic system is maintained in a state of permanent background activity. This fact is confirmed by the presence of fragments of clotting factors split, as a result of their activation. It should be noted that the concentrations of activation factors in the background state are extremely low and, therefore, experimental studies become very difficult. In [57] a system of ODEs (no diffusion) was considered to model the background state of blood coagulation. In these equations rates of enzymatic reactions were described by Michaelis-Menten approximations, whereas rates of interaction of enzymes with inhibitors were second order kinetic equations. Besides diffusion, the model also assumed the absence of platelets. Using this model the effects of deficiencies of factors II, V, VII, VIII, IX and X on the background state of hemostasis are also studied. Numerical simulations have shown that a decrease in concentration of FVII inhibited the activation of factors II, IX, X and protein C, whereas a decrease in the concentrations of factors VIII and IX had an insignificant effect on activation of factors II and X or protein C. These results suggest that only the extrinsic pathway of the background state is active, whereas FIX remains almost inactive, which is consistent with the fact that, in patients with hemophilia B (FIX deficiency) the level of FX activation is similar to that of healthy people. In our opinion these conclusions based on numerical results should be treated with caution, since the Hemophilia A and Hemophilia B are considered as the most important bleeding disorders (Sect. 3.3.1).

We referred to the Hageman Factor (FXII) deficiency at the end of Sect. 3.3.1. It is known that this deficiency has almost no effect on blood coagulation, see [58]. This paper deals with a first order nonlinear ODE system to study the conditions of FXII deficiency. Unlike other models, here the kinetic rate constants are taken from experimental studies only. Three positive feedbacks are considered: Thrombin activation of FV, FVIII and FXI. Solving the system by Runge-Kutta methods, numerical simulations show that the amount of Thrombin first rises, after activation, and then declines to background level zero. This is because blood coagulation is activated in the presence of background concentrations of initiation factors, provided that activated platelets are also present. Numerical simulations also show that about 200-fold decrease in the concentrations of FXIIa has no effect on the maximal Thrombin concentration but, on the contrary, a similar decrease on concentrations of FVIII or FIX implies a significant decrease on Thrombin concentration. As a consequence hemorrhage will take place.

Besides the conclusion that the absence of FXIIa is irrelevant for blood coagulation, simulations also show that when this deficiency occurs, the time interval required for Thrombin to reach its maximal concentration increases by a factor two (in the case of a 200-fold decrease in the concentration of FXIIa) or even by a factor three in the absence of activated platelets which, therefore, can be considered as responsible for the insensitivity of the blood coagulation system to the Hageman deficiency. The same conclusion applies to the positive feedback activating FXI. As already referred, this model does not take into account the effects of diffusion or blood flow. It may simulate Thrombin production in a closed cavity (hematoma).

Within the context of models accounting for protein deficiencies and diseases we refer to [90]. This paper addresses spatial aspects of clotting as a function of Factor IX deficiency (hemophilia B). It has been experimentally observed that clot formation and growth in hemophilia B plasma is substantially affected by the different levels of FIX deficiency. A detailed mathematical model consisting of a system of

27 ODEs derived in [64] has been adapted and numerical simulations have shown good agreement with the experimental results. The initiation and propagation phases of blood clotting is affected by severe deficiency in any of the components of the intrinsic pathway, FVIII, FIX or FXI, corresponding to hemophilia A, B or C, respectively. This study can also be found in [59]. We have already mentioned the paper [12] modelling the thrombocytopenia, which in perspective can provide an important tool to decide whether or not splenectomy is appropriate (see Sect. 3.3.1).

A model for Disseminated Intravascular Coagulation (DIC)

In the papers [32, 33] a space-dependent model has been formulated for the formation of mini thrombi in the blood flow. This process, known as Disseminated Intravascular Coagulation (DIC), is the result of abnormal delivery of TF into blood, following various pathological conditions. DIC may be lethal. Ultrasound inspection [92] reveals clouds of mini thrombi transported along the bloodstream. Besides the damage possibly caused by thrombi to various organs, DIC is harmful also because it drains from the patient's blood platelets and other thrombogenic elements with the possible consequence of spontaneous bleeding. References can be found in [32, 33].

The proposed model takes an oversimplified scheme of the coagulation cascade and, on the contrary, focuses in detail on the fibrin polymer formation. It is an ingenious combination of two fields for the first time put together in the mathematical literature: morphogenesis (describing pattern formation) and coagulation-fragmentation (describing the growth of polymer chains). From morphogenesis it borrows a system of just two reaction-diffusion equations for the evolution of thrombin and of an anticoagulant (representing the class of anticoagulants), in which the positive feedback is represented by the typical autocatalytic terms of morphogenesis. The approach of [32] is based on ordinary differential equations since a spatially homogeneous case is considered, while in [33] diffusion and convection are added. The development of the fibrin polymeric chain is described starting from the fibrinogen-thrombin interaction by means of the classical coagulation-fragmentation infinite differential system. The latter system is eventually replaced by the study of the first few moments of polymer concentrations M_0, M_1, M_2, M_3, defined as $M_k = \sum_{n=1}^{\infty} n^k F_n$, where F_n is the concentration of the chains consisting of precisely n monomers. A closure condition is necessary. The one adopted is $M_1 M_3 = M_2^2$ (see [32] for a justification). Numerical simulations show that the model does reproduce the micro thombi cloud formation.

3.4.3 Models incorporating mechanical and biochemical actions

None of the previous models includes the direct effects of mechanical loading on the process of thrombus formation and lysis, or the mechanical properties of blood itself and the forming clots. Most of them are also based in ODEs, neglecting the spatial distribution of reactants, catalysts and products, which is a simplifying assumption, since different reactions of blood coagulation take place in different regions. Moreover, practically all models concerned with hemodynamic factors and hemostasis do not capture all biochemical reactions that take place in flowing blood.

Fluid-platelets interaction models

A model that accounts for the influence of fluid motion and hemodynamic forces on the aggregation process, activation of platelets by chemicals, platelet-platelet and platelet-wall interactions, as well as the feedback role of aggregate growth on the fluid dynamics, has been considered in [23] (see also [24]). This model was developed to describe the formation of platelet thrombi in coronary-artery-sized blood vessels and is a follow up of earlier models of platelet aggregation (with no influence of the fluid dynamics), [21, 22], along with computational results, [64]. It gives an important contribution to better understand the interactions between local geometry, fluid dynamics, and aggregate growth. This continuum model involves two spatial scales, the microscale of the platelets and the macroscale of the vessel, using as a major tool the Immersed Boundary method (IB) introduced by Peskin (see [95]. More precisely, the microscopic scale tracks individual platelets, their detection and response to chemical activators, and mechanical interactions among the platelets, fluid and the vascular wall. The macroscopic scale tracks the dynamics of the same interactions on a larger scale; it follows the evolution of density functions that describe the distribution of nonactivated and activated platelets and of their links.

The model involves a large number of coupled nonlinear partial differential equations where the dynamics of the fluid is described by the Navier-Stokes equations; it involves a mix of Eulerian and Lagrangian communicating descriptions; steep spatial gradients appear due to the combination of rapid localized reactions and small diffusion coefficients; transport of platelets and chemicals needs to be confined to the portions of the domain inside of the immersed boundaries used to represent the vessel walls; the fluid – wall and fluid – platelet interactions can be stiff and present difficulties in achieving stable calculations. Appropriate numerical methods have been used to meet these challenges and numerical simulations have shown that it is possible to capture important behaviours in the platelet aggregation process during blood clotting, including vessel occlusion by thrombi growth.

This model does not include Thrombin which is an important platelet activator produced on the surface of activated platelets. Three-dimensional simulations at both micro- and macroscales are computationally very expensive.

Models incorporating the mechanical activation of platelets

Anand et al [2] developed a phenomenological mathematical model of the hemostatic system that takes into account biochemical, physiologic and rheological factors playing an important role in the formation, growth and lysis of blood clots (see also [3]). It models blood as a non-Newtonian fluid, namely as a shear-thinning viscoelastic fluid with a shear dependent relaxation time, and the clot as such a fluid with much higher viscosity. The extrinsic pathway of enzymatic cascade of reactions that leads to clot formation and growth was modelled by a set of 23 coupled convection-reaction-diffusion partial differential equations, and the system was closed by appropriate initial and flux boundary conditions, reflecting the injury to the vessel wall. A method for tracking the boundary between the growing clot and normal blood was introduced. Moreover, a criterion to quantify platelets activation due to prolonged exposure to shear stress was defined. Finally, the model also included convection-reaction-diffusion equations to account for fibrinolysis.

Numerical results for a 3D simplified version of this model, where the viscoelasticity of blood was not considered, can be found in [9]. Preliminary stability results have been obtained for this clot model in quiescent plasma (see [84]). In particular, a continuum of equilibria has been found for the kinetics of some of the most relevant Michaelis-Menten chemical reactions involved in the model and, using a nontangency-based Lyapunov criterion, semistability has been proved.

This model undergoes some oversimplifications and is far from being complete. In fact, biochemical factors have been partially included and, on the other hand, both the blood and the clot were treated as homogenized continua with a very similar constitutive structure, which does not allow to account for the rheology of their constituents. Nevertheless, the model provides a preliminary framework for the development of more comprehensive mathematical models that could better capture the relevant biochemical and mechanical processes of hemostasis and its regulation. In Weller [98] a model has been proposed in which thrombus growth is described as the result of the aggregation of activated platelets driven to the coagulation site by diffusion.

A model with a platelets activation front and a clotting front

In [10] a clotting model was presented based on the idea, suggested in [2], that platelets mechanical activation can be monitored by an activation number. Though the general principles adopted in [10] can be applied to any geometry, the paper was dealing with the special case of a clot growing in an arteriole of radius R in conditions of cylindrical symmetry, allowing a thoroughly mathematical investigation. The possibility of having an axisymmetric thrombus requires a neat separation of the time scales of the various concurrent processes, and this is a delicate aspect addressed in [10]. Hemodynamics was assumed to be described by a shear-thinning model in which the Cauchy stress tensor $\mathbf{T}$ is defined as

$$\mathbf{T} = \eta(\mathbf{D}, [FI_a])\mathbf{D}$$

where $\mathbf{D}$ is the shear rate tensor and

$$\eta(\mathbf{D}, [FI_a])\mathbf{D} = \hat{\eta}\,\eta_\infty([FI_a])\left[1 + \frac{\delta}{(1 + \alpha II_{\mathbf{D}}^2)^n}\right] \tag{3.4}$$

is the shear dependent viscosity. In (3.4) $\hat{\eta}$ is a reference viscosity, η_∞ is a dimensionless increasing function of the Fibrin concentration $[FI_a]$ (see [72]), δ is a dimensionless positive constant, $II_{\mathbf{D}} = \frac{1}{2}\sqrt{\text{tr } \mathbf{D}^2}$ is the second invariant of $\mathbf{D}$, α is a positive constant (measured in s^2), and n is a positive exponent. A nontrivial property of the constitutive law (3.4) is that the dependence on the second invariant may not be monotone, with interesting consequences on the fluid dynamics, that will not be dealt with here. The blood density ρ was taken constant.

The platelets activation by shear stress was described introducing a stress threshold σ_{thr} and supposing that activation depends on the exposure time to stress beyond that threshold. This was the motivation for defining the activation number as follows

$$A(t) = \left\{ \int_0^t \mathcal{H}\left(II_{\mathbf{T}} - \sigma_{thr}\right) e^k \left(\frac{II_{\mathbf{T}}}{\sigma_{thr}} - 1\right) d\tau \right\} \mathcal{H}\left([FI_a] - [FI_a]_{thr}\right) \qquad (3.5)$$

where $\mathcal{H}$ is the Heaviside function, $\mathbf{T} = \mathbf{T} - \frac{1}{3}(\text{tr } \mathbf{T})\mathbf{1}$, and k is a positive number. The factor $\mathcal{H}\left([FI_a] - [FI_a]_{thr}\right)$ keeps into account that a sufficient Fibrin concentration is needed to produce clotting. Following [2], clot formation was supposed to occur instantaneously when $A(t)$ exceeds a critical value A_{crit}.

Of course the evolution of $[FIa]$ is the consequence of the known chemical cascade involving numerous steps. A key simplification introduced in [10] was to single out Prothrombinase as pivotal element in the cascade. If the Prothrombinase concentration $[W]$ is supposed to be known as a function of time, then the only surviving section of the reaction-diffusion system involves just the concentrations $[FI]$, $[FIa]$, $[FII]$, $[FIIa]$, together with those of the fibrinolysis factors $[tPA]$, $[PLA]$, $[L2AP]$. The Prothrombinase concentration $[W]$ just enters the two balance equations for $[FII]$, $[FIIa]$, describing the Thrombin production, with a reaction term of the form $\frac{k_2[W][FII]}{K_{2m}+[FII]}$.

In [10] Prothrombinase is supposed to be formed in a scenario of *mild inflammation* of the blood vessel producing an increase of $[W]$ from 0 to some $[W]_\infty$, which is selected so to control the Thrombin production rate.

The following time scales are recognized to have an important role in the process: $t_f = \frac{\rho R^2}{\hat{\eta}}$ (fluid dynamics), $t_{diff} = \frac{R^2}{D}$ (largest diffusive time), t_{chem} (the time scale of Fibrin production), t_{clot} (the time scale of clot growth). The determination of these time scales relies on Table 3.1.

With those data and with the rate constants taken from [2] we have $t_f = 33s$, $t_{diff} = 1.3 \times 10^3 s$, and, assuming $[W]_\infty = 5 \times 10^{-3} nM$, then $t_{chem} = 10^4 s$. Thus we got the separation $t_f \ll t_{diff} \ll t_{chem}$. The heart beat time scale ($\approx 1s$) is comparatively short and plays no role. We remark that this is an abnormal situation, since when clotting has the purpose of wound sealing it has to be fast. Nevertheless, as we said, these conditions are imposed with the aim of keeping a simple geometry and can be released if the latter is not required (clearly leading to a much more complicated problem). The above chain of inequalities allows to justify the following approximation: *all concentrations in the biochemical system depend on time only, reducing it to an ODE system*. If in addition we want that *the blood flow is quasi steady during the whole process*, then we also need $t_f \ll t_{clot}$. The time scale t_{clot} can be defined as

Table 3.1. Parameters for determining time scales

R	D	ρ	$\hat{\eta}$
2×10^{-2} cm	3×10^{-7} cm^2 sec^{-1}	10^3 g cm^{-3}	1.2×10^{-2} Poise

the ratio $\frac{R}{V_c}$, where V_c is the typical speed of the clotting front. Thus its determination passes through the study of the motion of the front where $A(t) = A_{crit}$, which is precisely the clotting front. The latter is preceded by an activation front, characterized by the equality $\sigma = \sigma_{thr}$ (the shear stress becoming critical for platelet activation). Thus, from the mathematical point of view, the clot growth model is a problem with two free boundaries, governed by the flow equation and by the Fibrin production differential system.

For lack of space we cannot deal with the fluid dynamical problem, nor with the study of the free boundaries, which is performed in great detail in [10] and which is considerably complicated. We just mention that the final conclusion is that, in the framework of the selected parameters, t_{clot} can be estimated to be $120 s$, thus concluding that the whole scheme is self consistent.

3.5 Conclusions

The paper is very concise, due to space limitations. In the first part of the paper we have reviewed the complex mechanisms regulating the various stages of blood coagulation (primary hemostasis, secondary hemostasis, fibrinolysis). After a brief historical section illustrating the main discoveries, we have described the many elements entering the process and the bleeding disorders caused by their deficiency or dysfunctions. The recently formulated cell-based coagulation model has been illustrated in detail, also explaining why it replaced the previous 3-pathway cascade model, that has been used for over forty years, since 1964. Some space has been devoted to important illnesses such as Deep Vein Thrombosis, Thrombocytopenia, Disseminated Intravascular Coagulation, for which mathematical models have been proposed.

In the second part we have illustrated a selection of mathematical models, emphasizing different aspects. We could not point out which model is the best. First of all we have to say that the shift from the cascade to the cell-based model is rather recent and it poses a new challenge to mathematicians. Moreover, we remark that even models listing a "complete" set of reaction-diffusion equations miss to mention some basic feature, like e.g. the role of vWF or the volumetric contribution to thrombus growth by cells entrapment. In general the coupling of chemistry with mechanics is not frequently addressed. If the scenario depicted in most of the models may be incomplete, on the contrary there is sometimes a too scrupulous attention to chemical details, which can possibly be avoided since many reactions are very fast. Actually, opposite trends are identifiable: there are attempts to include as much as possible of the biochemistry (e.g. [23, 24, 85], while other models are extremely concise in order to emphasize specific aspects (e.g. [32, 33, 98]). Finally, there are phenomena not yet accounted for in mathematical models, like for instance the recently ascertained production of TF by platelets ([17, 63]) and the influence of blood slip at the vessel wall, contributing additional platelets to the coagulation site. Thus we conclude that, despite the considerable amount of literature on the subject, there is still much work to do in search of a mathematical model consistent with the modern views, including essential biochemical and mechanical features and allowing to incorporate bleeding disorders and related therapies.

Acknowledgements. Work partially supported by the Italian MIUR Project "Math. Models for Multicomponent Systems in Environmental and Medical Sciences", by the research center CE-MAT/IST through FCT's funding program and by the project PTDC/MAT/68166/2006. The authors are grateful to Dr. Jeremi Mizerski for his useful suggestions. Acknowledgments are also due to Jevgenija Pavlova, doctoral student at IST, for her assistance in the preparation of all the figures included in this paper.

References

[1] Anand M., Rajagopal K.R.: A shear-thinning viscoelastic fluid model for describing the flow of blood. Intern. Journal of Cardiovasc. Medicine and Sci. 4(2): low of blood, Intern. Journal of Cardiovasc. Medicine and Sci. 4(2): 59–68, 2004.

[2] Anand M., Rajagopal R., Rajagopal K.R.: A model incorporating some of the mechanical and biochemical factors underlying clot formation and dissolution in flowing blood. Journal of Theoretical Medicine 5(3–4): 183–218, 2003.

[3] Anand M., Rajagopal R., Rajagopal K.R.: A model for the formation, growth, and lysis of clots in quiescent plasma. a comparison between the effects of antithrombin III deficiency and protein C deficiency. Journal of Theoretical Biology 253: 725–738, 2008.

[4] Anning, S.T.: The historic aspects of venous thrombosis. Med. Hist. 1: 28–37, 1957.

[5] Baronciani L., Mannucci P.M.: The molecular basis of von willebrand disease. In D. Provan and J.G. Gribben (eds), Molecular Hematology, chapter 19, pp. 233–245. Wiley-Blackwell, third edition, 2010.

[6] Beltrami E. and Jesty J.: Mathematical analysis of activation thresholds in enzyme-catalyzed positive feedbacks: Application to the feedbacks of blood coagulation. Proc. Natl. Acad. Sci. USA 92: 8744–8748, 1995.

[7] Bernard J., Soulier J.P.: Sur une nouvelle variété de dystrophie thrombocytaire-hemor-rhagipau congénitale. Semin. Hop. 24: 3217–3223, 1948.

[8] Bertina R.M., Koeleman B.P., Koster T., et al.: Mutation in blood coagulation factor v associated with resistance to activated protein C. Nature 369: 64–67, 1994.

[9] Bodnár T., Sequeira A.: Numerical simulation of the coagulation dynamics of blood. Comp. Math. Methods in Medicine, 9(2): 83–104, 2008.

[10] Borsi I., Farina A., Fasano A., Rajagopal K.R.: Modelling the combined chemical and mechanical action for blood clotting. Math. Sci. and Appl. 9(2): 83–104, 2008.

[11] Brewer D.B.: Max Shultze (1865), G. Bizzozero (1882) and the discovery of the platelet. British Journal of Haematology 133: 251–258, 2006.

[12] Brugnano L., Di Patti F., Longo G.: An incremental mathematical model for immune thrombocytopenic purpura (ITP). Mathematical and Computer Modelling 42: 1299–1314, 2005.

[13] Caro C.G., Pedley T.J., Schroter R.C., Seed W.A.: The Mechanics of the Circulation. Oxford University Press, 1978.

[14] Clemetson K.J.: Platelets disorders. In D. Provan and J.G. Gribben (eds.), Molecular Hematology, chapter 20, pp. 246–258. Wiley-Blackwell, third edition, 2010.

[15] Dahlbäck B.: Blood coagulation and its regulation by anticoagulant pathways: genetic pathogenesis of bleeding and thrombotic diseases. Journal of Internal Medicine, 257: 209–223, 2005.

[16] Dahlbäck B., Hillarp A.: Molecular coagulation and thrombophilia. In D. Provan and J.G. Gribben (eds.) Molecular Hematology, chapter 17, pp. 208–218. Wiley-Blackwell, third edition, 2010.

[17] Davi G., Patrono C.: Platelets activation and atherothrombosis. N. Engl. J. Med. **357**: 2482–2494, 2007.

[18] Davie E.W., Ratnoff O.D.: Waterfall sequence for intrinsic blood clotting. Science **145**: 1310–1312, 1964.

[19] Eilersten K.-E., Osterud B.: The role of blood cells and their microparticles in blood coagulation. Biochem. Society Transactions **33**: 418–422, 2005.

[20] Fang J., Owens R.G.: Numerical simulations of pulsatile blood flow using a new constitutive model. Biorheology **43**: 637–660, 2006.

[21] Fogelson A.L.: A mathematical model and numerical method for studying platelet adhesion and aggregation during blood clotting. Journal of Computational Physics **56**: 111–134, 1984.

[22] Fogelson A.L.: Continuum models of platelet aggregation: formulation and mechanical properties. SIAM J. Appl. Math. **52**(4): 1089–1110, 1992.

[23] Fogelson A.L., Guy R.D.: Platelet-wall interactions in continuum models of platelet thrombosis: formulation and numerical solution. Mathematical Medicine and Biology **21**: 293–334, 2004.

[24] Fogelson A.L., Guy R.D.: Immersed-boundary-type models of intravascular platelet aggregation. Comp. Methods Appl. Mech. Engrg. **197**: 2087–2104, 2008.

[25] Fogelson A.L., Tania N.: Coagulation under flow: The influence of flow-mediated transport on the initiation and inhibition of coagulation. Pathophysiology of Haemostasis and Thrombosis **34**: 91–108, 2005.

[26] Formaggia L., Quarteroni A., Veneziani A.: The circulatory system: from case studies to mathematical modeling. In A. Quarteroni, L. Formaggia, and A. Veneziani (eds.), Integration of Complex Systems in Biomedicine, pp. 243–288. Springer, 2006.

[27] Frenette P.S., Johnson R.C., Hynes R.O., Wagner D.D.: Platelets role on stimulated endothelium in vivo: an interaction mediated by endothelial P-selection. Proc. Natl. Acad. Sci. USA **92**: 7450–7454, 1995.

[28] Gailani D., Zivelin A., Sinha D., Walsh P.N.: Do platelets synthesize factor XI? J. Thrombosis and Haemostasis **2**: 1709–1712, 2004.

[29] Gazzaniga V., Ottini L.: The discovery of platelets and their function. Vesalius **17**: 22–26, 2001.

[30] Giangrande P.L.F.: The molecular basis of hemophilia. In D. Provan and J.G. Gribben (eds.), Molecular Hematology, chapter 18. Wiley-Blackwell, third edition, 2010.

[31] Glanzmann E.: Hereditare hamorrhagische thrombasthenie: ein beitrag zur pathologie der bluttplatchen. J. Kinderkranken **88**: 113–141, 1918.

[32] Guria G.Th., Herrero M.A., Zlobina K.E.: A mathematical model of blood coagulation induced by activation sources. Discr. Cont. Dyn. Systems **25**(1): 175–194, 2009.

[33] Guria G.Th., Herrero M.A., Zlobina K.E.: Ultrasound detection of externally induced microthrombi cloud formation: a theoretical study. J. Eng. Math. **66**: 293–310, 2010.

[34] Hetland O., Brovold A.B., Holme R., Gaudernack G., Prydz H.: Thromboplastin (tissue factor) in plasma membranes of human monocytes. Biochem. J. **228**(3): 735–743, 1985.

[35] Hockin M.F., Jones K.C., Everse S.J., Mann K.G.: A model for the stoichiometric regulation of blood coagulation. The Journal of Biological Chemistry **277**(21): 18322–18333, 2002.

[36] Hoffman M.: Remodeling the blood coagulation cascade. Journal of Thrombosis and Thrombolysis **16**: 17–20, 2003.

[37] Hoffman M., Monroe D.: A cell-based model of hemostasis. Thrombosis and Haemostasis **85**: 958–965, 2001.

[38] Hoffman M., Monroe D.: Rethinking the coagulation cascade. Current Hematology Reports **4**: 391–396, 2005.

[39] Howell W.H., Holt E.: Two new factors in blood coagulation: heparin and pro-antithrombin. American Journal on Physiology **47**: 228–241, 1918.

[40] Jackson P., Mistry N., Yuan Y.: Platelets and the injured vessel wall – "rolling into action". TCM **10**: 192–197, 2000.

[41] Jones K.C., Mann K.G.: A model for the tissue factor pathway to thrombin. The Journal of Biological Chemistry **269**: 23367–23373, 1994.

[42] Kastrup C.J., Shen F., Runyon M.K., Ismagilov R.F.: Charaterization of the threshold response of initiation of blood clotting to stimulus patch size. Biophysical Journal **93**: 2969–2977, 2007.

[43] Kelton J.G., Warkentin T.E.: Heparin-induced thrombocytopenia. Blood **112**: 2607–2616, 2008.

[44] Khanin M.A., Semenov V.V.: A mathematical model of the kinetics of blood coagulation. Journal of Theoretical Biology **136**: 127–134, 1989.

[45] Kuharsky A.L., Fogelson A.L.: Surface-mediated control of blood coagulation: The role of binding site densities and platelet deposition. Biophysical Journal **80**: 1050–1074, 2001.

[46] Lawrence M.B., Springer T.A.: Leukocytes roll on a seletion at physiologic flow rates: distinction from and prerequisite for adhesion through integrins. Cell **65**: 859–873, 1991.

[47] Levine S.N.: Enzyme amplifier kinetics. Science **152**: 651–653, 1966.

[48] Macfarlane R.G.: An enzyme cascade in the blood clotting mechanism, and its function as a biochemical amplifier. Nature **202**: 498, 1964.

[49] Mann K.G.: Adding the vessel wall to Virchow's triad. Journal of Thrombosis and Haemostasis **4**: 58–59, 2006.

[50] Mann K.G., Brummel-Ziedins K., Orfeo T., Butenas S.: Models of blood coagulation. Blood Cells, Molecules and Diseases **36**: 108–117, 2006.

[51] Mann K.G., Butenas S., Brummel K.: The dynamics of thrombin formation. Arteriosclerosis Thrombosis and Vascular Biology **23**: 17–25, 2003.

[52] Martorana F., Moro A.: On the kinetics of enzyme amplifier systems with negative feedback. Mathematical Biosciences **21**: 77–84, 1974.

[53] Michaelis L., Menten M.L.: Die kinetik der invertinwirkung. Biochem. Z. **49**: 333–369, 1913.

[54] Morawitz P.: Die chemie der blutgerinnung. Ergebn. Physiol. **4**: 307–422, 1905.

[55] Moro A., Bharucha-Reid A.T.: On the kinetics of enzyme amplifier systems. Mathematical Biosciences **5**: 391–402, 1969.

[56] Naito K., Fujikawa K.; Activation of human blood coagulation Factor XI independent of Factor XII. J. Biolog. Chemistry **266**(12): 7353–7358, 1991.

[57] Obraztsov I.F., Kardakov D.V., Kogan A.E., Khanin M.A. A mathematical model of the background state of the blood coagulation system. Doklady Biochemistry and Biophysics **376**: 10–12, 2001.

[58] Obraztsov I.F., Kuz'min V.M., Khanin M.A.: Blood coagulation dynamics under the conditions of Hageman factor deficiency: A mathematical model. Doklady Biochemistry and Biophysics **386**: 248–250, 2002.

[59] Ovanesov M.V., Krasotkina J.V., Ul'yanova L.I., Abushinova K.V., Plyushch O.P., Domogatskii S.P., Vorob'ev A.I., Ataullakhanov. F.I.: Hemophilia A and B are associated with abnormal spatial dynamics of clot growth. Biochem. Biophys. Acta **1572**(1): 45–57, 2002.

[60] Owen Jr. C.A.: A History of Blood Coagulation. Mayo Foundation for Medical Education and Research, 2001.

[61] Owens R.G.: A new microstructure-based constitutive model for human blood. J. Non-Newtonian Fluid Mech. **140**: 57–70, 2006.

[62] Owren P.A.: Parahaemophilia. Haemorragic diathesis due to absence of a previously unknown clotting factor. Lancet **1**: 446–451, 1947.

[63] Panes O., Matus V., Saez C.G., Quiroga T., Pereira J., Mezzano D.: Human platelets synthesize and express functional tissue factor. Blood **109**, 5242–5250, 2007.

[64] Panteleev M.A., Ovanesov M.V., Kireev D.A., Shibeko A.M., Sinauridze E.I., Ananyeva N.M., Butylin A.A., Saenko E.L., Ataullakhanov F.I.: Spatial propagation and localization of blood coagulation are regulated by intrinsic and protein C pathways, respectively. Biophysical Journal **90**: 1489–1500, 2006.

[65] Peskin C.S.: The immersed boundary method. Acta Numerica **11**: 479–517, 2002.

[66] Podmore A., Smith M., Savidge G., Alhaq A.: Real-time quantitative PCR analysis of factor XI mRNA variants in human platelets. J. Thrombosis and Haemosthasis **2**: 1713–1719, 2004.

[67] Provan D., Gribben J.B. (eds.): Molecular Hematology. Wiley-Blackwell, third edition, 2010.

[68] Ratnoff O.D., Rosenblum J.M.: Role of Hageman factor in the initiation of clotting by glass: evidence that glass frees Hageman factor from inhibition. American Journal of Medicine **25**: 160–168, 1958.

[69] Reddik R.L., Griggs T.R., Lamb M.A., Brinkhous K.M.: Platelet adhesion on damaged coronary arteries: comparison in normal and von Willebrand disease swine. Proc. Natl. Acad. Sci. USA **79**: 5076–5079, 1982.

[70] Ribatti D., Crivellato E.: Giulio Bizzozero and the discovery of platelets. Leukemia Research **31**: 1339–1441, 2007.

[71] Riddel Jr. J.P., Aouizerat B.E., Miaskowski B.E., Lillicrap D.P.: Theories of blood coagulation. Journal of Pediatric Oncology Nursing **24**: 123–131, 2007.

[72] Riha P., Liao F., Stoltz J.F.: Effect of fibrin polymerization on flow properties of coagulation blood. Journal of Biological Physics **23**: 121–128, 1997.

[73] Roberts H.R.: Historical review. Oscar Ratnoff: His contributions to the golden era of coagulation research. British Journal of Haematology **122**: 180–192, 2003.

[74] Robertson A., Sequeira A., Kameneva M.V.: Hemorheology. In G. Galdi, R. Rannacher, A.M. Robertson, and S. Turek (eds.), Hemodynamical Flows: Modeling, Analysis and Simulation (Oberwolfach Seminars), pp. 63–120. Birkhäuser, 2008.

[75] Robertson A., Sequeira A., Owens R.: Rheological models for blood. In L. Formaggia, A. Quarteroni, and A. Veneziani (eds.), Cardiovascular Mathematics. Modeling and simulation of the circulatory system (MS&A), chapter 6, pp. 211–241. Springer-Verlag, 2009.

[76] Ruggeri Z.M.: Perspective series: Cell adhesion and vascular biology. Journal of Clin. Invest. **99**: 559–564, 1997.

[77] Ruysch F.: Thesaurus anatomicus. In Septimum, volume XXXIX. Amsterdam, Wolters, 1707.

[78] Sadler J.E.: Biochemistry and genetics of von Willebrand factor. Annu. Rev. Biochem. **67**: 395–424, 1998.

[79] Schmaier A.H., LaRusch G.: Factor XII: New life for an old protein. Thrombosis and Hemostasis **104**: DOI 10.1160/TH103–0171, 2010.

[80] Schmid-Schonbein H., Wells R.E.; Rheological properties of human erythrocytes and their influence upon the anomalous viscosity of blood. Ergeb. Physiol. **63**: 146–219, 1971.

[81] Schmidt A.: Neue untersuchungen ueber die fasserstoffesgerinnung. Pfluger's Archiv fur die gesamte Physiologie **6**: 413–538, 1872.

[82] Seegers W.H.: Blood clotting mechanisms: Three basic reactions. Annual Reviews in Physiology **31**: 269–294, 1969.

[83] Seligsohn U.: Factor XI in haemostasis and thrombosis: Past, present and future. J. Thromb. Haemost. **98**: 84–89, 2007.

[84] Sequeira A., Santos R.F., Bodnár T.: Blood coagulation dynamics: mathematical modeling and stability results. Math. Biosc. and Eng. **8**(2): 411-425, 2011.

[85] Shibeko A.M., Lobaneva E.S., Panteleev M.A., Ataullakhanov F.I.: Blood flow controls coagulation onset via the positive feedback of factor VII activation by factor Xa. BMC Systems Biology **4**: 5, 2010.

[86] Smith S.: The cell-based model of coagulation. Journal of Veterinary Emergency and Critical Care **19**: 3–10, 2009.

[87] Spreafico M., Peyvandi M.: Combined FV and FVIII deficiency. Hemophilia **14**: 1201–1208, 2008.

[88] Tangelder G.J., Slaaf D.W., Arts T., Reneman R.S.: Wall shear rates in arterioles in vivo: least estimates from platelet velocity profiles. Am. J. of Physiology **254**: H1059–H1064, 1988.

[89] Thurston G.B.: Viscoelasticity of human blood. Biophys. J. **12**: 1205–1217, 1972.

[90] Tokarev A.A., Krasotkina Y.V., Ovanesov M.V., Panteleev M.A., Azhigirova M.A., Volpert V.A., Ataullakhanov F.I., Butilin A.A.: Spatial dynamics of contact-activated fibrin clot formation in vitro and in silico in haemophilia B: Effects of severety and haemophilia B treatment. Mathematical Modeling of Natural Phenomena **1**(2): 124–137, 2006.

[91] Turitto V.T., Weiss H.J.: Red blood cells: their dual role in thrombus formation. Science **207**: 541–543, 1980.

[92] Uzlova S.G., Guria K., Guria G. Th.: Acoustic determination of early stages of intravascular blood coagulation. Phil. Trans. R. Soc. A **366**: 3649–3661, 2008.

[93] Virchow R.: Cellular Pathology. Churchill, 1860.

[94] Wagner. R. Erlauterungsfeln zur Physiologie unt Entwicklungsgeschichte. Leopold Voss, Leipzig, 1839.

[95] Wang N.-T., Fogelson A.L.: Computational methods for continuum models of platelet aggregation. Journal of Computational Physics **151**: 649–675, 1999.

[96] Xu C.Q., Zeng Y.J., Gregersen H.: Dynamic model of the role of platelets in the blood coagulation system. Medical Engineering & Physics **24**: 587–593, 2002.

[97] Yeleswarapu K., Kameneva M., Rajagopal K., Antaki J.: The flow of blood in tubes: theory and experiment. Mech. Res. Comm. **3**(25): 257–262, 1998.

[98] Weller F.F.: Platelet deposition in non-parallel flow. Influence of shear-stress and changes in surface relativity. J. Math. Biol. **57**: 333–359, 2008.

[99] Zwaal R.F., Comfurius P., Bevers E.M.: Surface exposure of phosphatidylserine in pathological cells. Cell Mol. Life Sci. **62**(9): 971–988, 2005.

4

Numerical simulation of electrocardiograms

Muriel Boulakia, Miguel A. Fernández, Jean-Frédéric Gerbeau, and
Nejib Zemzemi

Abstract. This chapter presents a concise overview of various mathematical and numerical problems raised by the simulation of electrocardiograms (ECGs). A model
for the propagation of the electrical activation in the heart and in the torso is proposed. Some of its mathematical properties are analyzed. This model is not aimed
at reproducing the complex phenomena taking place at the microscopic level. It has
been devised to produce realistic healthy ECGs, and some pathological ones, with
a reasonable level of complexity. Rather, it relies on various assumptions that are
carefully discussed through their impact on the ECGs. The coupling between the
heart and the torso is a critical numerical issue which is addressed. In particular, efficient coupling strategies based on explicit algorithms are presented and analyzed.
The chapter ends with some preliminary results of a reduced order model based on
the Proper Orthogonal Decomposition (POD) method.

Muriel Boulakia
Université Pierre et Marie Curie-Paris 6, UMR 7598, Laboratoire Jacques-Louis Lions, 75005 Paris,
France
e-mail: boulakia@ann.jussieu.fr

Miguel A. Fernández
INRIA Paris-Rocquencourt, 78153 Le Chesnay Cedex, France
e-mail: miguel.fernandez@inria.fr

Jean-Frédéric Gerbeau (✉)
INRIA Paris-Rocquencourt, 78153 Le Chesnay Cedex, France
e-mail: jean-frederic.gerbeau@inria.fr

Nejib Zemzemi
INRIA Paris-Rocquencourt and Laboratoire de mathématiques d'Orsay, Université Paris 11.
Current affiliation: University of Oxford, Computing laboratory, Oxford OX1 3QD, UK
e-mail: nejib.zemzemi@comlab.ox.ac.uk

Ambrosi D., Quarteroni A., Rozza G. (Eds.): Modeling of Physiological Flows.
DOI 10.1007/978-88-470-1935-5_4, © Springer-Verlag Italia 2012

4.1 Introduction

As for the skeletal muscles, the contraction of the heart is induced by an electrical stimulation. The electrocardiogram (ECG) is a noninvasive recording of this electrical activity. It is obtained from a standard set of skin electrodes (Fig. 4.1, left) and presented to the physician as the *12-lead ECG*, which is made of 12 graphs of the recorded voltage *vs.* time (Fig. 4.1, right). As explained for instance in [42], there is a direct connection between the ECG deflections and the electric state of the heart muscle: the first deflection, called the P-wave (see lead I in Fig. 4.1, right), corresponds to the depolarization of the atria; the group made of the second, third and fourth deflections, called the QRS-complex, corresponds to the depolarization of the ventricles; the last deflection, called the T-wave, corresponds to the repolarization of the ventricles.

The ECG is the most widely used clinical tool for the diagnosis of a broad range of cardiac conditions (see, e.g., [1, 27]). From the modelling standpoint, although the greatest challenge is the computation of the electrical activity of the heart, there are many reasons to try and compute the corresponding ECG: it can be seen as a way to assess the simulation since it can be easily submitted to the critical evaluation of a medical doctor; computer based simulations of ECGs can be a valuable tool for improving the understanding of some clinical signals; an ECG simulator can be useful in building a virtual database of pathological conditions, in order to test and

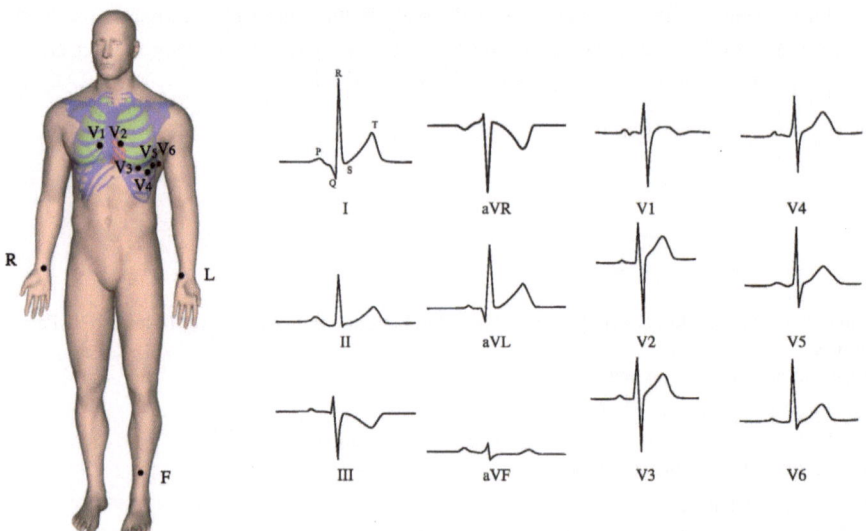

Fig. 4.1. *Left*: ECG electrodes location (black dots). *Right*: reproduction of a normal 12-lead ECG: standard leads (I, II, III), augmented leads (aVR, aVL, aVF) and chest leads (V1, V2, ..., V6). For example, lead I corresponds to the difference of potential between L and R, lead II to the difference between F and R, leads III to the difference between F and L (the other definitions can be found, e.g., in [7])

train medical devices (see [21]); last but not least, the simulation of realistic ECGs is a necessary step towards the development of patient-specific models from clinical ECG data, since it would be meaningless to address an inverse problem with a model unable to solve the direct one correctly.

As will be briefly shown in Sect. 4.2, the modelling of the cardiac electrical activity is a very complex and broad subject. The monographs [13, 53, 58, 61] give excellent introductions to the topic. The numerical simulation of ECGs using a whole-heart reaction-diffusion model has been addressed in many publications [7, 30, 33, 38, 50, 52, 67]. Among them, only a few [7, 50, 52] provide meaningful simulations of the complete 12-lead ECG. In [50, 52], simulations rely on either a monodomain approximation or a heart-torso decoupling approximation and a multi-dipole cardiac source representation (see [38, Sect. 4.2.4] and [28]). In publication [7], several simulations are based on a flexible coupled heart-torso model which allows to compare different modelling assumptions. More details about this model will be presented in Sect. 4.4.

In this presentation, we will focus on the numerical simulation of ECGs using a three-dimensional mathematical model fully based on partial/ordinary differential equations (PDE/ODE). The main ingredients of this model are standard: phenomenological cell dynamics, bidomain equations for the heart and a generalized Laplace equation for the torso. They are briefly overviewed in Sect. 4.2. A result on the existence of solution is presented in Sect. 4.3 for several classes of ionic models. In order to provide realistic ECG simulations, critical modelling aspects have to be discussed: heart-torso transmission conditions, cell heterogeneity, His bundle modelling, tissue anisotropy, etc. In Sect. 4.4, we will present different modelling options and show how they have been combined to get realistic ECGs. To complement the study, the impact on ECGs of alternative modelling choices is presented. Then, some decoupled time discretization schemes for the bidomain and heart-torso systems are analyzed in Sect. 4.5. Finally, some preliminary results obtained with a reduced-order model based on the Proper Orthogonal Decomposition (POD) method are presented in Sect. 4.6. This reduced-order model is about ten times as fast as the full-order model. Our results show that it is not always accurate and that further work is necessary to make it more reliable. Nevertheless, we will show that it seems to provide an interesting alternative to the full-order model in some particular cases, for example long-time simulations.

4.2 Mathematical modelling

The mathematical modelling of the ECG is known as the *forward problem* of electrocardiography (see [38]). It relies on three main ingredients: a model for the electrical activity of the heart, a model for the torso (extracardiac regions) and some specific heart-torso coupling conditions.

4.2.1 Isolated heart modelling

The bidomain equations, originally derived in [68], are the most widely accepted mathematical model of the macroscopic electrical activity of the heart (see, e.g., the monographs [53, 61]). This macroscopic model is based on the assumption that, at the cell scale, the cardiac tissue can be viewed as partitioned into two ohmic conducting media, separated by the cell membrane: the intracellular medium, made of the cardiac cells, and the extracellular one which represents the space between them. After a homogenization process (see [46, 49]), the intra- and extracellular domains are supposed to occupy the whole heart volume Ω_H (this also applies to the cell membrane). Hence, the averaged intra- and extracellular densities of current, j_i and j_e, the conductivity tensors, σ_i and σ_e, and the electric potentials, u_i and u_e, are defined in the whole heart domain Ω_H. The electrical charge conservation becomes

$$\operatorname{div}(j_i + j_e) = 0, \tag{4.1}$$

and the homogenized equation of the electrical activity of the cell membrane is given by

$$A_m\left(C_m \partial_t V_m + i_{\mathrm{ion}}(V_m, w)\right) + \operatorname{div}(j_i) = I_{\mathrm{app}}, \tag{4.2}$$

complemented with the Ohm's laws

$$j_i = -\sigma_i \nabla u_i, \quad j_e = -\sigma_e \nabla u_e.$$

Here V_m stands for the transmembrane potential, defined as $V_m \overset{\text{def}}{=} u_i - u_e$, A_m is a constant representing the rate of membrane area per volume unit and C_m the membrane capacitance per area unit. The reaction term $i_{\mathrm{ion}}(V_m, w)$ represents the ionic current across the membrane and I_{app} a given applied current stimulus. In general, the ionic variable w (possibly vector-valued) satisfies a system of ODEs of the type:

$$\partial_t w + g(V_m, w) = 0. \tag{4.3}$$

The definition of the functions g and I_{ion} depends on the cell membrane ionic model considered (see [53, 61] and the references therein). A very large number of ionic models have been proposed in the literature with different degrees of complexity and realism. The ionic models typically fall into one of the following categories (see [53, Chap. 3]): phenomenological (FitzHugh-Nagumo [26, 45], Aliev-Panfilov [2], Roger-McCulloch [57], van Capelle-Durrer [69], Fenton-Karma [24], Mitchell-Schaeffer [43]) or physiological (e.g., Beeler-Reuter [4], Luo-Rudy I[41], Luo-Rudy II [40], Noble-Varghese-Kohl-Noble [48], Djabella-Sorine [19]).

To sum up, the system of equations modelling the electrical activity within the heart (in terms of V_m and u_e) consists of a coupled system of ODEs, (4.3), a nonlinear reaction-diffusion equation, (4.2), and an elliptic equation, (4.1):

$$\begin{cases} \partial_t w + g(V_m, w) = 0 & \text{in} \quad \Omega_H \times (0, T), \\ \chi_m \partial_t V_m + I_{\mathrm{ion}}(V_m, w) - \operatorname{div}(\sigma_i \nabla V_m) - \operatorname{div}(\sigma_i \nabla u_e) = I_{\mathrm{app}} & \text{in} \quad \Omega_H \times (0, T), \quad (4.4) \\ -\operatorname{div}\left((\sigma_i + \sigma_e)\nabla u_e\right) - \operatorname{div}(\sigma_i \nabla V_m) = 0 & \text{in} \quad \Omega_H \times (0, T), \end{cases}$$

with appropriate boundary conditions

$$
\begin{cases}
\sigma_i \nabla V_m \cdot n + \sigma_i \nabla u_e \cdot n = 0 & \text{on} \quad \Sigma \times (0, T), \\
\sigma_e \nabla u_e \cdot n = 0 & \text{on} \quad \Sigma \times (0, T),
\end{cases}
\tag{4.5}
$$

and initial conditions $V_m|_{t=0} = V_m^0$ and $w|_{t=0} = w^0$. Here $(0, T)$ is the time interval of interest, $\chi_m \overset{\text{def}}{=} A_m C_m$, $I_{ion} \overset{\text{def}}{=} A_m i_{ion}$, the vector n stands for the outward unit normal to $\Sigma \overset{\text{def}}{=} \partial \Omega_H$ and V_m^0, w^0 are given initial data.

The boundary conditions $(4.5)_{1,2}$ state that the intra- and extracellular currents do not propagate outside the heart. While $(4.5)_1$ is a widely accepted condition (see, e.g., [35, 53, 61, 68]), the enforcement of $(4.5)_2$ is only justified under an isolated heart assumption (see [53, 61]). The coupled system of Eqs. (4.4)–(4.5) is often known in the literature as *isolated bidomain* model (see [15, 17, 61]).

Remark 1. The complexity of (4.4)–(4.5) can be reduced by using the so-called monodomain approximation:

$$
\begin{cases}
\chi_m \partial_t V_m + I_{ion}(V_m, w) - \text{div}\left(\sigma \nabla V_m\right) = I_{app} & \text{in} \quad \Omega_H, \\
\sigma \nabla V_m \cdot n = 0 & \text{on} \quad \Sigma,
\end{cases}
\tag{4.6}
$$

where $\sigma \overset{\text{def}}{=} \sigma_i (\sigma_i + \sigma_e)^{-1} \sigma_e$ is known as the bulk conductivity tensor. Note that (4.6) decouples the computation of V_m from that of u_e. Under the isolating condition $(4.5)_2$, problem (4.6) can be interpreted as the approximation of $(4.4)_2$ and $(4.5)_1$ of order zero with respect to a parameter $\varepsilon \in [0, 1]$; the latter measures the gap between the anisotropy ratios of the intra- and extracellular domains (see [14, 17] for details). Although several simulation analyses (see, e.g., [14, 51]) suggest that the monodomain approximation may be adequate for some propagation studies in isolated hearts, it cannot be applied in all situations since it neglects the extracellular feedback into V_m (see, e.g., [14, 22, 51]).

4.2.2 Coupling with the torso: ECG modelling

The myocardium is surrounded by a volume conductor, Ω_T, that contains all the extramyocardial regions (see Fig. 4.2). For convenience, Ω_T will be called the *torso*. It is commonly modelled as a passive conductor (generalized Laplace equation). A perfect electric heart-torso coupling, across the interface Σ, is generally assumed (see, e.g., [35, 53, 61, 68]):

$$
\begin{cases}
u_e = u_T & \text{on} \quad \Sigma, \\
\sigma_e \nabla u_e \cdot n = -\sigma_T \nabla u_T \cdot n_T & \text{on} \quad \Sigma,
\end{cases}
\tag{4.7}
$$

where σ_T stands for the conductivity tensor of the torso tissue and n_T for the outward unit normal to $\partial \Omega_T$.

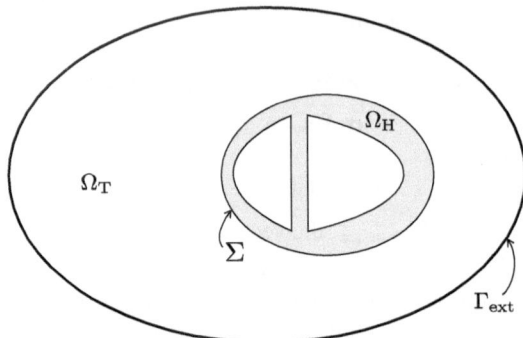

Fig. 4.2. Two-dimensional geometrical description: heart domain Ω_H, torso domain Ω_T (extramyocardial regions), heart-torso interface Σ and torso external boundary Γ_{ext}

The resulting coupled system can be formulated in terms of V_m, u_e, w and the torso potential u_T as follows (see, e.g., [53, 61]):

$$
\begin{cases}
\partial_t w + g(V_m, w) = 0, & \text{in } \Omega_H \times (0, T), \\
\chi_m \partial_t V_m + I_{ion}(V_m, w) - \text{div}(\boldsymbol{\sigma}_i \boldsymbol{\nabla} V_m) - \text{div}(\boldsymbol{\sigma}_i \boldsymbol{\nabla} u_e) = I_{app} & \text{in } \Omega_H \times (0, T), \\
- \text{div}((\boldsymbol{\sigma}_i + \boldsymbol{\sigma}_e) \boldsymbol{\nabla} u_e) - \text{div}(\boldsymbol{\sigma}_i \boldsymbol{\nabla} V_m) = 0 & \text{in } \Omega_H \times (0, T), \\
- \text{div}(\boldsymbol{\sigma}_T \boldsymbol{\nabla} u_T) = 0 & \text{in } \Omega_T \times (0, T), \\
\boldsymbol{\sigma}_T \boldsymbol{\nabla} u_T \cdot \boldsymbol{n}_T = 0 & \text{on } \Gamma_{ext} \times (0, T), \\
\boldsymbol{\sigma}_i \boldsymbol{\nabla} V_m \cdot \boldsymbol{n} + \boldsymbol{\sigma}_i \boldsymbol{\nabla} u_e \cdot \boldsymbol{n} = 0 & \text{on } \Sigma \times (0, T), \\
u_T = u_e & \text{on } \Sigma \times (0, T), \\
\boldsymbol{\sigma}_e \boldsymbol{\nabla} u_e \cdot \boldsymbol{n} = -\boldsymbol{\sigma}_T \boldsymbol{\nabla} u_T \cdot \boldsymbol{n}_T & \text{on } \Sigma \times (0, T),
\end{cases}
\tag{4.8}
$$

with the initial conditions $V_m|_{t=0} = V_m^0$, $w|_{t=0} = w^0$. The boundary condition on $\Gamma_{ext} \overset{\text{def}}{=} \partial \Omega_T \setminus \Sigma$ states that no current can flow from the external torso surface (see Fig. 4.2).

The coupled system of Eqs. (4.8) is often known in the literature as the *full bidomain* or *coupled bidomain* model (see, e.g., [14, 61]). It can be considered as the "gold standard" in the modelling of the ECG (see, e.g., [38, 53, 61]).

Remark 2. A common approach to reduce the computational complexity of (4.8) consists of uncoupling the computation of (w, V_m, u_e) from that of u_T, by neglecting the electrical torso-to-heart feedback (see, e.g., [14, 38, 50]). Thus, the coupling conditions (4.7) are replaced by

$$
\begin{cases}
u_T = u_e & \text{on } \Sigma, \\
\boldsymbol{\sigma}_e \boldsymbol{\nabla} u_e \cdot \boldsymbol{n} = 0 & \text{on } \Sigma,
\end{cases}
\tag{4.9}
$$

which amounts to working with an isolated bidomain model. The reliability of this approximation will be discussed later on. Moreover, isolating the heart might not be applicable to several studies. For instance this approximation is not physically acceptable while studying cardiac defibrillation mechanisms [66].

As will be shown in Sect. 4.4, this model has to be furnished by important modelling ingredients like cell heterogeneity, tissue anisotropy, modelling of fast-conducting networks (His bundle, Purkinje fibres) [6].

4.3 Mathematical analysis

Results on the existence and uniqueness of the solution for the isolated bidomain system (4.4) have been reported in a number of works. A first result for the bidomain model coupled with the FitzHugh-Nagumo ionic model, described below by Eq. (4.10), has been reported in [18]. The proof is based on a reformulation of the system of equations in terms of an abstract evolutionary variational inequality. The analysis for a simplified ionic model, namely $I_{ion}(V_m, w) = I_{ion}(V_m)$, has been addressed in [5]. In paper [9], existence, uniqueness and regularity of a local solution in time are proved for the bidomain model with a general ionic model by using a semi-group approach. Existence of a global in time solution of the bidomain problem is also proved in [9] for a wide class of ionic models – including FitzHugh-Nagumo (4.10), Aliev-Panfilov (4.11), and Roger-McCulloch (4.12) – through a compactness argument. As in the result presented below, uniqueness is only achieved for the FitzHugh-Nagumo ionic model. Finally, in [70], existence, uniqueness and some regularity results are proved with the Luo-Rudy ionic model [41]. All these papers deal with a model for the heart alone, without any coupling with the external medium.

Let us now present the result contained in [8], which addresses the mathematical analysis of the coupled heart-torso system (4.8). The well-posedness analysis of this system is obtained for an abstract class of two-variable ionic models including:

- the FitzHugh-Nagumo model [26, 45]:

$$I_{ion}(V_m, w) = kV_m(V_m - a)(V_m - 1) + w, \quad g(V_m, w) = -\varepsilon(\gamma V_m - w); \quad (4.10)$$

- the Aliev-Panfilov model [2]:

$$I_{ion}(V_m, w) = kV_m(V_m - a)(V_m - 1) + V_m w, \quad g(V_m, w) = \varepsilon(\gamma V_m(V_m - 1 - a) + w);$$
$$(4.11)$$

- the Roger-McCulloch model [57]:

$$I_{ion}(V_m, w) = kV_m(V_m - a)(V_m - 1) + V_m w, \quad g(V_m, w) = -\varepsilon(\gamma V_m - w); \quad (4.12)$$

- the Mitchell-Schaeffer model [43]:

$$I_{ion}(V_m, w) = \frac{w}{\tau_{in}} V_m^2 (V_m - 1) - \frac{V_m}{\tau_{out}},$$

$$g(V_m, w) = \begin{cases} \dfrac{1-w}{\tau_{open}} & \text{if } V_m \leq v_{gate}, \\[2mm] \dfrac{-w}{\tau_{close}} & \text{if } V_m > v_{gate}. \end{cases} \tag{4.13}$$

Here, $0 < a < 1$, k, ε, γ, τ_{in}, τ_{out}, τ_{open}, τ_{close} and $0 < v_{gate} < 1$ are given positive constants.

The next theorem asserts the existence of solutions for the coupled heart-torso model. It also provides the uniqueness of solutions if the ionic current is described by FitzHugh-Nagumo model.

Theorem 1. *Let $T > 0$, $I_{app} \in L^2(\Omega_H \times (0,T))$, $\sigma_i, \sigma_e \in L^\infty(\Omega_H)$ symmetric and uniformly definite positive, $w_0 \in L^2(\Omega_H)$ and $V_m^0 \in H^1(\Omega_H)$ be given data. Assume that I_{ion} and g are given by (4.10), (4.11), (4.12) or a regularized version of (4.13). Then, the heart-torso system (4.8) has a weak solution $V_m \in L^\infty(0,T;H^1(\Omega_H)) \cap H^1(0,T;L^2(\Omega_H))$, $w \in H^1(0,T;L^2(\Omega_H))$ and $u \in L^\infty(0,T;H^1(\Omega))$ with*

$$u \overset{\text{def}}{=} \begin{cases} u_e & \text{in} \quad \Omega_H, \\ u_T & \text{in} \quad \Omega_T, \end{cases}$$

and $\Omega \overset{\text{def}}{=} \Omega_T \cup \overline{\Omega_H}$. Moreover, for the FitzHugh-Nagumo model (4.10), the solution is unique.

The proof of Theorem 1 is reported in [8] and [72, Part II] and generalizes some of the arguments used for the analysis of the bidomain problem in [5, 9] to the case of the heart-torso coupling. The main idea consists in reformulating the bidomain system as a couple of degenerate reaction-diffusion equations and approximating the resulting heart-torso system by a suitably regularized problem in finite dimension; the latter is then analyzed through a Faedo-Galerkin/compactness procedure and a specific treatment of the non-linear terms. The heart-torso coupling is handled through an adequate definition of the Galerkin basis. Compared to models (4.10)–(4.12), the Mitchell-Schaeffer ionic model has a different structure that makes the existence proof slightly more involved. As shown in the next section, realistic ECG signals can be simulated with this ionic model.

4.4 ECG simulations

The aim of this section is twofold: first, provide realistic simulations of the 12-lead ECG based on the model given by system (4.8); secondly, discuss through numerical simulations the impact of various modelling options (e.g., uncoupling, monodomain, cell homogeneity, tissue anisotropy). More details can be found in [7].

Our reference mathematical model of ECG relies on (4.8) and the following additional modelling ingredients:

- the heart geometry only includes the ventricles (see Fig. 4.3). Note that this simplification prevents from computing the P-wave of the ECG;
- the "torso" geometry contains three regions: the lungs, the bones and the remaining extracardiac tissues (see Fig. 4.3). These regions are modelled with three different values of σ_T;
- the fast conduction system (His bundle and Purkinje fibres) is modelled by initializing the activation with a time-dependent external volume current I_{app}, acting on a thin sub-endocardial layer of left and right ventricles. The propagation speed of this external stimulus is a parameter of the model (see [7] for details);
- the dynamics of the cardiac cell's membrane are based on the Mitchell-Schaeffer ionic model (4.13);
- cells are heterogeneous in terms of *Action Potential Duration* (APD), which varies transmurally within the left ventricle. In practice, this amounts to consider a parameter τ_{close} in (4.13) which takes three different values in the left ventricle (τ_{close}^{endo} in the subendocardial region, τ_{close}^{intra} in the intracardial region and τ_{close}^{epi} in the subepicardial region) and another value τ_{close}^{rv} in the right ventricle. This is an important factor to obtain the T-wave with a correct polarity;
- the heart conductivities are anisotropic:

$$\boldsymbol{\sigma}_{i,e}(\boldsymbol{x}) \stackrel{def}{=} \sigma_{i,e}^t \boldsymbol{I} + (\sigma_{i,e}^l - \sigma_{i,e}^t)\boldsymbol{a}(\boldsymbol{x}) \otimes \boldsymbol{a}(\boldsymbol{x}),$$

where $\boldsymbol{a}(\boldsymbol{x})$ is a unit vector parallel to the local fibre direction and $\sigma_{i,e}^l$ and $\sigma_{i,e}^t$ are, respectively, the conductivity coefficients in the intra- and extra-cellular media, measured along the fibre and transverse directions.

4.4.1 Numerical approximation

Problem (4.8) can be cast into weak form as follows: for $t \in (0,T)$, find $w(\cdot,t) \in L^\infty(\Omega_H)$, $V_m(\cdot,t) \in H^1(\Omega_H)$, $u_e(\cdot,t) \in H^1(\Omega_H) \cap L_0^2(\Omega_H)$ and $u_T(\cdot,t) \in H^1(\Omega_T)$ with $u_e(\cdot,t) = u_T(\cdot,t)$ on Σ, such that

$$\int_{\Omega_H} \left(\partial_t w + g(V_m,w)\right)\xi = 0,$$

$$\int_{\Omega_H} \left(\chi_m \partial_t V_m + I_{ion}(V_m,w)\right)\phi + \int_{\Omega_H} \boldsymbol{\sigma}_i \boldsymbol{\nabla}(V_m + u_e) \cdot \boldsymbol{\nabla}\phi = \int_{\Omega_H} I_{app}\phi, \qquad (4.14)$$

$$\int_{\Omega_H} (\boldsymbol{\sigma}_i + \boldsymbol{\sigma}_e)\boldsymbol{\nabla} u_e \cdot \boldsymbol{\nabla}\psi + \int_{\Omega_H} \boldsymbol{\sigma}_i \boldsymbol{\nabla} V_m \cdot \boldsymbol{\nabla}\psi + \int_{\Omega_T} \boldsymbol{\sigma}_T \boldsymbol{\nabla} u_T \cdot \boldsymbol{\nabla}\zeta = 0$$

for all $(\xi,\phi,\psi,\zeta) \in L^2(\Omega_H) \times H^1(\Omega_H) \times (H^1(\Omega_H) \cap L_0^2(\Omega_H)) \times H^1(\Omega_T)$ with $\psi = \zeta$ on Σ. The weak formulation (4.14) is discretized in space using finite elements and in time using a semi-implicit scheme based on a backward difference formula (BDF).

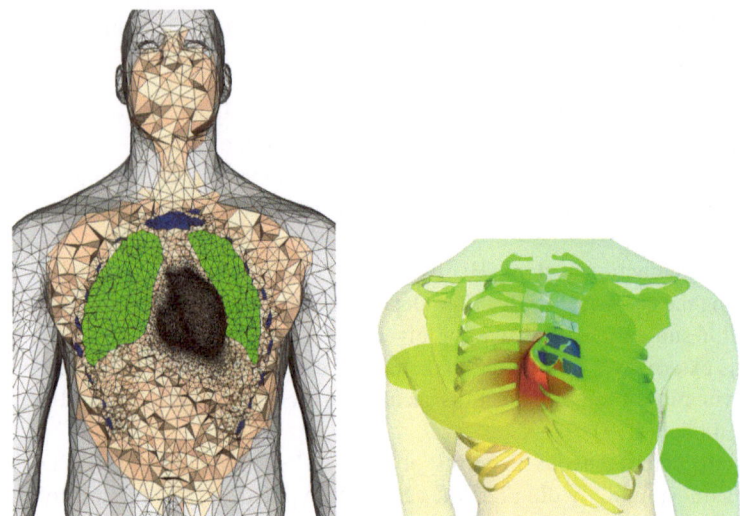

Fig. 4.3. *Left*: cut view of the heart-torso computational mesh. Heart (red), lungs (green), bone (blue) and remaining tissue (apricot). *Right*: posterior view and cut plane of the torso and heart potentials at time $t = 10$ ms

Let $N \in \mathbb{N}^*$ be a given integer and consider a uniform partition $\{[t_n, t_{n+1}]\}_{0 \leq n \leq N-1}$, with $t_n \stackrel{\text{def}}{=} n\tau$, of the time interval of interest $(0, T)$, with time-step size $\tau \stackrel{\text{def}}{=} T/N$. The notation $\partial_\tau x^{n+1}$ represents a (first or second order) backward difference formula, $\widetilde{x}^{n+1}$ the corresponding explicit extrapolation (i.e., x^n or $2x^n - x^{n-1}$) and $I_{\text{app}}^n \stackrel{\text{def}}{=} I_{\text{app}}(\cdot, t_n)$. The space $X_{\text{H},h}$ (resp. $X_{\text{T},h}$) is the internal approximation of $H^1(\Omega_{\text{H}})$ (resp. $H^1(\Omega_{\text{T}})$) made of continuous piecewise-affine functions. Moreover, we assume that the restrictions of $X_{\text{H},h}$ and $X_{\text{T},h}$ match at the interface Σ. The resulting fully discrete time advancing procedure reads as follows: for $0 \leq n \leq N-1$,

1. Ionic state: find $w^{n+1} \in X_{\text{H},h}$ such that

$$\int_{\Omega_{\text{H}}} \left(\partial_\tau w^{n+1} + g(\widetilde{V}_{\text{m}}^{n+1}, w^{n+1}) \right) \xi = 0 \qquad (4.15)$$

for all $\xi \in X_{\text{H},h}$.

2. Heart and torso potentials: find $V_{\text{m}}^{n+1} \in X_{\text{H},h}$, $u_{\text{e}}^{n+1} \in X_{\text{H},h} \cap L_0^2(\Omega_{\text{H}})$ and $u_{\text{T}}^{n+1} \in X_{\text{T},h}$ with $u_{\text{T}}^{n+1} = u_{\text{e}}^{n+1}$ on Σ, such that

$$\chi_{\text{m}} \int_{\Omega_{\text{H}}} \partial_\tau V_{\text{m}}^{n+1} \phi + \int_{\Omega_{\text{H}}} \boldsymbol{\sigma}_{\text{i}} \boldsymbol{\nabla}(V_{\text{m}}^{n+1} + u_{\text{e}}^{n+1}) \cdot \boldsymbol{\nabla}\phi$$

$$= \int_{\Omega_{\text{H}}} \left(I_{\text{app}}^{n+1} - I_{\text{ion}}(\widetilde{V}_{\text{m}}^{n+1}, w^{n+1}) \right) \phi,$$

$$\int_{\Omega_H} (\sigma_i + \sigma_e) \nabla u_e^{n+1} \cdot \nabla \psi + \int_{\Omega_H} \sigma_i \nabla V_m^{n+1} \cdot \nabla \psi + \int_{\Omega_T} \sigma_T \nabla u_T^{n+1} \cdot \nabla \zeta = 0,$$

(4.16)

for all $(\phi, \psi, \zeta) \in X_{H,h} \times (X_{H,h} \cap L_0^2(\Omega_H)) \times X_{T,h}$ with $\psi = \zeta$ on Σ.

At each time step, the linear problem (4.16) requires the coupled computation of the heart potentials (V_m^{n+1}, u_e^{n+1}) and the torso potential u_T^{n+1}. This coupling can be solved monolithically, that is, after full assembling of the whole system matrix (see, e.g., [38, Sects. 4.6 and 4.5.1] and [11, 61, 62]). Alas, this results in an increased number of unknowns with respect to the original bidomain system. Moreover, this procedure is less modular since the bidomain and torso equations cannot be solved independently. This shortcoming can be overcome by using a partitioned iterative procedure based on domain decomposition (see, e.g., [55, 65]). In this study, the heart-torso coupling is solved, in a partitioned fashion, via relaxed Dirichlet-Neumann preconditioned Richardson iterations. A related approach is adopted in [11] (see also [38, 53]), using an integral formulation of the torso equation.

As far as the discretization in time is concerned, alternative time-marching schemes will be discussed in Sect. 4.5.

4.4.2 Simulated ECG signals

Simulations have been carried out under the modelling hypotheses described at the beginning of this section and by solving Eqs. (4.15)–(4.16), as described in the previous paragraph. Fig. 4.4 shows some snapshots of the simulated body surface potentials. The corresponding 12-lead ECG signals are given in Fig. 4.5. Despite some minor flaws, the comparison with Fig. 4.1 (right) shows that the numerical ECGs obtained have the correct amplitudes, shapes and polarities, in all the twelve standard leads. To the best of our knowledge, this constitutes a breakthrough in the numerical simulations of ECGs with partial differential equations.

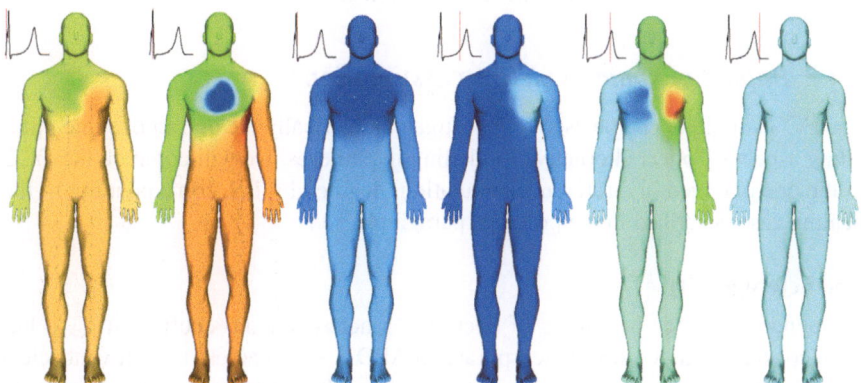

Fig. 4.4. Snapshots of the body surface potentials at times $t = 10, 32, 40$ ms (depolarization) and $t = 200, 250$ and 310 ms (repolarization), from left to right

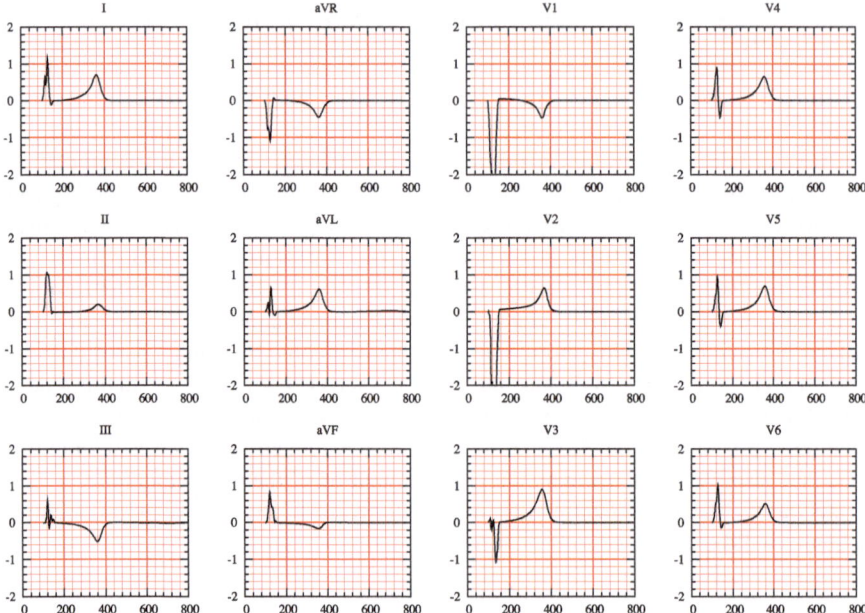

Fig. 4.5. Simulated normal 12-lead ECG signals

Numerical simulations have also been carried out for some pathological conditions like left or right bundle branch blocks (see [7] or [72] for details). The numerical ECG signals satisfy the typical criteria used by medical doctors to detect the pathology, and this without any recalibration of the model's parameters besides the natural modifications needed to model the disease (i.e. delayed activation in the right or left ventricle). This suggests that the ECG simulator has some predictive features.

Remark 3. We refer to [12, 21] for two examples of how the ECG simulator developed can be used in different contexts and applications.

4.4.3 Impact of some modelling assumptions

The ECG simulator can be used to investigate numerically the impact of some modelling aspects. To test alternative modelling hypotheses, we will compare the ECG to the one obtained in a reference simulation, denoted by **RS**, corresponding to the healthy case described in the previous paragraph.

Cell heterogeneity

As mentioned at the beginning of Sect. 4.4, a heterogeneous coefficient τ_{close} has been considered in the **RS** to incorporate an APD gradient across the left ventricle's transmural direction. To reduce the complexity of the model, it may be tempting to take a constant coefficient. But it has been observed that this simplification would affect the polarity of the T-wave in lead I (see [7]). Indeed, without transmural APD

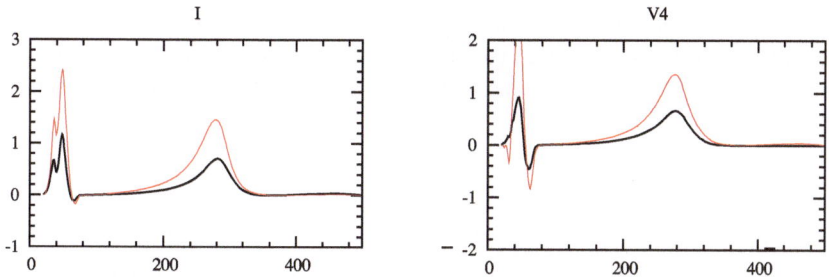

Fig. 4.6. Simulated ECG signals (leads I and V4) obtained using heart-torso full coupling (black) and uncoupling (red)

heterogeneity, the repolarization and depolarization waves travel in the same direction, which leads to the discordant polarity, between the QRS- and the T-waves, observed in lead I. On the contrary, unipolar leads (aVR, V1 and V4) present a similar polarity, irrespective of the ADP heterogeneity (see also [16]). As a result, as also noticed in [7, 33, 50, 51], transmural APD heterogeneity is a major ingredient in the simulation of a complete 12-lead ECG with physiological T-wave polarities.

Heart-torso uncoupling

Concerning the heart-torso uncoupling approximation (4.9), the comparison reported in Fig. 4.6 shows that heart-torso uncoupling compromises the accuracy of the ECG signals (notice the difference of amplitudes). Therefore uncoupling cannot be recommended in general, although it can be a reasonable choice to get qualitatively correct ECGs.

Monodomain approximation

Let us notice that, without the uncoupling assumption (4.9), approximation (4.6) becomes

$$
\begin{cases}
\chi_{\mathrm{m}}\partial_t V_{\mathrm{m}} + I_{\mathrm{ion}}(V_{\mathrm{m}}, w) - \mathrm{div}\left(\boldsymbol{\sigma}\boldsymbol{\nabla} V_{\mathrm{m}}\right) = I_{\mathrm{app}}, & \text{in} \quad \Omega_{\mathrm{H}}, \\
\boldsymbol{\sigma}\boldsymbol{\nabla} V_{\mathrm{m}} \cdot \boldsymbol{n} = -\mu\boldsymbol{\sigma}_{\mathrm{e}}\boldsymbol{\nabla} u_{\mathrm{e}} \cdot \boldsymbol{n}, & \text{on} \quad \Sigma,
\end{cases}
\tag{4.17}
$$

where $\mu \in (0,1)$ is a dimensionless parameter related to the local conductivities (see also [14]). Thus, we cannot derive anymore the usual monodomain model since the unknowns V_{m} and u_{e} are still coupled. In that case, solving this system does not reduce the computational complexity, compared to (4.8).

It seems more interesting, in terms of computational cost, to combine the monodomain approximation with the heart-torso uncoupling approximation (see, e.g. [30, 50]). Indeed, this yields a simplified mathematical model which allows a fully decoupled computation of V_{m}, u_{e} (which is equal to V_{m} up to a multiplicative constant) and u_{T}. In that case, the ECGs obtained are very similar to the ECGs obtained for the bidomain model combined with the heart-torso uncoupling hypothesis.

Remark 4. An alternative can be to neglect the boundary coupling in $(4.17)_2$ while keeping u_e and u_T fully coupled (see [52]). In a pure propagation framework (i.e., without extracellular pacing) numerical experiments suggest that this approach can provide accurate ECG signals.

Conclusion

According to the above discussion, we can conclude that cell heterogeneity has to be taken into account to get realistic ECGs. If we are only interested in a qualitatively reasonable ECG, heart-torso uncoupling can be considered, but for quantitatively precise results, it seems necessary to solve the coupled problem. It seems that replacing the bidomain equations by its monodomain approximation does not affect significantly the ECGs. Numerical simulations show that the geometries of the heart and the torso, the anisotropy in the heart and the heterogeneity of the conductivity in the torso also have an important impact on ECGs. We refer to [7, 72] for more details and further discussions about these effects and a sensitivity analysis.

4.5 Fully decoupled time-marching schemes

In this section, we introduce and analyze some time-marching schemes for the numerical approximation of the isolated bidomain model (Sect. 4.5.1) and the heart-torso system (Sect. 4.5.2). The particularity of these schemes is that they all allow for an uncoupled computation of the fields involved (ionic state, transmembrane potential, extracellular and torso potentials). The original ideas presented here were originally proposed in [25].

In what follows, the quantity $\partial_\tau x^n$ denotes the first-order backward difference $(x^n - x^{n-1})/\tau$.

4.5.1 Isolated bidomain model

The isolated bidomain system (4.4)-(4.5) can be cast into weak form as follows: for $t > 0$, find $w(\cdot,t) \in L^\infty(\Omega_H)$, $V_m(\cdot,t) \in H^1(\Omega_H)$ and $u_e(\cdot,t) \in H^1(\Omega_H) \cap L^2_0(\Omega_H)$, such that

$$\int_{\Omega_H} (\partial_t w + g(V_m, w))\xi = 0,$$

$$\int_{\Omega_H} (\chi_m \partial_t V_m + I_{ion}(V_m, w))\phi + \int_{\Omega_H} \sigma_i \nabla(V_m + u_e) \cdot \nabla\phi = \int_{\Omega_H} I_{app}\phi, \qquad (4.18)$$

$$\int_{\Omega_H} (\sigma_i + \sigma_e)\nabla u_e \cdot \nabla\psi + \int_{\Omega_H} \sigma_i \nabla V_m \cdot \nabla\psi = 0$$

for all $(\xi, \phi, \psi) \in L^2(\Omega_H) \times H^1(\Omega_H) \times (H^1(\Omega_H) \cap L^2_0(\Omega_H))$.

The rapid dynamics of the ODE system $(4.18)_1$, acting on the reaction terms $(4.18)_2$, detects a sharp propagating wavefront (see Fig. 4.3 (right)), which often requires fine resolutions in space and in time. Fully implicit time-marching is, therefore, extremely difficult to perform since it involves the resolution of a large system of non-linear equations at each time step (see, e.g., [10, 29, 44]). Attempts to reduce this computational complexity (without compromising numerical stability too much) consist in introducing some sort of explicit treatment within the time-marching procedure. For instance, by considering semi-implicit (see, e.g., Sect. 4.4 and [3, 15, 23, 39, 63]) or operator-splitting (see, e.g., [31, 64, 71]) schemes. All these approaches uncouple the ODE system (ionic state and non-linear reaction terms) from the electro-diffusive components (transmembrane and extracellular potentials). A few articles [3, 39, 63, 71] considered, without analysis, a decoupled (*Gauss-Seidel*-like) time-marching of the three fields.

It can be shown that the Gauss-Seidel and the Jacobi electro-diffusive splittings do not compromise the stability of the resulting scheme. They simply alter the energy norm, and the time-step restrictions are uniquely dictated by the semi-implicit treatment of the ODE system and the non-linear reaction terms. Let us consider the semi-discretization in time of (4.18) obtained by combining a first-order semi-implicit treatment of the ionic current with an explicit (Gauss-Seidel or Jacobi-like) treatment of the electro-diffusive coupling, as detailed in Algorithm 1.

Algorithm 1 Decoupled time-marching for the bidomain equation.

1. Ionic state: find $w^{n+1} \in L^\infty(\Omega_H)$ such that

$$\int_{\Omega_H} \left(\partial_\tau w^{n+1} + g(V_m^n, w^{n+1}) \right) \xi = 0$$

   for all $\xi \in L^2(\Omega_H)$;
2. Transmembrane potential: find $V_m^{n+1} \in H^1(\Omega_H)$ such that

$$\chi_m \int_{\Omega_H} \partial_\tau V_m^{n+1} \phi + \int_{\Omega_H} \sigma_i \nabla V_m^{n+1} \cdot \nabla \phi + \int_{\Omega_H} \sigma_i \nabla u_e^\star \cdot \nabla \phi$$
$$= \int_{\Omega_H} \left(I_{app}^{n+1} - I_{ion}(V_m^n, w^{n+1}) \right) \phi$$

   for all $\phi \in H^1(\Omega_H)$;
3. Extracellular potential: find $u_e^{n+1} \in H^1(\Omega_H) \cap L_0^2(\Omega_H)$,

$$\int_{\Omega_H} (\sigma_i + \sigma_e) \nabla u_e^{n+1} \cdot \nabla \psi + \int_{\Omega_H} \sigma_i \nabla V_m^\star \cdot \nabla \psi = 0$$

   for all $\psi \in H^1(\Omega_H) \cap L_0^2(\Omega_H)$;
4. Go to next time-step.

For $(u_e^\star, V_m^\star) = (u_e^{n+1}, V_m^{n+1})$, the unknown potentials V_m^{n+1} and u_e^{n+1} are implicitly coupled and, therefore, steps 2 and 3 of Algorithm 1 have to be performed simultaneously (as in Sect. 4.4). On the other hand, for $(u_e^\star, V_m^\star) = (u_e^n, V_m^{n+1})$ or $(u_e^\star, V_m^\star) = (u_e^n, V_m^n)$, the electro-diffusive coupling becomes explicit and therefore these steps can be performed separately: either sequentially (Gauss-Seidel) or in parallel (Jacobi).

Stability analysis

In order to render the analysis easier we make the following simplifying assumption on the structure of the ionic models (see [23, Sect. 3.2.2] and Remark 5 below):

$$
\begin{aligned}
I_{\mathrm{ion}}(V_{\mathrm{m}}, w) &\le C_I\big(|V_{\mathrm{m}}| + |w|\big), \\
g(V_{\mathrm{m}}, w) &\le C_g\big(|V_{\mathrm{m}}| + |w|\big)
\end{aligned}
\tag{4.19}
$$

for all V_{m}, w, and we set $\alpha \overset{\text{def}}{=} 1 + 3C_I + C_g$ and $\beta \overset{\text{def}}{=} C_I + 3C_g$. In what follows, the symbol $\lesssim$ indicates an inequality up to a multiplicative constant proportional to $e^{T/(1 - \tau \max\{\alpha, \beta\})}$.

Remark 5. Assumption (4.19) is not satisfied by the usual ionic models. Nevertheless, this simplification allows to motivate, from a theoretical point of view, the numerically-observed stability properties of the electro-diffusive splittings reported in Algorithm 1. Note that, for the ionic models (4.10)–(4.13) the assumption holds if one assumes *a priori* that V_{m} remains uniformly bounded, which allows to estimate empirically the constants C_I and C_g (see [23, Remark 3.1]).

The next result, proved in [25], establishes the energy-based stability of Algorithm 1, in terms of $u_{\mathrm{e}}^{\star}$ and $V_{\mathrm{m}}^{\star}$.

Theorem 2. *Assume that* (4.19) *holds and let* $(w^n, V_{\mathrm{m}}^n, u_{\mathrm{e}}^n)$ *be given by Algorithm 1. Then, under the condition*

$$
\tau < \frac{1}{\max\{\alpha, \beta\}},
\tag{4.20}
$$

we have:

- *For* $(u_{\mathrm{e}}^{\star}, V_{\mathrm{m}}^{\star}) = (u_{\mathrm{e}}^{n+1}, V_{\mathrm{m}}^{n+1})$:

$$
\begin{aligned}
&\|w^n\|_{0,\Omega_{\mathrm{H}}}^2 + \chi_{\mathrm{m}} \|V_{\mathrm{m}}^n\|_{0,\Omega_{\mathrm{H}}}^2 + 2\sum_{m=0}^{n-1} \tau \|\sigma_{\mathrm{e}}^{\frac{1}{2}} \nabla u_{\mathrm{e}}^{m+1}\|_{0,\Omega_{\mathrm{H}}}^2 \\
&\quad + 2\sum_{m=0}^{n-1} \tau \|\sigma_{\mathrm{i}}^{\frac{1}{2}} \nabla(V_{\mathrm{m}}^{m+1} + u_{\mathrm{e}}^{m+1})\|_{0,\Omega_{\mathrm{H}}}^2 \lesssim \|w^0\|_{0,\Omega_{\mathrm{H}}}^2 \\
&\quad\quad\quad\quad\quad\quad\quad\quad\quad + \chi_{\mathrm{m}} \|V_{\mathrm{m}}^0\|_{0,\Omega_{\mathrm{H}}}^2 + \sum_{m=0}^{n-1} \tau \|I_{\mathrm{app}}^{m+1}\|_{0,\Omega_{\mathrm{H}}}^2,
\end{aligned}
$$

with $1 \le n \le N$.

- *For* $(u_{\mathrm{e}}^{\star}, V_{\mathrm{m}}^{\star}) = (u_{\mathrm{e}}^n, V_{\mathrm{m}}^{n+1})$:

$$
\begin{aligned}
&\|w^n\|_{0,\Omega_{\mathrm{H}}}^2 + \chi_{\mathrm{m}} \|V_{\mathrm{m}}^n\|_{0,\Omega_{\mathrm{H}}}^2 + \tau \|\sigma_{\mathrm{i}}^{\frac{1}{2}} \nabla u_{\mathrm{e}}^n\|_{0,\Omega_{\mathrm{H}}}^2 \\
&\quad + 2\sum_{m=0}^{n-1} \tau \|\sigma_{\mathrm{e}}^{\frac{1}{2}} \nabla u_{\mathrm{e}}^{m+1}\|_{0,\Omega_{\mathrm{H}}}^2 + \sum_{m=0}^{n-1} \tau \|\sigma_{\mathrm{i}}^{\frac{1}{2}} \nabla(V_{\mathrm{m}}^{m+1} + u_{\mathrm{e}}^{m+1})\|_{0,\Omega_{\mathrm{H}}}^2 \\
&\quad \lesssim \|w^0\|_{0,\Omega_{\mathrm{H}}}^2 + \chi_{\mathrm{m}} \|V_{\mathrm{m}}^0\|_{0,\Omega_{\mathrm{H}}}^2 + \tau \|\sigma_{\mathrm{i}}^{\frac{1}{2}} \nabla u_{\mathrm{e}}^0\|_{0,\Omega_{\mathrm{H}}}^2 + \sum_{m=0}^{n-1} \tau \|I_{\mathrm{app}}^{m+1}\|_{0,\Omega_{\mathrm{H}}}^2,
\end{aligned}
$$

with $1 \le n \le N$.

- *For $(u_e^\star, V_m^\star) = (u_e^n, V_m^n)$:*

$$\|w^n\|_{0,\Omega_H}^2 + \chi_m \|V_m^n\|_{0,\Omega_H}^2 + \tau \|\sigma_i^{\frac{1}{2}} \nabla u_e^n\|_{0,\Omega_H}^2 + \tau \|\sigma_i^{\frac{1}{2}} \nabla V_m^n\|_{0,\Omega_H}^2$$

$$+ 2\sum_{m=0}^{n-1} \tau \|\sigma_e^{\frac{1}{2}} \nabla u_e^{m+1}\|_{0,\Omega_H}^2 \lesssim \|w^0\|_{0,\Omega_H}^2 + \chi_m \|V_m^0\|_{0,\Omega_H}^2$$

$$+ \tau \|\sigma_i^{\frac{1}{2}} \nabla V_m^0\|_{0,\Omega_H}^2 + \tau \|\sigma_i^{\frac{1}{2}} \nabla u_e^0\|_{0,\Omega_H}^2 + \sum_{m=0}^{n-1} \tau \|I_{app}^{m+1}\|_{0,\Omega_H}^2,$$

with $1 \le n \le N$.

Theorem 2 shows that electro-diffusive Gauss-Seidel and Jacobi splittings are energy-stable under condition (4.20), as for the unsplit case $(u_e^\star, V_m^\star) = (u_e^{n+1}, V_m^{n+1})$ (analyzed in [23]), but with slightly altered energy norms. As a result, stability is not compromised.

Remark 6. The proof of Theorem 2 does not depend on the time discretization considered in steps 1 and 2 of Algorithm 1. Indeed, we do not make use of any numerical dissipation produced by the scheme, apart from the one directly provided by the splitting. Therefore, the backward Euler quotients $\partial_\tau w^{n+1}$ and $\partial_\tau V_m^{n+1}$, can be safely replaced by a second-order backward difference formula, and then corrected (see, e.g., [59, 60]) to recover overall second-order accuracy.

Remark 7. The above stability result can be adapted, with minor modifications, to the case $(u_e^\star, V_m^\star) = (u_e^{n+1}, V_m^n)$. The full Jacobi splitting, obtained after replacing $I_{ion}(V_m^n, w^{n+1})$ by $I_{ion}(V_m^n, w^n)$ in step 2, could also be considered.

We conclude this subsection with a few numerical illustrations. The results reported in Fig. 4.7 (top) confirm that the electro-diffusive Gauss-Seidel and Jacobi splittings do not introduce additional constraints on the time-step size τ, as stated in Theorem 2. Fig. 4.7 (bottom) shows that the Coupled, the Gauss-Seidel and the Jacobi time-marching schemes all provide the expected first-order accuracy $\mathcal{O}(\tau)$ in time. Note that, at a given time-step size, Gauss-Seidel is slightly more accurate than Jacobi and that the Coupled scheme is slightly more accurate than Gauss-Seidel. This accuracy shifting could be related to the energy-norm weakening observed in the stability analysis.

4.5.2 Coupled heart-torso system

In this subsection, we propose a series of time-marching procedures for the heart-torso system (4.8) that allow a decoupled computation of the transmembrane, extracellular and torso potentials. The main idea consists in combining the bidomain splittings of the previous section with a specific explicit Robin-Robin treatment of the heart-torso coupling conditions. The proposed schemes are presented in Algorithm 2, with $\gamma > 0$ a dimensionless free parameter (fixed below) and where we have assumed that $\sigma_T|_\Sigma = \sigma_T^t I$ (without loss of generality).

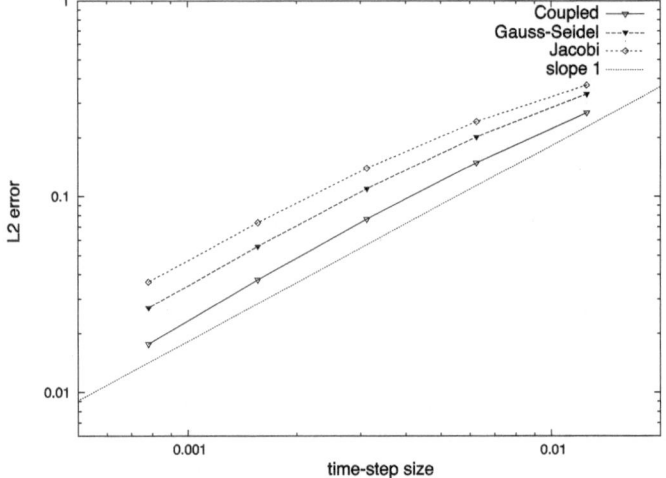

τ (ms)	Coupled	G-S	Jacobi
0.25	✓	✓	✓
0.50	✓	✓	✓
1.00	✓	✓	✓
1.25	✗	✗	✗
1.50	✗	✗	✗

Fig. 4.7. *Top*: stability sensitivity to time-step size (✗ indicates numerical instability). *Bottom*: time convergence history of the transmembrane potential error for Coupled $((u_e^\star, V_m^\star) = (u_e^{n+1}, V_m^{n+1}))$, Gauss-Seidel $((u_e^\star, V_m^\star) = (u_e^n, V_m^{n+1}))$ and Jacobi $(u_e^\star, V_m^\star) = (u_e^n, V_m^n)$ bidomain time-marching schemes

Note that the cardiac subproblem (steps 1–3) can be solved independently of the torso subproblem (step 4). In particular, the Gauss-Seidel-Robin algorithm $(u_e^\star, V_m^\star) = (u_e^n, V_m^{n+1})$, and the Jacobi-Robin algorithm $(u_e^\star, V_m^\star) = (u_e^n, V_m^n)$ lead to a fully decoupled computation of w^{n+1}, V_m^{n+1}, u_e^{n+1} and u_T^{n+1}. In other words, steps 1–4 are decoupled and can be performed sequentially.

Remark 8. The choices $(u_e^\star, V_m^\star) = (u_e^n, V_m^{n+1})$ or $(u_e^\star, V_m^\star) = (u_e^n, V_m^n)$ in Algorithm 2 allow for fully decoupled computation of w^{n+1}, V_m^{n+1}, u_e^{n+1} and u_T^{n+1} without the need to resort to monodomain and uncoupling approximations (see Sect. 4.4).

Remark 9. The interface coupling between steps 3 and 4 of Algorithm 2 corresponds to the following Robin-Robin-based explicit time-discretization of (4.7):

$$\boldsymbol{\sigma}_e \nabla u_e^{n+1} \cdot \boldsymbol{n} + \frac{\gamma \sigma_T^t}{h} u_e^{n+1} = -\boldsymbol{\sigma}_T \nabla u_T^n \cdot \boldsymbol{n}_T + \frac{\gamma \sigma_T^t}{h} u_T^n \quad \text{on } \Sigma,$$

$$\boldsymbol{\sigma}_T \nabla u_T^{n+1} \cdot \boldsymbol{n}_T + \frac{\gamma \sigma_T^t}{h} u_T^{n+1} = \boldsymbol{\sigma}_T \nabla u_T^n \cdot \boldsymbol{n}_T + \frac{\gamma \sigma_T^t}{h} u_e^{n+1} \quad \text{on } \Sigma.$$

(4.21)

Algorithm 2 Decoupled time-marching for the heart-torso system.

---

1. Ionic state: find $w^{n+1} \in X_{H,h}$ such that

$$\int_{\Omega_H} \left(\partial_\tau w^{n+1} + g(V_m^n, w^{n+1}) \right) \xi = 0$$

for all $\xi \in X_{H,h}$;

2. Transmembrane potential: find $V_m^{n+1} \in X_{H,h}$ such that

$$\chi_m \int_{\Omega_H} \partial_\tau V_m^{n+1} \phi + \int_{\Omega_H} \boldsymbol{\sigma}_i \nabla V_m^{n+1} \cdot \nabla \phi + \int_{\Omega_H} \boldsymbol{\sigma}_i \nabla u_e^\star \cdot \nabla \phi$$
$$= \int_{\Omega_H} \left(I_{app}^{n+1} - I_{ion}(V_m^n, w^{n+1}) \right) \phi$$

for all $\phi \in X_{H,h}$;

3. Extracellular potential: find $u_e^{n+1} \in X_{H,h}$ such that

$$\int_{\Omega_H} (\boldsymbol{\sigma}_i + \boldsymbol{\sigma}_e) \nabla u_e^{n+1} \cdot \nabla \psi + \int_{\Omega_H} \boldsymbol{\sigma}_i \nabla V_m^\star \cdot \nabla \psi + \frac{\gamma \sigma_T^t}{h} \int_\Sigma u_e^{n+1} \psi$$
$$= -\int_\Sigma \boldsymbol{\sigma}_T \nabla u_T^n \cdot \boldsymbol{n}_T \psi + \frac{\gamma \sigma_T^t}{h} \int_\Sigma u_T^n \psi$$

for all $\psi \in X_{H,h}$;

4. Torso potential: find $u_T^{n+1} \in X_{T,h}$

$$\int_{\Omega_T} \boldsymbol{\sigma}_T \nabla u_T^{n+1} \cdot \nabla \zeta + \frac{\gamma \sigma_T^t}{h} \int_\Sigma u_T^{n+1} \zeta = \int_\Sigma \boldsymbol{\sigma}_T \nabla u_T^n \cdot \boldsymbol{n}_T \zeta + \frac{\gamma \sigma_T^t}{h} \int_\Sigma u_e^{n+1} \zeta$$

for all $\zeta \in X_{T,h}$;

5. Go to next time-step.

Note that, since the (quasi-static) time discretization of $(4.8)_{3,4}$ does not generate numerical dissipation in time, a naive Dirichlet-Neumann explicit coupling, obtained by enforcing

$$\begin{cases} u_T^{n+1} = u_e^n & \text{on } \Sigma, \\ \boldsymbol{\sigma}_e \nabla u_e^{n+1} \cdot \boldsymbol{n} = -\boldsymbol{\sigma}_T \nabla u_T^{n+1} \cdot \boldsymbol{n}_T & \text{on } \Sigma, \end{cases}$$

might lead to numerical instability.

The next result, proved in [25], establishes the energy-based stability of Algorithm 2. There, $E_H^0(u_e^\star, V_m^\star)$ (resp. $E_H^n(u_e^\star, V_m^\star)$) denotes the discrete bidomain energy arising in the stability estimates provided by Theorem 2. For instance, in the case $(u_e^\star, V_m^\star) = (u_e^n, V_m^n)$, we have

$$E_H^0(u_e^\star, V_m^\star) \stackrel{\text{def}}{=}$$
$$\|w^0\|_{0,\Omega_H}^2 + \chi_m \|V_m^0\|_{0,\Omega_H}^2 + \tau \|\boldsymbol{\sigma}_i^{\frac{1}{2}} \nabla V_m^0\|_{0,\Omega_H}^2 + \tau \|\boldsymbol{\sigma}_i^{\frac{1}{2}} \nabla u_e^0\|_{0,\Omega_H}^2,$$

$$E_H^n\left(u_e^\star, V_m^\star\right) \stackrel{\text{def}}{=}$$

$$\left\|w^n\right\|_{0,\Omega_H}^2 + \chi_m \left\|V_m^n\right\|_{0,\Omega_H}^2 + \tau \left\|\sigma_i^{\frac{1}{2}} \nabla u_e^n\right\|_{0,\Omega_H}^2 + \tau \left\|\sigma_i^{\frac{1}{2}} \nabla V_m^n\right\|_{0,\Omega_H}^2$$

$$+ 2 \sum_{m=0}^{n-1} \tau \left\|\sigma_e^{\frac{1}{2}} \nabla u_e^{m+1}\right\|_{0,\Omega_H}^2,$$

and similarly for the rest.

Theorem 3. *Assume that the hypotheses of Theorem 2 hold and let* $(w^n, V_m^n, u_e^n, u_T^n)$ *be given by Algorithm 2. Then for* $\gamma \geq 2C_{ti}$ *and* $1 \leq n \leq N$ *the following estimate holds*

$$E_H^n\left(u_e^\star, V_m^\star\right) + \tau \frac{\gamma \sigma_T^t}{h} \left\|u_T^n\right\|_{0,\Sigma}^2$$

$$+ \sum_{m=0}^{n-1} \tau \left\|\sigma_T^{\frac{1}{2}} \nabla u_T^{m+1}\right\|_{0,\Omega_T}^2 + \frac{\gamma \sigma_T^t}{h} \sum_{m=0}^{n-1} \tau \left\|u_T^{m+1} - u_e^{m+1}\right\|_{0,\Sigma}^2$$

$$\lesssim E_H^0\left(u_e^\star, V_m^\star\right) + \tau \frac{\gamma \sigma_T^t}{h} \left\|u_T^0\right\|_{0,\Sigma}^2 + \tau \left\|\sigma_T^{\frac{1}{2}} \nabla u_T^0\right\|_{0,\Omega_T}^2 + \sum_{m=0}^{n-1} \tau \left\|I_{app}^{m+1}\right\|_{0,\Omega_H}^2.$$

Therefore, the explicit heart-torso coupling of Algorithm 2 is energy-stable provides $\tau = \mathcal{O}(h)$.

Remark 10. The proof of the above result does not make use of any numerical dissipation apart from the one directly provided by the explicit Robin-Robin splitting (4.21). Note that this is particularly well adapted to the heart-torso coupling (4.8), since the quasi-static elliptic equations (4.8)$_{3,4}$ do not generate numerical dissipation in time.

We conclude this subsection with a few numerical illustrations of the accuracy of Algorithm 2. Fig. 4.8 shows that the superior stability properties and computational cost reduction featured by the proposed Robin heart-torso decoupling schemes come with a price: a condition $\tau = \mathcal{O}(h^2)$ is required to guarantee an overall convergence rate $\mathcal{O}(h)$. Indeed, the penalty $1/h$ involved in the explicit Robin treatment introduces a non-standard coupling between the space and time discretizations, which has a consistency of $\mathcal{O}(\tau/h)$.

In spite of that, Fig. 4.9 shows that the proposed Robin splittings are able to provide accurate 12-lead ECG signals, both for a healthy and a pathological condition. Note that this is a major advantage with respect to the conventional heart-torso uncoupling approximation, which (for a similar computational cost) provides inaccurate ECG signals (see Fig. 4.6). Somehow the discretization error introduced by the Robin heart-torso decoupling is negligible with respect to the modelling error involved in the heart-torso uncoupling approximation.

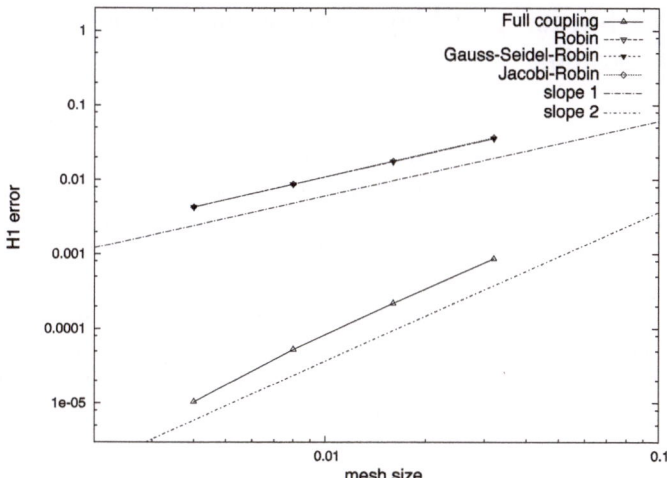

Fig. 4.8. Convergence history of the torso potential error for the full coupling ((4.15)-(4.16)), Robin ($(u_e^\star, V_m^\star) = (u_e^{n+1}, V_m^{n+1})$), Gauss-Seidel-Robin ($(u_e^\star, V_m^\star) = (u_e^n, V_m^{n+1})$) and Jacobi-Robin ($(u_e^\star, V_m^\star) = (u_e^n, V_m^n)$) with $\tau = \mathcal{O}(h^2)$

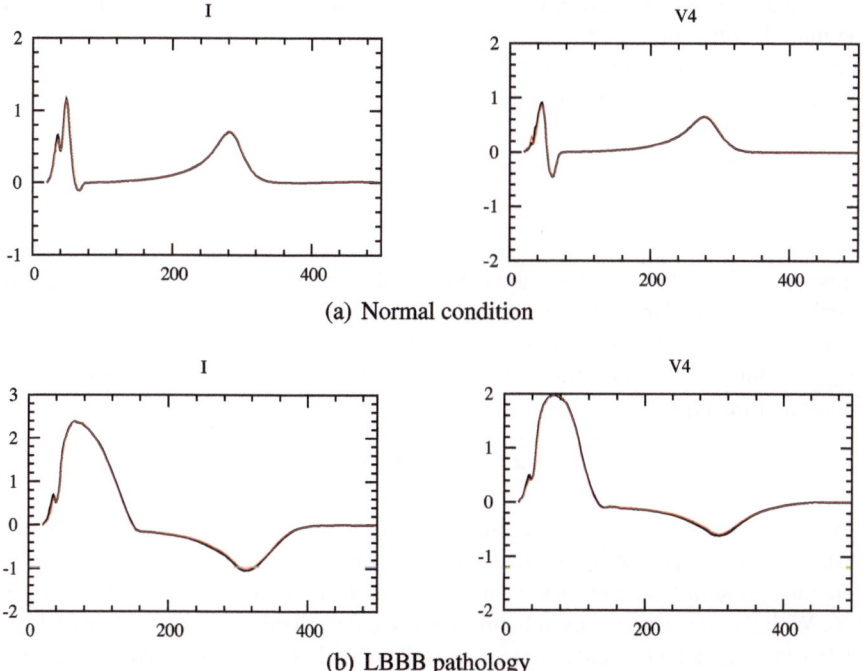

Fig. 4.9. Simulated ECG signals (leads I and V4) obtained using heart-torso full coupling (black) and the Jacobi-Robin scheme (red)

4.6 Reduced-order modelling by proper orthogonal decomposition

The model that has been proposed in the previous sections is based on a coupling between two expensive problems, the Poisson equation in the torso and the bidomain equations in the heart. Even with the uncoupled schemes that have been presented, the computational cost may be prohibitive in some situations. This is for example the case if an inverse problem has to be solved or if several heart beats have to be simulated. In those cases, a reduced-order model may be interesting. We propose some preliminary results obtained with a reduced order model based on the Proper Orthogonal Decomposition (POD) method. After briefly recalling the POD method, we present three examples of ECGs obtained with POD. All the simulations of this section assume a weak coupling with the torso.

4.6.1 POD in a nutshell

For convenience, we briefly recall some notions about POD in this paragraph. For more details about this well-known method, we refer the reader to [36, 56] for instance.

The basic idea is to replace the finite-element basis by a new basis that contains the main features of the expected solution. To generate this new basis a numerical simulation, or set of simulations, is run and some solutions $u(t_k)$, $1 \leq k \leq p$ (called "snapshots" in this context) are stored in a matrix $B = (u(t_1), \ldots, u(t_p)) \in \mathbb{R}^{N,p}$ where $N \geq p$ is the dimension of the finite-element basis. Then, the singular value decomposition (SVD) of this matrix is computed:

$$B = USV',$$

where $U \in \mathbb{R}^{N,N}$ and $V \in \mathbb{R}^{p,p}$ are orthogonal matrices, $S \in \mathbb{R}^{N,p}$ is the diagonal matrix of the singular values ordered by decreasing value.

The first N_{modes} POD basis functions $\{\Psi_i\}_{1 \leq i \leq N_{\text{modes}}}$ are then given by the first N_{modes} columns of U, and the POD Galerkin problem is solved by looking for a solution of the type

$$u = \sum_{i=1}^{N_{\text{modes}}} \alpha_i(t) \Psi_i.$$

The $N \times N$ sparse system of the finite-element method is thus replaced by a full system of size $N_{\text{modes}} \times N_{\text{modes}}$ with the POD method. To give a rough idea, it is generally possible to achieve a good accuracy for the problems at hand with $N_{\text{modes}} \approx 100$. With the time scheme used in this work and described by the procedure (4.15)–(4.16), the matrix is constant in time, since all the nonlinearities are treated explicitly. The matrix is therefore projected on the POD basis and factorized only once at the beginning of the computation. As a consequence, for the simulations presented in this section, the reduced-order model resolution is about one order of magnitude faster than the full order one.

4.6.2 Examples of POD simulations with different sets of parameters

The POD method has a practical interest only if a POD basis generated for a set of parameters can be used with other sets of parameters as well. In particular, if we have in mind to solve an inverse problem of parameter identification using POD, the parameters to be identified will take different values. It is therefore necessary to check that the accuracy of the POD basis is stable with respect to parameter perturbation. In this section, we study this issue for some parameters of the model.

Let us first consider the perturbation of the parameter τ_{close} which corresponds to the characteristic closing time of ionic channels. As detailed in the beginning of Sect. 4.4, this parameter takes four different constant values. We focus on two of these values, $\tau_{\text{close}}^{\text{epi}}$ and $\tau_{\text{close}}^{\text{rv}}$. A POD basis of 80 vectors is first constructed with $(\tau_{\text{close}}^{\text{epi}}, \tau_{\text{close}}^{\text{rv}}) = (80, 80)$. This basis is sufficient to get an excellent accuracy if the *same* experiment is run with the reduced-order model. More interestingly, it still gives quite good results when $(\tau_{\text{close}}^{\text{epi}}, \tau_{\text{close}}^{\text{rv}})$ are significantly modified. For example, if we take $(\tau_{\text{close}}^{\text{epi}}, \tau_{\text{close}}^{\text{rv}}) = (90, 120)$, Fig. 4.10 (left) shows a comparison of the first lead of the ECGs obtained with the full model and the reduced model. The QRS is in excellent agreement, while the T-wave is slightly underestimated, which is not surprising since τ_{close} mainly affects the repolarization phase. It is interesting to note that for the values $(\tau_{\text{close}}^{\text{epi}}, \tau_{\text{close}}^{\text{rv}}) = (80, 80)$, the T-wave of the ECG is negative whereas it is positive with $(90, 120)$. It is therefore particularly satisfactory to obtain the correct T-wave orientation with the reduced-order model after perturbing these coefficients.

Similar tests have been done with τ_{in}, C_{m}, A_{m} and $\tau_{\text{close}}^{\text{rv}}$. These parameters were chosen since it was observed in [7] that the ECGs are particularly sensitive to them. We have run simulations taking $\tau_{\text{in}} = 0.8$, $C_{\text{m}} = 10^{-3}$, $A_{\text{m}} = 200$, $\tau_{\text{close}}^{\text{rv}} = 120$ and we have compared the ECGs obtained with the full-order model and with the reduced-order model corresponding to the basis generated with $\tau_{\text{in}} = 1.5$, $C_{\text{m}} = 2 \times 10^{-3}$, $A_{\text{m}} = 100$, $\tau_{\text{close}}^{\text{rv}} = 70$. The results are reported in Fig. 4.10 (right). We see that the use of POD induces a temporal shift and a difference of amplitudes. On the first lead, the wave orientation is preserved, but for some other leads changes of orientation have been noticed.

The results get even worse with other parameters, like for example those governing the initial activation: if we use a POD basis obtained from a reference simulation with an initial activation in the septum to run a reduced order simulation with an initial activation at the apex, the results are totally wrong, as shown in Fig. 4.11.

The last two examples show that POD approximation has to be used very carefully. In particular, it is necessary to examine to what extent the parameters chosen to compute the POD basis have to be close to the parameters used in the simulation. This issue is particularly important to address the problem of parameter identification. Practical techniques to improve these results will be proposed in a forthcoming publication.

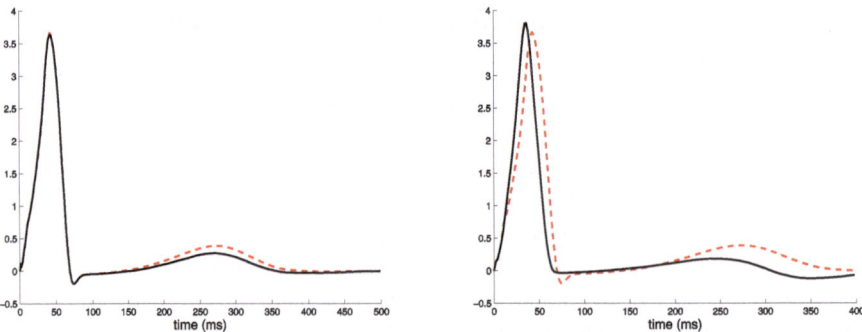

Fig. 4.10. *Left:* First lead of the ECGs with $(\tau_{\text{close}}^{\text{epi}}, \tau_{\text{close}}^{\text{rv}}) = (90, 120)$. Comparison between the full model (dotted red line) and the reduced model (solid black line) with a POD basis generated with $(\tau_{\text{close}}^{\text{epi}}, ECGt\,\tau_{\text{close}}^{\text{rv}}) = (80, 80)$. *Right:* First lead of the ECGs with $(\tau_{\text{in}}, C_{\text{m}}, A_{\text{m}}, \tau_{\text{close}}^{\text{rv}}) = (0.8, 10^{-3}, 200, 120)$. Comparison between the full model (dotted red line) and the reduced model (black) with a POD basis generated with $(\tau_{\text{in}}, C_{\text{m}}, A_{\text{m}}, \tau_{\text{close}}^{\text{rv}}) = (1.5, 2 \times 10^{-3}, 100, 70)$

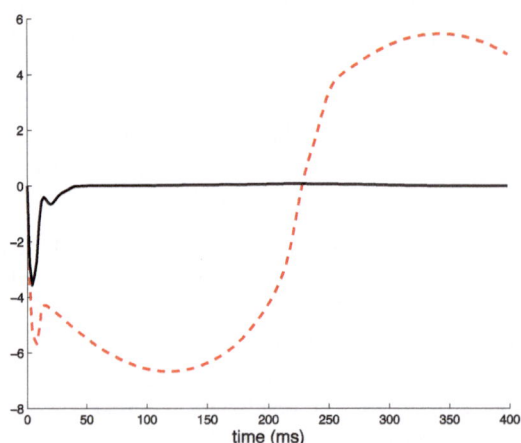

Fig. 4.11. First leads of ECGs obtained for an initial stimulation in the apex. Comparison between the full model (dotted red line) and the reduced model (black) with the POD basis obtained from a simulation with an initial activation on the septum

4.6.3 Examples of POD simulations for several heart beats

We conclude this section with an example where POD seems to give promising results. The problem consists in computing many heart beats and increasing progressively the heart rate. This kind of simulation allows to generate a curve known as *the restitution curve*, which represents the action potential duration (APD) versus the preceding diastolic interval (DI). For more details about the restitution curve and its link with cardiac arrhythmia, we refer the reader to [54] for instance.

Fig. 4.12 shows the ECGs during eleven beats with an increasing heart rate, obtained with the full- and the reduced-order models. The POD basis is only based

Firsrt lead : I

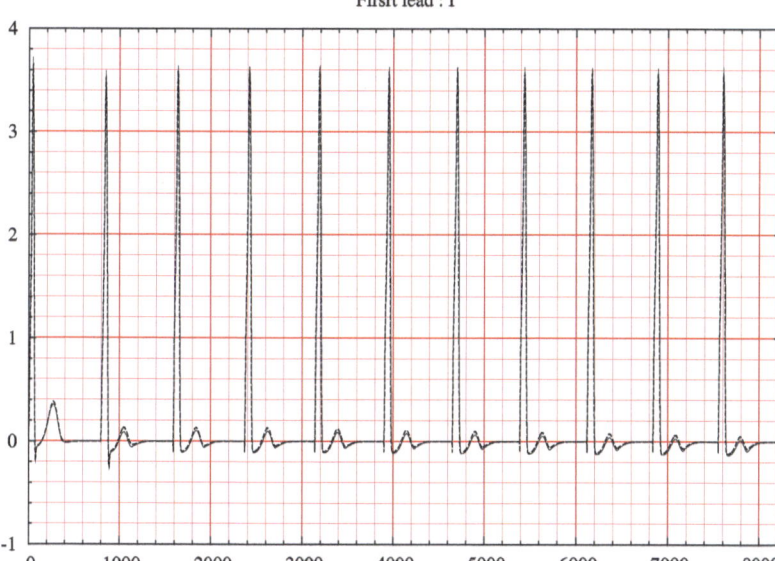

Fig. 4.12. First lead of ECGs for a simulation with several beats. Comparison between the full order model (full line) and the POD reduced order model (dotted line)

on the snapshots obtained from the first heart beat. In Fig. 4.13 (top) we see, as expected, that the first beats perfectly match. Fig. 4.13 (bottom) shows that, at the eleventh beat, the ECGs are still very close. Hence, the POD reduced-order model seems to be reliable to compute long time simulations with variable heart rates. It offers therefore an interesting option to compute efficiently restitution curves.

4.7 Conclusions

In this overview, various aspects of the modelling of cardiac electrophysiology were considered: mathematical analysis, numerical simulation and numerical analysis. The main goal was to present a strategy to compute 12-lead ECGs with a model fully based on PDEs and ODEs and depending on a moderate number of parameters. The results turned out to be satisfactory in physiological conditions, and for some pathologies. It is nevertheless clear that many aspects could be improved. For example, the atria should be included to get P-waves; His bundle and Purkinje fibres could be explicitly modelled; ionic models could be based on physiological models instead of phenomenological ones. But it is worth keeping in mind that these improvements would come with many additional parameters and would require data that are extremely difficult to obtain clinically.

The ECG simulator presented in this chapter has already been used in various applications: optimization of pacing sites [20], ECG reconstruction from electrograms

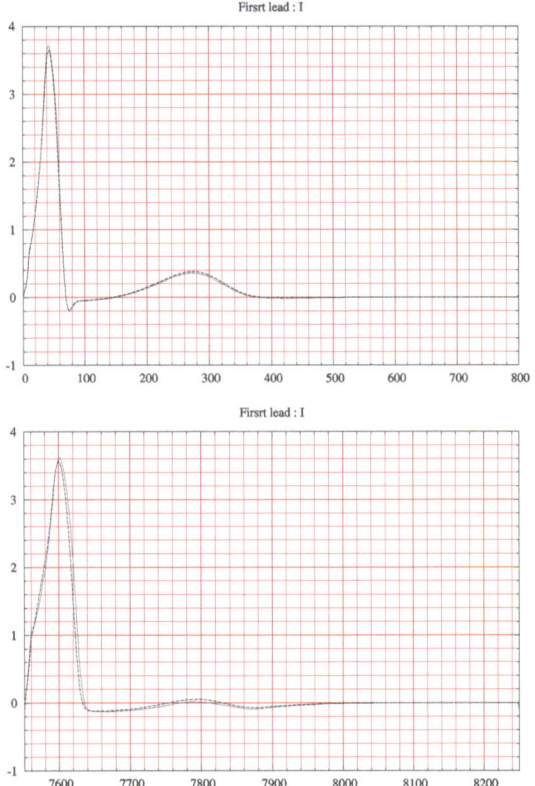

Fig. 4.13. Zoom of the previous figure. Top: first beat, Bottom: 11th beat

[21]. It can naturally be used with a model of the mechanics of the myocardium. Such a coupled system would allow to complete the ECG signals with important mechanical indicators, like the mean pressure in the cavities and the stroke volumes. Preliminary tests have been presented in [12], neglecting the mechano-electric feedback, i.e., the action of the deformation on the electrical activity (see, e.g., [32, 34, 37, 47] for an overview on the cardiac electro-mechanical coupling). Future extensions of this work will include the effect of myocardium deformation in the electrical computations.

As already mentioned, an interesting feature of the proposed ECG model is the low number of parameters it relies on. Therefore it may be a good candidate to address the inverse problem of electrocardiography (the reconstruction of the cardiac electrical activity from the body's surface potential) through parameter estimation. From this perspective, it is important to reduce the complexity of the forward problem in order to keep the computational cost of the inverse problem reasonable. We have presented preliminary results obtained with a reduced-order model based on proper orthogonal decomposition (POD). When the parameters related to the ionic current are perturbed or when the heart rate varies, this reduced-order model seems to give reasonable results. But it is unfortunately very poor when the initial activation site is modified. To be more stable, the reduced-order model has to be defined on larger sets of experiments. This point, as well as the application to inverse problems, will be addressed in future works.

Acknowledgements. The authors wish to thank Michel Sorine for many fruitful discussions, and Elisa Schenone for the simulations shown in Figs. 4.12 and 4.13.

References

[1] Aehlert B.: ECGs Made Easy. Mosby Jems, Elsevier, third edition, 2006.

[2] Aliev R.R., Panfilov A.V.: A simple two-variable model of cardiac excitation. Chaos, Solitons & Fractals **3**(7): 293–301, 1996.

[3] Austin T.M., Trew M.L., Pullan A.J.: Solving the cardiac bidomain equations for discontinuous conductivities. IEEE Trans. Biomed. Eng. **53**(7): 1265–72, 2006.

[4] Beeler G., Reuter H.: Reconstruction of the action potential of ventricular myocardial fibres. J. Physiol. (Lond.) **268**: 177–210, 1977.

[5] Bendahmane M., Karlsen K.H.: Analysis of a class of degenerate reaction-diffusion systems and the bidomain model of cardiac tissue. Netw. Heterog. Media **1**(1): 185–218 (electronic), 2006.

[6] Bordas R., Grau V., Burton R.A.B., Hales P., Schneider J.E., Gavaghan D., Kohl P., Rodriguez B.: Integrated approach for the study of anatomical variability in the cardiac purkinje system: from high resolution MRI to electrophysiology simulation. In Engineering in Medicine and Biology Society (EMBC), 2010 Annual International Conference of the IEEE, pp. 6793–6796. IEEE, 2010.

[7] Boulakia M., Cazeau S., Fernández M.A., Gerbeau J.-F., Zemzemi N.: Mathematical modeling of electrocardiograms: a numerical study. Ann. Biomed. Eng. **38**(3): 1071–1097, 2010.

[8] Boulakia M., Fernández M.A., Gerbeau J.-F., Zemzemi N.: A coupled system of PDEs and ODEs arising in electrocardiograms modelling. Applied Math. Res. Exp. 2008(abn002): **28**, 2008.

[9] Bourgault Y., Coudière Y., Pierre C.: Existence and uniqueness of the solution for the bidomain model used in cardiac electrophysiology. Nonlinear Anal. Real World Appl. **10**(1): 458–482, 2009.

[10] Bourgault Y., Ethier M., Le Blanc V.G.: Simulation of electrophysiological waves with an unstructured finite element method. M2AN Math. Model. Numer. Anal. **37**(4): 649–661, 2003.

[11] Buist M., Pullan A.: Torso coupling techniques for the forward problem of electrocardiography. Ann. Biomed. Eng. **30**(10): 1299–1312, 2002.

[12] Chapelle D., Fernández M.A., Gerbeau J.-F., Moireau P., Sainte-Maire J., Zemzemi N.: Numerical simulation of the electromechanical activity of the heart. In N. Ayache, H. Delingette, and M. Sermesant (eds.), Functional Imaging and Modeling of the Heart, volume 5528 of Lecture Notes in Computer Science, pp. 357–365. Springer, 2009.

[13] Clayton O., Bernus R.H., Cherry E.M., Dierckx H., Fenton F.H., Mirabella L., Panfilov A.V., Sachse F.B., Seemann G., Zhang H.: Models of cardiac tissue electrophysiology: Progress, challenges and open questions. Progress in Biophysics and Molecular Biology **104**(1–3): 22–48, 2011.

[14] Clements J., Nenonen J., Li P.K.J., Horacek B.M.: Activation dynamics in anisotropic cardiac tissue via decoupling. Ann. Biomed. Eng. **32**(7): 984–990, 2004.

[15] Colli Franzone P., Pavarino L.F.: A parallel solver for reaction-diffusion systems in computational electrocardiology. Math. Models Methods Appl. Sci. **14**(6): 883–911, 2004.

[16] Colli Franzone P., Pavarino L.F., Scacchi S., Taccardi B.: Effects of anisotropy and transmural heterogeneity on the T-wave polarity of simulated electrograms. In N. Ayache, H. Delingette,

and M. Sermesant (eds,), Functional Imaging and Modeling of the Heart, volume 5528 of Lecture Notes in Computer Science, pp. 513–523. Springer, 2009.

[17] Colli Franzone P., Pavarino L.F., Taccardi B.: Simulating patterns of excitation, repolarization and action potential duration with cardiac bidomain and monodomain models. Math. Biosci. **197**(1): 35–66, 2005.

[18] Colli Franzone P., Savaré G.: Degenerate evolution systems modeling the cardiac electric field at micro- and macroscopic level. In Evolution equations, semigroups and functional analysis (Milano, 2000), volume 50 of Progr. Nonlinear Differential Equations Appl., pp. 49–78. Birkhäuser, Basel, 2002.

[19] Djabella K., Sorine M.: Differential model of the excitation-contraction coupling in a cardiac cell for multicycle simulations. In EMBEC'05, volume 11, pp. 4185–4190, Prague, 2005.

[20] Dumas L., El Alaoui L.: How genetic algorithms can improve a pacemaker efficiency. In GECCO '07: Proceedings of the 2007 GECCO conference companion on Genetic and evolutionary computation, pp. 2681–2686, New York, NY, USA, 2007. ACM.

[21] Ebrard G., Fernández M.A., Gerbeau J.-F., Rossi F., Zemzemi N.: From intracardiac electrograms to electrocardiograms. models and metamodels. In N. Ayache, H. Delingette, and M. Sermesant (eds.), Functional Imaging and Modeling of the Heart, volume 5528 of Lecture Notes in Computer Science, pp. 524–533. Springer, 2009.

[22] Efimov I.R., Gray R.A., Roth B.J.: Virtual electrodes and deexcitation: new insights into fibrillation induction and defibrillation. J. Cardiovasc. Electrophysiol. **11**(3): 339–353, 2000.

[23] Ethier M., Bourgault Y.: Semi-implicit time-discretization schemes for the bidomain model. SIAM J. Numer. Anal. **46**(5): 2443–2468, 2008.

[24] Fenton F., Karma A.: Vortex dynamics in three-dimensional continuous myocardium with fiber rotation: Filament instability and fibrillation. Chaos **8**(1): 20–47, 1998.

[25] Fernández M.A., Zemzemi N.: Decoupled time-marching schemes in computational cardiac electrophysiology and ECG numerical simulation. Math. Biosci. **226**(1): 58–75, 2010.

[26] FitzHugh R.: Impulses and physiological states in theoretical models of nerve membrane. Biophys. J. **1**: 445–465, 1961.

[27] Goldberger A.L.: Clinical Electrocardiography: A Simplified Approach. Mosby-Elsevier, 7th edition, 2006.

[28] Gulrajani R.M.: Models of the electrical activity of the heart and computer simulation of the electrocardiogram. Crit. Rev. Biomed. Eng. **16**(1): 1–6, 1988.

[29] Hooke N., Henriquez C.S., Lanzkron P., Rose D.: Linear algebraic transformations of the bidomain equations: implications for numerical methods. Math. Biosci. **120**(2): 127–145, 1994.

[30] Huiskamp G.: Simulation of depolarization in a membrane-equations-based model of the anisotropic ventricle. IEEE Trans. Biomed. Eng. **5045**(7): 847–855, 1998.

[31] Keener J. P., Bogar K.: A numerical method for the solution of the bidomain equations in cardiac tissue. Chaos **8**(1): 234–241, 1998.

[32] Keldermann R.H., Nash M.P., Panfilov A.V.: Modeling cardiac mechano-electrical feedback using reaction-diffusion-mechanics systems. Physica D: Nonlinear Phenomena **238**(11–12): 1000–1007, 2009.

[33] Keller D.U.J., Seemann G., Weiss D.L., Farina D., Zehelein J., Dössel O.: Computer based modeling of the congenital long-qt 2 syndrome in the visible man torso: From genes to ECG. In Proceedings of the 29th Annual International Conference of the IEEE EMBS, pp. 1410–1413, 2007.

[34] Kerckhoffs R.C.P., Healy S.N., Usyk T.P., McCulloch A.D.: Computational methods for cardiac electromechanics. Proc. IEEE **94**(4): 769–783, 2006.

[35] Krassowska W., Neu J.C.: Effective boundary conditions for syncitial tissues. IEEE Trans. Biomed. Eng. **41**(2): 143–150, 1994.

[36] Kunisch K., Volkwein S.: Galerkin proper orthogonal decomposition methods for parabolic problems. Numerische Mathematik **90**(1): 117–148, 2001.

[37] Lab M.J., Taggart P., Sachs F.: Mechano-electric feedback. Cardiovasc. Res. **32**: 1–2, 1996.

[38] Lines G.T., Buist M.L., Grottum P., Pullan A.J., Sundnes J., Tveito A.: Mathematical models and numerical methods for the forward problem in cardiac electrophysiology. Comput. Visual. Sci. **5**(4): 215–239, 2003.

[39] Lines G.T., Grøttum P., Tveito A.: Modeling the electrical activity of the heart: a bidomain model of the ventricles embedded in a torso. Comput. Vis. Sci. **5**(4): 195–213, 2003.

[40] Luo C., Rudy Y.: A dynamic model of the cardiac ventricular action potential. i. simulations of ionic currents and concentration changes. Circ. Res. **74**(6): 1071–1096, 1994.

[41] Luo C.H., Rudy Y.: A model of the ventricular cardiac action potential. depolarisation, repolarisation,and their interaction. Circ. Res. **68**(6): 1501–1526, 1991.

[42] Malmivuo J., Plonsey R.: Bioelectromagnetism. Principles and applications of bioelectric and biomagnetic fields. Oxford University Press, New York, 1995.

[43] Mitchell C.C., Schaeffer D.G.: A two-current model for the dynamics of cardiac membrane. Bulletin Math. Bio. **65**: 767–793, 2003.

[44] Murillo M., Cai X.-C.: A fully implicit parallel algorithm for simulating the non-linear electrical activity of the heart. Numer. Linear Algebra Appl. **11**(2-3): 261–277, 2004.

[45] Nagumo J., Arimoto S., Yoshizawa S.: An active pulse transmission line simulating nerve axon. Proceedings of the IRE **50**(10): 2061–2070, 1962.

[46] Neu J.C., Krassowska W.: Homogenization of syncytial tissues. Crit. Rev. Biomed. Eng. **21**(2): 137–199, 1993.

[47] Nickerson D., Smith N.P., Hunter P.: New developments in a strongly coupled cardiac electromechanical model. Europace **7** Suppl 2: 118–127, 2005.

[48] Noble D., Varghese A., Kohl P., Noble P.: Improved guinea-pig ventricular cell model incorporating a diadic space, ikr and iks, and length- and tension-dependent processes. Can. J. Cardiol. **14**(1): 123–134, 1998.

[49] Pennacchio M., Savaré G., Colli Franzone P.: Multiscale modeling for the bioelectric activity of the heart. SIAM Journal on Mathematical Analysis **37**(4): 1333–1370, 2005.

[50] Potse M., Dubé B., Gulrajani M.: ECG simulations with realistic human membrane, heart, and torso models. In Proceedings of the 25th Annual International Conference of the IEEE EMBS, pp. 70–73, 2003.

[51] Potse M., Dube B., Richer J., Vinet A., Gulrajani R.M.: A comparison of monodomain and bidomain reaction-diffusion models for action potential propagation in the human heart. IEEE Trans. Biomed. Eng. **53**(12): 2425–2435, 2006.

[52] Potse M., Dubé B., Vinet A.: Cardiac anisotropy in boundary-element models for the electrocardiogram. Med. Biol. Eng. Comput. **47**: 719–729, 2009.

[53] Pullan A.J., Buist M.L., Cheng L.K.: Mathematically modelling the electrical activity of the heart: From cell to body surface and back again. World Scientific Publishing Co. Pte. Ltd., Hackensack, NJ, 2005.

[54] Qu Z., Xie Y., Garfinkel A., Weiss J.N.: T-wave Alternans and Arrhythmogenesis in Cardiac Diseases. Frontiers in Physiology **1**(154): 1, 2010.

[55] Quarteroni A., Valli A.: Domain decomposition methods for partial differential equations. Numerical Mathematics and Scientific Computation. The Clarendon Press Oxford University Press, 1999.

[56] Rathinam M., Petzold L.R.: A new look at proper orthogonal decomposition. SIAM Journal on Numerical Analysis **41**(5): 1893–1925, 2004.

[57] Roger J.M., McCulloch A.D.: A collocation-Galerkin finite element model of cardiac action potential propagation. IEEE Trans. Biomed. Engr. **41**(8): 743–757, 1994.

[58] Sachse F.B.: Computational Cardiology: Modeling of Anatomy, Electrophysiology, and Mechanics. Springer, 2004.

[59] Skouibine N., Trayanova K., Moore P.: A numerically efficient model for simulation of defibrillation in an active bidomain sheet of myocardium. Math. Biosci. **166**(1): 85–100, 2000.

[60] Stetter H.J.: The defect correction principle and discretization methods. Numer. Math. **29**: 425–443, 1978.

[61] Sundnes J., Lines G.T., Cai X., Nielsen B.F., Mardal K.-A., Tveito A.: Computing the electrical activity in the heart. Springer, 2006.

[62] Sundnes J., Lines G.T., MardalK.A., Tveito A.: Multigrid block preconditioning for a coupled system of partial differential equations modeling the electrical activity in the heart. Comput. Methods Biomech. Biomed. Engin. **5**(6): 397–409, 2002.

[63] Sundnes J., Lines G.T., Tveito A.: Efficient solution of ordinary differential equations modeling electrical activity in cardiac cells. Math. Biosci. **172**(2): 55–72, 2001.

[64] Sundnes J., Lines G.T., Tveito A.: An operator splitting method for solving the bidomain equations coupled to a volume conductor model for the torso. Math. Biosci. **194**(2): 233–248, 2005.

[65] Toselli A., Widlund O.: Domain decomposition methods – algorithms and theory, volume 34 of Springer Series in Computational Mathematics. Springer, Berlin, 2005.

[66] Trayanova N.: Defibrillation of the heart: insights into mechanisms from modelling studies. Experimental Physiology **91**(2): 323–337, 2006.

[67] Trudel M.-C., Dubé B., Potse M., Gulrajani R.M., Leon L.J.: Simulation of qrst integral maps with a membrane-based computer heart model employing parallel processing. IEEE Trans. Biomed. Eng. **51**(8): 1319–1329, 2004.

[68] Tung L.: A bi-domain model for describing ischemic myocardial D–C potentials. PhD thesis, MIT, USA, 1978.

[69] van Capelle F.H., Durrer D.: Computer simulation of arrhythmias in a network of coupled excitable elements. Circ. Res. **47**: 453–466, 1980.

[70] Veneroni M.: Reaction-diffusion systems for the macroscopic bidomain model of the cardiac electric field. Nonlinear Anal. Real World Appl. **10**(2): 849–868, 2009.

[71] Vigmond E.J., Weber dos Santos R., Prassl A.J., Deo M., Plank G.: Solvers for the cardiac bidomain equations. Progr. Biophys. Molec. Biol. **96**(1–3): 3–18, 2008.

[72] Zemzemi N.: Étude théorique et numérique de l'activité électrique du cœur: Applications aux électrocardiogrammes. PhD thesis, Université Paris XI, 2009. http://tel.archives-ouvertes.fr/tel-00470375/en/.

5

Mathematical and numerical methods for reaction-diffusion models in electrocardiology

Piero Colli-Franzone, Luca F. Pavarino, and Simone Scacchi

Abstract. This paper presents a review of current mathematical and numerical models of the bioelectrical activity in the ventricular myocardium, describing cardiac cells excitability and the action-potential propagation in cardiac tissue. The degenerate reaction-diffusion system called the Bidomain model is introduced and interpreted as macroscopic averaging of a cellular model on a periodic assembling of myocytes. The main theoretical results for the cellular and Bidomain models are given. Various approximate models based on some relaxed approaches are also considered, such as Monodomain and eikonal-curvature models. The main numerical methods for the Bidomain and Monodomain models are then reviewed. In particular, we focus on isoparametric finite elements, semi-implicit time discretizations and a parallel iterative solver based on a multilevel Schwarz preconditioned conjugate gradient method. The Bidomain solver is finally applied to the study of the excitation processes generated by virtual electrode response in 3D orthotropic blocks of myocardial tissue.

5.1 Introduction

Electrocardiology deals with the investigation of intracardiac bioelectric phenomena and the evolution of cardiac potential fields at the body surface is one of the main purposes of Electrocardiology. Clinic Electrocardiography deals with the de-

Piero Colli-Franzone
University of Pavia, Department of Mathematics, via Ferrata 1, 27100 Pavia, Italy
e-mail: colli@imati.cnr.it

Luca F. Pavarino
University of Milano, Department of Mathematics, via Saldini 50, 20133 Milano, Italy
e-mail: luca.pavarino@unimi.it

Simone Scacchi
University of Milano, Department of Mathematics, via Saldini 50, 20133 Milano, Italy
e-mail: simone.scacchi@unimi.it

Ambrosi D., Quarteroni A., Rozza G. (Eds.): Modeling of Physiological Flows.
DOI 10.1007/978-88-470-1935-5_5, © Springer-Verlag Italia 2012

tection and interpretation of the morphology of the usual electrocardiograms (ECG) at a few points on the body surface or from the evolution of body surface maps (see [135, 136, 160] for a survey). The information content of ECGs and body maps is limited, due to the strong signal attenuation and smoothing associated with current conduction from heart to thorax; thus it is a difficult task to detect from these signals detailed information on pathological heart states. The scientific ground of Electrocardiology is the so called *Forward Problem of Electrocardiology*, that studies the bioelectric cardiac sources and conducting media in order to derive the potential field. Of considerable interest are also the so called *Inverse Problems of Electrocardiography* in terms of potentials or cardiac sources (see e.g. the reviews [54, 96, 112]). In this paper, we will focus on the modelling and simulation techniques for describing the bioelectrical activity in the myocardium at the macroscopic level which are the basic tools for the formulation of the Forward and Inverse Problems in terms of cardiac sources.

In the past few decades, experimental electrophysiology has been increasingly supported by the mathematical and numerical models of computational electrocardiology. These models provide essential quantitative tools to integrate the increasing knowledge of bioelectrochemical phenomena occurring at several time and space scales, from microscopic models of ionic channels and currents in the cellular membrane, to macroscopic models of anisotropic cardiac tissue and organ, see e.g. [17, 18, 30, 63, 96, 113, 133, 160]. These coupled multiscale models are then validated by comparing simulated results with experimental *in vitro* and *in vivo* data. At the next step, these electrophysiological models need to be coupled and integrated with mechanical model of tissue deformation, hemodynamical models of cardiac blood flow and, more in general, of the cardiovascular system. This complex integrative effort is the current focus of several research projects, such as the EC sponsored Virtual Physiological Human (VPH) Initiative, see [63]. Ultimately, the integration of these models should provide new tools enabling the biomedical community to link genetic and proteomic databases to anatomy and functions at the cellular, tissue and organ levels.

In the rest of this paper, we investigate the main mathematical models of the cardiac bioelectric activity in Sect. 2, covering ionic membrane models, cardiac cell arrangements, the Bidomain model, together with the main theoretical results on their mathematical analysis. We also introduce the approximate Eikonal and Monodomain models. In Sect. 3 we review some numerical methods for discretizing the Monodomain and Bidomain models in space and time, and build a parallel Bidomain solver based on a multilevel Schwarz preconditioned conjugate gradient method. In Sect. 4 we report some numerical results illustrating the parallel scalability of the resulting Bidomain solver. In Sect. 5 we apply the solver to three-dimensional Bidomain simulations of anode make mechanisms of cardiac excitation generated by a unipolar extracellular pulse.

5.2 Mathematical models of the cardiac bioelectric activity

The bioelectric activity of the cellular membrane of a myocyte is described by the time course of the transmembrane potential, i. e. the potential jump v across the cellular membrane surface separating the intra- *(i)* and extracellular *(e)* media, usually called *action potential*. The whole process of action-potential generation and propagation is due to ionic membrane currents and to the electrotonic diffusion in the *(i)* and *(e)* conducting media.

Starting from the sino-atrial node, which acts as a pacemaker, a front-like variation of the transmembrane potential v spreads first in the atria and then reaches the ventricles through the Purkinje network, with a very fast transition from the resting value v_r to the plateau value v_p. This phase constitutes the excitation or depolarization phase, and it is followed by fast and slow repolarization phases with a subsequent return to the initial state. The time profile of the transmembrane potential v may depend in general on the position $\mathbf{x}$ and on the local state of the heart; the whole bioelectric cycle lasts about 300 msec in the human heart. Moreover, the fibre structure strongly affects both the excitation and repolarization processes and in particular is the main cause of the anisotropic conductivity in the cardiac tissue, see [71, 130].

5.2.1 Ionic current membrane models

The electrical behavior of the membrane is represented by a circuit consisting of a capacitor connected in parallel with a resistor, modelling the various ionic channels regulating the selective and independent ionic fluxes through the membrane. The total transmembrane current I_m is given by

$$I_m = C_m \frac{dv}{dt} + I_{ion} = I_{app}, \tag{5.1}$$

with I_{ion} the ionic membrane current, C_m the membrane capacity and I_{app} the applied current per unit area of the membrane surface.

Most mathematical models of the ionic currents are based on appropriate extensions of the formalism introduced by Hodgkin and Huxley in [59]. Current progress in molecular biology continues to produce more detailed data and knowledge on the dynamics of the ionic fluxes through the cellular membrane, see e. g. the review papers [17, 81, 113]. The general form of the ionic current is given by

$$I_{ion}(v, w, c) = \sum_{k=1}^{P} g_k(c) \prod_{j=1}^{M} w_j^{p_{j_k}} (v - v_k(c)) + I_0(v, c), \tag{5.2}$$

where $g_k(c)$, $v_k(c) = (RT/zF) \log \frac{c_k^e}{c_k^i}$ and $c_k^{i,e}$ are the maximal conductance of the ion channel, the Nernst potential and the intra- and extracellular ion concentrations for the k-th ionic species, respectively, and p_{j_k} are integers. In (5.2), we have split the ionic current as the sum of a term related to ionic fluxes modulated by the gating dynamics and a time-independent term $I_0(v,c)$. The gating variables $w := (w_1, \ldots, w_M)$ regulate the conductances of the various ionic fluxes and $c := (c_1, \ldots, c_Q)$ are variables regulating the intracellular concentrations of the various ions. The dynamics of the gating variables w is given by a first-order kinetic model, while the ionic concentrations c satisfy differential equations associated to ion channels, pumps and exchanger currents that are carrying the same ionic species.

Simplified models of lower complexity, with only 1 or 2 gating variables and no ionic concentrations, have been proposed and employed for analytical and numerical studies. The simplest model with only one gating variable w is the FitzHugh-Nagumo (FHN) model [50] and its variants [1, 101], the latter yielding better approximation of a typical cardiac action potential. A simplified ionic model with two gating variables ($M = 2$) was proposed in [48, 49], and for human ventricular cell see [137].

5.2.2 The anisotropic macroscopic Bidomain model

The cardiac ventricular tissue is conceived at a macroscopic level as an arrangement of cardiac fibres organized in toroidal layers nested within the ventricular wall and rotating counterclockwise from epi- to endocardium (see e.g. [130]). More recently, [71, 72] have shown that this fibre structure has an additional laminar organization modelled as a set of muscle sheets running radially from epi- to endocardium. Therefore, at any point $\mathbf{x}$, it is possible to identify a triplet of orthonormal principal axes $\mathbf{a}_l(\mathbf{x})$, $\mathbf{a}_t(\mathbf{x})$, $\mathbf{a}_n(\mathbf{x})$, with $\mathbf{a}_l(\mathbf{x})$ parallel to the local fibre direction, $\mathbf{a}_t(\mathbf{x})$ and $\mathbf{a}_n(\mathbf{x})$ tangent and orthogonal to the radial laminae, respectively, and both transversal to the fibre axis [38, 72]. Denoting by $\sigma_l^{i,e}$, $\sigma_t^{i,e}$, $\sigma_n^{i,e}$ the conductivity coefficients in the intra and extracellular media measured along the corresponding directions $\mathbf{a}_l, \mathbf{a}_t, \mathbf{a}_n$, the anisotropic conductivity tensors $D_i(\mathbf{x})$ and $D_e(\mathbf{x})$ related to the *orthotropic anisotropy* of the media are given by

$$D_{i,e}(\mathbf{x}) = \sigma_l^{i,e}\, \mathbf{a}_l(\mathbf{x})\mathbf{a}_l^T(\mathbf{x}) + \sigma_t^{i,e}\, \mathbf{a}_t(\mathbf{x})\mathbf{a}_t^T(\mathbf{x}) + \sigma_n^{i,e}\, \mathbf{a}_n(\mathbf{x})\mathbf{a}_n^T(\mathbf{x}). \qquad (5.3)$$

From the macroscopic point of view the cardiac tissue is represented as a *bidomain* (see e.g. [22, 53, 57, 69, 76, 92, 103, 145]), i. e. the superposition of two anisotropic continuous media, the intra- and extracellular media (i) and (e), coexisting at every point of the tissue and separated by a distributed continuous cellular membrane. Let Ω_H and $\Gamma_H = \partial\Omega_H$ denote the volume and the heart surface, respectively. The intra and extracellular electric potentials u_i, u_e in the anisotropic Bidomain model are described by a reaction-diffusion system coupled with a system of ODEs for ionic gating variables w and ion concentrations c. Let $I_{app}^{e,i}$ be an applied extracellular and intracellular current per unit volume, satisfying the compatibility condition $\int_{\Omega_H}(I_{app}^e + I_{app}^i)\,dx = 0$, and denote by $\mathbf{j}_{i,e} = -D_{i,e}\nabla u_{i,e}$ the intra- and extracellular current density. Due to the current conservation law, it follows

$$\text{div}\,\mathbf{j_i} = -J_m + I_{app}^i, \quad \text{div}\,\mathbf{j_e} = J_m + I_{app}^e, \quad J_m = \chi I_m = c_m\frac{\partial v}{\partial t} + i_{ion}(v,w,c),$$

where χ is the ratio of membrane area per tissue volume and J_m, $c_m = \chi C_m$ and $i_{ion} = \chi I_{ion}$ are the transmembrane current, the membrane capacitance and the ionic membrane current per unit volume, respectively. Then the *anisotropic Bidomain model* in term of the potential unknowns $u_i(\mathbf{x},t)$, $u_e(\mathbf{x},t)$, $v(\mathbf{x},t) = u_i(\mathbf{x},t) - u_e(\mathbf{x},t)$, the gating variables $w(\mathbf{x},t)$ and ion concentrations $c(\mathbf{x},t)$ can be written as

$$\begin{cases} c_m \dfrac{\partial v}{\partial t} - \mathrm{div}(D_i \nabla u_i) + i_{ion}(v,w,c) = I^i_{app} & \text{in } \Omega_H \times (0,T) \\[2mm] -c_m \dfrac{\partial v}{\partial t} - \mathrm{div}(D_e \nabla u_e) - i_{ion}(v,w,c) = I^e_{app} & \text{in } \Omega_H \times (0,T) \\[2mm] \dfrac{\partial w}{\partial t} - R(v,w) = 0, \quad \dfrac{\partial c}{\partial t} - S(v,w,c) = 0 & \text{in } \Omega_H \times (0,T). \end{cases} \tag{5.4}$$

The Reaction-Diffusion (R-D) system (5.4) uniquely determines v, while the potentials u_i and u_e are defined only up to the same time-dependent additive constant relating to the reference potential. Until now the Bidomain model has been formulated in terms of the potential fields u_i and u_e, but it can be equivalently expressed in terms of the transmembrane and extracellular potentials $v(\mathbf{x},t)$ and $u_e(\mathbf{x},t)$; in fact, adding the two evolution equations of system (5.4) and substituting $u_i = v + u_e$, we obtain an elliptic equation in the unknown (v, u_e) which, coupled with one of the evolution equations, gives the following equivalent formulation of the anisotropic Bidomain model

$$\begin{cases} -\mathrm{div}((D_i + D_e)\nabla u_e) - \mathrm{div}(D_i \nabla v) = I^i_{app} + I^e_{app} & \text{in } \Omega_H \times (0,T) \\[2mm] c_m \dfrac{\partial v}{\partial t} + i_{ion}(v,w,c) - \mathrm{div}(D_i \nabla v) - \mathrm{div}(D_i \nabla u_e) = I^i_{app} & \text{in } \Omega_H \times (0,T). \end{cases} \tag{5.5}$$

The system must be supplemented by the initial conditions

$$v(\mathbf{x},0) = v_0(\mathbf{x}), \quad w(\mathbf{x},0) = w_0(\mathbf{x}), \quad c(\mathbf{x},0) = c_0(\mathbf{x}) \text{ in } \Omega_H$$

and by boundary conditions. In the case of an insulated heart surface Γ_H, both the the intra- and extracellular current vector are tangent to the interface, i. e. $\mathbf{n}^T \mathbf{j_i} = \mathbf{n}^T \mathbf{j_e} = 0$, where $\mathbf{n}$ denotes the outward normal with respect to Ω_H.

In the non-insulated case, we must couple the macroscopic Bidomain model of the cardiac tissue with the description of the current conduction in the extracardiac medium in order to relate the noninvasive potential measurements on the body surface to the bioelectric cardiac source currents. Let us denote by Ω_0, D_0, $\mathbf{j_0} = -D_0 \nabla u_0, u_0$, the extracardiac volume, the conductivity tensor, the current density and the extracardiac potential respectively, and by $\Gamma_0 = \partial \Omega_0 \setminus \Gamma_H$ the body surface. Disregarding, for instance, the presence of extracardiac applied currents, no current sources lie outside the heart, thus div $\mathbf{j_0} = 0$ in Ω_0. Moreover, the body is embedded in the air, which is an insulated medium, hence $\mathbf{n}^T \mathbf{j_0} = 0$ on Γ_0. Current conservation on the heart interface Γ_H requires that $\mathbf{n}^T (\mathbf{j_i} + \mathbf{j_e})) = \mathbf{n}^T \mathbf{j_0}$, where $\mathbf{n}$ denotes the outward normal to Ω_H, and the zero intracellular flux condition $\mathbf{n}^T \mathbf{j_i} = 0$. In terms of potentials, the system must be coupled with the following elliptic problem

$$\begin{cases} \operatorname{div} D_0(\mathbf{x})\nabla u_0(\mathbf{x},t) = 0 \ \ \text{in } \Omega_0, & \mathbf{n}^T D_0 \nabla u_0(\mathbf{x},t) = 0 \ \text{on } \Gamma_0 \\ u_0(\mathbf{x},t) = u_e(\mathbf{x},t) \ \text{on } \Gamma_H, & \mathbf{n}^T D_0(\mathbf{x})\nabla u_0 = \mathbf{n}^T D_e(\mathbf{x})\nabla u_e \ \ \text{on } \Gamma_H. \end{cases}$$

Defining

$$\widehat{D} = \widehat{D}(\mathbf{x}) = \begin{cases} D_i(\mathbf{x}) + D_e(\mathbf{x}), \ \mathbf{x} \in \Omega_H \\ D_0(\mathbf{x}), \qquad\quad \mathbf{x} \in \Omega_0 \end{cases} \qquad u(\mathbf{x},t) = \begin{cases} u_e(\mathbf{x},t), \ \mathbf{x} \in \Omega_H \\ u_0(\mathbf{x},t), \ \mathbf{x} \in \Omega_0, \end{cases}$$

then the extracellular and extracardiac potential field u satisfies the following elliptic problem

$$\begin{cases} \operatorname{div} \widehat{D}\nabla u(\mathbf{x},t) = \begin{cases} \operatorname{div} \mathbf{J}_v(\mathbf{x},t) \ \ \text{in } \Omega_H \\ 0 \qquad\qquad \text{in } \Omega_0 \end{cases} \\[2mm] [\![\, u(\mathbf{x},t) \,]\!]_{\Gamma_H} = 0, \quad [\![\, \mathbf{n}^T \widehat{D}\nabla u(\mathbf{x},t) \,]\!]_{\Gamma_H} = \mathbf{n}^T \mathbf{J}_v(\mathbf{x},t) \\[2mm] \mathbf{n}^T D_0 \nabla u(\mathbf{x},t) = 0 \qquad\qquad\qquad \text{on } \Gamma_0, \end{cases} \qquad (5.6)$$

where $\mathbf{J}_v(\mathbf{x},t) = -D_i\nabla v(\mathbf{x},t)$ and $[\![\, \Phi \,]\!]_S = \Phi_{S^+} - \Phi_{S^-}$ denotes the jump through a surface S, $\Phi|_{S^\pm}$. We remark that the right-hand sides $\operatorname{div} \mathbf{J}_v(\mathbf{x},t)$ and $\mathbf{n}^T \mathbf{J}_v(\mathbf{x},t)$ act as *impressed* current and *current source density*. Thus, if we assume known the transmembrane potential distribution $v(\mathbf{x},t)$ is known, the above elliptic problem fully characterizes the extracellular and extracardiac field $u(\mathbf{x},t)$ up to an additive constant.

5.2.3 Modelling cardiac cells arrangements

The Bidomain model introduced in the previous section will now be derived from a family of cellular problems by using homogenization techniques, which will solidly justify its macroscopic meaning. The cardiac tissue is composed of a collection of elongated cardiac cells having roughly a cylindrical form with diameter $d_c \approx 10\,\mu\text{m}$ and length $l_c \approx 100\,\mu\text{m}$. The cells are coupled together in end-to-end, mainly, and also in side-to-side apposition by gap junctions [62, 114]. The end-to-end contacts produce the fibres structure of the cardiac muscle, whereas the presence of lateral junctions establishes a connection between the elongated fibres. The cellular structure of the cardiac tissue can be roughly viewed as composed by two ohmic conducting media Ω_i (intracellular space) and Ω_e (extracellular space) separated by the active membrane $\Gamma_m := \partial\Omega_i \cap \partial\Omega_e$. The effects of the microstructure on the current flow are described by the conductivity tensors $\Sigma_i(\mathbf{x})$, $\Sigma_e(\mathbf{x})$ reflecting the local variations of conductances because of the presence of structural intra- and extracellular inhomogeneities of resistance associated with e. g. gap junctions, connective tissue, collagen, blood vessel. Due to the presence of gap junctions connecting the cardiac cells end-to-end and side-to-side, Ω_i and Ω_e are regarded as two simply-connected open sets of $\mathbb{R}^3$.

Let u_i, u_e be the intra- and extracellular potentials and $\mathbf{J}_{i,e} = -\Sigma_{i,e}\nabla u_{i,e}$ their current densities. Let v_i, v_e denote the unit exterior normals to the boundary of Ω_i and Ω_e respectively, satisfying $v_i = -v_e$ on Γ_m. Let $v := u_i|_{\Gamma_m} - u_e|_{\Gamma_m}$ denote the transmembrane potential, evidencing that Γ_m is a discontinuity surface for the potential field. Under quasi-stationary conditions (see [93]), due to the current conservation law, the normal current flux through the membrane is continuous, i.e. $\mathbf{J_i} \cdot v_i = \mathbf{J_e} \cdot v_i$, and this common flux equals the total transmembrane current I_m per unit area (5.1). If no currents are applied to the intra- and extracellular spaces, since the only active source elements lie on the membrane Γ_m, in terms of potentials we have:

$$-\operatorname{div}(\Sigma_i \nabla u_i) = 0, \quad \text{in } \Omega_i \qquad -\operatorname{div}(\Sigma_e \nabla u_e) = 0 \quad \text{in } \Omega_e \qquad (5.7)$$

$$I_m = -v_i^T \Sigma_i \nabla u_i = v_e^T \Sigma_e \nabla u_e = C_m \frac{\partial v}{\partial t} + I_{ion} \quad \text{on } \Gamma_m, \qquad (5.8)$$

where we supplement the equations in (5.7) with homogeneous Neumann boundary conditions for u_i, u_e, assuming that the cellular aggregate is embedded in an insulated medium, and we assign for (5.8) degenerate initial conditions.

We consider two characteristic length scales for the electric potentials u_i, u_e: a *micro* scale related to the typical cell dimensions $\{d_c, l_c\}$ and a *macro* scale defined below. Define $\bar{\mu} = \bar{\mu}_i + \bar{\mu}_e$, with $\bar{\mu}_i$, $\bar{\mu}_e$ being the average eigenvalues of $\Sigma_i(\mathbf{x})$, $\Sigma_e(\mathbf{x})$ over the cells' arrangement. Let R_m be an estimate of the passive membrane resistance near the equilibrium point v_r, i. e. the resting transmembrane potential. Then, we introduce the membrane time constant $\tau_m = R_m C_m$, the length scale unit $\Lambda = \sqrt{l_c \bar{\mu} R_m}$ and the dimensionless parameter $\varepsilon = l_c/\Lambda$, i. e. the ratio between the *micro* and the *macro* length constants. We consider the following scalings of the space and time variables $\widehat{\mathbf{x}} = \mathbf{x}/\Lambda$, $\widehat{t} = t/\tau_m$ and define $u_i^\varepsilon(\widehat{\mathbf{x}}, \widehat{t}) = u_i(\Lambda\widehat{\mathbf{x}}, \tau_m\widehat{t})$ and analogously u_e^ε, w^ε, c^ε. Rescaling the Eqs. (5.7) and (5.8) in the intra- and extracellular media and omitting the superscripts $\widehat{}$ of the dimensionless variables, we obtain the following model:

Dimensionless cellular model P^ε. Let $\Omega := \Omega_i^\varepsilon \cup \Omega_e^\varepsilon \cup \Gamma_m^\varepsilon$ be the dimensionless cardiac tissue volume, Γ_m^ε the surface cellular membrane, $\sigma_{i,e} = \Sigma_{i,e}/\bar{\mu}$ and $\mathscr{I}_{ion} = I_{ion} R_m$ the dimensionless conductivity matrices and ionic membrane current, respectively. Then, the full Reaction-Diffusion system associated with the cellular model in dimensionless form can be formulated as follows: the vector $(u_i^\varepsilon, u_e^\varepsilon, w^\varepsilon, c^\varepsilon)$, with $v^\varepsilon = u_i^\varepsilon - u_e^\varepsilon$, satisfies the problem

$$-\operatorname{div}\sigma_i^\varepsilon(x)\nabla u_i^\varepsilon = 0 \quad \text{in } \Omega_i^\varepsilon, \qquad -\operatorname{div}\sigma_e^\varepsilon(x)\nabla u_e^\varepsilon = 0 \quad \text{in } \Omega_e^\varepsilon \qquad (5.9)$$

$$I_m^\varepsilon = -\sigma_i^\varepsilon(x)n^T\nabla u_i^\varepsilon = -\sigma_e^\varepsilon(x)n^T\nabla u_e^\varepsilon$$

$$= \varepsilon\,[\,\frac{\partial v^\varepsilon}{\partial t} + \mathscr{I}_{ion}(v^\varepsilon, w^\varepsilon, c^\varepsilon)\,], \quad \text{on } \Gamma_m^\varepsilon \qquad (5.10)$$

$$\frac{\partial w^\varepsilon}{\partial t} - \mathscr{R}(v^\varepsilon, w^\varepsilon) = 0, \qquad \frac{\partial c^\varepsilon}{\partial t} - \mathscr{S}(v^\varepsilon, w^\varepsilon, c^\varepsilon) = 0, \quad \text{on } \Gamma_m^\varepsilon \qquad (5.11)$$

$$\mathbf{n}^T \nabla u_i^\varepsilon = 0 \quad \text{on} \quad \partial \Omega_i^\varepsilon \setminus \Gamma_m^\varepsilon \quad \text{and} \quad \mathbf{n}^T \nabla u_e^\varepsilon = 0 \quad \text{on} \quad \partial \Omega_e^\varepsilon \setminus \Gamma_m^\varepsilon, \tag{5.12}$$

$$v^\varepsilon(\mathbf{x},0) = v_0^\varepsilon(\mathbf{x}), \quad w^\varepsilon(\mathbf{x},0) = w_0^\varepsilon(\mathbf{x}), \quad c^\varepsilon(\mathbf{x},0) = c_0^\varepsilon(\mathbf{x}) \quad \text{su } \Gamma_m^\varepsilon, \tag{5.13}$$

where $\mathscr{R} = \tau_m R$, $\mathscr{S} = \tau_m S$ and $n = v_i = -v_e$ denotes the normal to Γ_m^ε pointing towards Ω_e^ε. We note that the variables $v^\varepsilon, w^\varepsilon, c^\varepsilon$ and I_m^ε are defined only on the membrane surface Γ_m^ε. Problem $\mathbf{P}^\varepsilon$ is not a standard parabolic homogenization problem and its main difficulties are associated with the fact that the evolution term depends explicitly on ε, it is *degenerate* and that the boundaries of $\Omega_{i,e}^\varepsilon$ could be quite irregular.

Formal two-scale homogenization. The two-scale method of homogenization (see [12, 82, 118]) can be applied to the previous current conservation equations (5.9), (5.10), where we assume that the cells are distributed according to an ideal periodic organization similar to a regular lattice of interconnected cylinders. We denote by E_i, $E_e := \mathbb{R}^3 \setminus \overline{E}_i$, two open, connected and periodic reference subsets of $\mathbb{R}^3$ with common Lipschitz boundary $\Gamma_m := \partial E_i \cap \partial E_e$, by Y the elementary periodicity region, composed of the intra- and extracellular volumes $Y_{i,e} = Y \cap E_{i,e}$ representing a reference volume box containing a cellular configuration Y_i with cell membrane surface $S_m = \Gamma_m \cap Y_i$. The physical region Ω occupied by the cardiac tissue is decomposed into the intra- and extracellular domains $\Omega_{i,e}^\varepsilon$, obtained as $\Omega_i^\varepsilon = \Omega \cap \varepsilon E_i$, $\Omega_e^\varepsilon = \Omega \cap \varepsilon E_e$. The common boundary $\Gamma_m^\varepsilon = \partial \Omega_i^\varepsilon \cap \partial \Omega_e^\varepsilon = \Omega \cap \varepsilon \Gamma_m$ models the cellular membrane. Since the cardiac tissue exhibits a number of significant inhomogeneities, such as those related to cell-to-cell communications, the conductivity tensors are considered dependent on both slow and fast variables, i.e. $\sigma_{i,e}(x, \frac{x}{\varepsilon})$. The dependence of σ_i on $\frac{x}{\varepsilon}$ models the inclusion of gap-junction effects. We then define the rescaled symmetric conductivity matrices $\sigma_{i,e}^\varepsilon(x) = \sigma_{i,e}(x, \frac{x}{\varepsilon})$, where $\sigma_{i,e}(x,\xi) : \Omega \times E_{i,e} \to \mathbb{M}^{3 \times 3}$ are continuous functions satisfying the usual uniform ellipticity and periodicity conditions.

The following macroscopic bidomain model can be formally derived by taking the average of a cellular model in the periodic case, see for details [27, 86].

Dimensionless averaged model P. For a periodic network of interconnected cells, let $\beta = |S_m|/|Y|$ be the ratio between the surface membrane and the volume of the reference cell and let $\beta_{i,e} = |Y_{i,e}|/|Y|$. Then the governing dimensionless equations of the macroscopic intra and extracellular potentials of zero order in ε are given by

$$\begin{cases} \operatorname{div} D_i(\mathbf{x}) \nabla_x u_i = \beta \left(\dfrac{\partial v}{\partial t} + \mathscr{I}_{ion}(v, w, c) \right) \\[2mm] \operatorname{div} D_e(\mathbf{x}) \nabla_x u_e = -\beta \left(\dfrac{\partial v}{\partial t} + \mathscr{I}_{ion}(v, w, c) \right) \\[2mm] v = u_i - u_e, \quad \dfrac{\partial w}{\partial t} - \mathscr{R}(v, w) = 0, \quad \dfrac{\partial c}{\partial t} - \mathscr{S}(v, w, c) = 0, \end{cases} \tag{5.14}$$

where the effective conductivity tensors are given by

$$D_{i,e}(\mathbf{x}) = \frac{1}{|Y|} \int_{Y_{i,e}} \sigma_{i,e}(\mathbf{x},\xi) \left\{ I - \nabla_\xi (\mathbf{w}^{i,e})^T \right\} d\xi \qquad (5.15)$$

and $\mathbf{w}^{i,e} = (w_1^{i,e}, w_2^{i,e}, w_3^{i,e})^T$ are solutions of the following cellular problems

$$\begin{cases} \operatorname{div}_\xi \sigma_{i,e}(\mathbf{x},\xi) \nabla_\xi w_k^e = 0, & \text{in } Y_{i,e} \\ \mathbf{n}_\xi^T \sigma_{i,e}(\mathbf{x},\xi) \nabla_\xi w_k^{i,e} = n_{\xi_k}, & \text{on } S, \quad k = 1,2,3. \end{cases} \qquad (5.16)$$

The tensors $D_{i,e}$ are symmetric and positive definite matrices.

The abstract variational framework of the cellular and averaged models share the same structural properties, see [27]. It was shown in [27] that both the cellular and the averaged models are well posed, assuming the FitzHugh-Nagumo membrane model [50], where the ionic current is a cubic-like function in v, linear in the recovery variable w, i.e. $\mathscr{I}_{ion}(v,w) = g(v) + \alpha w$, and $\mathscr{R}(v,w) = \eta v - \gamma, \alpha, \eta, \gamma > 0$.

Theorem 1 (model $\mathbf{P}^\varepsilon$ well-posedness). *Assume Γ^ε regular, suppose that the initial data satisfy $(v_0^\varepsilon, w_0^\varepsilon) \in L^2(\Gamma_m^\varepsilon) \times L^2(\Gamma_m^\varepsilon)$ and define the quotient space $\mathbb{V}^\varepsilon = \{H^1(\Omega_i^\varepsilon) \times H^1(\Omega_e^\varepsilon)\} / \{(\gamma,\gamma) : \gamma \in \mathbf{R}\} \times L^2(\Gamma_m^\varepsilon)$. Then there exists a unique solution $U^\varepsilon = (u_i^\varepsilon, u_e^\varepsilon, w^\varepsilon) \in C^0(]0,T]; \mathbb{V}^\varepsilon)$ of the variational formulation of Problem $\mathbf{P}^\varepsilon$ with $\partial v^\varepsilon / \partial t, \; \partial w^\varepsilon / \partial t \in L^2(0,T; L^2(\Gamma_m^\varepsilon))$, and $u_-^\varepsilon := (u_i^\varepsilon, u_e^\varepsilon)$ solution of the differential problem $\mathbf{P}^\varepsilon$ in the standard distributional sense.*

Theorem 2 (model $\mathbf{P}$ well-posedness). *Let $(v_0, w_0) \in L^2(\Omega) \times L^2(\Omega)$ and define the quotient space $\mathbb{V} := \{H^1(\Omega) \times H^1(\Omega)\} / \{(\gamma,\gamma) : \gamma \in \mathbf{R}\} \times L^2(\Omega)$. Then there exists a unique solution $U = (u_i, u_e, w) \in C^0(]0,T]; \mathbb{V})$ of the variational formulation of the averaged Problem $\mathbf{P}$ with $\partial v / \partial t, \; \partial w / \partial t \in L^2(0,T; L^2(\Omega))$ and $u_- := (u_i, u_e)$ solution of the differential problem $\mathbf{P}$ in the standard distributional sense.*

Existence results for global solutions of the averaged Bidomain model with other simplified membrane models have been obtained using different techniques in [10, 14, 15]. Extension of the well-posedness results for the cellular and the averaged models with more complex ionic current membrane dynamics have been obtained recently in [146, 147], both for the classical Hodgkin-Huxley model of the nerve action potential [59] and for the Luo-Rudy Phase I model [74].

The derivation of the Bidomain model based on the two-scale method is only formal. The following convergence result for the Bidomain model with FHN gating has been developed in [86] using homogenization techniques and Γ−convergence theory.

Theorem 3 (convergence result for the averaged model with FHN dynamics).
Let us assume that $v_0^\varepsilon = u_{i,0}^\varepsilon - u_{e,0}^\varepsilon, w_0^\varepsilon$ are converging to $v_0 = u_{i,0} - u_{e,0}, w_0$ in the following "distributional" sense,

$$\lim_{\varepsilon \downarrow 0} \varepsilon \left(\int_{\Gamma_m^\varepsilon} v_0^\varepsilon \, \zeta \, d\gamma, \int_{\Gamma_m^\varepsilon} w_0^\varepsilon \, \zeta \, d\gamma \right)$$
$$= \beta \left(\int_\Omega v_0 \, \zeta \, dx, \int_\Omega w_0 \, \zeta \, dx \right), \quad \forall \zeta \ \ \forall \zeta \in C_0^\infty(\Omega),$$

and the following properties of the energy-like functionals hold

$$\lim_{\varepsilon \downarrow 0} \varepsilon \left(\int_{\Gamma_m^\varepsilon} |v_0^\varepsilon|^2 \, d\gamma, \int_{\Gamma_m^\varepsilon} |w_0^\varepsilon|^2 \, d\gamma \right) = \beta \left(\int_\Omega |v_0|^2 \, dx, \int_\Omega |w_0|^2 \, dx \right)$$

$$\limsup_{\varepsilon \downarrow 0} \left(\sum_{i,e} \int_{\Omega_{i,e}^\varepsilon} \sigma_{i,e}^\varepsilon |\nabla u_{(i,e),0}^\varepsilon|^2 dx + \varepsilon \int_{\Gamma_m^\varepsilon} G(v_0^\varepsilon) d\gamma \right) < +\infty,$$

where G is a positive primitive of the cubic-like function g in the FHN model. Let $\Omega_0 \subset\subset \Omega$ be a reference open subdomain with positive measure, $u^\varepsilon = (u_i^\varepsilon, u_e^\varepsilon), v^\varepsilon = u_i^\varepsilon - u_e^\varepsilon, w^\varepsilon$ and $u_\_ = (u_i, u_e), v = u_i - u_e, w$ be the solutions of the cellular and averaged models with $\int_{\Omega_0 \cap \Omega^\varepsilon} u_e^\varepsilon dx = 0$, and $\int_{\Omega_0 \cap \Omega} u_e dx = 0$, respectively. Then for every time $t \in [0,T]$

$$(u_{i,e}^\varepsilon, v^\varepsilon, w^\varepsilon) \rightarrow (u_{i,e}, v, w) \quad as \ \varepsilon \downarrow 0,$$

in the following distributional sense,

$$\lim_{\varepsilon \downarrow 0} \varepsilon \left(\int_{\Gamma_m^\varepsilon} v^\varepsilon \, \zeta \, d\gamma, \int_{\Gamma_m^\varepsilon} w^\varepsilon \, \zeta \, d\gamma \right) = \beta \left(\int_\Omega v \, \zeta \, dx, \int_\Omega w \, \zeta \, dx \right), \quad \forall \zeta \in C_0^\infty(\Omega),$$

$$\lim_{\varepsilon \downarrow 0} \int_{\Omega_{i,e}^\varepsilon} u_{i,e}^\varepsilon(x) \zeta(x) \, dx = \beta_{i,e} \int_\Omega u_{i,e}(x) \zeta(x) \, dx, \quad \forall \zeta \in C_0^\infty(\Omega).$$

Moreover, there exist extensions $\mathcal{T}_i^\varepsilon u_i^\varepsilon, \mathcal{T}_e^\varepsilon u_e^\varepsilon$ of $u_i^\varepsilon, u_e^\varepsilon$ the whole domain Ω, solution of the cellular problem $\mathbf{P}^\varepsilon$, which converge in $L^2(0,T; H_{loc}^1(\Omega))$ to the unique solution $(u_i, u_e) \in \mathbb{V}$ of the averaged model $\mathbf{P}$.

We note that, at present, a rigorous derivation of homogenized interface conditions at the cardiac tissue boundary in contact with a conducting medium is still missing.

5.2.4 Approximate modelling of the cardiac bioelectric activity by reduced models

In the Bidomain model (5.4), the transmembrane potential v during the excitation phase of the heartbeat exhibits a steep propagating layer spreading throughout the myocardium with a thickness of about 0.5 mm. At every point of the cardiac domain, this upstroke phase lasts about 1 msec. Therefore, the simulation of the excitation process requires the numerical solution of problems with small space and time steps (of the order of 0.1 mm and 0.01 msec). This fact constrains 3-D simulations to limited blocks with dimensions of a few centimetres and with small duration of the evolution, see e.g. [23, 29, 58, 148] and the recent survey [149]. For large scale simulations involving the whole ventricles, computer memory and time requirements become excessive, and less demanding approximations have been developed, such as Eikonal and Monodomain models.

Eikonal models

We note that during the excitation phase of the heart beat the main feature, at a macroscopic level, is the excitation wavefront configuration and its motion. In order to describe the excitation sequence and avoid the high computational costs of the full Bidomain model, in [8, 64, 66, 67] the *eikonal – curvature models* were developed. With these models, the simulation of the activation sequence in large volumes of cardiac tissue has become computationally practical, since they do not require a fine spatial and temporal resolution.

During the excitation phase, the Bidomain model in dimensionless form (see [22] for details on the suitable scaling) can be written as a singularly perturbed R–D system

$$\frac{\partial v^\varepsilon}{\partial t} + \frac{1}{\varepsilon} g(v^\varepsilon) - \varepsilon \operatorname{div}(D_i \nabla u_i^\varepsilon) = 0 \ , \ -\frac{\partial v^\varepsilon}{\partial t} - \frac{1}{\varepsilon} g(v^\varepsilon) - \varepsilon \operatorname{div}(D_e \nabla u_e^\varepsilon) = 0, \quad (5.17)$$

where the dimensionless parameter ε is of the order $10^{-3} - 10^{-2}$, $v^\varepsilon = u_i^\varepsilon - u_e^\varepsilon$ and g is a scaled cubic–like form of an instantaneous ionic current-voltage relationship $i_{ion}(v)$ related to the excitation phase. We denote by $v_r < v_{th} < v_p$ the three zeroes of g representing the resting, threshold and excited transmembrane values respectively, and we assume that $\int_{v_r}^{v_p} g(v) dv < 0$. Given the previous singular perturbation structure, $u_\varepsilon^i, u_\varepsilon^e$ diffuse quite slowly, while the reaction takes place much faster; hence, the development of a moving layer associated with a traveling wavefront solution is to be expected. Assuming that the excitation propagates in fully recovered tissue, a monotonic temporal behavior of v is expected, thus we define the *activation time* as the time instant $t = \psi(\mathbf{x})$ at which $v(\mathbf{x}, t) = (v_r + v_p)/2$. Then the excitation wavefront $S_\varepsilon(t)$ is represented by the level surface of the activation time at the time instant t, i.e.:

$$S_\varepsilon(t) = \{\mathbf{x} \in \Omega_H, \ \psi(\mathbf{x}) = t\}. \quad (5.18)$$

We then introduce the indicatrix function $\Phi(\mathbf{x}, \xi) = \sqrt{q(\mathbf{x}, \xi)}$, where

$$q(\mathbf{x}, \xi) = \frac{\left(\xi^T D_i(\mathbf{x})\xi\right)\left(\xi^T D_e(\mathbf{x})\xi\right)}{\xi^T (D_i(\mathbf{x}) + D_e(\mathbf{x}))\xi} = \frac{1}{q_i(\mathbf{x}, \xi)^{-1} + q_i(\mathbf{x}, \xi)^{-1}} \quad (5.19)$$

is the conductivity of the bulk medium at a point $\mathbf{x}$ along the unit vector ξ, expressed as the harmonic mean of the quadratic forms $q_{i,e}(\mathbf{x}, \xi) = \xi^T D_{i,e}(\mathbf{x})\xi$ associated with the conductivity tensors $D_{i,e}$. The bulk medium is composed by coupling in series the media (i) and (e).

Exploiting the singular perturbation approach, we can derive formally anisotropic geometric evolution laws capturing the asymptotic behavior of traveling wavefront solutions of the R–D system (5.17) (see [8, 9, 64, 66, 67]). The propagation of the wavefront $S_\varepsilon(t)$ behaves, up to $\mathscr{O}(\varepsilon^2)$ terms, as a hypersurface $S(t)$, propagating with the anisotropic geometric law with velocity $\theta(v)$, along the Euclidean unit vector v normal to the wavefront and oriented toward the resting tissue, (see e.g. [8], Appendix B), given by:

$$\theta(v) = \Phi(\mathbf{s}, v)(c - \varepsilon \operatorname{div} \Phi_\xi(\mathbf{s}, v)). \qquad (5.20)$$

Equations of this type are also called *eikonal-curvature models* since $K_\Phi = \operatorname{div} \mathbf{n}_\Phi = \operatorname{div} \Phi_\xi(\mathbf{s}, v))$ is the *anisotropic* mean curvature with respect to a suitable Finsler metric; see [8, 9] for definitions and details. Moreover, c is the velocity of a 1-D traveling wave solution , i.e. (c,a) is the unique bounded solution of the eigenvalue problem:

$$a'' + c\, a' - g(a) = 0 \quad a(-\infty) = v_p, \quad a(\infty) = v_r, \quad a(0) = (v_p + v_r)/2.$$

An eikonal model, equivalent to (5.20) up to second-order terms of ε and called *eikonal-diffusion* equation, is also formally derived in [21, 22] and used in studies of the anisotropic 3-D propagation of the excitation wavefronts, see e.g. [23, 28, 139].

The rigorous justification of the connection between the evolution of a suitable level set of v and the surface flowing under a geometric evolution law remains to our knowledge an open problem. A partial rigorous characterization of the anisotropic curvature term was obtained for the stationary Bidomain model in [2].

Relaxed linear and nonlinear anisotropic Monodomain model

Many large scale simulations have been performed using the so-called Monodomain model, in order to avoid the high computational costs of the full Bidomain model. This model is a *relaxed Bidomain model* described by a system of a parabolic equation coupled with an elliptic equation, but, unlike the Bidomain model, the former evolution equation is fully uncoupled to the elliptic one. An anisotropic Monodomain model consists in a single parabolic reaction-diffusion equation for v, with conductivity tensor $D_m(\mathbf{x})$ and applied current I_{app}^m, coupled with gating and ionic concentration systems and with the elliptic problem:

$$
\begin{cases}
c_m \dfrac{\partial v}{\partial t} - \operatorname{div}(D_m(\mathbf{x})\nabla v) + i_{ion}(v, w, c) = I_{app}^m & \text{in } \Omega_H \times (0, T) \\[2ex]
\dfrac{\partial w}{\partial t} - R(v, w) = 0, \quad \dfrac{\partial c}{\partial t} - S(v, w, c) = 0 & \text{in } \Omega_H \times (0, T) \\[2ex]
\mathbf{n}^T(D_m(\mathbf{x})\nabla v = 0 & \text{in } \Gamma_H \times (0, T) \\
v(\mathbf{x}, 0) = v_0(\mathbf{x}), \quad w(\mathbf{x}, 0) = w_0(\mathbf{x}), \quad c(\mathbf{x}, 0) = c_0(\mathbf{x}) & \text{in } \Omega_H. \\[2ex]
-\operatorname{div}((D_i + D_e)\nabla u_e) = \operatorname{div}(D_i \nabla v) + I_{app}^i + I_{app}^e & \text{in } \Omega_H, \\
-\mathbf{n}^T(D_i + D_e)\nabla u_e = \mathbf{n}^T D_i \nabla v & \text{on } \Gamma_H.
\end{cases}
$$

$$(5.21)$$

The evolution equation determines the distribution of $v(\mathbf{x},t)$, then the extracellular potential distribution u_e is derived by solving the elliptic boundary value problem. It is well known that if the two media have the same anisotropy ratio $\frac{\sigma_l^e}{\sigma_l^i} = \frac{\sigma_t^e}{\sigma_t^i} = \frac{\sigma_n^e}{\sigma_n^i} = \lambda$, i. e. $D_e(\mathbf{x}) = \lambda D_i(\mathbf{x})$ with constant λ, then the Bidomain system reduces to the Monodomain model (see [19, 36, 57]) with conductivity tensor and applied current given by

$$D_m(\mathbf{x}) = \lambda D_i(\mathbf{x})/(1+\lambda), \qquad I_{app}^m = \left(I_{app}^i - \lambda I_{app}^e\right)/(1+\lambda). \tag{5.22}$$

We remark that this is not a physiological case, as clearly follows from well established cardiac experimental evidence. Derivations of a reduced Bidomain model which does not make such an assumption of *a priori equal anisotropy* assumption have been developed in [19, 36, 65, 102], disregarding some source terms, i.e. the projections of the total current flowing in the two media $J_{tot} = \mathbf{j_i} + \mathbf{j_e}$ along and across fibre directions $\mathbf{a}_t(\mathbf{x})$, $\mathbf{a}_n(\mathbf{x})$. In these derivations, the Monodomain model is characterized by the following anisotropic conductivity tensor and applied current:

$$D_m = D_e D^{-1} D_i, \qquad I_{app}^m = \left(I_{app}^i \sigma_l^e - I_{app}^e \sigma_l^i\right)/(\sigma_l^e + \sigma_l^i), \tag{5.23}$$

where $D = D_i + D_e$. We remark that it is possible to rescale the reaction-diffusion equation (5.21) of the Monodomain model, obtaining a singular perturbation problem as in the Bidomain model (5.17). Performing formal asymptotic expansions (see [8], Appendix A), it is possible to derive the anisotropic evolution law of the wavefront. Up to terms of order $O(\varepsilon^2)$, the normal velocity $\theta^m(v)$ is given by

$$\theta^m(v) = \Phi^m(\mathbf{s}, v)(c - \varepsilon \operatorname{div} \Phi_\xi^m(\mathbf{s}, v)), \tag{5.24}$$

presenting the same structure of (5.20), but for the Monodomain model the *indicatrix* is $\Phi^m(\mathbf{x},\xi) = \sqrt{q_m(\mathbf{x},\xi)}$. This *indicatrix* does not coincide with the one derived from the Bidomain model (5.19), since it is now defined in terms of the harmonic mean of the conductivity tensors, i.e.

$$q_m(\mathbf{x},\xi) = \xi^T D_m(\mathbf{x},\xi)\xi \quad \text{where} \quad D_m(\mathbf{x}) = D_e D^{-1} D_i. \tag{5.25}$$

Therefore the wavefront motion of the Monodomain model characterized by (5.25) is different at the zero order from that of the Bidomain excitation wavefront, compare (5.19) with (5.25).

In order to obtain an anisotropic Monodomain model, having the same anisotropic law of motion of the excitation wavefront of the Bidomain model, the following nonlinear diffusion term, i.e. a nonlinear conductivity tensor, was introduced in [8] in the parabolic equation:

$$D_m(\mathbf{x},\xi) := \left(\frac{\xi^T D_i \xi}{\xi^T D \xi}\right)^2 D_e(\mathbf{x}) + \left(\frac{\xi^T D_e \xi}{\xi^T D \xi}\right)^2 D_i(\mathbf{x}). \tag{5.26}$$

In [8], it was derived formally that the eikonal-curvature equation associated with this Monodomain model coincides with that of the Bidomain model given by (5.20) and (5.19). We remark that the following relation holds between (5.26) and (5.19): $D_m(\mathbf{x}, \xi)\xi = \Phi_\xi(\mathbf{x}, \xi)\Phi(\mathbf{x}, \xi)$. Moreover, for the FHN membrane model and a suitable convexity property of Φ, in [8] it was shown that the approximation provided by the eikonal-curvature equation can be justified rigorously by estimating the distance between the evolution of a suitable level set of v, generated by the Monodomain model, and the surface flowing under the geometric law. We remark that the linear and nonlinear conductivity tensors considered above both reduce to (5.22) in the case of equal anisotropy ratio of the two media. Finally, we have focused so far on the use of reduced models for the excitation phase, but the relaxed linear and nonlinear Monodomain model could also be used as a reduced model in all phases of the heartbeat, see for a recent validations [36, 94].

5.3 Numerical approximation of the Monodomain and Bidomain models

Time discretization. The time discretization of the Bidomain and Monodomain equations can be performed using either implicit, semi-implicit or explicit schemes, requiring accordingly vector updates or the solution of a nonlinear or linear system. Fully implicit methods in time have been considered in [60, 77, 78, 79, 80, 121, 158]. The advantage of implicit methods is that they do not require stability constraints on the choice of the time step, but they are very expensive, because at each time step one has to solve a nonlinear system. A good compromise between stability and efficiency is obtained using linear implicit methods, studied e. g. in [20, 40, 79, 158], which require at each time step the solution of 2–4 linear systems. Semi-implicit methods have been studied in [29, 30, 46, 68, 119, 127, 151]. Finally, explicit schemes, in particular the Forward Euler method, have been used by [17, 84, 97, 138]. For a detailed comparative study on the stability and accuracy of several Bidomain time discretizations (implicit, semi-implicit, explicit) we refer to the recent work [46]. However, the most popular technique is based on operator splitting, hence on separating the diffusion operator, related to conduction in the media, from the reaction operator, related to the ionic current, gating and ionic concentrations dynamics. The advantage of splitting methods is to allow different numerical schemes for the diffusion and the reaction terms in order to maximize computational efficiency and eliminate complex dependency between variables. The disadvantage is a loss of accuracy, because the simultaneous dependency between variables is neglected. For the parabolic-elliptic formulation of the Bidomain model, a further splitting method consists in solving sequentially the elliptic equation, followed by the parabolic one. About operator splitting techniques, we refer e. g. to the studies [75, 90, 94, 98, 132, 133, 134, 148, 149, 150]. Finally we mention the approximation results related to semi-discrete approximation in time of the Bidomain model with the FHN membrane model. In [5] *a priori* and *a posteriori* error estimates have been developed for the implicit Euler scheme.

Space discretization. Many different approaches have been developed for the space discretization of the Bidomain and Monodomain equations. Finite differences were studied in [56, 84, 94, 99, 115, 143, 156, 157, 161]. Finite elements have been widely used, see e. g. [20, 29, 51, 89, 119, 148]. The finite-volume method, which has the advantage of conserving local flux, was developed in [11, 39, 55, 142]. Finally, some researchers recently investigated the spectral element method, see [13]. In order to reduce the computational load of the Bidomain system, also adaptive remeshing techniques have been developed, see e. g. [6, 7, 16, 20, 40, 141, 152]; these techniques have proved successful for problems of moderate size and are currently under investigation. Interested readers can find numerical approximation schemes and numerical simulations based on eikonal approaches in [64, 66, 67, 83] and in [24, 25, 28, 139], using mesh-size of the order of 1 mm.

Semi-implicit scheme with FEM. We briefly describe here the approximation for the anisotropic Bidomain system (5.4), obtained applying the finite element method in space and a semi-implicit method in time. In the following, we will denote vectors representing finite element functions by boldface symbols, for both product space functions and their components.

Let $\mathscr{T}^h$ be a uniform hexahedral triangulation of Ω and V^h the associated space of trilinear finite elements. We obtain a semidiscrete problem by applying a standard Galerkin procedure. Choosing a finite element basis $\{g_i\}$ for V^h, we denote by

$$\mathrm{M} = \{\mathrm{m}_{rs} = \int_\Omega g_r g_s \mathrm{dx}\}, \qquad \mathrm{A}_{i,e} = \{\mathrm{a}_{rs}^{i,e} = \int_\Omega (\nabla g_r)^T D_{i,e} \nabla g_s \mathrm{dx}\},$$

the symmetric mass matrix and stiffness matrices, respectively, and by $\mathrm{I}_{ion}^h, \mathrm{I}_{app}^h$ the finite element interpolants of I_{ion} and I_{app}, respectively.

We sketch now a semi-implicit time discretization of the Bidomain system, using for the diffusion term the implicit Euler method, while the nonlinear reaction term I_{ion} is treated explicitly. The implicit treatment of the diffusion terms appearing in the Monodomain or Bidomain models is essential in order to adapt the time step according to the stiffness of the various phases of the heart-beat. The ODE system for the gating variables is discretized by the semi-implicit Euler method and the explicit Euler method is applied for solving the ODE system for the ion concentrations. As a consequence, the full evolution system is decoupled by first solving the gating and ion concentrations system (given the potential $\mathbf{v}_M^n$ at the previous time-step)

$$\mathbf{w}^{n+1} - \Delta t\, R(\mathbf{v}_M^n, \mathbf{w}^{n+1}) = \mathbf{w}^n, \qquad \mathbf{c}^{n+1} = \mathbf{c}^n + \Delta t\, S(\mathbf{v}_M^n, \mathbf{w}^{n+1}, \mathbf{c}^n),$$

and then solving for $\mathbf{u}^{n+1} = (\mathbf{u}_i^{n+1}, \mathbf{u}_e^{n+1})$ the system

$$\mathscr{A}\mathbf{u}^{n+1} = \mathscr{F} \qquad\qquad (5.27)$$

where

$$\mathscr{A} := \left(\gamma \begin{bmatrix} \mathrm{M} & -\mathrm{M} \\ -\mathrm{M} & \mathrm{M} \end{bmatrix} + \begin{bmatrix} \mathrm{A}_i & 0 \\ 0 & \mathrm{A}_e \end{bmatrix}\right), \qquad \gamma := \frac{c_m}{\Delta t}, \qquad (5.28)$$

and

$$\mathscr{F} = \begin{pmatrix} F^n \\ -F^n \end{pmatrix} = \begin{pmatrix} M[-I_{ion}^h(\mathbf{v}_M^n, \mathbf{w}^{n+1}, \mathbf{c}^{n+1}) + I_{app}^h] \\ M[\ \ I_{ion}^h(\mathbf{v}_M^n, \mathbf{w}^{n+1}, \mathbf{c}^{n+1}) - I_{app}^h] \end{pmatrix}, \qquad (5.29)$$

where $\mathbf{v}_M^n = \mathbf{u}_i^n - \mathbf{u}_e^n$. As in the continuous model, $\mathbf{v}_M^n$ is uniquely determined, while $\mathbf{u}_i^n$ and $\mathbf{u}_e^n$ are determined only up to the same additive time-dependent constant chosen by imposing the condition $\mathbf{1}^T M \mathbf{u}_e^n = 0$. Hence, at each time step we have to solve the large linear system (5.27), that, as shown in [29], is very ill-conditioned and increases considerably the computational costs of the simulations.

5.3.1 Linear solvers for the Bidomain model

The advantage of using semi-implicit and/or operator-splitting methods is that they only require the solution of linear systems at each time step. In order to devise efficient iterative solvers for these linear systems, many different preconditioners have been proposed: diagonal preconditioners [126], SSOR preconditioners [87], Block Jacobi (BJ) preconditioners [29, 75, 148], Block triangular Monodomain-based preconditioners [52], Multigrid preconditioners [3, 73, 75, 88, 89, 90, 131, 149, 150], and Multilevel Schwarz preconditioners [85, 120]. These studies have shown that Multigrid and Multilevel Schwarz methods yield efficient solvers for the discrete Bidomain equations and improve considerably the performance of the simplest diagonal or block-diagonal preconditioner on several processors.

5.3.1.1 Functional reformulation of the semi-discrete Bidomain system

In this section, we introduce the notations and basic results that we will need in the next sections in order to construct a two-level Schwarz preconditioner for the Bidomain system and to prove its scalability. We will reformulate (5.27) as an elliptic problem and we will adapt to it the standard abstract Schwarz theory described e. g. in [129, 140]. In the rest of the section, we will denote by $\mathbf{v}, \mathbf{u}$ generic functions in the product space U^h defined below and we will denote their intra- and extracellular components by $\mathbf{v} = (\mathbf{v}_i, \mathbf{v}_e)$, $\mathbf{u} = (\mathbf{u}_i, \mathbf{u}_e)$.

Let us introduce the spaces

$$U^h = V^h \times \widetilde{V}^h, \quad \text{with} \quad \widetilde{V}^h = \{\mathbf{u}_e \in V^h : \mathbf{1}^T M \mathbf{u}_e = \int_\Omega \mathbf{u}_e dx = 0\}, \qquad (5.30)$$

equipped with the inner product $((\mathbf{u}, \mathbf{v})) := \int_\Omega \nabla \mathbf{u}_i \cdot \nabla \mathbf{v}_i + \int_\Omega \nabla \mathbf{u}_e \cdot \nabla \mathbf{v}_e + \int_\Omega (\mathbf{u}_i - \mathbf{u}_e)(\mathbf{v}_i - \mathbf{v}_e)$ and with the induced norm $|||\cdot|||$.

We introduce the continuous and elliptic, with respect to the $|||\cdot|||$-norm, bilinear form $a_{bid}(\cdot, \cdot) : U^h \times U^h \to R$ defined by

$$a_{bid}(\mathbf{u}, \mathbf{v}) := \int_\Omega D_i \nabla \mathbf{u}_i \cdot \nabla \mathbf{v}_i + \int_\Omega D_e \nabla \mathbf{u}_e \cdot \nabla \mathbf{v}_e + \gamma \int_\Omega (\mathbf{u}_i - \mathbf{u}_e)(\mathbf{v}_i - \mathbf{v}_e),$$

and the continuous linear form $\mathbf{f} : U^h \times U^h \to R$ defined as $(\mathbf{f}, \mathbf{v}) = (f, \mathbf{v}_i) - (f, \mathbf{v}_e)$ $\forall \mathbf{v} \in U^h$. We define the linear operator $\mathscr{A} : U^h \to U^h$ as

$$((\mathscr{A}\mathbf{u}, \mathbf{v})) = a_{bid}(\mathbf{u}, \mathbf{v}) \quad \forall \mathbf{v} \in U^h,$$

and $\mathscr{F} \in U^h$ as $((\mathscr{F}, \mathbf{v})) = (\mathbf{f}, \mathbf{v})$, where $(\cdot, \cdot)$ represents the standard L^2-inner product.

With these definitions, we can give the following equivalent formulation of the discrete bidomain system (5.27): given $f \in L^2(\Omega)$, find $\mathbf{u} \in U^h$ such that

$$a_{bid}(\mathbf{u}, \mathbf{v}) = (\mathbf{f}, \mathbf{v}), \quad \forall \mathbf{v} \in U^h, \text{or in term of linear operators } \mathscr{A}\mathbf{u} = \mathscr{F}. \quad (5.31)$$

5.3.1.2 The two-level Additive Schwarz preconditioner

In this section, we construct and analyze a two-level overlapping Additive Schwarz (AS) method for problem (5.31). Let $\mathscr{T}^0 = \mathscr{T}^H$ be a coarse shape-regular triangulation of Ω consisting of N nonoverlapping hexahedral subdomains Ω_m, $m = 1, ..., N$ of diameter H_m and let $H = \max_m H_m$. Let $\mathscr{T}^1 = \mathscr{T}^h$ be a fine shape-regular triangulation nested in $\mathscr{T}^0$, consisting of hexahedral elements τ_j, $j = 1, ..., N_e$ of diameter h_j and let $h = \max_j h_j$. An overlapping partition of Ω is then constructed using the standard technique of adding to each subdomain Ω_m all the fine elements $\tau_j \in \mathscr{T}^1$ within a distance δ from its boundary $\partial\Omega_m$. We denote by Ω'_m the overlapping subdomain obtained by such extensions of Ω_m. With each subdomain Ω'_m, we associate the following finite-element spaces

$$V_m := \{\mathbf{u}_i \in V^h : \mathbf{u}_i(x) = 0 \; x \in \Omega \backslash \Omega'_m\} \quad \text{and} \quad U_m := V_m \times V_m.$$

Let V^H be the space of trilinear finite elements associated to the coarse triangulation $\mathscr{T}_0$ and define

$$\widetilde{V}^H := \{\mathbf{u}_e \in V^H : \int_\Omega \mathbf{u}_e = 0\}, \quad U_0 = U^H := V^H \times \widetilde{V}^H.$$

We remark that $U_0 \subset U^h$, whereas U_m is not a subset of U^h, $m = 1, ..., N$. Define the interpolation operators $\mathbf{I}_m : U_m \to U^h$, $m = 1, ..., N$, as given

$$\mathbf{u} = (\mathbf{u}_i, \mathbf{u}_e) \in U_m, \quad \mathbf{I}_m \mathbf{u} = (\mathbf{I}_{m,i}\mathbf{u}, \mathbf{I}_{m,e}\mathbf{u}) := \left(\mathbf{u}_i - \int_\Omega \mathbf{u}_e, \mathbf{u}_e - \int_\Omega \mathbf{u}_e\right).$$

The operator $\mathbf{I}_0 : U_0 \to U^h$ is simply the embedding operator. Defining the projection operators

$$\widetilde{\mathbf{T}}_m : U^h \to U_m \quad \text{by } a_{bid}(\widetilde{\mathbf{T}}_m\mathbf{u}, \mathbf{v}) = a_{bid}(\mathbf{u}, \mathbf{I}_m\mathbf{v}) \quad \forall \mathbf{v} \in U_m,$$

and $\mathbf{T}_m = \mathbf{I}_m\widetilde{\mathbf{T}}_m$, we can define the AS operator as

$$T_{AS} = \mathbf{B}_{AS}\mathscr{A} := \mathbf{T}_0 + \mathbf{T}_1 + ... + \mathbf{T}_N. \quad (5.32)$$

It is easy to see that the matrix form of this operator is

$$T_{AS} = \mathbf{B}_{AS}\mathscr{A},$$

where $\mathscr{A}$ is the Bidomain stiffness matrix (5.28) and $\mathbf{B}_{AS}$ is the AS preconditioner

$$\mathbf{B}_{AS} = \sum_{m=0}^N B_m = R_m^T A_m^{-1} R_m. \quad (5.33)$$

Here, for $m = 1, ..., N$, R_m are boolean restriction matrices and A_m the local stiffness matrices for the Bidomain problems restricted to the subdomain Ω'_m, while for

$m = 0$ R_0 is the fine-to-coarse restriction matrix and A_0 is the coarse Bidomain stiffness matrix associated with the coarse space $U_0 = U^H$. More general projection-like operators $\mathbf{T}_m$ associated with approximate bilinear forms and inexact local solvers could be used as well, see [129, 140]. The use of the $\mathbf{B}_{AS}$ preconditioner (5.33) for the iterative solution of the Bidomain discrete system (5.31) can also be regarded as a way for replacing (5.31) with the preconditioned system

$$\mathbf{T}_{AS}\mathbf{u} = \mathbf{g},$$

where $\mathbf{g} = \mathbf{B}_{AS}\mathbf{f}$, which can be accelerated by a Krylov subspace method.

The following result extend the classical overlapping Schwarz analysis to the condition number of $\mathbf{T}_{AS}$, see [85].

Theorem 4. *The condition number of the 2-level additive Schwarz operator $\mathbf{T}_{AS}$ defined in (5.32) is bounded by*

$$\kappa_2(\mathbf{T}_{AS}) \leq C\left(1 + \frac{H}{\delta}\right),$$

with C independent of h, H, δ.

5.3.1.3 The multilevel Additive Schwarz preconditioner

The method described in the previous section uses two levels with a coarse and a fine mesh. The smaller is H, the smaller are the local problems, while the algebraic linear system corresponding to the problem in U_0 becomes larger. We can then apply recursively this two-level technique to the coarse problem and obtain an additive multilevel method for the Bidomain system. We refer to the original work of Dryja and Widlund [41], Zhang [159] and Dryja, Sarkis and Widlund [42] for an overview of these methods for scalar elliptic problems, including multilevel diagonal scaling variants and different choices of coarse spaces.

Let us consider $L \geq 2$ rather than just two mesh levels: on each level $k = 0, 1, \ldots, L-1$, Ω is discretized with a shape-regular mesh $\mathcal{T}^k$ with elements of characteristic size h_k. $\mathcal{T}^k$ is a refinement of $\mathcal{T}^{k-1}$, with level $L-1$ being the finest, hence $h_{L-1} = h$ and $h_0 = H$. On each level $k = 0, \ldots, L-1$, we define a finite-element space V^{h_k}, with $V^{h_{L-1}} = V^h$ and $V^{h_0} = V^H$, and the spaces $\widetilde{V}^{h_k}$, U^{h_k} as in the previous section. We introduce $L-1$ sets of overlapping subdomains $\{\Omega'_{km}\}_{m=1}^{N_k}$ for $k = 1, \ldots, L-1$, such that on each level there is an overlapping decomposition $\Omega = \bigcup_{m=1}^{N_k} \Omega'_{km}$ and we denote with δ_k the overlap at level k. As in [159], we make the following assumption about the sets $\{\Omega'_{km}\}$.

Assumption 5.3.1. *On each level the decomposition $\Omega = \bigcup_{m=1}^{N_k} \Omega'_{km}$ satisfies the following:*

(a) $\partial\Omega'_{km}$ aligns with the boundaries of level k elements, i.e. Ω'_{km} is the union of level k elements. In addition, diameter$(\Omega'_{km}) = O(h_{k-1})$.

(b) The subdomains $\{\Omega'_{km}\}_{m=1}^{N_k}$ form a finite covering of Ω, with a covering constant N_c, i.e. we can color $\{\Omega'_{km}\}_{m=1}^{N_k}$ using at most N_c colors in such a way that subdomains of the same color are disjoint.

(c) There exists a partition of unity $\{\theta_{km}\}$, associated with $\{\Omega'_{km}\}_{m=1}^{N_k}$, which satisfies

$$\sum_m \theta_{km} = 1, \quad with \ \theta_{km} \in H_0^1(\Omega_{km}) \cap C^0(\Omega_{km}), \ 0 \le \theta_{km} \le 1,$$

$$|\nabla \theta_{km}|_{L^\infty} \le C/\delta_k \quad and \quad \|\theta_{km} - \bar{\theta}_{km}\|_{L^\infty(K)} \le C(h_k/\delta_k), \tag{5.34}$$

where $K \in \mathcal{T}^k$ and $\bar{\theta}_{km}$ is the average of θ_{km} on K.

For $k = 1, ..., L-1$ and $m = 1, ..., N_k$, we define the spaces

$$V_{km} := V^{h_k} \cap H^1(\Omega'_{km}), \quad U_{km} = V_{km} \times V_{km}.$$

Let $\mathbf{I}_{km}$ be the interpolation operators as in the two-level case. Defining the projections $\tilde{\mathbf{T}}_{km} : U^h \to U_{km}$ by

$$a_{bid}(\tilde{\mathbf{T}}_{km}\mathbf{u}, \mathbf{v}_k) = a_{bid}(\mathbf{u}, \mathbf{I}_{km}\mathbf{v}_k) \quad \forall \mathbf{v}_k \in U_{km},$$

and $\mathbf{T}_{km} = \mathbf{I}_{km}\tilde{\mathbf{T}}_{km}$, we can introduce the Multilevel Additive Schwarz (MAS) operator

$$\mathbf{T}_{MAS} = \mathbf{B}_{MAS}\mathscr{A} := \mathbf{T}_0 + \sum_{k=1}^{L-1} \sum_{m=1}^{N_k} \mathbf{T}_{km}. \tag{5.35}$$

As mentioned before, approximate local solvers $\tilde{\mathbf{T}}_{km}$ could be used as well. The preconditioner $\mathbf{B}_{MAS}$ is given by

$$\mathbf{B}_{MAS} = R_0^T A_0^{-1} R_0 + \sum_{k=1}^{L-1} \sum_{m=1}^{N_k} R_{km}^T A_{km}^{-1} R_{km},$$

where A_{km} is the local stiffness matrix of the subdomain Ω'_{km} at level k and R_{km} is the restriction matrix from the finest level to subdomain Ω'_{km}, see [129] for more details. The following result for the multilevel case has been proved in [85].

Theorem 5. *The condition number of the multilevel additive Schwarz operator $\mathbf{T}_{MAS}$ defined in (5.35) is bounded:*

$$\kappa_2(\mathbf{T}_{MAS}) \le C \max_{k=1,...,L-1} \left(1 + \frac{h_{k-1}}{\delta_k}\right),$$

where C is a constant independent of the mesh sizes h_k and the number of levels L.

5.4 Numerical results: parallel scalability

The numerical experiments have been performed on the Linux Cluster IBM BCX of the Cineca Consortium (www.cineca.it), with 5120 processors. Our FORTRAN code is based on the parallel library PETSc, from the Argonne National Laboratory [4].

Domain geometry. We solve the Bidomain system on a domain Ω that is either a slab or the image of a Cartesian slab using ellipsoidal coordinates, yielding a portion of a truncated ellipsoid (see Fig. 5.1). This second choice has been used in cardiac simulations with idealized ventricular geometries, see e. g. [29]. These two choices also allow us to test the performance of our multilevel preconditioner in presence of severe domain deformations.

Conductivity coefficients and parameter calibration. The values of the conductivity coefficients used in the numerical tests are

$$\sigma_l^i = 3 \cdot 10^{-3}, \ \sigma_t^i = 3.1525 \cdot 10^{-4}, \ \sigma_n^i = 3.1525 \cdot 10^{-5},$$
$$\sigma_l^e = 2 \cdot 10^{-3}, \ \sigma_t^e = 1.3514 \cdot 10^{-3}, \ \sigma_n^e = 6.757 \cdot 10^{-4} \ (Siemens \ cm^{-1}).$$

and the capacitance per unit volume is set to $c_m = 1$ mF cm^{-3}. The values of the coefficients and parameters in the LR1 model are given in the original paper [74].

Fine and coarse meshes. For both types of domains (Cartesian slabs and truncated ellipsoids), we denote the Cartesian mesh used by $\mathscr{T} = \mathscr{T}_i \times \mathscr{T}_j \times \mathscr{T}_k$, indicating the number of elements in each coordinate direction. This notation applies to both fine and coarse meshes. When we scale up the mesh by a factor c, for brevity we define $c\mathscr{T} = c\mathscr{T}_i \times c\mathscr{T}_j \times c\mathscr{T}_k$, i.e. the number of elements in $c\mathscr{T}$ is c^3 times the number of elements in $\mathscr{T}$.

Stimulation site, initial and boundary conditions. The depolarization process is started by applying a stimulus of $I_{app}^i = 200$ mA/cm^3, $I_{app}^e = -200$ mA/cm^3 lasting 1 ms on the face of the domain modelling the endocardial surface. The initial conditions are at resting values for all the potentials and LR1 gating variables, while the boundary conditions are for insulated tissue.

Linear solver. At each time step, we solve iteratively the discrete Bidomain system (5.27) by PCG with the two-level AS preconditioner. The PCG's initial guess is the discrete solution of the previous time step and the stopping criterion is a 10^{-4} reduction of the residual norm on all levels except the coarse one, where it is 10^{-8} instead. In all runs, we report the PCG condition number, extreme eigenvalues, iteration counts and cpu timings (all in seconds). The extreme eigenvalues are computed by the standard Lanczos procedure during the PCG iteration (see e.g. [140]), yielding close approximations of the exact extreme eigenvalues.

Scaled speedup on ellipsoidal domains. We consider a scaled speedup test on deformed ellipsoidal domains, that represent a severe test for the two-level AS solver. The local size of each subdomain on the finest mesh is kept fixed at the value $32 \times 32 \times 32$ (before adding the overlap) and the number of subdomains (hence processors) is increased from 8 to 1024, forming increasing portions of ellipsoidal domains Ω as shown in Fig 5.1. The fine mesh is chosen proportionally to the coarse mesh as $\mathscr{T}_1 = 16\mathscr{T}_0$ so as to keep the local mesh size on each processor fixed at $32 \times 32 \times 32$, and the overlap size is $\delta = h$. The simulation is run for 3 time steps of 0.05 ms during the depolarization phase.

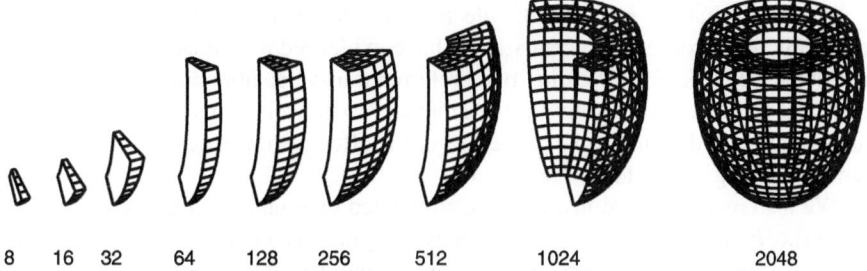

8   16   32   64   128   256   512   1024   2048

Fig. 5.1. Ellipsoidal domains for scaled speedup Test 2 (Table 5.1), decomposed in 8, 16, ..., 2048 subdomains, each one composed of $32 \times 32 \times 32$ finite elements (not shown)

Table 5.1. Test 1. Scaled speedup test on ellipsoidal domains, 3 time steps with $\tau = 0.05$ ms. Condition number κ_2, extreme eigenvalues $\lambda_{max}, \lambda_{min}$, iteration counts it and cpu timings in seconds (*time*) of the PCG solver with AS preconditioner

procs.	$\mathcal{T}_0$	$\mathcal{T}_1 = 16\mathcal{T}_0$	d.o.f.	$\kappa_2 = \lambda_{max}/\lambda_{min}$	it	$time$
8	$2 \times 4 \times 8$	$32 \times 64 \times 128$	0.5M	336.50=20.19/0.06	50.33	14.25
16	$2 \times 8 \times 8$	$32 \times 128 \times 128$	1.1M	196.86=13.78/0.07	44.33	13.12
32	$2 \times 16 \times 8$	$32 \times 256 \times 128$	2.2M	134.12=10.73/0.08	40.67	11.38
64	$2 \times 32 \times 8$	$32 \times 512 \times 128$	4.4M	93.89=8.45/0.09	37.00	11.95
128	$4 \times 32 \times 8$	$64 \times 512 \times 128$	8.6M	109.87=8.79/0.08	38.00	15.06
256	$8 \times 32 \times 8$	$128 \times 512 \times 128$	17.1M	89.89=8.09/0.09	36.33	17.73
512	$16 \times 32 \times 8$	$256 \times 512 \times 128$	34.1M	87.89=7.91/0.09	36.00	17.57
1024	$32 \times 32 \times 8$	$512 \times 512 \times 128$	67.9M	88.11=7.93/0.09	36.00	21.03
2048	$64 \times 32 \times 8$	$1024 \times 512 \times 128$	135.7M	88.67=7.98/0.09	36.00	26.54

Table 5.1 reports the condition number κ_2, the extreme eigenvalues $\lambda_{max}, \lambda_{min}$, and the iteration counts it. and cpu timings of the PCG solver with AS preconditioner. The results clearly show the scalability of the AS operator, since all the reported quantities appear to approach constant values as the number of subdomains (hence processors) increases.

5.5 Numerical results: Bidomain simulations

The previous Bidomain solver has been used for studying the effects of subendocardial ischemic regions on the epicardial surface [31], for assessing the information content of the controversial hybrid monophasic action potentials [32], for studying the effects of cellular heterogeneities on the spatial distributions of repolarization time [35, 37] and for validating extracellular markers of repolarization time [122, 123]. We have recently applied this parallel Bidomain solver to study the three dimensional details of the mechanisms of cardiac excitation elicited by unipolar extracellular stimuli of anodal and cathodal type, see [33, 34].

We recall that the macroscopic electrical properties of the cardiac muscle are markedly anisotropic and recent studies have evidenced a laminar organization of the fibres (see e.g. [71]), yielding two preferred transverse fibre directions, one tangent and the other orthogonal to the laminae, respectively. This geometrical fibre architecture yields orthotropic properties of the effective macroscopic conductivity tensors, see e.g. [61, 144].

We also recall that the heart can be stimulated by a unipolar extracellular electrode (i.e. $I_{app}^i = 0$ and $I_{app}^e \neq 0$ in (5.4)) by applying a cathodal ($I_{app}^e < 0$) or anodal ($I_{app}^e > 0$) pulse in four distinct ways: by turning a negative current on (cathode make) or off (cathode break), or by turning a positive current on (anode make) or off (anode break). In [104, 105] the bidomain model with unequal anisotropic ratio was first proposed and used to establish that the stimulation by a unipolar extracellular cathodal or anodal electrode produces a characteristic transmembrane pattern called *virtual electrodes response*. Only bidomain models with unequal anisotropy ratios of the intra- and extracellular media are able to generate regions known as *virtual electrodes*, see e.g. [124, 153]. The effects and features of make and break excitation mechanism, have been explained in terms of the underlying *virtual electrodes polarization*, and have been subsequently investigated by simulation studies in [45, 91, 100, 106, 107, 108, 109, 110, 116, 117, 128] and by experimental studies in [43, 44, 125, 154]; see also the recent surveys [111, 155].

Bidomain simulations of anode make mechanism. Here we focus on the 3D investigation of the excitation mechanisms associated with an anodal stimulation, which can be generated only in tissues with unequal anisotropy ratio.

The cardiac domain H considered is a Cartesian slab of dimensions $0.96 \times 0.96 \times 0.32\,\mathrm{cm}^3$, modelling a portion of the left ventricular wall and we use as cardiac membrane model the Luo-Rudy LRd model, see [47], augmented with the funny and electroporation currents; see [33] for more details. The conductivity coefficients used are given in Table 5.2 and yield local conductivity velocities of $0.065, 0.03, 0.015\,\mathrm{cm/ms}$ along the fibres, and tangent and orthogonal to the laminae, respectively. These values are in partial agreement with the conductivity coefficient estimates of [61] derived from experimental measurements. The same table reports also the associated unequal anisotropy ratios $\rho_{lt}^{i,e}, \rho_{tn}^{i,e}$. In order to compare the predicted response in unequal and equal anisotropy tissues we also considered an equal anisotropy tissue with conductivity coefficients given in the same Table 5.2 that yield the same conservative values of conductivity velocities, i.e. $0.065, 0.03, 0.015\,\mathrm{cm/ms}$. The cardiac fibres are parallel on each intramural plane parallel to the epicardium and rotate linearly with depth of $120°$ starting from $-45°$ on the epicardium. In order to investigate the 3D shape of the virtual electrode response, we apply an anodal stimulus to a slab initially at rest. Tha anodal stimulus is applied with a duration of 10 *ms* and an amplitude of 0.0648 *mA* in a small region of dimensions $0.06 \times 0.06 \times 0.03\,\mathrm{cm}^3$ (i.e. $I_{app}^i = 0\,\mathrm{mA/cm}^3$, $I_{app}^e = 600\,\mathrm{mA/cm}^3$) at the center of the epicardial face.

Figs. 5.2 and 5.3 show the epicardial distribution of the intracellular potential (top), extracellular potential (middle) and transmembrane potentials (bottom) 2 ms after the anodal stimulus elicited in a slab without and with transmural fibre rotation, respectively. In each figure, the panels on the right column refer to a tissue

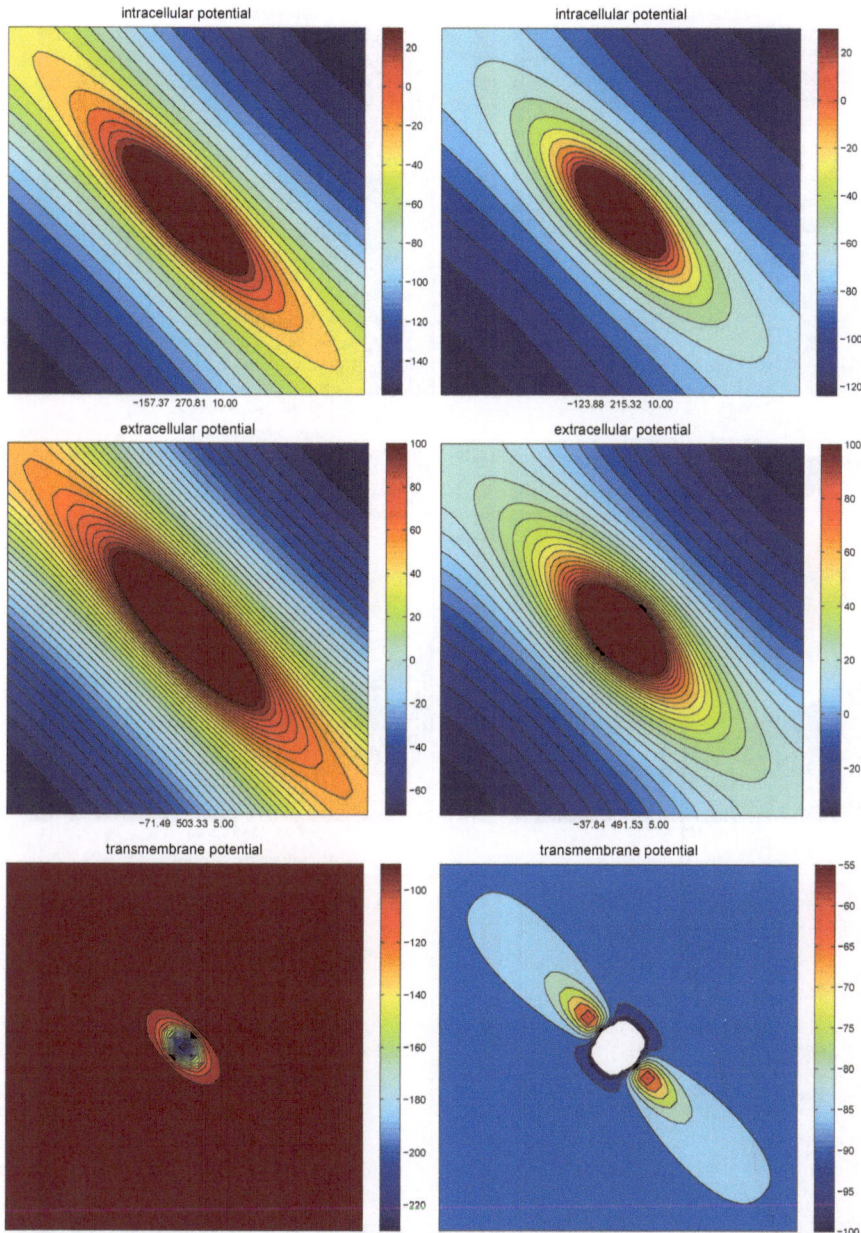

Fig. 5.2. Slab without transmural fibre rotation. Epicardial distribution of the intracellular potential (top), extracellular potential (middle) and transmembrane potentials (bottom) 2 ms after an anodal stimulus. Right panels: tissue with the unequal anisotropy ratios of Table 5.2. Left panels: tissue with equal anisotropy ratio. Below each panel the minimum, maximum and step of the displayed map are reported and the colorbar denotes the range of values of the displayed equipotential lines

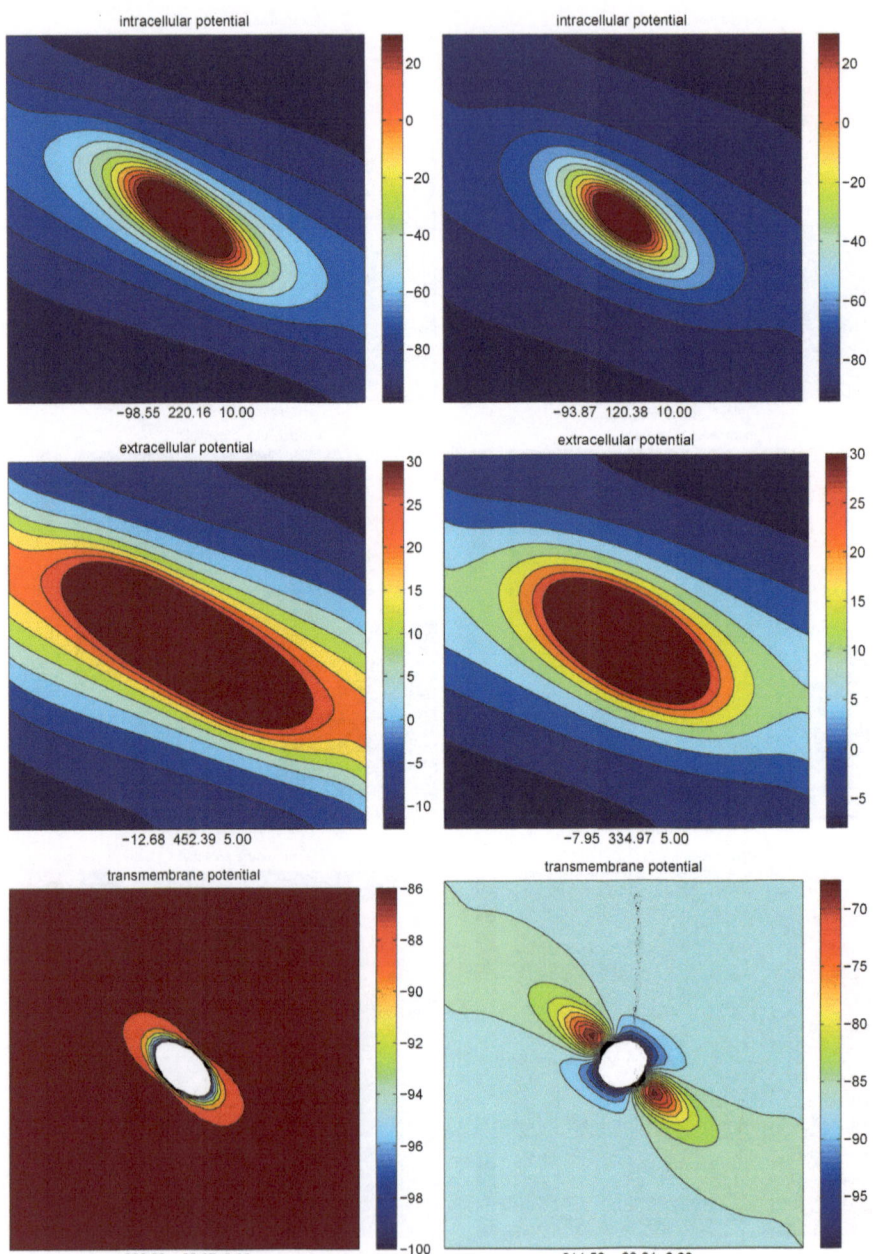

Fig. 5.3. Slab with transmural fibre rotation. Epicardial distribution of the intracellular potential (top), extracellular potential (middle) and transmembrane potentials (bottom) 2 ms after an anodal stimulus. Right panels: tissue with the unequal anisotropy ratios of Table 5.2. Left panels: tissue with equal anisotropy ratio. Below each panel the minimum, maximum and step of the displayed map are reported and the colorbar denotes the range of values of the displayed equipotential lines

Table 5.2. Conductivity coefficients of the intra and extracellular media in *Siemenscm*$^{-1}$ and anisotropy ratios $\rho_{lt}^{i,e} = \sigma_l^{i,e}/\sigma_t^{i,e}$, $\rho_{tn}^{i,e} = \sigma_t^{i,e}/\sigma_n^{i,e}$

	Unequal anisotropy				
medium intra	σ_l^i 2.31724e-3	σ_t^i 2.43504e-4	σ_n^i 5.69083e-5	ρ_{lt}^i 9.51622	ρ_{tn}^i 4.27889
medium extra	σ_l^e 1.54483e-3	σ_t^e 1.04385e-3	σ_n^e 3.7221e-4	ρ_{lt}^e 1.47993	ρ_{tn}^e 2.80447
	Equal anisotropy				
medium intra	σ_l^i 1.34511e-3	σ_t^i 3.36278e-4	σ_n^i 8.40695e-5	ρ_{lt}^i 4	ρ_{tn}^i 4
medium extra	σ_l^e 5.35282e-3	σ_t^e 1.3382e-3	σ_n^e 3.34551e-4	ρ_{lt}^e 4	ρ_{tn}^e 4

with the unequal anisotropy ratios of Table 5.2, while the panels on the left column refer to a tissue slab with equal anisotropy ratio. In the left panel of Fig. 5.2, the elliptical intra- and extracellular potential patterns have the same major to minor semi-axis ratio, yielding a transmembrane potential pattern (bottom left panel) with more rounded elliptical lines around the anodal electrode. Conversely, in a slab with unequal anisotropy ratio, the intra- and extracellular equipotential contour lines, displayed in the top and middle right panels of Fig. 5.2, have elliptical shapes with different ratio of the two semi-axes. The difference between these two epicardial patterns yields a transmembrane potential distribution with the typical virtual electrode response (Fig. 5.2, bottom right panel). In fact, the tissue within and around the anodal electrode is negatively polarized (hyperpolarized) and it exhibits an epicardial *dog-bone* shape, developing perpendicularly to the epicardial fibre direction of $-45°$. Two regions of positive polarization (depolarization), i.e. two virtual cathodes, develop along the fibre direction adjacent to the concave parts of the epicardial virtual anode boundary.

The same anodal stimulation was applied to a tissue slab with transmural fibre rotation. In case of unequal anisotropy ratio, see the right panels of Fig. 5.3, the epicardial transmembrane potential pattern exhibits a *virtual electrode response* with a twisted hyperpolarized *dog-bone* shaped region, which is not symmetric with respect to the epicardial fibre direction because of the counterclockwise fibre rotation from epicardium to endocardium. In fact, the equipotential surfaces inside the virtual cathode volumes, shown in Fig. 5.4, are shaped as two horns pointing counterclockwise when proceeding from the upper (epicardial) face to the lower (endocardial) face. On intramural sections parallel to the epicardial face, the equipotential lines inside the virtual anode preserve the same *dog-bone* shape as on the epicardium but with a counterclockwise rotation, thus reducing their area and yielding a twisted tote bag; see Fig. 5.4. On the contrary, the same anodal stimulus applied to a slab with equal anisotropy ratios and transmural fibre rotation does not yield virtual cathodal

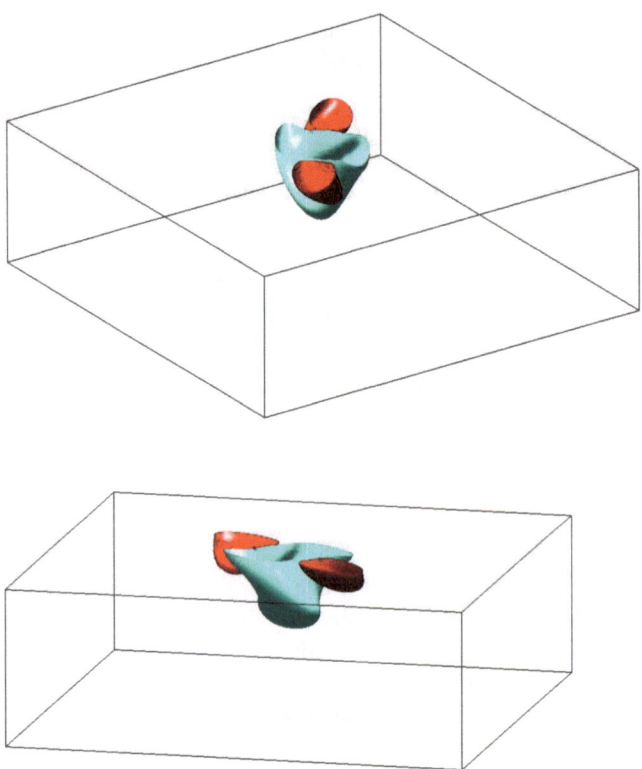

Fig. 5.4. Slab with transmural fibre rotation. Anodal stimulation applied to a slab with unequal anisotropy ratio. Two different views of the isopotential surfaces of the transmembrane potential distribution of values −88 and −75 mV, 2 ms after the beginning of the stimulation

regions but only an slightly twisted elliptical hyperpolarized region centered at the stimulation site; see Fig. 5.3, bottom left panel. This confirms that *virtual electrodes* can only be generated by bidomain models with unequal anisotropy ratios of the intra- and extracellular media. The formation of the two virtual cathodes elicited by the anodal stimulation explains why an anodal stimulation is able to excite the myocardium, as shown by the excitation sequence shown in Fig. 5.5 and described below.

Excitation wavefront sequence elicited by anode make mechanism. The isochrones of activation time on the epicardial surface and transmural diagonal sections displayed in Fig. 5.5 show clearly that the first-triggered areas are the two epicardial sites located at the center of the virtual cathodes, generating two distinct excitation wavefronts propagating outward and inward along the diagonal parallel to the fibre direction. When the excitation isochrones reach points lying on a boundary of the virtual anode, a block of the inward propagation takes place since this region is inexcitable until the stimulus is turned off. Therefore, from about 2 to 6 ms, two distinct activation wavefronts propagate outwards on the epicardial surface, moving faster

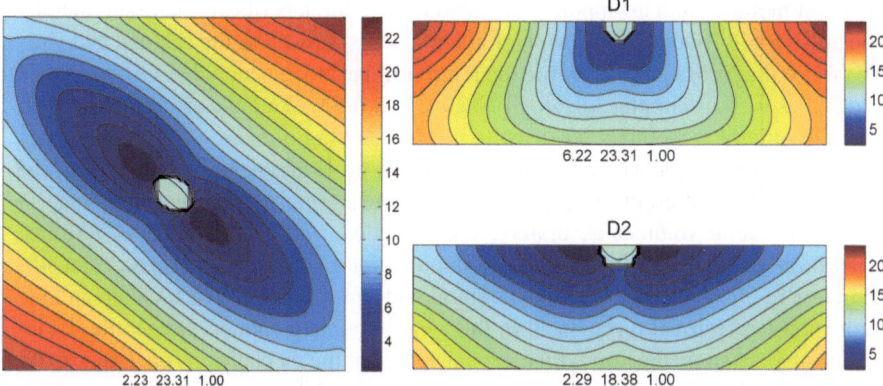

Fig. 5.5. Isochrones of activation time on the epicardial surface (left) and the two transmural diagonal sections (right) elicited by anode make stimulation in tissue with the unequal anisotropy ratios (Fig. 5.3, right column). Below each panel the minimum, maximum and step of the displayed map are reported

along fibres, and also across the slab thickness, and subsequently collide. These two 3D wavefronts are shaped as open surfaces, each having a closed curve as rim lying on the epicardial surface. At about 12 ms, the central epicardial area of the virtual anode is suddenly activated, since it becomes excitable only after the stimulus is turned off. Subsequently, only a unique large wavefront remains.

The epicardial and intramural excitation pathways start from the virtual cathodes, proceed turning around the boundary of the virtual anode volume and point toward its epicardial center. These pairs of pathways, coming from the virtual cathodes, cause the merging of the two wavefronts by forming an activated tissue volume bounded by two intramural surfaces, one consisting of a wavefront by moving toward the epicardial boundary and intramurally toward the endocardial face, and the other being the conduction block surface surrounding the inexcitable region, which is suddenly activated after the stimulus is turned off. From the inspection of the transmural isochrones pattern displayed in Fig. 5.5, we observe that the major transmural effect of the virtual electrode polarization is the propagation of two wavefronts, separated by the transmural part of the virtual anode region, that merge about 5 ms after the beginning of epicardial excitation at a 1 *mm* depth.

5.6 Conclusions

We have presented the main mathematical models used in computational electrocardiology to describe the complex multiscale structure of the bioelectrical activity of the heart, from the microscopic activity of ion channels of the cellular membrane to the macroscopic properties of the anisotropic propagation of excitation and recovery fronts in the whole heart. We have described how reaction-diffusion systems can be rigorously derived from microscopic models of cellular aggregates by homoge-

nization methods and asymptotic expansions. The models of cardiac tissue include
the anisotropic Bidomain and Monodomain models, as well as the Eikonal and various relaxed approximations, while the ionic cellular models include Luo-Rudy-type
models as well as simpler FitzHugh-Nagumo variants. We have also presented advanced numerical methods for discretizing and solving numerically these complex
models on three-dimensional domains, using adaptive and parallel techniques. The
resulting solvers are able to reproduce complete normal heartbeat phenomena in
large ventricular volumes accurately, simulating e.g. various potential waveforms,
activation and recovery fronts, and action potential dispersion. The solvers can also
simulate re-entry phenomena such as spiral and scroll waves, their breakup and the
transition to electrical turbulence. In this work, we have applied the parallel Bidomain solver to study the three-dimensional details of the mechanisms of cardiac excitation elicited by unipolar extracellular stimuli of anodal type, yielding new insights
on the origin of excitation based on the virtual electrode polarization. Current work
is investigating the role of inhomogeneities of the tissue and heterogeneity of the
cellular membrane properties, due e.g. to ischemia, and the coupling of electrocar-
diological models with mechanical and fluid dynamic models, with the future goal
of integrating them with cardiovascular and circulatory models.

Acknowledgements. The authors would like to thank Bruno Taccardi for introducing them to the
field of Mathematical Physiology and for many stimulating discussions.

References

[1] Aliev R.R., Panfilov A.V.: A simple two-variable model of cardiac excitation. Chaos Sol.
Fract. **7**: 293–301, 1996.
[2] Ambrosio L., Colli Franzone P., Savaré G.: On the asymptotic behaviour of anisotropic
energies arising in the cardiac bidomain model. Interface Free Bound. **2**(3): 213–266, 2000.
[3] Austin T.M., Trew M.L., Pullan A.J.: Solving the cardiac Bidomain equations for discontinuous conductivities. IEEE Trans. Biomed. Eng. **53**(7): 1265–1272, 2005.
[4] Balay S., Buschelman K., Gropp W.D., Kaushik D., Knepley M., Curfman McInnes L.,
Smith B.F., Zhang H.: PETSc Users Manual.Tech. Rep. ANL-95/11 – Revision 2.1.5, Argonne National Laboratory, 2002.
[5] Bassetti F.: Variable time-step discretization of degenerate evolution equations in Banach
space. Numer. Funct. Anal. Optim. **24**(3–4): 391–426, 2003.
[6] Belhamadia Y.: A Time-Dependent Adaptive Remeshing for Electrical Waves of the Heart.
IEEE Tran. Biomed. Eng. **55**(2): 443–452, 2008.
[7] Belhamadia Y., Fortin A., Bourgault Y.: Towards accurate numerical method for monodomain models using a realistic heart geometry. Math. Biosci. **220**(2): 89–101, 2009.
[8] Bellettini G., Colli Franzone P., Paolini M.: Convergence of front propagation for
anisotropic bistable reaction-diffusion equations. Asymp. Anal. **15**: 325–358, 1997.
[9] Bellettini G., Paolini M.: Anisotropic motion by mean curvature in the context of Finsler
geometry. Hokkaido Math. J. **25**: 537–566, 1996.
[10] Bendahmane M., Karlsen K.H.: Analysis of a class of degenerate reaction–diffusion systems
and the bidomain model of cardiac tissue. Netw. Heterog. Media **1**(1): 185–218, 2006.
[11] Bendahmane M., Karlsen K.H.: Convergence of a finite volume scheme for the bidomain
model of cardiac tissue. Appl. Numer. Math. **59**(9): 2266–2284, 2009.

[12] Bensoussan A., Lions J.-L., Papanicolaou G.: Asymptotic Analysis for Periodic Structures. North-Holland, Amsterdam, 1978.

[13] Bordas R., Carpentieri B., Fotia G., Maggio F., Nobes R., Pitt-Francis J., Southern J.: Simulation of cardiac electrophysiology on next-generation high-performance computers. Phil. Trans. R. Soc. A **367**(1895): 1951–1969, 2009.

[14] Boulakia M., Fernandez M.A., Gerbeau J.-F., Zemzemi N.: A coupled system of PDEs of ODEs arising in electrocardiograms models. Appl. Math. Res. Express. AMRX, 2: Art. ID abn002, 28, 2008.

[15] Bourgault Y., Coudiere Y., Pierre C.: Existence and uniqueness of the solution for the bidomain model used in cardiac electrophysiology. Nonlinear Anal.-Real World Appl. **10**(1): 458–482, 2009.

[16] Cherry E.M., Greenside H.S., Henriquez C.S.: Efficient simulation of three-dimansional anisotropic cardiac tissue using an adaptive mesh refinement method. Chaos **13**: 853–865, 2003.

[17] Clayton R.H., Panfilov A.V.: A guide to modelling cardiac electrical activity in anatomically detailed ventricles. Progr. Biophys. Molec. Biol. **96**: 19–43, 2008.

[18] Clayton R.H., et al.: Models of cardiac tissue electrophysiology: Progress, challenges and open questions. Progr. Biophys. Molec. Biol. **104**: 22–48, 2011.

[19] Clements J.C., Nenonen J., Li P.K.J., Horacek B.M.: Activation dynamics in anisotropic cardiac tissue via decoupling. Ann. Biomed. Eng. **32**(7): 984–990, 2004.

[20] Colli Franzone P., Deuflhard P., Erdmann B., Lang J., Pavarino L.F.: Adaptivity in space and time for reaction-diffusion systems in Electrocardiology. SIAM J. Sci. Comput. **28**(3): 942–962, 2006.

[21] Colli Franzone P., Guerri L., Rovida S.: Wavefront propagation in an activation model of the anisotropic cardiac tissue: Asymptotic analysis and numerical simulations. J. Math. Biol. **28**: 121–176, 1990.

[22] Colli Franzone P., Guerri L., Tentoni S.: Mathematical modeling of the excitation process in myocardial tissue: Influence of fibre rotation on wavefront propagation and potential field. Math. Biosci. **101**: 155–235, 1990.

[23] Colli Franzone P., Guerri L.: Spread of excitation in 3-D models of the anisotropic cardiac tissue I: Validation of the eikonal approach. Math. Biosci. **113**: 145–209, 1993.

[24] Colli Franzone P., Guerri L., Pennacchio M., Taccardi B.: Spread of excitation in 3-D models of the anisotropic cardiac tissue II: Effects of fibre architecture and ventricular geometry. Math. Biosci. **147**: 131–171, 1998.

[25] Colli Franzone P., Guerri L., Pennacchio M., Taccardi B.: Spread of excitation in 3-D models of the anisotropic cardiac tissue III: Effects of ventricular geometry and fibre structure on the potential distribution. Math. Biosci. **151**: 51–98, 1998.

[26] Colli Franzone P., Guerri L., Pennacchio M., Taccardi B.: Anisotropic mechanisms for multiphasic unipolar electrograms. Simulation studies and experimental recordings. Ann. Biomed. Eng. **28**: 1–17, 2000.

[27] Colli Franzone P., Savaré G.: Degenerate evolution systems modeling the cardiac electric field at micro and macroscopic level. In Evolution equations, Semigroups and Functional Analysis, A. Lorenzi and B. Ruf Editors, pp. 49–78, Birkhäuser, 2002.

[28] Colli Franzone P., Guerri L., Taccardi B.: Modeling ventricular excitation: axial and orthotropic effects on wavefronts and potentials. Math. Biosci. **188**: 191–205, 2004.

[29] Colli Franzone P., Pavarino L.F.: A parallel solver for reaction-diffusion systems in computational electrocardiology. Math. Mod. Meth. Appl. Sci. **14**(6): 883–911, 2004.

[30] Colli Franzone P., Pavarino L.F., Savarè G.: Computational Electrocardiology: mathematical and numerical modeling, in Complex Systems in Biomedicine, A. Quarteroni et al. (eds.), Springer, pp. 187–241, 2006.

[31] Colli Franzone P., Pavarino L.F., Scacchi S.: Dynamical effects of myocardial ischemia in anisotropic cardiac models in three dimensions. Math. Mod. Meth. Appl. Sci. **17**(12): 1965–2008, 2007.

[32] Colli Franzone P., Pavarino L.F., Scacchi S., Taccardi B.: Monophasic action potentials generated by bidomain modeling as a tool for detecting cardiac repolarization times. Am. J. Physiol. (Heart Circ. Physiol.) **293**: H2771–H2785, 2007.

[33] Colli Franzone P., Pavarino L.F., Scacchi S.: Exploring anodal and cathodal make and break cardiac excitation mechanisms in a 3D anisotropic Bidomain model. Math. Biosci. **230**(2): 96–114, 2011.

[34] Colli Franzone P., Pavarino L.F., Scacchi S.: Anode make and break excitation mechanisms and strength–interval curves: bidomain simulations in 3D rotational anisotropy. In FIMH 2011, D.N. Metaxas and L. Axel (eds.), LNCS 6666: 1–10, Springer-Verlag Berlin Heidelberg 2011.

[35] Colli Franzone P., Pavarino L.F., Scacchi S., Taccardi B.: Modeling ventricular repolarization: effects of transmural and apex-to-base heterogeneities in action potential durations. Math. Biosci. **214**(1–2): 140–152, 2008.

[36] Colli Franzone P., Pavarino L.F., Taccardi B.: Simulating patterns of excitation, repolarization and action potential duration with cardiac Bidomain and Monodomain models. Math. Biosci. **197**: 35–66, 2005.

[37] Colli Franzone P., Pavarino L.F., Taccardi B.: Effects of transmural electrical heterogeneities and electrotonic interactions on the dispersion of cardiac repolarization and action potential duration: A simulation study. Math. Biosci. **204**(1): 132–165, 2006.

[38] Costa K.D., May-Newman K., Farr D., O'Dell W.G., McCulloch A.D., Omens J.H.: Three-dimensional residual strain in midanterior canine left ventricle. Am. J. Physiol. (Heart Circ. Physiol.) **42**: H1968–H1976, 1997.

[39] Coudiere Y., Pierre C.: Stability and convergence of a finite volume method for two systems of reaction-diffusion equations in electro-cardiology. Nonlinear Anal.-Real World Appl. **7**(4): 916–935, 2006.

[40] Deuflhard P., Erdmann B., Roitzsch R., T G.: Lines. Adaptive finite element simulation of ventricular fibrillation dynamics. Comput. Visual. Sci. **12**(5): 201–205, 2009.

[41] Dryja M., Widlund O.B.: Multilevel additive methods for elliptic finite element problems. In Parallel algorithms for partial differential equations (Kiel, 1990). Notes Numer. Fluid Mech. **31**: 58–69, 1991.

[42] Dryja M., Sarkis M.V., Widlund O.B.: Multilevel Schwarz methods for elliptic problems with discontinuous coefficients in three dimensions. Numer. Math. **72**(3): 313–348, 1996.

[43] Efimov I.R., Gheng Y., Van Eagoner D.R., Mazgalev T., Tchou P.J.: Virtual electrode-induced phase singularity: a basic mechanism of defibrillation failure. Circ. Res. **82**: 918–925, 1998.

[44] Efimov I.R., Gray R.A., Roth B.J.: Virtual electrodes and deexcitation: new insights into fibrillation induction and defibrillation. J. Cardiovasc Electrophysiol. **11**: 339–353, 2000.

[45] Entcheva E., Eason J., Efimov I.R., Cheng Y., Malkin R., Clayton F.: Virtual electrode effects in transvenous defibrillation-modulation by structure and interface: evidence from bidomain simulations an optical mapping. J. Cardiovasc. Electrophysiol. **9**: 949–961, 1998.

[46] Ethier M., Bourgault Y.: Semi-implicit time-discretization schemes for the Bidomain model. SIAM J. Numer. Anal. **46**(5): 2443–2468, 2008.

[47] Faber G.M., Rudy Y.: Action potential and contractility changes in [Na+](i) overloaded cardiac myocytes: a simulation study. Biophys. J. **78**(5): 2392–2404, 2000.

[48] Fenton F.H., Cherry E.M., Hastings H.M., Evans S.J.: Multiple mechanisms of spiral wave breakup in a model of cardiac electrical activity. Chaos **12**(3): 852–892, 2002.

[49] Fenton F.H., Karma A.: Vortex dynamics in three-dimensional continuous myocardium with fibre rotation: filament instability and fibrillation. Chaos **8**: 20–47, 1998.

[50] FitzHugh R.: Impulses and physiological states in theoretical models of nerve membrane. Biophys. J. **1**: 445–466, 1961.

[51] Fischer G., Tilg B., Modre R., Huiskamp G., Fetzer J., Rucker W., Wach P.: A bidomain model based BEM-FEM coupling formulation for anisotropic cardiac tissue. Ann. Biomed. Eng. **28**(10): 1229–1243, 2000.

[52] Gerardo Giorda L., Mirabella L., Nobile F., Perego M., Veneziani A.: A model-based block-triangular preconditioner for the Bidomain system in electrocardiology. J. Comp. Phys. **228**(10): 3625–3639, 2009.

[53] Geselowitz D.B., Miller W.T.: A bidomain model for anisotropic cardiac muscle. Ann. Biomed. Eng. **11**: 191–206, 1983.

[54] Gulrajani R.M., Roberge F.A., Savard P.: The inverse problem of electrocardiography. In Comprehensive Electrocardiology, P.W. Macfarlane and T.T.V. Lawrie (eds.), I: ch. 9, pp. 237–288, Pergamon Oxford, 1989.

[55] Harrild D.M., Henriquez C.S.: A finite volume model of cardiac propagation. Ann. Biomed. Eng. **28**(2): 315–334, 1997.

[56] Heidenreich E.A., Rodriguez J.F., Gaspar F.J., Doblaré M.: Fourth-order compact schemes with adaptive time step for monodomain reaction–diffusion equations. J. Comput. Appl. Math. **216**(1): 39–55, 2008.

[57] Henriquez C.S.: Simulating the electrical behavior of cardiac tissue using the bidomain model. Crit. Rev. Biomed. Eng. **21**: 1–77, 1993.

[58] Henriquez C.S., Muzikant A.L., Smoak C.K.: Anisotropy, fibre curvature, and bath loading effects on activation in thin and thick cardiac tissue preparations: Simulations in a three-dimensional bidomain model. J. Cardiovasc. Electrophysiol. **7**(5): 424–444, 1996.

[59] Hodgkin A., Huxley A.: A quantitative description of membrane current and its application to conduction and excitation in nerve. J. Physiol. (Lond.) **117**: 500–544, 1952.

[60] Hooke N.: Efficient simulation of action potential propagation in a bidomain. Ph. D. Thesis, Duke Univ., Dept. of Comput. Sci., 1992.

[61] Hooks D.A., Trew M.L.: Construction and validation of a plunge electrode array for three-dimensional determination of conductivity in the heart. IEEE Trans. Biomed. Eng. **55**(2): 626–635, 2008.

[62] Hoyt R.H., Cohen M.L., Saffitz J.E.: Distribution and three-dimensional structure of inter-cellular junctions in canine myocardium. Circ. Res. **64**: 563–574, 1989.

[63] Hunter P. et al.: A vision and strategy for the virtual physiological human in 2010 and beyond. Phil. Trans. R. Soc. A **368**: 2595–2614, 2010.

[64] Keener J.P.: An eikonal-curvature equation for action potential propagation in myocardium. J. Math. Biol. **29**: 629–651, 1991.

[65] Keener J.P.: Direct activation and defibrillation of cardiac tissue, J. Theor. Biol. **178**: 313–324, 1996.

[66] Keener J.P., Panfilov A.V.: Three-Dimensional propagation in the heart: the effects of geometry and fibre orientation on propagation in myocardium. In Cardiac Electrophysiology: From Cell to Bedside, D.P. Zipes and J. Jalife (eds.), W.B. Sounders Co, Philadelphia, pp. 335-347, 1995.

[67] Keener J.P., Panfilov A.V.: The effects of geometry and fibre orientation on propagation and extrcellular potentials in myocardium. In Computational Biology of the Heart, A.V. Panfilov, and A.V. Holden (eds.), John Wiley & Sons, New York, Chapter 8, pp. 235–258, 1997.

[68] Keener J.P., Bogar K.: A numerical method for the solution of the bidomain equations in cardiac tissue. Chaos **8**: 234–241, 1998.

[69] Keener J.P., Sneyd J.: Mathematical Physiology. Springer-Verlag, New York 1998.

[70] Krassowska W., Neu J.C.: Effective boundary conditions for syncytial tissue. IEEE Trans. Biomed. Eng. **41**: 143–150, 1994.

[71] LeGrice I.J., Smaill B.H., Chai L.Z., Edgar S.G., Gavin J.B., Hunter P.J.: Laminar structure of the heart: ventricular myocyte arrangement and connective tissue architecture in the dog. Am. J. Physiol. (Heart Circ. Physiol.) **269**(38): H571–H582, 1995.

[72] LeGrice I.J., Smaill B.H., Hunter P.J.: Laminar structure of the heart: a mathematical model. Am. J. Physiol. (Heart Circ. Physiol.) **272**(41): H2466–H2476, 1997.

[73] Linge S., Sundnes J., Hanslien M., Lines G.T., Tveito A.: Numerical solution of the bidomain equations. Phil. Trans. R. Soc. A **367**(1895): 1931–1950, 2009.

[74] Luo C., Rudy Y.: A model of the ventricular cardiac action potential: depolarization, repolarization, and their interaction. Circ. Res. **68**(6): 1501–1526, 1991.

[75] Mardal K.-A., Nielsen B.F., Cai X., Tveito A.: An order optimal solver for the discretized bidomain equations. Numer. Linear Algebra Appl. **14**(2): 83–98, 2007.

[76] Miller W.T., Geselowitz D.B.: Simulation studies of the electrocardiogram I. The normal heart. Circ. Res. **43**(2): 301–315, 1978.

[77] Murillo M., Cai X.C.: A fully implicit parallel algorithm for simulating the non-linear electrical activity of the heart. Numer. Linear Algebr. Appl. **11**(2–3): 261–277, 2004.

[78] Munteanu M., Pavarino L.F.: Implicit parallel solvers in computational electrocardiology. in Applied Analysis and Differential Equations, O. Carja and I. I. Vrabie (eds.), World Scientific, pp. 255–266, 2007.

[79] Munteanu M., Pavarino L.F.: Decoupled Schwarz algorithms for implicit discretization of nonlinear Monodomain and Bidomain systems. Math. Mod. Meth. Appl. Sci. **19**(7): 1065–1097, 2009.

[80] Munteanu M., Pavarino L.F., Scacchi S.: A scalable Newton-Krylov-Schwarz method for the Bidomain reaction-diffusion system. SIAM J. Sci. Comput. **31**(5): 3861–3883, 2009.

[81] Noble D., Rudy Y.: Models of cardiac ventricular action potentials: iterative interaction between experiment and simulation. Phil. Trans. R. Soc. A **359**: 1127–1142, 2001.

[82] Oleinik O.A., Shamaev A.S., Yosifian G.A.: Mathematical problems in elasticity and homogenization. North-Holland, Amsterdam, 1992.

[83] Osher S., Fedkin R.: Level set methods and dynamic implicit surfaces. Applied Mathematical Sciences, vol. 153, Springer-Verlag, New York, 2003.

[84] Panfilov A.V.: Spiral breakup as a model ov ventricular fibrillation. Chaos **8**: 57–64, 1998.

[85] Pavarino L.F., Scacchi S.: Multilevel additive Schwarz preconditioners for the Bidomain reaction-diffusion system. SIAM J. Sci. Comput. **31**(1): 420–443, 2008.

[86] Pennacchio M., Savarè G., Colli Franzone P.: Multiscale modeling for the electrical activity of the heart. SIAM J. Math. Anal. **37**(4): 1333–1370, 2006.

[87] Pennacchio M., Simoncini V.: Efficient algebraic solution of reaction–diffusion systems for the cardiac excitation process. J. Comput. Appl. Math. **145**(1): 49–70, 2002.

[88] Pennacchio M., Simoncini V.: Algebraic multigrid preconditioners for the bidomain reaction-diffusion system. Appl. Numer. Math. **59**(12): 3033–3050, 2009.

[89] Plank G., Burton R.A.B., Hales P., Bishop M., Mansoori T., Bernabeu M.O., Garny A., Prassl A.J., Bollendorsff C., Mason F., Mahmood F., Rodriguez B., Grau V., Schneider J.E., Gavaghan D., Kohl P.: Generation of histo-anatomically representative models of the individual heart: tools and application. Phil. Trans. R. Soc. A **367**(1895): 2257–2292, 2009.

[90] Plank G., Liebmann M., Weber dos Santos R., Vigmond E.J., Haase G.: Algebraic Multigrid Preconditioner for the Cardiac Bidomain Model. IEEE Trans. Biomed. Eng. **54**(4): 585–596, 2007.

[91] Plank G., Prassl A., Hofer E., Trayanova N.A.: Evaluation intramural virtual electrodes in the myocardial wedge preparation: simulations of experimental conditions. Biophys. J. **94**: 1904–1915, 2008.

[92] Plonsey R.: Bioelectric sources arising in excitable fibres (Alza lecture), Ann. Biomed. Eng. **16**: 519–546, 1988.

[93] Plonsey R., Heppner D.: Consideration of quasi-stationarity in electrophysiological systems. Bull. Math. Biophys. **29**: 657–664, 1967.

[94] Potse M., Dubè B., Richer J., Vinet A., Gulrajani R.: A comparison of Monodomain and Bidomain reaction–diffusion models for action potential propagation in the human heart. IEEE Trans. Biomed. Eng. **53**(12): 2425–2434, 2006.

[95] Potse M., Vinet A., Opthof T., Coronel R.: Validation of simple model for the morphology of the T wave in unipolar electrograms. Am. J. Physiol. HeartCirc. Physiol. **297**: H792–H801, 2009.

[96] Pullan A.J., Buist M.L., Cheng L.K.: Mathematical Modelling and Electrical Activity of the Heart: From Cell to Body Surface and back Again. World Scientific, Singapore, 2005.

[97] Puwal S., Roth B.J.: Forward Euler stability of the bidomain model of cardiac tissue. IEEE Trans. Biomed. Eng. **54**(5): 951–953, 2007.

[98] Qu Z., Garfinkel A.: An advanced algorithm for solving partial differential equation in cardiac conduction. IEEE Trans. Biomed. Eng. **46**: 1166–1168, 1999.

[99] Quan W., Evans S.J., Hastings H.M.: Efficient integration of a realistic two-dimensional cardiac tissue model by domain decomposition. IEEE Trans. Biomed. Eng. **45**: 372–385, 1998.

[100] Ranjan R., Tomaselli G.F., Marban E.: A novel mechanism of anode-break stimulation predicted by bidomain modeling. Circ. Res. **84**: 153–156, 1999.

[101] Rogers J.M., McCulloch A.D.: A collocation-Galerkin finite element model of cardiac action potential propagation. IEEE Trans. Biomed. Eng. **41**: 743–757, 1994.

[102] Roth B.J.: Approximate analytic solutions to the bidomain equations with unequal anisotropy ratio. Phys. Rev. E **55**(2): 1819–1826, 1997.

[103] Roth B.J.: How the anisotropy of the intracellular and extracellular conductivities influence stimulation of cardiac muscle. J. Math. Biol. **30**: 633–646, 1992.

[104] Roth B.J., Wikswo J.P. Jr.: Electrical stimulation of cardiac tissue: a bidomain model with active membrane properties. IEEE Trans. Biomed. Eng. **41**(3): 232–240, 1994.

[105] Roth B.J.: A mathematical model of make and break electrical stimulation of cardiac tissue by a unipolar anode or cathode. IEEE Trans. Biomed. Eng. **42**: 1174–1184, 1995.

[106] Roth B.J.: Strength-Interval curve for cardiac tissue predicted using the bidomain model. J. Cardiovasc. Electrophysiol. **7**: 722–737, 1996.

[107] Roth B.J.: Nonsustained reentry following successive stimulation of cardiac tissue through a unipolar electrode. J. Cardiovasc. Electrophysiol. **8**: 768–778, 1997.

[108] Roth B.J., Lin S.-F., Wikswo J.P. Jr.: Unipolar stimulation of cardiac tissue. J. Electrocardiol. **31**(Suppl.): 6–12, 1998.

[109] Roth B.J., Chen J.: Mechanism of anode break excitation in the heart: the relative influence of membrane and electrotonic factors. J. Biol. Systems **7**(4): 541–552, 1999.

[110] Roth B.J., Patel S.G.: Effects of elevated extracellular potassium ion concentration on anodal excitation of cardiac tissue. J. Cardiovasc. Electrophysiol. **14**: 1351–1355, 2003.

[111] Janks D.L., Roth B.J.: The bidomain theory of pacing. In Cardiac Bioelectric Therapy, Efimov I.R., Kroll M.W. and Tchou (eds.), Ch. 2.1, 63–83, Springer Science+Business Media, LLc, 2009.

[112] Rudy Y., Oster H.S.: The electrocardiographic inverse problem. Crit. Rev. Biomed. Eng. **20**: 25–45, 1992.

[113] Rudy Y., Silva J.R.: Computational biology in the study of cardiac ion channels and cell electrophysiology. Quart. Rev. Biophys. **39**(1): 57–116, 2006.

[114] Saffitz J.E., Kanter H.L., Green K.G., Tolley T.K., Beyer E.C.: Tissue-specific determinants of anisotropic conduction velocity in canine atrial and ventricular myocardium. Circ. Res. **74**: 1065–1070, 1994.

[115] Saleheen H.I., Ng K.T.: A new three-dimensional finite-difference bidomain formulation for inhomogeneous anisotropic cardiac tissues. IEEE Trans. Biomed. Eng. **45**(1): 15–25, 1998.

[116] Sambelashvili A., Efimov I.R.: Dynamics of virtual electrode-induced scroll-wave reentry in a 3D bidomain model. Am. J. Physiol Heart Circ. Physiol. **287**: H1570–H1581, 2004.

[117] Sambelashvili A., Nikolski V.P., Efimov I.R.: Virtual electrode theory explains pacing threshold increase caused by cardiac tissue damage. Am. J. Physiol Heart Circ. Physiol. **286**: H2183–H2194, 2004.

[118] Sanchez-Palencia E., Zaoui A.: Homogenization Techniques for Composite Media. Lectures Notes in Physics, volume 272. Springer-Verlag; Berlin; 1987

[119] Sanfelici S.: Convergence of the Galerkin approximation of a degenerate evolution problem in electrocardiology. Numer. Meth. Part. Diff. Eq. **18**: 218–240, 2002.

[120] Scacchi S.: A hybrid multilevel Schwarz method for the bidomain model. Comp. Meth. Appl. Mech. Engrg. **197**(45–48): 4051–4061, 2008.

[121] Scacchi S.: A multilevel hybrid Newton-Krylov-Schwarz method for the Bidomain model of electrocardiology. Comp. Meth. Appl. Mech. Engrg. **200**(5–8): 717–725, 2011.

[122] Scacchi S., Colli Franzone P., Pavarino L.F., Taccardi B.: A reliability analysis of cardiac repolarization time markers. Math. Biosci. **219**(2): 113–128, 2009.

[123] Scacchi S., Colli Franzone P., Pavarino L.F., Taccardi B.: Computing cardiac recovery maps from electrograms and monophasic action potentials under heterogeneous and ischemic conditions. Math. Mod. Meth. Appl. Sci. **20**(7): 1089–1127, 2010.

[124] Sepulveda N.G., Roth B.J., Wikswo J.P. Jr.: Current injection into a two-dimensional anisotropic bidomain. Biophys. J. **55**: 987–999, 1989.

[125] Sidorov V.Y., Woods M.C., Baudenbacher P., Baudenbacher F.: Examination of stimulation mechanism and strength-interval curve in cardiac tissue. Am. J. Physiol Heart Circ. Physiol. **289**: H2602–H2615, 2005.

[126] Skouibine K.B., Krassowska W.: Increasing the computational efficiency of a bidomain model of defibrillation using a time-dependent activating function. Ann. Biomed. Eng. **28**: 772–780, 2000.

[127] Skouibine K.B., Trayanova N., Moore P.: A numerically efficient model for the simulation of defibrillation in an active bidomain sheet of myocardium. Math. Biosci. **166**(1): 85–100, 2000.

[128] Skouibine K.B., Trayanova N.A., Moore P.: Anode/cathode make and break phenomena in a model of defibrillation. IEEE Trans. Biomed. Eng. **46**(7): 769–777, 1999.

[129] Smith B.F., Bjørstad P., Gropp W.D.: Domain Decomposition: Parallel Multilevel Methods for Elliptic Partial Differential Equations, Cambridge University Press, 1996.

[130] Streeter D.: Gross morphology and fibre geometry in the heart. In Handbook of Physiology. Vol. 1, Sect. 2, pp. 61–112. R.M. Berne (ed.), Williams & Wilkins, 1979.

[131] Sundnes J., Lines G.T., Mardal K.A., Tveito A.: Multigrid Block Preconditioning for a Coupled System of Partial Differential Equations Modeling the Electrical Activity in the Heart. Comput. Meth. Biomech. Biomed. Eng. **5**(6): 397–409, 2002.

[132] Sundnes J., Lines G.T., Tveito A.: An operator splitting method for solving the bidomain equations coupled to a volume conductor model for the torso. Math. Biosci. **194**(2): 233–248, 2005.

[133] Sundnes J., Lines G.T., Cai X., Nielsen B.F., Mardal K.-A., Tveito A.: Computing the electrical activity of the heart, Springer, 2006.

[134] Sundnes J., Nielsen B.F., Mardal K.A., Lines G.T., Mardal K.A., Tveito A.: On the computational complexity of the bidomain and the monodomain models of electrophysiology. Ann. Biomed. Engrg. **34**(7): 1088–1097, 2006.

[135] Taccardi B., Punske B., Lux R., MacLeod R., Ershler P., Dustman T., Vyhmeister Y.: Usefull lesson from body surface mapping on body. J. Cardiovasc. Electrophysiol. **9**(7): 773–786, 1998.

[136] Taccardi B., Punske B.B.: Body surface Potential Mapping In Cardiac Electrophysiology. From cell to Bedside., D. Zipes and J. Jalife (eds.), W.B. Saunders Company, Philadelphia, PA, 4th Edition, pp. 803–811, 2004.

[137] Ten Tusscher K.H.W.J., Panfilov A.V.: Cell model for efficient simulation of wave propagation in human ventricular tissue under normal and pathological conditions. Phys. Med. Biol. **51**: 6141–6156, 2006.

[138] Ten Tusscher K.H.W.J., Panfilov A.V.: Modelling of the ventricular conduction system. Progr. Biophys. Molec. Biol. **96**(1–3): 152–170, 2008.

[139] Tomlinson K.A., Hunter P.J., Pullan A.J.: A finite element method for an eikonal equation model of myocardial excitation wavefront propagation. SIAM J. Appl. Math. **63**(1): 324–350, 2002.

[140] Toselli A., Widlund O.B.: Domain Decomposition Methods: Algorithms and Theory. Computational Mathematics, Vol. 34. Springer-Verlag, Berlin, 2004.

[141] Trangenstein J.A., Kim C.: Operator splitting and adaptive mesh refinement for the Luo-Rudy I model. J. Comput. Phys. **196**: 645–679, 2004.

[142] Trew M., Le Grice I., Smaill B., Pullan A.: A finite volume method for modeling discontinuous electrical activation in cardiac tissue. Ann. Biomed. Eng. **33**(5): 590–602, 2005.

[143] Trew M., Smaill B., Bullivant D., Hunter P., Pullan A.: A generalized finite difference method for modeling cardiac electrical activation on arbitrary, irregular computational meshes. Math. Biosci. **198**(2): 169–189, 2005.

[144] Trew M.L., Caldwell B.J., Sands G.B., Hooks D.A., Tai D.C.-S., Austin T.M., LeGrice I.J., Pullan A.J., Smaill B.H.: Cardiac electrophysiology and tissue structure: bridging the scale gap with a joint measurement and modelling paradigm. Exp. Physiol. **91**(2): 355–370, 2006.

[145] Tung L.: A bidomain model for describing ischemic myocardial D.C. potentials. Ph.D. dissertation M.I.T., Cambridge, MA, 1978.

[146] Veneroni M.: Reaction-diffusion systems for the microscopic cellular model of the cardiac action potential. Math. Meth. Appl. Sci. **29**: 1631–1661, 2006.

[147] Veneroni M.: Reaction-diffusion systems for the macroscopic Bidomain model of the cardiac action potential. Nonlinear Anal.-Real World Appl. **10**(2): 849–868, 2009.

[148] Vigmond E.J., Aguel F., Trayanova N.A.: Computational techniques for solving the bidomain equations in three dimensions. IEEE Trans. Biomed. Eng. **49**(11): 1260–1269, 2002.

[149] Vigmond E.J., Weber dos Santos R., Prassl A.J., Deo M., Plank G.: Solvers for the cardiac bidomain equations. Progr. Biophys. Molec. Biol. **96**: 3–18, 2008.

[150] Weber dos Santos R., Plank G., Bauer S., Vigmond E.J.: Parallel multigrid preconditioner for the cardiac bidomain model. IEEE Trans. Biomed. Eng. **51 (11)**: 1960–1968, 2004.

[151] Whiteley J.P.: An efficient numerical technique for the solution of the monodomain and bidomain equations. IEEE Trans. Biomed. Eng. **53**(11): 2139–2147, 2006.

[152] Whiteley J.P.: Physiology Driven Adaptivity for the Numerical Solution of the Bidomain Equations. Ann. Biomed. Eng. **35**(9): 1510–1520, 2007.

[153] Wikswo J.P. Jr.: Tissue anisotropy, the cardiac bidomain, and the virtual cathode effect, In Cardiac Electrophysiology: from cell to bedside, (2nd ed), Zipes D.P. and Jalife J. (eds.): 348–361, WB Saunders, Philadelphia, 1994.

[154] Wikswo J.P. Jr., Lin L.-F., Abbas R.A.: Virtual electrodes in cardiac tissue: a common mechanism for anodal and cathodal stimulation. Biophys. J. **69**: 2195–2210, 1995.

[155] Wiskwo J.P. Jr., Roth B.J.: Virtual electrode theory of pacing. In Cardiac Bioelectric Therapy Efimov I.r., Kroll M.W. and Tchou (eds.), Ch. 4.3, 283–330, Springer Science+Business Media, LLc, 2009.

[156] Xie F., Qu Z.L., Yang J., Baher A., Weiss J.N., Garfinkel A.: A simulation study of the effects of cardiac anatomy in ventricular fibrillation. J. Clin Invest. **113**: 686–693, 2004.

[157] Xu A., R M.: Guevara. Two forms of spiral-wave reentry in an ionic model of ischemic ventricular myocardium. Chaos **8**(1): 157–174, 1998.

[158] Ying W.J., Rose D.J., Henriquez C.S.: Efficient Fully Implicit Time Integration Methods for Modeling Cardiac Dynamics. IEEE Trans. Biomed. Eng. **55**(12): 2701–2711, 2008.

[159] Zhang X.: Multilevel Schwarz methods. Numer. Math. **63**(4): 521–539, 1992.

[160] Zipes D.P., Jalife J. (eds.): Cardiac Electrophysiology: From Cell to Bedside, 5th ed., Saunders, 2009.

[161] Zozor S., Blanc O., Jacquemet V., Virag N., Vesin J., Pruvot E., Kappenberger L., Henriquez C.: A numerical scheme for modeling wavefront propagation on a monolayer of arbitrary geometry. IEEE Trans. Biomed. Eng. **50**(4): 412–420, 2003.

6

Structurally motivated damage models for arterial walls. Theory and application

Anne M. Robertson, Michael R. Hill, and Dalong Li

6.1 Introduction

The mechanical integrity of the arterial wall is vital for the health of the individual. This integrity is in turn dependent on the state of the central load bearing components of the wall: collagen fibres, elastic fibres and smooth muscle. Of these, the elastic fibres, composed largely of the protein elastin, are viewed as responsible for the highly elastic behaviour of the wall at low loads [92]. The collagen fibres are recruited under increasing extension, leading to a highly nonlinear behaviour of the arterial wall [117]. They are responsible for the structural integrity of the wall at elevated physiological loads. Changes in the quantity, distribution, orientation and mechanical properties of these components (the microstructure) are known to occur as part of a healthy response to changing stimuli (e.g. growth and remodelling) as well as during pathological and damage processes in disease and aging. For example, degradation of the elastic fibres is linked to pathological conditions including cerebral aneurysms [12, 15, 20, 65], dissection aneurysms [101], arteriosclerosis [11, 44, 86, 113, 114], and complications from balloon angioplasty [84]. Age related arterial stiffening is attributed to degradation of the elastic fibres, possibly from fatigue failure [11, 30]. The subject of arterial damage is addressed in Sect. 6.4.

The mechanical behaviour of the arterial wall is modelled from several perspectives. Phenomenological models are based on bulk measurements of the mechanical response of the arterial wall, while structural theories directly integrate information

Anne M. Robertson (✉)
University of Pittsburgh, Pittsburgh, PA 15261 USA
e-mail: rbertson@pitt.edu

Michael R. Hill
University of Pittsburgh, Pittsburgh, PA 15261 USA
e-mail: mrh54@pitt.edu

Dalong Li
Ansys Inc., Canonsburg, PA USA
e-mail: dalong.li@ansys.com

Ambrosi D., Quarteroni A., Rozza G. (Eds.): Modeling of Physiological Flows.
DOI 10.1007/978-88-470-1935-5_6, © Springer-Verlag Italia 2012

on the tissue composition, structure, and load carrying mechanisms of individual components. In doing so, they provide more insight into the function and mechanics of tissue components, at the expense of requiring constitutive data for each of the elements. For example, recent experimental and modelling studies have addressed the mechanical behaviour of isolated arterial elastin [42, 43, 112]. In between these two extremes are so called structurally motivated models, which bring in an intermediate level of structural information such as the fibrous nature of the tissue without directly incorporating the response of individual components. For example, the mechanical behaviour of individual fibres is not prescribed. Mixture theories have been developed to describe the growth and mechanical response of individual wall constituents [59]. These models have shown great promise in studies of aneurysm development, vasospasm, and remodelling under altered loads [37, 38, 58, 60].

Lanir appears to be the first to develop a three dimensional microstructural model for fibre reinforced soft tissues that directly includes fibre orientation [67]. Though Lanir formulated this structural model more than 25 years ago [66, 67], these models have recently received increasing attention. This surge in interest arises partially due to improvements in imaging and computational technology as well as the desire to model more complex behaviour of the artery such as subfailure damage [74, 87, 88], growth, and remodelling [59, 60, 94].

One of the most significant challenges in applying microstructural models is acquiring experimental values for the structural components. With few exceptions, most structural information about the orientation and distribution of wall components are obtained using destructive techniques such as fixation, followed by sectioning and staining for the particular constituent of interest. As a result, multiple tissue samples and tissue fixation steps are typically needed to explore the wall structure under different loading conditions. For example, in the most detailed study available on collagen fibre alignment in cerebral vessels during loading, Finlay, McCullough and Canham evaluated vessels that were embedded in paraffin and sectioned [28].

Sacks developed an approach using small angle light scattering (SALS) to non-destructively measure collagen fibre orientation [98] which he then directly incorporated into Lanir's structural model [97]. While this approach was used effectively in a number of applications including bovine pericardium, it is not suitable for thicker tissues samples such as the intact arterial wall. Further, this approach does not distinguish between fibre types or provide information about the distribution of orientation through the thickness of the tissue.

Recent advances in multi-photon microscopy (MPM) provide an opportunity to nondestructively acquire quantitative structural information during the loading process of interest for thicker tissues such as cerebral vessels. Using MPM, elastin and collagen can be imaged without staining or fixation at greater depths than previously possibly [123]. We recently developed an MPM compatible uniaxial system (UA-MPM system) which enables quantitative assessment of the microstructure, simultaneous with mechanical loading experiments [49, 50]. This makes it possible to correlate microstructure (including some changes due to damage) and mechanical response in a single sample over a range of loading conditions and time points. A major advantage of using a MPM is the ability to obtain images of arterial collagen and elastin without the destructive processing typical of previous approaches.

In this work, we consider damage in arteries, with particular emphasis on cerebral vessels. A brief introduction to the structure of the healthy arterial wall is given in the Sect. 6.2. In Sect. 6.3, an overview of structural constitutive models for the arterial wall are discussed, with particular emphasis on modelling fibre recruitment and distribution of fibre orientation. This section ends with a discussion of the UA-MPM device and its application to the healthy arterial wall. In Sect. 6.4, we turn attention to damage in the arterial wall, beginning with a discussion of sources of damage and clinical relevance of this subject. We then discuss a continuum damage theory for the arterial wall. This section ends with a brief overview of the application of the UA-MPM system to study damage in cerebral vessels. Finally, in Sect. 6.5, we consider the an application of the continuum damage model to the subject of cerebral angioplasty.

6.2 Background – structure of the undamaged arterial wall

The healthy artery wall consists of three concentric layers which, moving outward from the lumen, are 1) the tunica intima, including endothelial cells and a fenestrated sheet of elastin fibres called the internal elastic lamina (IEL); the tunica media, containing mostly smooth muscle cells, some elastin fibres and collagen fibres; and 3) the tunica adventitia, composed mainly of a network of type I collagen fibres and fibroblasts, Fig. 6.1a. An external elastic lamina (EEL) is found interior to the adventitia. It should be noted that the microstructure varies with location in the arterial tree, age and disease state. For example, in cerebral arteries the EEL is absent and nearly all the elastin is confined in the IEL. Further, the microstructure of the wall can change in response to changes in stimuli through growth, remodelling or damage.

Structure/function relationship in healthy arteries

It has long been recognized that the passive mechanical response of arteries is largely due to the collagen and elastin found in the arterial wall, and in some cases smooth muscle cells [6, 90, 93]. The typical stress strain curve from mechanical tests of arterial walls, for example under uniaxial stretch, is highly nonlinear, characterized by high flexibility at low loads (the toe region) and rapidly increasing stiffness at higher loads. The toe region has been conjectured to arise from the loading of elastic fibres and the highly nonlinear response at increasing loads to the recruitment of collagen fibres [11, 90, 92]. Samila et al. morphologically observed the gradual unfolding of crimped collagen fibres in human carotid artery strips that were stretched uniaxially [100]. It is believed that collagen fibres contribute little resistance during this unfolding process, [92]. As loading intensifies, the collagen contribution takes on an increasing role as seen by the heightened vessel stiffness. If the elastic fibres are significantly damaged, the region of high flexibility at low loads will be lost and the mechanical properties of the vessel will change significantly.

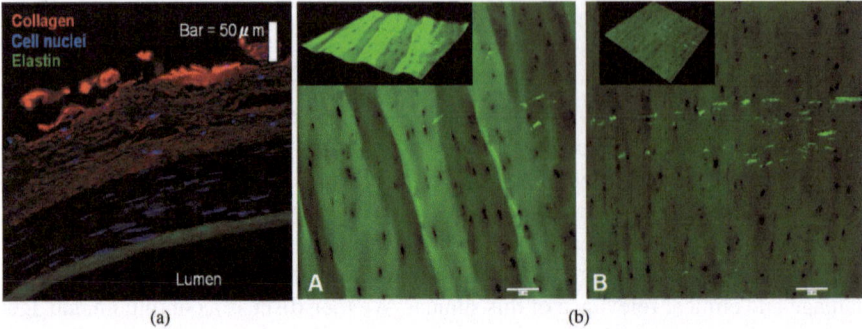

Fig. 6.1. (a) Fluorescence microscopy images of cross sectional preparations of the human left vertebral artery (cerebral), fixed at 30 % stretch. Immunohistochemical staining of the arterial wall reveals elastin (green) localized in the internal elastic lamina, cell nuclei (blue, DAPI stain), collagen fibres (red). (b) 3D reconstruction of confocal microscopy image slices taken of a human anterior cerebral artery (ACA) revealing the autofluorescent internal elastic lamina, under (A) zero stretch and (B) 30 % strain. All three scale bars = 50 microns

Collagen in the arterial wall

Collagen molecules in the arterial wall are assembled into fibrils that are packed together to form collagen fibres. In Type 1 collagen, thick fibrils ($\sim$ 74 nm) are densely packed to form collagen fibres (2–10 mm in diamter) that are packed together in bundles [80]. In contrast, Type III collagen is formed as individual fibres (0.5–1.5 mm in diameter), composed of loosely packed thin fibrils ($\sim$ 45 nm). The mechanical behaviour of the vascular wall is affected by the orientation and unloaded geometry of these fibres as well as their distribution across the wall layers. These features vary with location in the vasculature as welll as in health and disease.

Elastic fibres in the arterial wall

Mammalian elastic fibres (0.2–1.5 µm) are composed of a core of twisted, rope-like structures of highly cross-linked elastin protein surrounded by fibrillin microfibrils (10–12 nm in diameter) [80, 96]. The microfibrils are believed to contribute to elastogenesis and elastic fibre function [104, 111].

In cerebral vessels the elastic fibres are nearly all localized in a fenestrated (windowed) sheet of elastic fibres, the IEL (Fig. 6.1b), [63]. SEM has revealed the presence of folds in the IEL, projecting into the lumen and running parallel to the longitudinal axis of the artery [68]. This folded state has been found under zero applied loading and the folds can be "pulled out" by applying loads perpendicular to the axial direction of the vessel, Fig. 6.1b. The main function of the fenestrae in the IEL appears to be the enhancement of passage of water, nutrients and electrolytes across the wall. However, the size and number of these fenestrae change in time and have been shown to affect the mechanical properties of the vessel [40].

6.3 Structurally motivated constitutive models – elastic regime

In this section, we summarize recent structural constitutive models for the elastic behaviour of arteries. A number of references are available on this subject (e.g. [54]). Here, we focus on models which explicitly include fibre orientation that will be used in the discussion of damage in the next section. For simplicity of discussion, we will not include the residual stress or the viscoelastic effects. Further, we concentrate on the passive response of the arterial wall and in particular will not discuss the active response due to smooth muscle. A review of models for active contributions from smooth muscle cells can be found in [54] including early phenomenological modelling by Rachev and Hayashi [89] and by Zulliger et al. [125]. Humphrey and collaborators include an anisotropic contribution from passive smooth muscle in their constitutive models of the arterial wall (e.g. [110]) in addition to an active response.

In this work, the arterial wall is treated as an incompressible, hyperelastic multi-mechanism material with independent passive contributions from collagen and elastin and smooth muscle. The stress tensor will be decomposed into a constraint response that does no work as well as an extra stress contribution $\underline{\sigma}_E$ that can do work,

$$\underline{\sigma} = -p\underline{I} + \underline{\sigma}_E \tag{6.1}$$

where p is the Lagrange multiplier arising from the constraint of incompressibility. Assuming the material response to be hyperelastic it then follows that,

$$\underline{\sigma} : \underline{D} = \underline{\sigma}_E : \underline{D} = \dot{W} \tag{6.2}$$

where W is the strain energy per unit volume in the unloaded configuration κ_0, $\underline{D}$ is the symmetric part of the velocity gradient, and the overdot is used to denote the material derivative. We consider the strain energy to consist of the sum of strain energies from an isotropic mechanism W_{iso} and an anisotropic mechanism W_{aniso}

$$W = W_{iso} + W_{aniso} \tag{6.3}$$

and then use (6.2) and (6.3) to define corresponding contributions to the Cauchy stress tensor $\underline{\sigma}$,

$$\underline{\sigma}_E = \underline{\sigma}_{iso} + \underline{\sigma}_{aniso} \tag{6.4}$$

with

$$\underline{\sigma}_{iso} : \underline{D} = \dot{W}_{iso}, \qquad \underline{\sigma}_{aniso} : \underline{D} = \dot{W}_{aniso}. \tag{6.5}$$

Both collagen fibres and passive smooth muscle will contribute to the anisotropic mechanism. These mechanisms will in general have different constitutive responses and different unloaded configurations. The arterial collagen is in a wavy or crimped state in the unloaded artery and is gradually recruited to load bearing as the vessel is strained. Hence, multiple reference configurations will be needed to identify the kinematic state for recruitment. Further, damage to the elastin and collagen fibres may occur independently and prior to failure of the arterial wall. The explicit treatment of these multiple mechanisms makes it possible to model damage and failure of the elastic fibres separately from that of collagen fibres [118, 119]. The layered nature of the wall can be handled by idealizing the wall as a membrane in which the effects of the three layers are collapsed into one surface. At the other extreme, we can consider the wall as a three layered composite with a separate constitutive equation of the form (6.3) for each layer. The first approach is taken in Sect. 6.3.5 while the second is used in the angioplasty study in Sect. 6.5.

6.3.1 Isotropic mechanism

In the cerebral artery, the elastin is largely confined to a single elastic lamina which we model as isotropic. In this section, it is not necessary to exclude the possibility that other structural components contribute to the isotropic mechanism. In discussions of damage to individual components (Sect. 6.4) and in the discussion of the multi-layered wall models introduced in Sect. 6.5, we will be more specific about this issue. While the IEL is known to be in a wavy or buckled state when the arterial wall is unloaded, Fig. 6.1, for simplicity, we assume the load necessary to reach this uncrimped state is negligible compared with the load levels of interest. We therefore assume the unloaded configuration for the isotropic mechanism corresponds to the unloaded configuration of the body, κ_0, Fig. 6.2. The strain energy of this isotropic mechanism will then be a function of the deformation gradient relative to κ_0 denoted as $\underline{F}_0$. After imposing invariance requirements, the incompressibility condition and material isotropy, we can write the strain energy as a function of the first and second principal invariants of $\underline{C}_0 = \underline{F}_0^T \underline{F}_0$ or alternatively of $\underline{b}_0 = \underline{F}_0 \underline{F}_0^T$, since they have the same principal invariants,

$$W_{iso} = W_{iso}(I_0, II_0), \quad \text{where} \quad I_0 = \text{tr}\,\underline{b}_0, \quad II_0 = \frac{1}{2}\left((\text{tr}\,\underline{b}_0)^2 - \text{tr}\,\underline{b}_0^2\right). \quad (6.6)$$

The third principal invariant is a constant (one) since the motion is necessarily iso-choric for incompressible materials and therefore does not arise in (6.6). It then follows from (6.5), (6.6) (see, e.g. Spencer [108]) that the Cauchy stress tensor can be written as a function of $\underline{b}_0$,

$$\underline{\sigma}_{iso} = 2\left[\frac{\partial W_{iso}}{\partial I_0}\underline{b}_0 - \frac{\partial W_{iso}}{\partial II_0}\underline{b}_0^{-1}\right]. \quad (6.7)$$

Defining the second Piola-Kirchhoff tensor $\underline{S}$ through $\underline{S} = \underline{F}_0^{-1}\underline{\sigma}\underline{F}_0^{-T}$, it follows from (6.7) that,

$$\underline{S} = -\bar{p}\underline{C}_0^{-1} + \underline{S}_E, \quad \underline{S}_E = \underline{S}_{iso} + \underline{S}_{aniso},$$
$$\underline{S}_{iso} = 2\left[\left(\frac{\partial W_{iso}}{\partial I_0} + I_0\frac{\partial W_{iso}}{\partial II_0}\right)\underline{I} - \frac{\partial W_{iso}}{\partial II_0}\underline{C}_0\right], \quad (6.8)$$

where the Lagrange multipliers p and $\bar{p}$ introduced in (6.1) and (6.8) differ by a factor $2II_0\,\partial W_{iso}/\partial II_0$.

6.3.2 Kinematics of the anisotropic mechanism

In general, the collagen fibres will display a distribution of fibre crimp. Namely, there will not be a *single* discrete strain associated with the initiation of load bearing of all collagen fibres. For the moment, we consider an arbitrary fibre family at an arbitrary material point in the body. We denote the configuration where this fibre is recruited to load bearing (activated) as κ_a, Fig. 6.2. For simplicity of the ensuing discussion, we refer to the time at which this configuration is reached as t_a. The length and orientation of an infinitesimal material segment of fibre in κ_a will be denoted as dS_a and $\underline{a}_a$, respectively. Under an affine transformation, this material element will be mapped back to a material element in κ_0 which we denote as $dS_0\underline{a}_0$, so that,

$$dS_a\underline{a}_a = \underline{F}_0(t_a)dS_0\underline{a}_0. \tag{6.9}$$

The stretch of a material element in $\kappa(t)$ relative to its length is κ_0 is denoted by λ_0 so that,

$$\lambda_0 = ds/dS_0, \quad \text{and} \quad \lambda_0^2 = \underline{C}_0(t):(\underline{a}_0 \otimes \underline{a}_0). \tag{6.10}$$

For definiteness, we take $\underline{a}_0$ and $\underline{a}_a$ to be unit vectors. Based on histological information, it is clear the fibre is in a buckled state in κ_0. Therefore, the direction $\underline{a}_0$ defined in (6.9) is not the local orientation of the fibre, Fig. 6.2 and dS_0 is not the fibre length in κ_0. We denote the stretch of the underlying material element during the deformation from κ_0 and κ_a as λ_a,

$$\lambda_a = dS_a/dS_0, \quad \text{and} \quad \lambda_a^2 = \underline{C}_0(t_a):(\underline{a}_0 \otimes \underline{a}_0). \tag{6.11}$$

The value of λ_a will be treated as a material property of the fibre reinforced material defining the necessary material stretch for the fibre to become uncrimped. In Sect. 6.3.5, we will discuss means of acquiring an estimate of λ_a from imaging data. In principal, a theory could be developed to relate the geometry of the fibre in κ_0 to λ_a. We do not address this issue here. In fact, one of the advantages of the current approach arises from the fact that the constitutive contribution of collagen is independent of this uncrimping process. Rather, the response of the collagen fibres depends only on the deformation of collagen relative to the reference

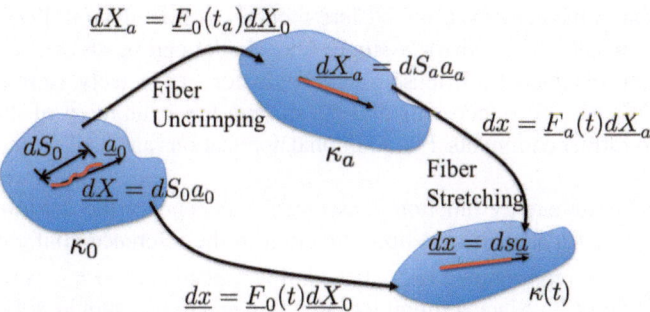

Fig. 6.2. Schematic of configurations used for the fibre recruitment constitutive model

configuration κ_a, not κ_0. This is discussed in more detail in Sect. 6.3.3. There-fore, while the mapping of the fibre back to κ_0 might be of interest, it is not nec-essary other than serving as a means of relating the undeformed configuration for the collagen fibre to the unloaded configurations of other mechanisms such as the IEL.

The deformation of an infinitesimal material element from $\underline{dX}_a = dS_a\underline{a}_a$ in κ_a to $\underline{dx} = ds\underline{a}$ in the current configuration $\kappa(t)$ is described through

$$\underline{dx} = \underline{F}_a(t)\underline{dX}_a \tag{6.12}$$

where $\underline{F}_a$ is the deformation gradient tensor in $\kappa(t)$ relative to κ_a. The correspond-ing right Cauchy stretch tensor is then $\underline{C}_a = \underline{F}_a^T\underline{F}_a$. The actual (true) stretch of the material fibre element during the deformation from κ_a to $\kappa(t)$ will be denoted as λ_t and is therefore,

$$\lambda_t = ds/dS_a = \underline{C}_a : (\underline{a}_a \otimes \underline{a}_a). \tag{6.13}$$

In some cases, it may be convenient to relate the stretch at time t relative to κ_a to that relative to κ_0. It follows from (6.10), (6.11) and (6.13) that

$$\lambda_t = \lambda_0/\lambda_a, \qquad \text{and} \qquad \lambda_t^2 = \frac{1}{\lambda_a^2}\underline{C}_0 : (\underline{a}_0 \otimes \underline{a}_0). \tag{6.14}$$

6.3.3 Anisotropic mechanism – discrete fibre models (N-fibre models)

In this section, we consider N-fibre models, whereby each material point in the body is assumed to contain N families of fibres, each with a distinct direction and re-cruitment stretch. The contributions of these individual fibre families in the mate-rial are directly incorporated into the constitutive model. The theory presented here closely follows that in given in [108, 109] and applications to the arterial wall in-troduced in [52] with the exception that we consider the recruitment of fibre family i at some activation stretch λ_{ai} that may be larger than one. A discussion of acti-vation stretch in the context of the N-fibre model was first given in [71]. As noted by Spencer, results obtained for a single fibre model can be shown to be equiv-alent to those obtained by Ericksen and Rivlin for transversely isotropic elastic materials [26]. See also Sects. 6.7 and 6.8 of [53] for a summary of the work of Spencer and further comments. For additional applications and background material see [7].

Here, the strain energy function is assumed to depend on the deformation gra-dient relative to κ_{ai} as well as a fibre direction in the reference configuration. To ensure invariance requirements are satisfied, this dependence on $\underline{F}_a^{(i)}$ is replaced by a dependence on $\underline{C}_a^{(i)}$. Since a flipping of the fibre end to end should not play a role (i.e. $\underline{a}_0$ and $-\underline{a}_0$ represent the same fibre), the dependence on $\underline{a}_a^{(i)}$ is assumed to be even, (for example, through $\underline{a}_a^{(i)} \otimes \underline{a}_a^{(i)}$),

$$W_{aniso} = \sum_{i=1}^{N} W_{aniso}^{(i)}(\underline{C}_a^{(i)}, \underline{a}_a^{(i)} \otimes \underline{a}_a^{(i)}). \tag{6.15}$$

We define the strain energy function to be zero when the true stretch of the ith fibre family, $\lambda_t^{(i)}$, is less than one. The activation stretch $\lambda_{ai}^{(i)}$ of a fibre family i will be given as part of the constitutive equation. In writing (6.15), we have implicitly neglected any contributions to the strain energy arising from coupled effects between the fibres (see [108] for a formulation with coupling between the fibres). We now assume the only anisotropy in the material is due to the fibre orientation as characterized by $\underline{a}_a^{(i)}$. In this case, $W_{aniso}^{(i)}$ is a scalar valued isotropic tensor function of the tensor $\underline{C}_a^{(i)}$ and $\underline{a}_a^{(i)} \otimes \underline{a}_a^{(i)}$. Mathematically, this implies that without loss in generality the strain energy can be written as a function of the integrity basis of $\underline{C}_a^{(i)}$ and $\underline{a}_a^{(i)} \otimes \underline{a}_a^{(i)}$. As in [35], we assume, $W_{aniso}^{(i)}$ only depends on one member of this basis $(IV_a^{(i)})$:

$$W_{aniso}^{(i)} = W_{aniso}^{(i)}(\lambda_t^{(i)}), \qquad \text{where} \qquad IV_a^{(i)} = \underline{C}_a^{(i)} : (\underline{a}_a^{(i)} \otimes \underline{a}_a^{(i)}) = \lambda_t^{(i)2}. \tag{6.16}$$

It follows from (6.5) that

$$\underline{\sigma}_{aniso} = 2\underline{F}_0 \left(\sum_{i=1}^{N} \frac{dW_{aniso}^{(i)}(\lambda_t^{(i)})}{d\lambda_t^{(i)}} \frac{\partial \lambda_t^{(i)}}{\partial \underline{C}_0} \right) \underline{F}_0^T, \qquad \text{so that,}$$

$$\underline{S}_{aniso} = 2 \left(\sum_{i=1}^{N} \frac{dW_{aniso}^{(i)}(\lambda_t^{(i)})}{d\lambda_t^{(i)}} \frac{\partial \lambda_t^{(i)}}{\partial \underline{C}_0} \right), \tag{6.17}$$

and therefore using (6.14) the Cauchy and second Piola-Kirchhoff stress tensors can be obtained

$$\underline{\sigma}_{aniso} = \sum_{i=1}^{N} \lambda_t^{(i)} \frac{dW_{aniso}^{(i)}(\lambda_t^{(i)})}{d\lambda_t^{(i)}} \underline{a}^{(i)} \otimes \underline{a}^{(i)},$$

$$\underline{S}_{aniso} = \sum_{i=1}^{N} \frac{1}{\lambda_a^{(i)2}} \frac{1}{\lambda_t^{(i)}} \frac{dW_{aniso}^{(i)}(\lambda_t^{(i)})}{d\lambda_t^{(i)}} \underline{a}_0^{(i)} \otimes \underline{a}_0^{(i)}. \tag{6.18}$$

In this case, the, a two fibre model with homogeneous fibre properties of the form

$$W_{aniso}^{(i)} = \sum_{i=1}^{N} H(\lambda_t^{(i)} - 1) \frac{\eta}{2\gamma} \left(e^{\gamma(\lambda_t^{(i)2} - 1)^a} - 1 \right), \tag{6.19}$$

with $a = 2$ is frequently used for fibres in structurally motivated constitutive models [52, 71, 110]. In this case, the corresponding Cauchy stress tensor is

$$\underline{\sigma}_{aniso} = 2\eta \sum_{i=1}^{N} H(\lambda_t^{(i)} - 1) \lambda_t^{(i)2} (\lambda_t^{(i)2} - 1) e^{\gamma(\lambda_t^{(i)2} - 1)^2} \underline{a}^{(i)} \otimes \underline{a}^{(i)}. \qquad (6.20)$$

The symbol H in (6.19) and (6.20) is the unit step function. Namely, the anisotropic contribution from fibre family i is zero if $\lambda_t^{(i)} \leq 1$. In [110], (6.19) is used for both collagen and smooth muscle, through $\lambda_\alpha^{(i)} = 1$ in that work.

6.3.4 Anisotropic mechanism – fibre distribution models

In this section, we consider materials for which the fibres do not have a small number of distinct orientations in the reference configuration. Rather, they appear to be dispersed about distinct directions. With this in mind, we define an *orientation density function* (ODF) denoted by $\rho(\underline{M}_0)$ which characterizes the three-dimensional distribution of fibre angles in the reference configuration κ_0, Fig. 6.3, (see, e.g. [35]). In the most general case, $\underline{M}_0$ is an arbitrary unit vector which is characterized by two Euler angles $\Theta \in [0, \pi]$ and $\Phi \in [-\pi/2, \pi/2]$ and can be written in terms of a three-dimensional Cartesian coordinate system with basis $\{\underline{e}_1, \underline{e}_2, \underline{e}_3\}$,

$$\underline{M}_0(\Theta, \Phi) = \sin\Theta\cos\Phi\underline{e}_1 + \sin\Theta\sin\Phi\underline{e}_2 + \cos\Theta\underline{e}_3, \qquad (6.21)$$

so that we can also write $\rho = \rho(\Theta, \Phi)$. The orientation density function is non-negative and defined such that $\rho(\Theta, \Phi)d\Omega$ represents the normalized number of fibres in κ_0 with orientations in the range $[(\Theta, \Theta + d\Theta), (\Phi, \Phi + d\Phi)]$, where $d\Omega = \sin\Theta d\Theta d\Phi$. For definiteness, $\rho(\Theta, \Phi)$ will be defined to satisfy the normalization condition

$$1 = \frac{1}{2\pi} \int_\Omega \rho(\Theta, \Phi)d\Omega. \qquad (6.22)$$

The range of Φ arises from the restriction that the distribution function be insensitive to fibre reflections $(\rho(\underline{M}_0) = \rho(-\underline{M}_0))$. Two limiting cases of fibre distribution will

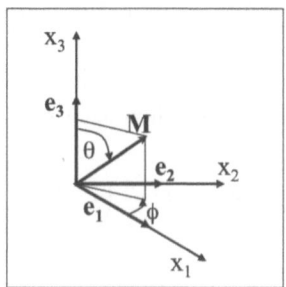

Fig. 6.3. Schematic of geometric variables used in dispersion model for collagen fibre distribution

be considered here. If at an arbitrary point, a coordinate axis can be chosen such that ρ is independent of Φ, then the material is transversely isotropic at that point and (6.22) reduces to,

$$\textbf{Conical Splay} \qquad \rho = \rho(\Theta), \qquad 1 = \frac{1}{2}\int_0^\pi \rho(\Theta)\sin\Theta d\Theta. \qquad (6.23)$$

For example, Gasser and Holzapfel [35] considered a transversely isotropic π-periodic von Mises orientation density function for collagen fibres, suitably modified so the normalization condition given in (6.23) was satisfied,

$$\rho(\Theta) = 4\sqrt{\frac{b}{2\pi}}\,\frac{\exp[b(\cos(2\Theta)+1)]}{\operatorname{erfi}(\sqrt{2b})}, \qquad (6.24)$$

where $\operatorname{erfi}(x) = -i\operatorname{erf}(x)$ is the imaginary error function. Other distributions such as Gaussian, Gamma [99] and Bingham [2] have also been considered for arterial collagen. The latter has been considered in the context of a microsphere-based constitutive model developed using a computational homogenization scheme [1]. For materials idealized as having a planar splay of distributions, we assume that suitable coordinates can be chosen such that the tangent to all fibres at an arbitrary material point lie in the plane $\Theta = \pi/2$. In this case, the two-dimensional orientation distribution function ρ_{2D} and normalization condition can be written as,

$$\textbf{Planar Splay} \qquad \rho = 2\delta[\Theta - \pi/2]\rho_{2D}(\Phi), \qquad 1 = \frac{1}{\pi}\int_{-\pi/2}^{\pi/2}\rho_{2D}(\Phi)\,d\Phi$$
$$(6.25)$$

where $\delta[f]$ is the well known generalized function referred to as the "delta Dirac function".

Integral fibre distribution models

Lanir [67] appears to be the first to introduce fibre dispersion directly into a strain energy function. Following this work, we assume a distribution of fibre angles exists at each arbitrary point in reference configuration κ_0. For simplicity, most models of the arterial wall, assume either an isotropic, transversely isotropic or planar distribution of fibres. The total strain energy is assumed to arise from the combined effect of fibres with a distribution of orientations as W_f'. Further, at each orientation, the fibres may have a distribution $D(\lambda_a)$ of critical stretches λ_a at which load bearing commences. Namely, $D(\lambda_a)$ is the recruitment distribution function. The minimum and maximum critical stretches will be denoted by λ_{a1} and λ_{a2}, respectively. As in an earlier work, we consider the possibility that λ_{a1} may be greater than one, (e.g. [119]),

$$W_{ansio} = \frac{1}{2\pi}\int_\Omega \rho(\Theta,\Phi)\,W_f'(\lambda_0)\,d\Omega \qquad \text{where}$$
$$W_f' = \mathrm{H}(\lambda_0 - \lambda_{a1})\int_{\lambda_{a1}}^{\lambda_0} D(\lambda_a)\,W_f(\lambda_t)\,d\lambda_a \qquad (6.26)$$

where λ_0 will, in general, depend on the Euler angles and $\lambda_t = \lambda_0/\lambda_a$. Using similar steps as those used to obtain (6.18), it then follows from (6.5) and (6.26) that

$$\sigma_{aniso} = \underline{F}_0 \left[\frac{1}{2\pi} \int_\Omega \rho(\Theta, \Phi) \frac{1}{\lambda_0} \frac{dW_f'(\lambda_0)}{d\lambda_0} \underline{M}_0 \otimes \underline{M}_0 \, d\Omega \right] \underline{F}_0^T \qquad (6.27)$$

where

$$\frac{dW_f'(\lambda_0)}{d\lambda_0} = \mathsf{H}(\lambda_0 - \lambda_{a1}) \int_{\lambda_{ai}}^{\lambda_0} D(\lambda_a) \frac{dW_f(\lambda_t)}{d\lambda_t} \frac{1}{\lambda_a} \, d\lambda_a, \qquad (6.28)$$

and we have used the normalization $W_F = 0$ for $\lambda_t = 1$.

If the material is idealized as having **a single, discrete recruitment stretch** λ_a, the strain energy function in (6.27) reduces to,

$$W_{aniso} = \frac{1}{2\pi} \int_\Omega \rho(\Theta, \Phi) \mathsf{H}(\lambda_t - 1) W_f(\lambda_t) \, d\Omega. \qquad (6.29)$$

The corresponding Cauchy stress tensor can directly be calculated from (6.29) and (6.5),

$$\sigma_{aniso} = \underline{F}_0 \left[\frac{1}{2\pi} \int_\Omega \rho(\Theta, \Phi) \mathsf{H}(\lambda_t - 1) \frac{dW_f(\lambda_t)}{d\lambda_t} \frac{1}{\lambda_t \lambda_a^2} \underline{M}_0 \otimes \underline{M}_0 \, d\Omega \right] \underline{F}_0^T. \qquad (6.30)$$

Generalized Structure Tensor (GST) models

There are few closed form solutions for materials described by (6.27) and (6.28) or even (6.30). Further, computationally it can be resource consuming to evaluate integrals such as (6.30) at each point in the body for each deformation level. To address this issue, Freed et al. [32] and Gasser and Holzapfel [35] introduced an alternative approach using a structure tensor. Here, as in [71], we consider a general structure tensor model for materials with fibre recruitment at a finite stretch, though we take a slightly different approach. We start with the assumption that the strain energy of the anisotropic mechanism is dependent on a weighted average of fibre orientations. In the case of finite stretch recruitment,

$$W_{aniso} = \mathsf{H}(\overline{\lambda}_t - 1) W_f(\overline{\lambda}_t) \qquad (6.31)$$

where a single representative recruitment angle λ_a is used and $\overline{\lambda}_t$ is an average true stretch defined as,

$$\overline{\lambda}_t^2 = \frac{1}{\lambda_a^2} \frac{1}{2\pi} \int_\Omega \rho(\Theta, \Phi) \lambda_0^2 \, d\Omega. \qquad (6.32)$$

It follows from (6.5), (6.31) and (6.32), that

$$\sigma_{aniso} = \underline{F}_0 \left[\mathsf{H}(\overline{\lambda}_t - 1) \frac{1}{\overline{\lambda}_t \lambda_a^2} \frac{dW_f(\overline{\lambda}_t)}{d\overline{\lambda}_t} \underline{H}_0 \right] \underline{F}_0^T \qquad (6.33)$$

where $\underline{H}_0$ is called the generalized structure tensor,

$$\underline{H}_0 = \frac{1}{2\pi} \int_\Omega \rho(\Theta, \Phi) \underline{M}_0 \otimes \underline{M}_0 \, d\Omega \qquad (6.34)$$

and can be seen to be an average of the structure tensor $\underline{M}_0 \otimes \underline{M}_0$, weighted by the distribution of fibre angles. As noted in [22], the average stretch can be calculated using the structure tensor. From (6.32) and (6.34),

$$\overline{\lambda}_t^2 = \frac{1}{\lambda_a^2} \underline{C}_0 : \underline{H}_0. \qquad (6.35)$$

An important advantage of the Generalized Structure Tensor approach is that $\underline{H}_0$ can be calculated a priori since it depends on the material structure in κ_0. Further, the structure tensor simplifies greatly for the distributions in (6.23) and (6.25). However, as discussed below, there are clear limitations to this approach.

Generalized Structure Tensor for materials with conical splay

For the special case where the fibres have rotational symmetry about a mean referential direction aligned with $\overline{a}_0$ (e.g. direction $\underline{e}_3$ in Fig. 6.3), we can use (6.21) and (6.23) with (6.34) to show,

Conical Splay $\underline{H}_0 = \kappa \underline{I} + (1 - 3\kappa) \overline{a}_0 \otimes \overline{a}_0, \quad \kappa = \frac{1}{4} \int_0^\pi \rho(\Theta) \sin^3 \Theta \, d\Theta.$

(6.36)

When there is no preferred orientation, ρ is equal to one and $\kappa = 1/3$, reducing $\underline{H}_0$ to one third of the identity tensor. It follows from (6.35) and (6.36) that the average stretch can be written with respect to the invariants of $\underline{C}_0$ and $\overline{a}_0 \otimes \overline{a}_0$

$$\overline{\lambda}_t^2 = \frac{1}{\lambda_a^2} \underline{C}_0 : \underline{H}_0 = \frac{1}{\lambda_a^2} \left(\kappa I_0 + (1 - 3\kappa) \overline{IV}_0 \right), \qquad \text{where}$$

$$I_0 = \mathrm{tr} \underline{C}_0, \qquad \overline{IV}_0 = \underline{C}_0 : (\overline{a}_0 \otimes \overline{a}_0). \quad (6.37)$$

A commonly used strain energy function with a conical splay generalized structure tensor is [35, 54]

$$W_{aniso} = \mathrm{H}(\lambda_t - 1) \frac{\eta}{\gamma} \left[e^{\left(\gamma(\overline{\lambda}_t^2 - 1)^2 \right)} - 1 \right], \qquad (6.38)$$

where $\lambda_a = 1$.

For the case where the fibre distribution has rotational symmetry, the structure tensor can be written with respect to a single *dispersion parameter* which represents the fibre distribution in an integral sense. This parameter can either be thought of as a material parameter, determined directly from the experimental data, or calculated, from experimental knowledge of $\rho(\Theta)$. If the first approach is taken, it is desirable to know what range of κ is allowable. Li and Robertson [71] proved that $\kappa \in (0, 1/2]$.

As noted by Holzapfel and Ogden [54], the limiting case of $\kappa = 1/2$ corresponds to an isotropic distribution of fibres in a plane perpendicular to $\bar{a}_0$. They found unphysical results using a von Mises fibre distribution for pressure inflation at some values of $\kappa > 1/3$. It should be noted that as κ approaches $1/2$ and fibre orientation tends towards a planar distribution, there will be no fibres in the $\bar{a}_0$ direction to resist loading. In this case, it would be more appropriate to consider an alternate $\bar{a}_0$ or use the planar splay discussed below.

The generalized structural tensor approach has received a great deal of attention for modelling biological tissues such as pulmonary aveoli [115], abdominal aneurysms [95] and cerebral arteries [71]. Pandolfi and Holzapfel modelled the human cornea using a slight generalization of the structure tensor model given in (6.38) to include two average fibre directions rather than one [51].

Generalized Structure Tensor for materials with planar splay

The use of generalized structure tensors for a planar distribution of fibres has also been considered previously [22, 55] If we assume the planar distribution of fibre angles displays a symmetry to reflections about a line tangent to direction $\bar{a}_0$ in this plane (e.g. such that $\rho(\Phi) = \rho(-\Phi)$ in Fig. 6.3, then it follows from (6.21), (6.25) and (6.34) that,

Planar Splay with Symmetry $\underline{H}_0 = \kappa_{2D}\underline{I}_{2D} + (1 - 2\kappa_{2D})\bar{a}_0 \otimes \bar{a}_0$

$$\kappa_{2D} = \frac{1}{\pi}\int_{-\pi/2}^{\pi/2}\rho_{2D}(\Phi)\sin^2\Phi\,d\Phi. \tag{6.39}$$

Namely, as for the case of conical splay, the structure tensor depends only on one material parameter. When the fibres are isotropically distributed in the plane, ρ_{2D} is equal to one and $\kappa_{2D} = 1/2$, reducing $\underline{H}_0$ to one half of the planar identity tensor.

Following similar approaches to that used in [71], it is straightforward to show that $\kappa_{2D} \in [0, 1]$.

Comparison of Generalized Structure Tensor (GST) model with integral fibre distribution model

The generalized structure tensor strain energy function given in (6.31) will be recovered from the integral strain energy function in (6.29) if the angle dependent fibre strain energy in (6.29) is approximated by the same function, evaluated at the representative average stretch. Therefore, in some sense, (6.31) can be considered as an approximation of the integral fibre distribution model (6.29). However, there is an additional difference, in that the condition for fibre loading $\lambda_t \geq \lambda_a$ is only met in an averaged sense for the generalized structure tensor model. Efforts have recently been made to try to quantify the magnitude of the differences in predictions of these two models [22, 27] though controversy remains [55]. Cortes et al. have compared the Piolo-Kirchhoff stress from the GST model with the results of the integral model using material constants chosen for the supraspinatus tendon [22]. Results were compared for uniaxial tension, biaxial loading and simple shear. The formulations were found to be equivalent for equal-biaxial stretch of materials with planar distributions. For other loading conditions and fibre orientations, it was recommended that κ be substantially less 0.1 to ensure errors in the GST model are less than 10 %.

6.3.5 Microstructural analysis

In this section, we discuss methods that have recently been developed to non-destructively evaluate collagen and elastin fibre orientation simultaneous with uni-axial loading experiments.

Multiphoton Microscopy (MPM) – 2PE and SHG

Using multiphoton microscopy, elastin can be imaged due to its intrinsic fluorescence under two- photon excitation (2-PE). The same microscope can be used to simultaneously image collagen using its second harmonic generation (SHG).

Two photon excitation (2PE) microscopy is so named because of the employment of two-photon excitation. Briefly, two photons excite a flourophore in the test sample, resulting in the emission of a fluorescence photon. The wavelength of the emitted photon depends on the type of fluorophore. In 2PE MPM, the emitted photons from laser scanned surfaces are collected in a photomultiplier tube to generate images which can then be 3D reconstructed. Compared with confocal microscopy, 2PE microscopy has deeper tissue penetration (up to 1 mm) more efficient light detection and reduced phototoxicity. Elastin is well imaged by 2PE excited intrinsic fluorescence without exogenous stains [121].

Second harmonic imaging (SHG) microscopy makes use of a nonlinear optical phenomena in which photons interacting with a nonlinear material are combined, forming new photons with twice the energy (so twice the frequency and half the wavelength). The variation in contrast in the image arises due to the variability in the specimen's ability to form second harmonic light. In particular, SHG enables direct imaging of some anisotropic biological structures such as collagen without staining or fixation [24]. As for 2PE microscopy, infrared light can be used, making it possible to generate 3D images of specimen regions relatively deep in the tissue.

Uniaxial Multi-Photon Microscopy system (UA-MPM)

A uniaxial mechanical testing device was custom designed to operate in conjunction with a multiphoton microscope (MPM) for coincident stress-strain analysis and laser scan imaging of biotissues, Fig. 6.4, [49, 50]. During testing, arterial segments are uniaxially loaded in the device under applied strain. Utilizing the second harmonic generation (SHG) and two-photon excited autofluorescence (TPEA), endogenous fibrillar collagen and elastin fibres are visualized. Image stacks are then 3D reconstructed (Metamorph, Danaher, Washington, DC, USA and Imaris software, Bit-plane, Switzerland).

In the UA-MPM system, data acquisition and motion control is performed with National Instruments (NI, Austin, TX, USA) software and hardware. A stepper mo-

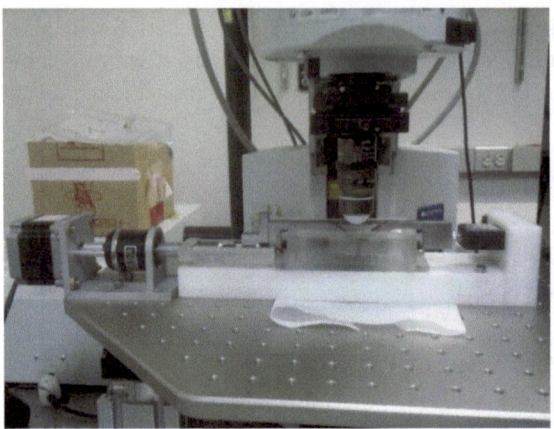

Fig. 6.4. Horizontal uniaxial mechanical testing device mounted on the stage of the multi-photon microscope

tor and driver from Applied Motion Products (Watsonville, CA, USA) connected to a Velmex (Bloomfield, NY, USA) linear slide is used to apply extension, and a MDB-5 5-lb. capacity load cell (Transducer Techniques, Temecula, CA, USA) is used to measure force. A custom clamping system is used to remove the specimen under applied strain. An acrylic bath is constructed with a low profile, so the submerged tissue is accessible by the MPM lens for imaging during mechanical testing. The MPM is an an Olympus FV1000 MPE (Olympus, Tokyo, Japan) equipped with a Spectra-Physics DeepSee Mai Tai Ti-Sapphire laser (Newport, Mountain View, CA, USA) and a 1.12NA 25x MPE lens.

Tissue assessment using the UA-MPM system

In this illustrative example, we describe the application of the UA-MPM system to evaluate circumferential samples of the left common carotid arteries from New Zealand white rabbits and then use this data in a structurally motivated model for the wall. Here, the arterial wall is modelled as a single layer multi-mechanism material with isotropic and anisotropic mechanisms. Attention is focused on fibres contributions from the medial layer. The isotropic mechanism is modelled as a Neo-Hookean material and the anisotropic mechanism is modelled using a generalized structure tensor approach with conical splay, Eq. (6.23). Earlier studies suggest collagen recruitment can be approximated as commencing at at a finite stretch and taking place over a narrow stretch range [50]. Therefore, in this work, we approximate the collagen recruitment stretch λ_a as a discrete, finite value. In particular, using (6.23) and (6.36)

$$W = W_{iso} + W_{aniso}, \quad \text{with} \quad W_{iso} = \eta_{iso}(I_0 - 3),$$

$$W_{aniso} = \mathsf{H}(\overline{\lambda}_t - 1)\frac{\eta}{\gamma}\left[e^{\left(\gamma(\overline{\lambda}_t^2 - 1)\right)} - 1\right]$$

where

$$\overline{\lambda}_t^2 = \frac{1}{\lambda_a^2}\underline{C}_0 : \underline{H}_0, \qquad \underline{H}_0 = \kappa\underline{I} + (1-3\kappa)\overline{a}_0 \otimes \overline{a}_0,$$

$$\kappa = \frac{1}{4}\int_0^\pi \rho(\Theta)\sin^3\Theta\,d\Theta. \quad (6.40)$$

Material parameters λ_a, $\overline{a}_0$ and $\rho(\Theta)$ can be obtained from fibre images and used to calculate $\underline{H}_0$ and κ. Constants η_{iso}, η and γ can be obtained from the nonlinear stretch-stress data. It is straightforward to consider an integral fibre representation as in (6.26) and (6.29) rather than the GST model [50], or to use planar splay rather than conical splay.

Samples of left common carotid arteries were obtained from New Zealand white rabbits and tested in the UA-MPM device (Fig. 6.4) using a previously developed protocol [49, 50]. Briefly, a sample of artery was removed from its source, opened longitudinally, manually cut into circumferential strips with a "dogbone" shape, and then placed in the UA-MPM system for extension tests. Prior to testing, the thickness (H) and width (W) of the unloaded tissue sample were measured five times with calipers and averaged to obtain cross sectional area.

After preconditioning, force and stretch values were recorded under quasi-static loading conditions, Fig. 6.5. Force data recorded from the load cell was converted to Cauchy stress assuming an isochoric deformation. Stretch λ was measured directly from displacement in the uniaxial device. For a given deformation of the arterial strip, we then obtained averaged Cauchy stress as a function of sample stretch Fig. 6.5. The collagen fibres were imaged at six different stretch values during this process Fig. 6.6.

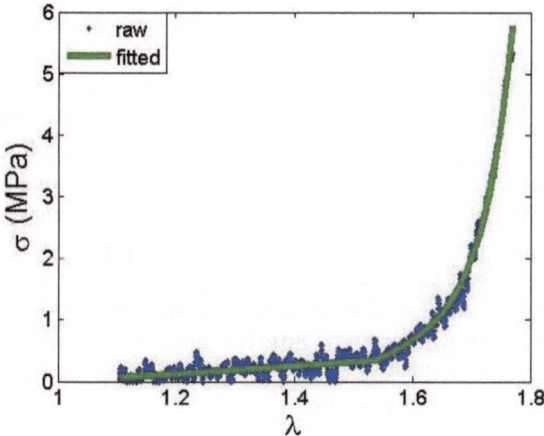

Fig. 6.5. Raw and fitted data from uniaxial extension of Sample 01 of rabbit carotid artery. Corresponding MPM-UA images are shown in Fig. 6.6

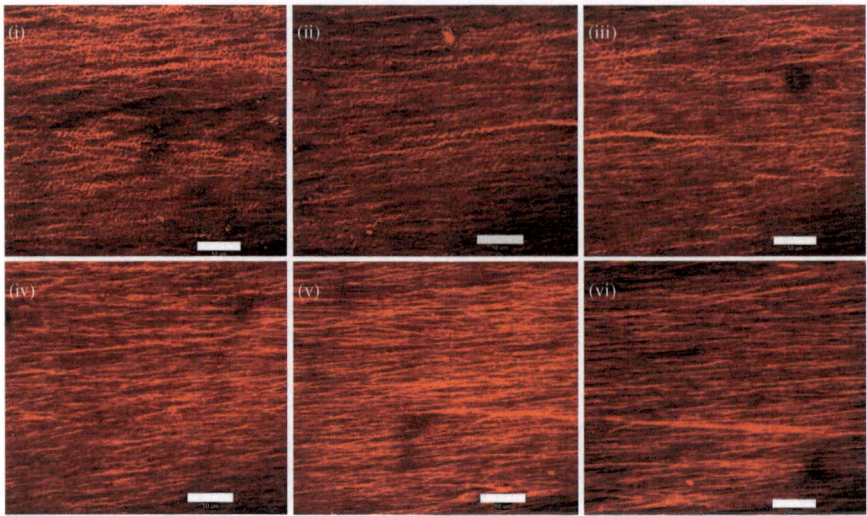

Fig. 6.6. MPM Images of collagen at various stretches: (i) 1.10, (ii) 1.33, (iii) 1.44, (iv) 1.55, (v) 1.66, and (vi) 1.77 (Bars = 50 μm). Sample 1

Recruitment stretch

In Sample 01, all visible fibres appear to be wavy at stretch of 1.1. Between, stretch values of 1.44 and 1.55, the collagen crimp appears to be nearly if not completely eliminated. The slope of the stress stretch curve visibly increases in this range, consistent with recruitment of collagen Fig. 6.5. These results suggest medial collagen fibres are recruited at a finite stretch over a relatively small stretch range [3]. Further analysis is being performed to evaluate this hypothesis [50].

We have recently developed techniques to quantify the degree of crimp and these methods are being used to quantify the fibre recruitment process, including an assessment of the validity of the discrete recruitment approximation [50]. Briefly, a threshold level for each fibre is selected manually. Then skeletonization and edge detecting algorithms are used to trace a curve passing along a fibre segment, Fig. 6.7, MATLAB (The Mathworks, Inc.). A commonly used measure of tortuosity is used to quantify crimp [10, 45], defined as the ratio of fibre arc length (L) to the linear distance of the fibre (L_0), Fig. 6.7. The linear distance was obtained as the intersection of the curve and the line with minimal distance to the curve (obtained using linear least squares), (Fig. 6.7). Tortuosity was computed at six different stretches for Sample 02 (Fig. 6.8a). A decrease in average tortuosity can be seen between data at stretches of 1.7 and 1.8. The corresponding stretch-stress curve as well as the slope of this curve is shown in (Fig. 6.8b). The slope of the stress stretch curve begins to increase rapidly in this stretch range, Fig. 6.8a, consistent with the decreased tortuosity during collagen recruitment.

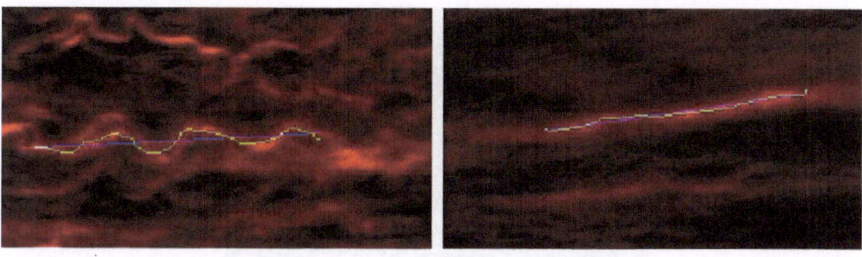

Fig. 6.7. Images of medial collagen in a rabbit carotid artery wall taken with the UA-MPM system under circumferential stretches of 1.4 (left) and 2.0 (right) and analyzed with a custom MATLAB program to give tortuosity values of 1.1 and 1.0, respectively

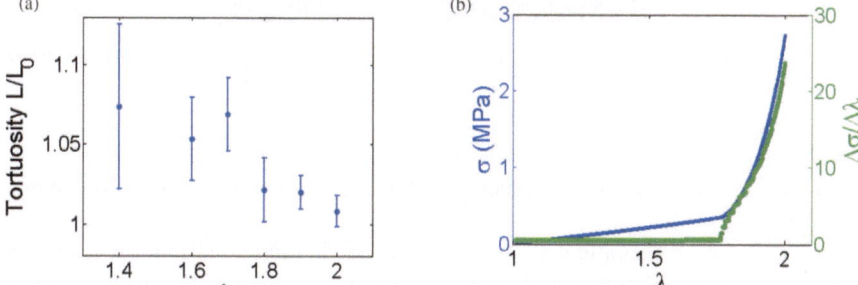

Fig. 6.8. Results for sample 2 of rabbit artery under uniaxial loading showing (a) Mean tortuosity (L/L_0) as a function of circumferential stretch, λ, (error bars = standard deviation) and (b) Stress versus stretch as well as slope of stress versus stretch

Collagen fibre distribution

A custom program was used to evaluated fibre angle distribution using a method introduced by Courtney for tissue-engineered constructs [23]. To avoid artifacts from fibre waviness, collagen fibre orientation was measured in a configuration (stretch state) with little fibre crimp. The orientation distribution of fictitious uncrimped fibres in κ_0 that are related to the actual fibres in $\kappa_{(t)}$ through an affine transformation was obtained using a pull back operation. Consistent with previously reported results, we found the distribution of projected fibre orientations in the media to be approximately symmetrically distributed about the circumferential direction.

Therefore, as in [35], a single mean reference direction $\overline{a}_0$ was used, aligned with the circumferential direction. In this way, material constants $\lambda_a, \overline{a}_0$, and $\rho(\Theta)$ were obtained from images of the collagen fibres. The parameters $\underline{H}_0$ and κ could then be calculated from these values using (6.40). $\underline{C}_0$ was calculated from the applied tissue stretch and used to obtain $\overline{\lambda}_t$ from (6.40). The material constants $\eta_{iso}, \eta_{ansio}, \gamma_{aniso}$ were determined using nonlinear regression analysis of the fit of (6.40) to the measured data, Fig. 6.5. Data fit the conic splay model well for the illustrative sample (Fig. 6.5), with $R^2 = 0.96$. Material parameters were obtained as $\eta_{iso} = 196$ kPa, η

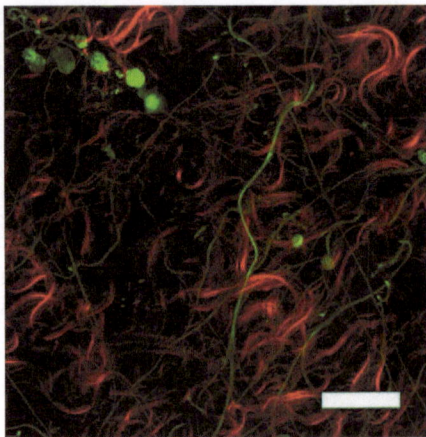

Fig. 6.9. Adventitia of a rabbit carotid artery imaged with the UA-MPM system, revealing elastin fibres (green) and collagen fibres (red), bar = 50 μm

= 566 kPa, and γ = 11.3. The thickness was 99 μm, width was 3.2 mm, initial length of 3.7 mm.

This illustrative example was focused on the medial fibre distribution. A similar approach is currently being used for a comprehensive study of the distinct contributions of medial and adventitial collagen. Fig. 6.9 shows a representative image of the adventitia obtained using the MPM-UA system.

6.4 Damage to the arterial wall

Mechanical damage can take the form of creation and growth of microvoids or microcracks. These are discontinuities in a medium that are often modelled as continuous on a larger scale [69]. In the arterial wall, damage can occur to both the collagen and elastin. However, while collagen naturally turns over in a health artery, the mature arterial wall appears to be incapable of generating functional elastic fibres. Further, damage to elastic fibres and the IEL are associated with a number of significant diseases. For these reasons, much of the attention in this section is directed at damage to the elastic fibres and the IEL. In this section, we discuss the physical nature of elastin damage in the arterial wall as well as theoretical models of damage to both collagen and elastic fibres/IEL. In Sect. 6.5, we consider the application of these models to to cerebral angioplasty.

6.4.1 Physical nature of damage

In the case of the IEL, damage in the form of tears and fragmentation has been reported. As early as the 1920s, Reuterwall reported tears in human IELs that have been termed "Reuterwall's tears" by Hassler [47]. These are common in older (>30

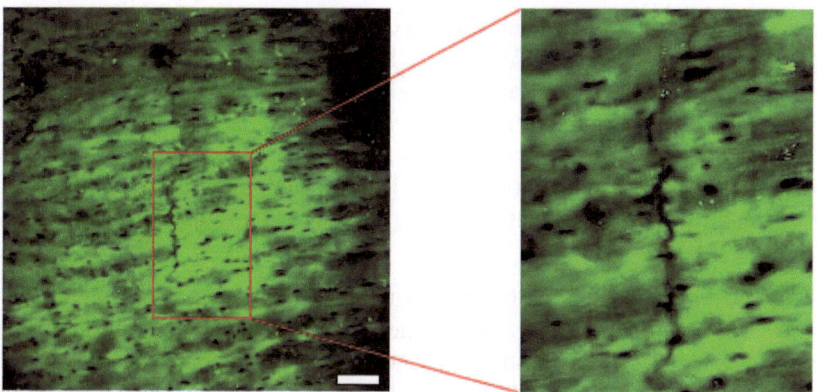

Fig. 6.10. Multiphoton microscopy images from en face preparations of the basilar artery in a human, fixed at 30 % strain, depicting autofluorescent elastin, by utilizing two-photon emission (2PE) spectroscopy and revealing the fenestrated internal elastic lamina with a crack oriented in the direction of applied load (circumferential direction, vertical in image). Bar = 50microns

years) individuals, and they are generally found in larger arteries, such as basilar and internal carotid arteries, as transverse gaps 700-3000 µm in length [47]. We are able to assess these cracks using the UA-MPM system, Fig. 6.10.

Cracks have also been seen in experimental arteriovenous fistulas created between the common carotid artery and jugular vein [9]. In this latter case, no evidence of elastolytic activity was found, so the cause was hypothesized to be due to direct over-stressing (acute rupture) or from fatigue-type wear, discussed below. Histological examination of the IEL from common carotid arteries subjected to longitudinal [9] and circumferential [49] uniaxial failure tests have also shown damage to the IEL in the form of mechanically-induced tears. Elastic fibres have been reported to naturally fray and fragment over time [11]. As these fibres are progressively damaged and possibly fail, the mechanical load will be transferred to the stiffer collagen fibres [41, 92, 117, 119], leading to arterial stiffening.

Sources of damage

Low cycle fatigue at high loads. A study of the effect of balloon angioplasty on bovine carotid arteries revealed that mechanically-induced damage under high loads resulted in a shift in the mechanical response curve [92]. Elastic fibre damage was not assessed histologically, but the results were applied to a continuum damage model for arteries. Additionally, tests have been performed on "dogbone" shaped specimens from the human aorta to evaluate ultimate stress and extension ratio at failure during circumferential and axial loading [79], though neither a continuum damage framework nor histological techniques were employed. Scott et al. have demonstrated damage in cerebral vessels under low cycle pressure inflation loading conditions [102]. The tension/stretch curves displayed a loss of the toe region after three cycles of loading to approximately 200 mmHg.

High cycle fatigue. The arterial wall is exposed to over 3.0×10^9 cycles of loading during a 75 year lifetime. In 1976, O'Rourke hypothesized this high cycle fatigue was responsible for the frayed appearance of the elastic fibres in aged arteries. Fractal analysis has been performed on histological sections of elastin lamellae in arteries [61], with results indicating that elastin fatigue occurred due to increased cardiac cycles over the same time period [3]. Constituent-based structural models have been developed to account for aging [31, 124] and fatigue related degradation of the IEL [74].

Biochemical degradation of elastin. In vitro studies in rabbit carotid arteries have demonstrated qualitative changes in the mechanical properties of the wall after enzymatic damage to arterial elastin [31]. Chemically induced damage to the IEL has also been hypothesized to arise from self induced enzymatic damage (e.g. [104]). For example, mechanical factors such as altered wall shear stress and wall shear stress gradient may trigger a signaling cascade in the endothelial cells lining blood vessels, causing them to initiate the degradation of the IEL seen in cerebral aneurysm formation.

6.4.1.1 Lack of repair of damaged elastin

Deposition of arterial elastin begins in vivo and peaks early in post natal development, driven by hemodynamic factors The creation of elastic fibres in adulthood is believed to be negligible (see, e.g. [111]) Using tritium-labeled valine, Davis showed that no elastin turnover or growth occurs in the mouse aorta during adulthood [25]. Shapiro et al. estimated the longevity of elastic fibres in the lung in to be on the order of the human lifetime, using aspartic acid racemization and 14C turnover [103]. These results suggest the detrimental effects of damage to the elastic fibres will not, in general, be repaired. The inability of cell in the arterial wall to reactivate the many genes in the proper ratios and sequences required for normal fibre assembly [105] may prevent its deposition later in life as well as elastic fibre repair.

6.4.1.2 Medical impact of damage to the IEL

Age-associated arterial stiffening has been conjectured to arise as a result of fatigue damage to elastic fibres that then results in load transfer to the (stiffer) collagen fibres [85]. The stiffened vessels are less capable of storing elastic energy during the cardiac cycle resulting in an increase in pressure during diastole and reduced pressure through diastole. Increases in atherosclerosis, arterial weakening, and detrimental cardiac changes have been attributed to these changes in pressure waveform [86]. This stiffening leads to an increase in the arterial wave speed [86] and wave reflections. Studies have indicated that increased arterial wave reflections are predictors of severe cardiovascular events, and therapeutic attempts to reduce these wave reflections to improve prognosis have been suggested [114].

Damage to the elastin in the IEL is associated with pathological disorders such as cerebral aneurysms [4, 12, 13, 14, 19, 92, 102], spontaneous cervical artery dissection aneurysms [8, 101], and complications associated with balloon angioplasty [57].

6.4.2 Structurally motivated constitutive models – damage regime

Motivated by a need to model failure of the IEL separately from failure of the other components of the arterial wall, a nonlinear, inelastic, isotropic, dual-mechanism constitutive equation was developed for cerebral artery tissue damage [119]. This model was subsequently extended to include the collagen derived anisotropy [71], subfailure isotropic damage to the IEL [74] and collagen damage [70]. While a scalar damage function was used for collagen in this later work, the resulting damage was anisotropic. Both discontinuous and continuous damage were considered as well as enzymatic damage. Balzani et al. previously considered discontinuous damage to the anisotropic collagen fibres using a scalar damage function [5]. Damage to the isotropic mechanism was not considered. A method was used to include the effect of the residual stresses in the unloaded configuration. Circumferential overstretching of atherosclerotic arteries was modelled.

In this section, we briefly provide the theoretical background for the model. In the next section, we apply this model to the analysis of damage in arteries. Continuum damage models are introduced to model the progressive deterioration in mechanical properties on a scale where it is suitable to homogenize the individual cracks. We restrict attention to isothermal theory and by way of example consider materials for which the undamaged response is hyperelastic. It is straightforward to generalize this approach to rate-type models. In the following discussion, a single mechanism is considered. Similar approaches follow directly for multi-mechanisms materials [70, 74] and multi-mechanism damage will be used in Sect. 6.5. In this work, attention is confined to isotropic damage where a scalar damage variable can be used. This can be extended to anisotropic damage.

Following earlier work on continuum damage mechanics, (e.g. [107]), we begin by postulating the existence of a Helmholtz strain energy function per unit volume in κ_0 (W) that depends on $\underline{C}_0$ as well as a scalar damage variable (d) We take a strain space based approach, assuming the dependence on d can be explicitly written as,

$$W = (1-d)W^o(\underline{C}_0) \tag{6.41}$$

where W^o is the effective strain energy of the hypothetical undamaged material subject to the conditions

$$W^o(\underline{I}_0) = 0, \qquad W^o(\underline{C}_0) \geq 0. \tag{6.42}$$

The internal variable d is defined to be in the range $[0, 1]$ with zero corresponding to no damage and one to total damage. The factor $(1-d)$ is called the **reduction factor** (for obvious reasons). This form was first proposed by Kachanov [62] to model creep rupture of metals. The Clausius-Planck inequality is imposed by requiring the internal dissipation $\mathscr{D}_{in}$ be non-negative for all times and material points in the body,

$$\mathscr{D}_{in} = -\dot{W} + \underline{\sigma} : \underline{D} \geq 0 \tag{6.43}$$

where $\underline{D}$ is the symmetric part of the velocity gradient and the over dot signifies the material derivative. It then follows from (6.41)

$$\dot{W} = -\dot{d}W^o + 2(1-d)\left(F_0\frac{\partial W^o}{\partial C_0}F_0^T\right):\underline{D} \tag{6.44}$$

and therefore, from (6.43),

$$\dot{W}^o\dot{d} + \left(\underline{\sigma} - 2(1-d)F_0\frac{\partial W^o}{\partial C_0}F_0^T\right):\underline{D} \geq 0 \tag{6.45}$$

so that,

$$\underline{\sigma} = (1-d)\underline{\sigma}^o, \qquad \underline{\sigma}^o = 2\left(F_0\frac{\partial W^o}{\partial C_0}F_0^T\right) \tag{6.46}$$

with the additional requirement that,

$$\mathcal{D}_{in} = W^o\dot{d} \geq 0. \tag{6.47}$$

The quantity $\underline{\sigma}^o$ is the effective Cauchy stress tensor for the hypothetical undamaged material. Defining f as the thermodynamic conjugate to d through

$$\mathcal{D}_{in} = f\dot{d} \tag{6.48}$$

we see immediately from (6.41), (6.43), (6.47) and (6.48) that,

$$f = W^o = -\frac{\partial W}{\partial d} \quad \text{and} \quad f\dot{d} \geq 0. \tag{6.49}$$

The quantity f is termed the thermodynamic force and drives the damage evolution, [78]. The evolution equation for f (or equivalently W^o) can then be calculated from (6.49),

$$\dot{f} = \underline{\sigma}^o : \underline{D}. \tag{6.50}$$

The quantity $\dot{f}$ is therefore the rate of work by the effective stresses per unit volume of the body, (*stress power*).

Damage modes

We now turn attention to the cause of damage and in particular to the definition of a damage evolution equation. As in [74], we consider three modes of damage with damage variables denoted as d_j with $j = 1, 2, 3$. Two of these modes are purely mechanical following earlier work [78, 107]. As elaborated on below, the third mode of damage arises due to enzymatic degradation of the wall, arising for example from elastase activity [74]. We quantify damage accumulation through the variable α_j and assume $d_j(\alpha_j)$ is a monotonically increasing and smooth function with

$$d_j(\alpha_j) \in [0, 1] \quad \text{for all} \quad \alpha_j \in [\alpha_{js}, \alpha_{jf}] \tag{6.51}$$

and initial conditions $d_j(0) = 0$. As discussed below, the rate of damage accumulation will be different for different modes. In particular, modes one and two will defined a functions of the strain energy the material has experienced while the rate of accumulation in mode three will depend on the history of exposure to hemodynamic wall shear stress. An example of the functional dependence $d_j(\alpha_j)$ is,

$$
d_j(\alpha_j) = \begin{cases} 0, & \alpha_j < \alpha_{js} \\ D_j[\alpha_j(t)], & \alpha_{js} \le \alpha_j < \alpha_{jf} \\ 1, & \alpha_{jf} \le \alpha_j \end{cases} \qquad \text{and}
$$

$$
D_j[\alpha_j(t)] = 1 - \frac{1 - e^{c_j(1-\alpha_j/\alpha_{jf})}}{1 - e^{c_j(1-\alpha_{js}/\alpha_{jf})}}. \quad (6.52)
$$

Here, j is the damage mode, α_{js} is the critical value of α_j for the start damage mode j, and α_{jf} is the critical value for complete failure.

Discontinuous damage mode

When a rubber specimen is loaded uniaxially in tension, unloaded and then reloaded, the applied stress necessary to achieve a given level of strain decreases in the following loading cycles. This phenomena is termed *stress softening*. It is also often referred to as the *Mullins effect*, so named due to the early studies by Mullins on stress softening in rubber materials with imbedded particles [82, 83].

Several microstructual explanations have been given to explain this phenomena. It seems likely that multiple mechanisms are involved, (e.g. [39]). For example, it has been conjectured that one mechanism for stress softening is the breakage of bonds between the filler particles and surrounding matrix during previous loading cycles. Damage of this kind has been modelled by setting the current value of the accumulation variable α_1, equal to the maximum effective strain energy the material has experienced [107],

$$
\alpha_1(t) = \max_{s \in [0,t]} W^o(s). \quad (6.53)
$$

The corresponding evolution equation for α_1 is then

$$
\dot{\alpha}_1 = \begin{cases} \dot{W}^o & \text{if } W^o = \alpha_1 \text{ and } \dot{W}^o > 0 \\ 0 & \text{otherwise} \end{cases} \quad (6.54)
$$

with the initial condition $\alpha_1(0) = 0$. Namely, damage only accumulates when the effective strain energy increases beyond the previous maximum.

Continuous damage

Damage has also been found to increase during cyclic loading with effective strain energies below the maximum value obtained during the prior history of loading. To address this phenomena, Miehe [78] introduced a contribution to damage evolution arising as the arc length of the effective strain energy. Following this work, the second mechanical damage variable d_2 is defined as a function of accumulated equivalent strain through the accumulation variable $\alpha_2(t)$,

$$\alpha_2(t) = \int_0^t \left| \frac{dW^o}{ds} \right| ds. \tag{6.55}$$

The corresponding evolution equation is,

$$\dot{\alpha}_{02} = \left| \dot{W}^o(t) \right| \tag{6.56}$$

with the initial condition $\alpha_2(0) = 0$. For this damage mode, α_2 accumulates continuously as the material deforms.

Enzymatic damage

We also consider the possibility that damage is accumulated due to enzymatic (or other chemical) damage. For example, in arteries, abnormally high wall shear stress under suitable wall shear stress gradients has been associated with damage to the internal elastic laminae of arteries [33, 46, 64, 81]. Animal studies have shown that this degradation can be induced by exposing bifurcations to abnormal hemodynamic loads. In particular, the combination of elevated wall shear stress WSS and positive elevated wall shear stress gradient $WSSG$ has been associated with elastin degradation in native and nonnative bifurcations [76, 77]. The degradation has been found to be progressive in that damage increases with exposure time [81]. The degree of pre-aneurysmal type change was found to be dose dependent. Motivated by these results, the following form for α was proposed [74],

$$\alpha_3(t) = f(WSS, WSSG) \tag{6.57}$$

where the appropriate definition of $WSSG$ is discussed in [120]. This form of α_3 is purposely left quite general, but will be similarly chosen to be an invariant quantity. A particular example is

$$\alpha_3 = \frac{1}{T} \int_0^t H(\zeta)H(\eta)(\zeta + b\eta)ds \quad \text{where}$$
$$\zeta = \frac{WSS - WSS_T}{WSS_T}, \quad \eta = \frac{WSSG - WSSG_T}{WSSG_T}. \tag{6.58}$$

The corresponding evolution equation is

$$\dot{\alpha}_3 = \frac{1}{T} H(\zeta(t))H(\eta(t)) \left(\zeta(t) + b\eta(t) \right) \tag{6.59}$$

where b and T are material constants. Eq. (6.59) satisfies the criterion that damage does not increase unless WSS is sufficiently elevated above a threshold value and the WSSG is elevated above a positive threshold value. Further, the evolution Eq. (6.59) satisfies the condition that damage increases with exposure time and dosage. More data is needed to determine whether the rate of accumulation should increase with increased η (amount by which WSSG exceeds the threshold value) or whether it is only necessary that η be positive, in which case b can be set to zero. Coupling between the damage modes can be included. For example (6.59) can be modified to the rate of enzymatic damage is depemdent on the strain state of the tissue.

Effect of combined damage modes

Following Miehe 1995 [78], when more than one damage mode are involved, we can define the damage parameter as the sum of the damage parameters for the various modes so that the reduction factor given in (6.41) becomes in [78],

$$(1-d) = (1-d_1-d_2-d_3) \tag{6.60}$$

with (6.51). In this case, we must impose the additional restriction that condition that $d_1 + d_2 + d_3 \leq 1$. Namely, the combined damage must not exceed complete damage for the material. An alternative choice is to consider a multiplicative combination of these two modes,

$$(1-d) = (1-d_1)(1-d_2)(1-d_3). \tag{6.61}$$

In this case, there is no additional restriction beyond (6.51).

The damage metrics $\alpha_1(t)$, $\alpha_2(t)$ are clearly invariant to superposed rigid body motions. The form of α_3 defined in (6.57) is purposely left quite general, but will similarly be an invariant quantity.

6.4.3 Damage assessment using the UA-MPM device

Two human basilar arteries were obtained from cadaver circles of Willis, cut into segments, and tested in the UA-MPM device, using the previously described protocol [49], as mentioned in Sect. 6.3.5. Uniaxial damage experiments were performed with the UA-MPM system on artery segments until total large scale mechanical tears of the IEL were confirmed. We modelled the undamaged arterial wall as a single layer multi-mechanism material, and discontinuous damage was modelled as described in Sect. 6.4. Thus, the strain energy function was modified to give

$$W = (1-d_1(\alpha))W_{iso} + W_{aniso} \tag{6.62}$$

with W_{iso} given as in Eq. 6.40. The scalar-valued isotropic damage parameter for discontinuous damage d_1 was given by (6.52), with the damage accumulation metric $\alpha_1(t)$ given by Eq. (6.53). We assumed the anisotropic response was dominated by circumferentially-oriented medial collagen fibres. W_{aniso} was defined as in (6.19) with $N = 1$. Collagen recruitment occurred when the circumferential stretch ratio exceeded the activation stretch, λ_a.

Data were fit to the discontinuous damage model via nonlinear regression to obtain material and damage parameters (Fig. 6.11). At high strain, the intima region split perpendicular to the circumferential loading direction and peeled away from the media, as observed from visual inspection and images taken from a CCD camera (Fig. 6.12). Multi-photon images revealed a cat's eye-shaped longitudinal tear in the internal elastic lamina, which pulled back under strain to reveal mostly

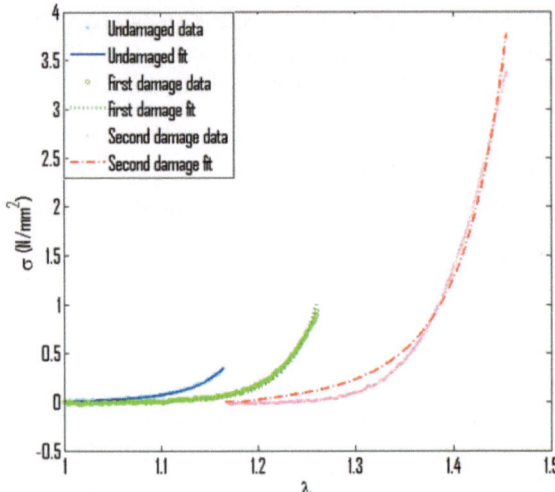

Fig. 6.11. Damage results for a strip of basilar artery tissue loaded in the circumferential direction. Data were fit to the discontinuous acute rupture model

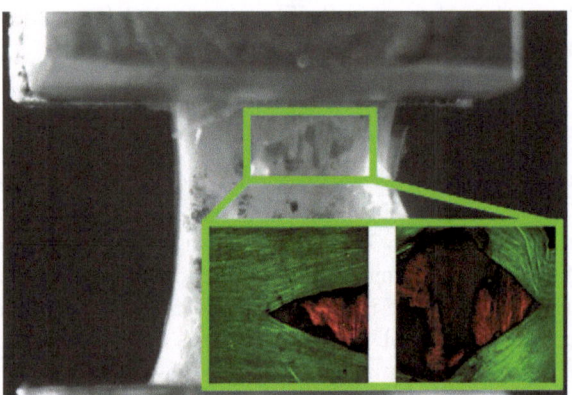

Fig. 6.12. Damage from circumferential testing is evident on the macroscale towards the upper region of the tissue, shown here clamped at a stretch of approximately 1.45. Inset: Multi-photon image from MetaMorph depicting the damage shown in the left image. The elastin (green) appears torn and pulled away from the underlying medial collagen (red), bar = 50 μm

circumferentially oriented collagen in the media (Fig. 6.12). Values for material parameters were: $\lambda_a = 1.16 \pm 0.08$, $\eta_{iso} = 69.0 \pm 85.9$ kPa, $\gamma_{iso} = 17.8 \pm 14.5$, $\eta_{aniso} = 677 \pm 615$ kPa, $\gamma_{aniso} = 5.6 \pm 2.4$. Values for damage parameters were $\lambda_s = 1.31 \pm 0.08$, $c_1 = 2.9 \pm 1.5$, $\alpha_{1f} = 28.9 \pm 33.5$. Note that λ_s, the stretch at the start of elastin damage, was used to compute α_{1s} in Eq. (6.52). R^2 values were typically close to unity for each specimen.

The damage observed in these experiments is on a scale larger than what can be reasonably considered in continuum damage mechanics. The continuum damage model may be suitable for the other types of damage, cyclic fatigue and enzymatic degradation. Evidence of a uniform frayed appearance has been observed in aged arteries [30]. We are currently conducting cyclic fatigue studies to further explore this damage mode. Similar to the acute rupture tests presented here, resulting data will be fit to the continuum damage model described above.

6.5 Applications of the damage model to cerebral angioplasty

Cerebral percutaneous transluminal angioplasty (PTA) is an important interventional neurovascular technique for the treatment of atherosclerotic obstructions and va-sospasm in cerebral vessels. During this procedure, a balloon is inflated within the cerebral vessels in an attempt to improve cerebral perfusion and as a result reduce long-term stroke and death [48, 57, 116]. The primary mechanical mechanism of lumen enlargement by PTA is explained as the overstretching of the arterial wall [17]. Vessel wall damage is observed at the site of PTA including intimal damage (endothelial damage, subendothelial destruction, fractured IEL) and medial changes (damaged myocytes, loss of dense bodies, gap in the extracellular matrix, disorga-nized collagen fibres) [16, 18, 21, 57, 122]. In addition to local structural damages such as areas of disruption and dissection throughout the vessel layers, larger scale structural damage can occur including partial tears of the intima or media and even vessel rupture [21, 57].

Arterial wall damage caused by angioplasty was studied in common carotid, iliac, and femoral arteries of mongrel dogs, [122]. Damage was found to be dose depen-dent. After 25 % inflation, the wall exhibited localized fractures and stretching of the IEL and damage to the inner one third of the media. At 50 % inflation, exten-sive damage to the IEL, dissection of the media, distorted SMC and disorganized collagen fibres through more than one-half of the media were reported. Damage was largest in the inner layers and increased outwards. While the repair of smooth muscle cells and collagen in the media was visible at six months, the IEL showed no signs of recovery.

It is important to develop realistic computational tools for studying this disease. Sidorov [106] simulated balloon angioplasty with a homogeneous and isotropic multi-mechanism arterial model developed by [119]. Material anisotropy, dam-age and heterogeneity were not included in that study. Gasser and Holzapfel [34] modelled plaque fissure and dissection during angioplasty using an anisotropic and elastoplastic material formulation with two arterial layers [36, 52]. Arterial injury was analyzed indirectly via the distribution of a plastic hardening variable.

The damage to the arterial wall following PTA has not been rigorously inves-tigated theoretically and numerically. In this section we summarize recent results using the multi-mechanism framework discussed in Sect. 6.3 with the damage the-ory in Sect. 6.4 to model cerebral angioplasty, [71, 74]. An extension of the theory is made to include anisotropic damage to collagen fibres so medial injury can be modelled, [70, 72, 73, 75]. A three-layer heterogeneous vessel wall model is used in which each layer of the wall is defined as a structural multi-mechanism material. A three-dimensional computational model of artery and balloon is formulated to simulate the artery-balloon interaction during cerebral angioplasty. Due to the extensive computational requirements for contact analysis with nonlinear materials, an axisymmetric geometry and simplified balloon model are used. We do not consider the role of SMC or of residual stresses. The loading is broken into multiple load steps including vessel inflation-tension, balloon deployment, artery-balloon contact and balloon deflation. Future experimental work is needed to provide guidance in the selection of damage parameters.

Arterial wall model

To represent the heterogeneous histological structure of cerebral arteries [29], we consider the wall as composed of three distinct layers: the IEL, media and adventitia, Fig. 6.13. The unloaded geometry is idealized as a circular cylinder with constant wall thickness.

Based on the histology studies of cerebral arteries [28, 91], it is assumed that all elastin is concentrated in the IEL, the fibres in the media are near the circumferential direction while the fibre families in the adventitia are dispersed about two fibre directions. All layers will be modelled as incompressible. The intima will be assumed to be composed solely of an isotropic mechanism. Since we expect that in cerebral arteries, the mechanical properties of this layer will be dominated by the response of the IEL, we simply refer to this layer as the IEL. In diseased arteries with intimal thickening, it may be appropriate to generalize the model to include intimal collagen fibres. The medial and adventitial layers will be modelled fibre-reinforced composites with an isotropic mechanism arising from the ground matrix and an anisotropic mechanism arising from the collagen fibres. The anisotropic mechanism is assumed to arise from a pair of helically wound fibres oriented symmetrically with respect to the circumferential direction, Fig. 6.13. A two fibre model will be used for the media and a generalized structure tensor approach with two fibre directions will be used for the adventitia. In both cases, it is assumed the two families are oriented symmetrically with respect to the circumferential direction of the artery, Fig. 6.13.

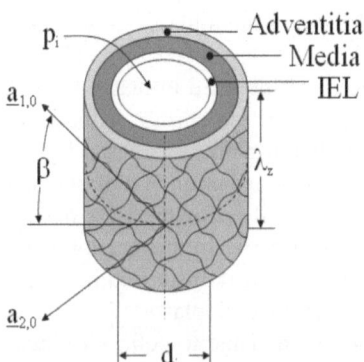

Fig. 6.13. Cylindrical multi-layer arterial wall model with the IEL, media and adventitia layers

So, for example,

$$\bar{a}_0^{(1)} = \cos\beta\underline{e}_\theta + \sin\beta\underline{e}_z, \qquad \bar{a}_0^{(2)} = \cos\beta\underline{e}_\theta - \sin\beta\underline{e}_z. \qquad (6.63)$$

Since acute vessel injury following PTA has been found to be limited to the intima and media layers [57, 122], we only consider damage to the mechanisms in the intima and media. The damage to each layer will be modelled using the isotropic structural damage model introduced in Sect. 6.3. Here, only mechanical damage is considered since the time scale is too short for a secondary response of hemodynamic loads to be important [72]. Further, due the low number of loading cycles and large magnitude of the load, only discontinuous mechanical damage is considered.

For all layers, the isotropic contributions will be assumed to be of the form,

$$W_{iso} = (1 - d(\alpha_{iso})) W_{iso}^o(I_0), \text{ where } \begin{cases} W_{iso}^o(I_0) = \dfrac{\eta_{iso}}{2\gamma_{iso}} \left(e^{(\gamma_{iso}(I_0 - 3))} - 1\right) \\ \alpha_{iso} = \max_{s \in [0,t]} W_{iso}^o(s) \\ d(\alpha_{iso}) \text{ is given in Eq. (6.52)} \end{cases} \qquad (6.64)$$

where η_{iso} and γ_{iso} are layer dependent material constants.

The components of the anisotropic mechanism in the medial layer are modelled using a two fibre model. It is assumed the fibre families in the two directions have the same material properties, differing only in their fibre orientation. As introduced in (6.19),(6.41),(6.53),

$$W_{aniso} = \sum_{i=1}^{2} \left(1 - d(\alpha_{aniso}^{(i)})\right) W_{aniso}^o(\lambda_t^{(i)}), \qquad \text{where}$$

$$\begin{cases} W_{aniso}^o(\lambda_t^{(i)}) = \dfrac{\eta_{aniso}}{2\gamma_{aniso}} \left(e^{\left(\gamma_{aniso}(\lambda_t^{(i)2} - 3)\right)} - 1\right) \\ \lambda_t^{(i)} = \dfrac{1}{\lambda_a^2} \underline{C}_0 : \underline{a}_0^{(i)} \otimes \underline{a}_0^{(i)} \\ \alpha_{aniso}^{(i)} = \max_{s \in [0,t]} \lambda_t^{(i)} \\ d(\alpha_{aniso}) \text{ is given in Eq. (6.52)} \end{cases} \qquad (6.65)$$

where $\eta_{aniso}, \gamma_{aniso}, \lambda_a$, and β are material constants associated with the families of fibres. The components of the anisotropic mechanism in the adventitial layer are modelled using a generalized structure tensor approach. As introduced in (6.36),(6.38) are defined as,

$$W_{aniso} = \sum_{i=1}^{2} \left(1 - d^{(i)}\right) W_{aniso}^o(\bar{\lambda}_t^{(i)}),$$

where

$$
\begin{cases}
W^o_{aniso}(\overline{\lambda}_t^{(i)}) = \dfrac{\eta_{aniso}}{2\gamma_{aniso}} \left(e^{\left(\gamma_{aniso}(\overline{\lambda}_t^{(i)2}-3)\right)} - 1 \right) \\[2mm]
\overline{\lambda}_t^{(i)} = \dfrac{1}{\lambda_a^2} \underline{C}_0 : \underline{H}_0^{(i)} \\[2mm]
\underline{H}_0^{(i)} = \kappa \underline{I} + (1-3\kappa)\overline{\underline{a}}_0^{(i)} \otimes \overline{\underline{a}}_0^{(i)} \\[2mm]
d^{(i)} = \begin{cases} 1 & for\ \lambda^{(i)} < \lambda_a, \\ 0 & for\ \lambda^{(i)} \geq \lambda_a, \end{cases}
\end{cases}
\qquad (6.66)
$$

where this model has similar material constants as for the medial anisotropic mechanism with the addition of κ, the measure of fibre dispersion.

Further, each of the three mechanisms in the intima and media have three constants associated with the damage model, Eq. 6.52, α_s, α_f, c. Representative material parameters for arterial layers used in the balloon-artery interaction are shown in Table 6.1.

Angioplasty model

In this section, cerebral PTA is simulated using a general purpose finite element code ANSYS 14.0 PREVIEW 1 (ANSYS, Inc., Canonsburg, PA, USA) in which the damage model of Sect. 6.4 was implemented using user subroutines. The numerical implementation of the inelastic constitutive model in an implicit finite element code requires the derivation of its stress response and elasticity tensor as well as a reformulation of the constitutive equation as a slightly compressible material. This reformulation requires some care because of the multiple reference configurations used in the model. Details of the finite element formulation can be found in [70, 75].

Table 6.1. Representative material parameters for three arterial layers in balloon-artery interaction

Intima-isotropic mechanism

$\eta_{iso}(KPa)$	γ_{iso}	c	$\alpha_s(KPa)$	$\alpha_f(KPa)$
4.55	0.565	2.1	80.0	2100

Media-isotropic mechanism

$\eta_{iso}(KPa)$	γ_{iso}	c	$\alpha_s(KPa)$	$\alpha_f(KPa)$
4.55	0.565	1.8	100.0	1800

Media-anisotropic mechanism

$\eta_{aniso}(KPa)$	γ_{aniso}	λ_a^2	β	c	$\alpha_s(KPa)$	$\alpha_f(KPa)$
125.0	1.88	1.54	7°	0.006	1.5	6.0

Adventitia-isotropic		Adventitia-anisotropic				
$\eta_{iso}(KPa)$	γ_{iso}	$\eta_{aniso}(KPa)$	γ_{aniso}	λ_a^2	κ	β
2.27	1.13	500.0	7.53	1.5	0.20	56°

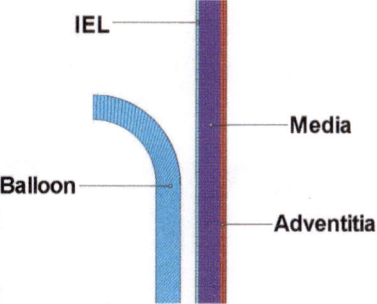

Fig. 6.14. Balloon and multi-layer artery model for cerebral angioplasty simulation. 3D solid 185 elements are used to discretize the artery and balloon. The arterial inner surface and balloon outer surface are meshed with surface-surface contact elements, 170 and 173

A three-dimensional axisymmetric angioplasty model is created to simulate the interaction between a balloon and the multi-layer artery wall. Due to fore-aft symmetry of the model, only part of the arterial segment and balloon are included in the model, Fig. 6.14. A clinically relevant choice of intracerebral single-lumen angioplasty catheters is 2 mm in diameter by 10 mm in length [21]. Most angioplasty lesions treated by PTA are less than 10 mm long (usually 2 to 4 mm), so very short balloons are necessary. In the computational model, the unloaded internal diameter of the artery is set to 2.5 mm (3.816 mm at the physiological loading state), with a thickness of 125 μm and a length of 10 mm. The balloon has a external diameter of 1.8 mm and a length of 5 mm. The IEL, media and adventitia occupy 1/10, 6/10 and 3/10 of the wall thickness respectively.

A surface contact strategy is used to simulate the interaction between artery and balloon during PTA. The axisymmetric model includes 17000 3D solid elements, 2600 surface-surface contact elements and four solid materials for the balloon, IEL, media and adventitia.

Loading states

Four representative deformation states are considered, Fig. 6.15. In State A, the artery is inflated to a transmural pressure $\Delta p = 13.33 KPa$ with an axial stretch $\lambda_z = 1.1$. This generates the arterial physiological deformation state before PTA which is in a purely elastic regime. In State B, the balloon is deployed to contact and then further deployed to State C, where the internal diameter has been increased by 130% from its value in State A. Radial displacement loads were applied on the balloon to reach States B and C. The inelastic damage and injury of arterial tissues take place during this oversized dilatation process from State B to C. Finally, the ballon is unloaded to bring the artery back to the physiological state after PTA (State D). At this final unloading state, a residual unloaded deformation is observed, extending beyond the area contacted by the PTA. This non-homogeneous residual strain arises from nonrecoverable inelastic damage of the IEL and media induced by the supraphysiological dilatation loads.

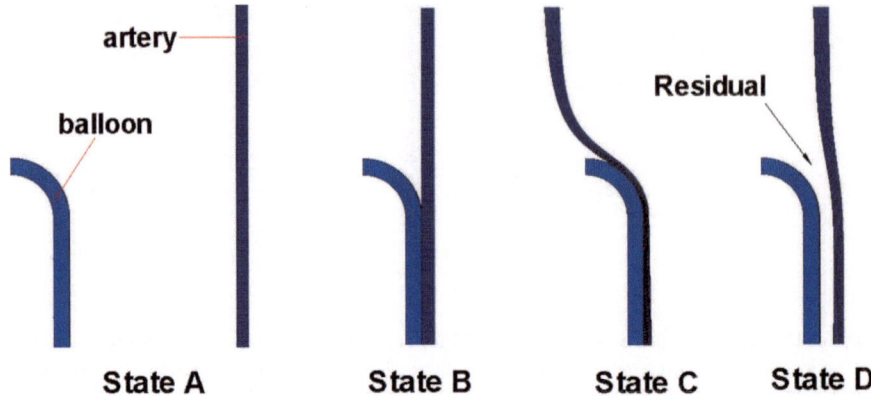

Fig. 6.15. Deformation states of artery and balloon during multi-step cerebral angioplasty simulation. State A: arterial physiological state before angioplasty (transmural pressure $\Delta p = 13.33 KPa$ and axial stretch $\lambda_Z = 1.1$); State B: initial contact of balloon and artery after balloon deploys; State C: maximum balloon inflation with arterial dilatation to 130% of the internal diameter; State D: arterial physiological state after angioplasty displaying residual deformation in parts of the vessel

Results

The distributions of arterial tissue damage in the IEL and media layers are shown in Figs. 6.16 and 6.17 for two different balloon inflation levels during PTA: 120% oversized dilatation state (an intermediate state between States B and C), and 130% oversized dilatation state (State C in Fig. 6.15). The maximum arterial damage can be seen at the outer region of the balloon-artery contact. As the deformation is increased from 120% to 130%, the damage variable increases in magnitude and the damaged region extends further outside the balloon contact area. At 120% dilatation the maximum elastin damage in the IEL is $d_{0E} = 0.27$, the maximum ground matrix damage in the media is $d_{0M} = 0.22$, and the maximum collagen damage in the media is $d_{iM} = 0.17$. For further dilatation to 130% (maximum balloon inflation), arterial damage further accumulates to the following maximum values: $d_{0E} = 0.71$, $d_{0M} = 0.49$ and $d_{iM} = 0.27$.

Fig. 6.18 shows the distributions of the von Mises stresses in the IEL, media and adventitia layers at the 120% oversized dilation level. The largest arterial stresses are found in regions corresponding to the highest IEL and media damage shown previously in Fig. 6.16. Compressive radial Cauchy stresses are also seen in highly damaged regions of the IEL and media (not shown). Fig. 6.19 similarly shows the distribution of von Mises stresses in the IEL, media and adventitia layers after the completion of angioplasty (State D). The maximum stress values after unloading have shifted to regions where residual deformation remains. The minimum von-Mises stresses is seen in highly damaged regions of the IEL after unloading.

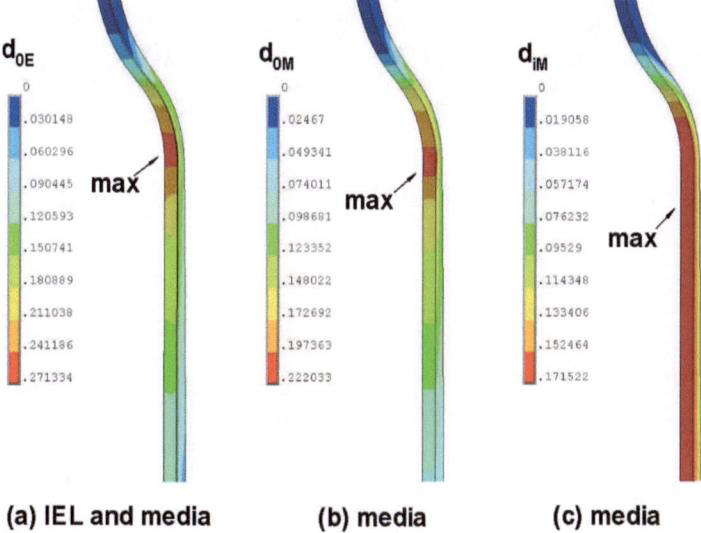

(a) IEL and media **(b) media** **(c) media**

Fig. 6.16. Damage distributions in the arterial layers at 120 % oversized dilatation state (an intermediate state between States B and C in Fig. 6.15). The arrows indicate the locations of the maximum tissue damage: (a) maximum elastin damage in the IEL $d_{0E} = 0.27$; (b) maximum ground matrix damage in the media $d_{0M} = 0.22$; (c) maximum collagen damage in the media $d_{iM} = 0.17$

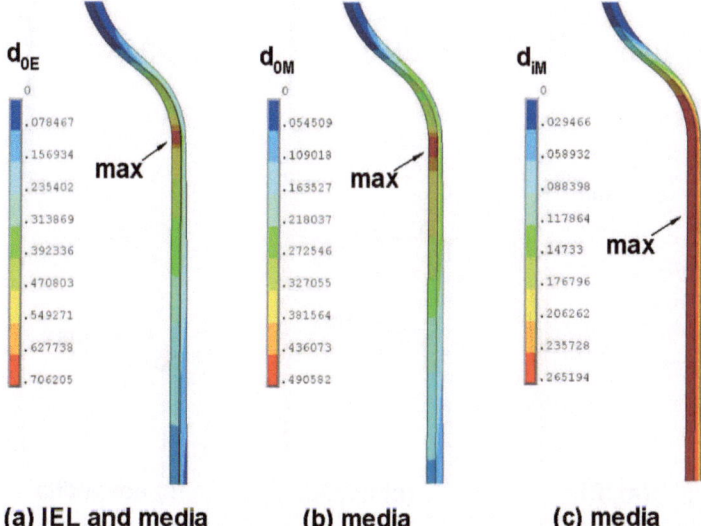

(a) IEL and media **(b) media** **(c) media**

Fig. 6.17. Damage distributions in the arterial layers at 130 % oversized dilatation state (State C in Fig. 6.15). The arrows indicate the locations of the maximum tissue damage: (a) maximum elastin damage in the IEL $d_{0E} = 0.71$; (b) maximum ground matrix damage in the media $d_{0M} = 0.49$; (c) maximum collagen damage in the media $d_{iM} = 0.27$

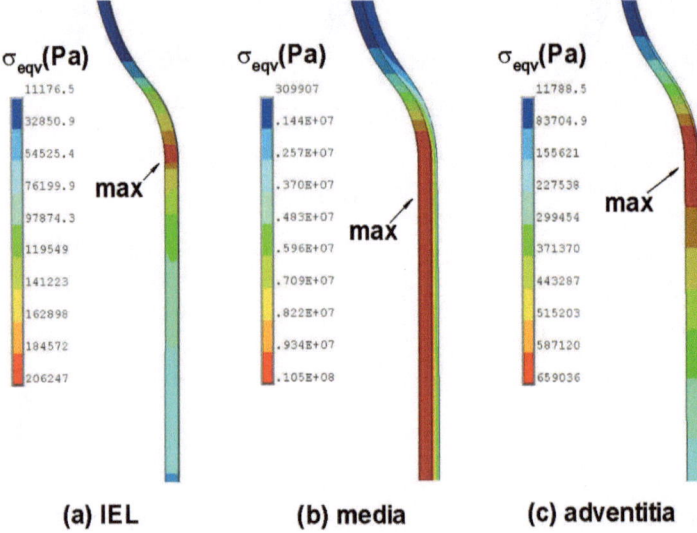

Fig. 6.18. Distributions of the von Mises stresses in the IEL, media and adventitia layers at 120% oversized dilatation state (an intermediate state between States B and C in Fig. 6.15). The arrows indicate the locations of the maximum stresses

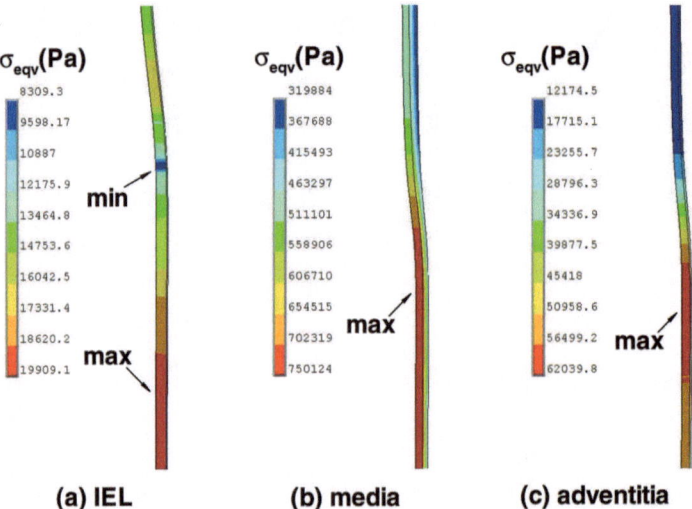

Fig. 6.19. Distributions of the von Mises stresses in the IEL, media and adventitia layers under normal arterial loads, but after angioplasty (State D). The arrows indicate the locations of the maximum stresses

Discussion

Using representative constitutive equations and material parameters cerebral angioplasty was modelled using an acute damage model. Large tissue damage and an increase in unloaded vessel diameter modelled demonstrated. Further investigation is needed to assess the modelling idealizations. Currently, we use a rigid walled balloon controlled by displacement loads. For simplicity, we have not included arterial plaque. For some applications, this idealization may also need to be relaxed. The current model does not incorporate arterial residual stresses, which may change wall stress distributions. Further we have only considered the passive behaviour of arteries. The contribution of smooth muscle cells will have to be included to study the active response of arterial tissues.

To further verify and refine the current constitutive model, in-vitro and in-vivo studies are in great needed. For example, additional experimental data for the layer-specific responses of cerebral arteries are needed. The inelastic response of the IEL and media must be investigated to develop the functional forms of the damage criteria. Experimental data are also required for a quantitative validation of the computational results, especially the relationship between loading and residual stretch following PTA. Due to the large number of material parameters utilized in the model, a detailed sensitivity analysis should be carried out in future studies.

We also recognize that in some vessels, the assumption of isotropy many not be appropriate. For example, Holzapfel et al. [56] analyzed the mechanical properties of nonstenotic human coronary arteries with intimal thickening and found the intima to be the stiffest wall layer and to display. The intima layer was highly anisotropic with increased stiffness in the longitudinal direction. They conjectured that since this stiffness was seen for larger loads, it is likely due to orientation to collagen fibres. For the long lasting effects of PTA, tissue mechanobiology including degeneration, repair, growth and remodelling will be important features to be modelled.

References

[1] Alastrué V., Martínez M.A., Doblaré M., Menzel A.: Anisotropic micro-sphere-based finite elasticity applied to blood vessel modelling. Journal of the Mechanics and Physics of Solids **57**(1): 178–203, 2009.

[2] Alastrué V., Sáez, P., Martínez, M.A., Doblaré, M.: On the use of the Bingham statistical distribution in microsphere-based constitutive models for arterial tissue. Mech Res Commun **37**(8): 700–706, 2010.

[3] Avolio A., Jones D., Tafazzoli-Shadpour M.: Quantification of alterations in structure and function of elastin in the arterial media. Hypertension **32**(1): 170–175, 1998.

[4] Baker C.J., Fiore A., Connolly E.S., Baker K.Z., Solomon R.A.: Serum elastase and alpha-1-antitrypsin levels in patients with ruptured and unruptured cerebral aneurysms. Neurosurgery **37**(1): 56–61; discussion 61–52, 1995.

[5] Balzani D., Schröder J., Gross D.: Simulation of discontinuous damage incorporating residual stresses in circumferentially overstretched atherosclerotic arteries. Acta Biomater **2**(6): 609–618, 2006.

[6] Bergel D.H.: The static elastic properties of the arterial wall. The Journal of Physiology **156**(3): 445–457, 1961.

[7] Betten J.: Formulation of anisotropic constitutive equations. In: J.P. Boehler (ed.) Applications of Tensor Functions in Solid Mechanics, CISM Courses and Lectures, 292, pp. 227–250. International Center for Mechanical Sciences, Springer-Verlag, 1984.

[8] Brandt T., Morcher M., Hausser I.: Association of cervical artery dissection with connective tissue abnormalities in skin and arteries. Frontiers of neurology and neuroscience 20: 16–29, 2005.

[9] Broom N., Ramsey G., Mackie R., Martins B., Stehbens W.: A new biomechanical approach to assessing the fragility of the internal elastic lamina of the arterial wall. Connective Tissue Research 30(2): 143–155, 1993.

[10] Bullitt E., Lin N., Smith J., Zeng D., Winer E., Carey L., Lin W., Ewend M.: Blood vessel morphological changes depicted with mr angiography during treatment of brain metastases: a feasibility study. Radiology 40: 824–830, 2007.

[11] Busby D.E., Burton A.C.: The effect of age on the elasticity of the major brain arteries. Canadian journal of physiology and pharmacology 43: 185–202, 1965.

[12] Cajander S., Hassler O.: Enzymatic destruction of the elastic lamella at the mouth of the cerebral berry aneurysm? Acta Neruol Scand 53: 171–181, 1976.

[13] Campbell G., Roach M.: The use of ligament efficiency to model fenestrations in the internal elastic lamina of cerebral arteries. I–modelling scheme. J Biomech 16: 875–882, 1983.

[14] Campbell G., Roach M.: The use of ligament efficiency to model fenestrations in the internal elastic lamina of cerebral arteries. II–analysis of the spatial geometry. J Biomech 16: 883–91, 1983.

[15] Campbell G., Roach M.: A physical model for the formation of evaginations: a prospective precursor to the creation of saccular aneurysms. Stroke 15: 642–52, 1984.

[16] Castaneda-Zuniga W.R., Amplatz K., Laerum F., Formanek A., Sibley R., Edwards J., Vlodaver Z.: Mechanics of angioplasty: an experimental approach. RadioGraphics 1(3): 1–14 (1981)

[17] Castaneda-Zuniga W.R., Formanek A., Tadavarthy M., Vlodaver Z., Edwards J.E., Zollikofer C., Amplatz K.: The mechanism of balloon angioplasty. Radiology 135(3): 565–571, 1980.

[18] Chavez L., Takahashi A., Yoshimoto T., Su C.C., Sugawara T., Fujii Y.: Morphological changes in normal canine basilar arteries after transluminal angioplasty. Neurol Res 12(1): 12–16 (1990)

[19] Chyatte D., Reilly J., Tilson M.D.: Morphometric analysis of reticular and elastin fibres in the cerebral arteries of patients with intracranial aneurysms. Neurosurgery 26(6): 939–943, 1990.

[20] Connolly E.S.J.,Fiore A.J., Winfree C.J., Prestigiacoma C.J., Goldman J.E., Solomon R.A.: Elastin degradation in the superficial temporal arteries of patients with intracranial aneurysms reflects changes in plasma elastase. Neurosurgery 40(5): 903–908; discussion 908–909, 1997.

[21] Connors J.J., Wojak J.C.: Percutaneous transluminal angioplasty for intracranial atherosclerotic lesions: evolution of technique and short-term results. J Neurosurg 91(3): 415–423, 1999.

[22] Cortes D.H., Lake S.P., Kadlowec J.A., Soslowsky L.J., Elliott D.M.: Characterizing the mechanical contribution of fibre angular distribution in connective tissue: comparison of two modeling approaches. Biomech Model Mechanobiol 9: 651–658, 2010.

[23] Courtney T., Sacks M., Stankus J., Guan J., Wagner W.: Design and analysis of tissue engineering scaffolds that mimic soft tissue mechanical anisotropy. Biomaterials 27: 3631–3638, 2006.

[24] Cox G., Kable E.: Second-harmonic imaging of collagen. In: D.J. Taatjes, B.T. Mossman (eds.) Cell Imaging Techniques: Methods and Protocols, Methods in Molecular Biology, vol. 319, pp. 15–35 (2006)

[25] Davis E.C.: Stability of elastin in the developing mouse aorta: a quantitative radioautographic study. Histochemistry and Cell Biology 100(1): 17–26, 1993.

[26] Ericksen J.E., Rivlin R.S.: Large elastic deformations of homogeneous anisotropic materials. J. Rat. Mech. Anal. **3**: 281–301, 1954.

[27] Federico S., Herzog W.: Towards an analytical model of soft biological tissues. Journal of biomechanics **41**(16): 3309–3313, 2008.

[28] Finlay H., McCullough L., Canham P.: Three-dimensional collagen organization of human brain arteries at different transmural pressures. J. Vasc. Res. **32**: 301–312, 1995.

[29] Finlay H.M., McCullough L., Canham P.B.: Three-dimensional collagen organization of human brain arteries at different transmural pressures. J. Vasc. Res. **32**: 301–312, 1995.

[30] Fonck E., Feigl G.G., Fasel J., Sage D., Unser M., Rufenacht D.A., Stergiopulos N.: Effect of aging on elastin functionality in human cerebral arteries. Stroke **40**(7): 2552–2556, 2009.

[31] Fonck E., Prod'hom G., Roy S., Augsburger L., Rufenacht D.A., Stergiopulos N.: Effect of elastin degradation on carotid wall mechanics as assessed by a constituent-based biomechanical model. American Journal Of Physiology. Heart And Circulatory Physiology **292**(6): H2754–2763, 2007.

[32] Freed A., Einstein D., Vesely I.: Invariant formulation for dispersed transverse isotropy in aortic heart valves. Biomechanics and Modeling in Mechanobiology **4**(2): 100–117, 2005.

[33] Gao L., Hoi Y., Swartz D.D., Kolega J., Siddiqui A., Meng H.: Nascent aneurysm formation at the basilar terminus induced by hemodynamics. Stroke; A Journal Of Cerebral Circulation **39**(7): 2085–2090, 2008.

[34] Gasser C.T., Holzapfel G.: Modeling plaque fissuring and dissection during balloon angioplasty intervention. Annals of Biomedical Engineering **35**(5): 711–723, 2007.

[35] Gasser C.T., Ogden R.W., Holzapfel G.A.: Hyperelastic modelling of arterial layers with distributed collagen fibre orientations. Journal Of The Royal Society, Interface / The Royal Society **3**(6): 15–35, 2006.

[36] Gasser T.C., Holzapfel G.: A rate-independent elastoplastic constitutive model for biological fibre-reinforced composites at finite strains: continuum basis, algorithmic formulation and finite element implementation. Computational Mechanics **29**(4-5): 340–360 , 2002.

[37] Gleason R.L., Humphrey J.: A 2D constrained mixture model for arterial adaptations to large changes in flow, pressure and axial stretch. Mathematical Medicine And Biology: A Journal Of The IMA **22**(4): 347–369, 2005.

[38] Gleason R.L., Taber L.A., Humphrey J.D.: A 2-d model of flow-induced alterations in the geometry, structure, and properties of carotid arteries. Journal of Biomechanical Engineering **126**(3): 371–381, 2004.

[39] Goktepe S., Miehe C.: A micro-macro approach to rubber-like materials. part iii: The microsphere model of anisotropic mullins-type damage. Journal of the Mechanics and Physics of Solids **53**(10): 2259–2283, 2005.

[40] Gonzalez J., Briones A., Starcher B., Conde M., Somoza B., Daly C., Vila E., McGrath I., Arribas S.: Influence of elastin on rat small artery mechanical properties. Exp Physiol **90**: 463–8, 2005.

[41] Greenwald S.E.: Ageing of the conduit arteries. The Journal of pathology **211**(2): 157–172, 2007.

[42] Gundiah N., Ratcliffe M.B., Pruitt L.: Determination of strain energy function for arterial elastin: Experiments using histology and mechanical tests. Journal of Biomechanics **40**(3): 586–594, 2007.

[43] Gundiah N., Ratcliffe M.B., Pruitt L.A.: The biomechanics of arterial elastin. Journal of the Mechanical Behavior of Biomedical Materials **2**(3): 288–296, 2009.

[44] Hadjinikolaou L., Kotidis K., Galinanes M.: Relationship between reduced elasticity of extracardiac vessels and left main stem coronary artery disease. European heart journal **25**(6): 508–513, 2004.

[45] Hart W., Goldbaum M., Cote B., Kube P., Nelson M.: Measurement and classification of retinal vascular tortuosity. International Journal of Medical Informatics **53**: 239–252, 1999.

[46] Hashimoto N., Kim C., Kikuchi H., Kojima M., Kang Y., Hazama F.: Experimental induction of cerebral aneurysms in monkeys. Journal of Neurosurgery **67**(6): 903–905, 1987.

[47] Hassler O.: Morphological studies on the large cerebral arteries, with reference to the aetiology of subarachnoid haemorrhage. Acta psychiatrica Scandinavica **154**: 1–145, 1961.

[48] Higashida R.T., Halbach V.V., Dowd C.F., Dormandy B., Bell J., Hieshima G.B.: Intravascular balloon dilatation therapy for intracranial arterial vasospasm: patient selection, technique, and clinical results. Neurosurg Rev **15**(2): 89–95, 1992.

[49] Hill M., Robertson A.M.: Combined histological and mechanical evaluation of isotropic damage to elastin in cerebral arteries. In: 6th World Congress on Biomechanics. Singapore, 2010.

[50] Hill M., Robertson A.: Abrupt recruitment of medial collagen fibres in the rabbit carotid artery – SBC2011-5341. Proceedings of the ASME 2011 Summer Bioengineering Conference (SBC2011), June 22–25, Nemacolin Woodlands Resort. Farmington, PA, USA, 2 pages, 2011.

[51] Holzapfel A.P.G.A.: Three-dimensional modeling and computational analysis of the human cornea considering distributed collagen fibril orientations. Journal of Biomechanical Engineering **130**(6): 061,006–061,012, 2008.

[52] Holzapfel G., Gasser T., Ogden R.: A new constitutive framework for arterial wall mechanics and a comparative study of material models. Journal of Elasticity **61**(1–3): 1–48, 2000.

[53] Holzapfel G.A.: Nonlinear Solid Mechanics A Continuum Approach for Engineering. J. Wiley & Sons, 2000.

[54] Holzapfel G.A., Ogden R.W.: Constitutive modelling of arteries. Proceedings of the Royal Society A: Mathematical, Physical and Engineering Sciences **466**(2118): 1551–1597, 2010.

[55] Holzapfel G.A., Ogden R.W.: Modelling the layer-specific three-dimensional residual stresses in arteries, with an application to the human aorta. Journal of The Royal Society Interface **7**(46): 787–799, 2010.

[56] Holzapfel G.A., Sommer G., Gasser C.T., Regitnig P.: Determination of layer-specific mechanical properties of human coronary arteries with nonatherosclerotic intimal thickening and related constitutive modeling. Am. J. Physiol. Heart Circ. Physiol. **289**: H2048–H2058, 2005.

[57] Honma Y., Fujiwara T., Irie K., Ohkawa M., Nagao S.: Morphological changes in human cerebral arteries after pta for vasospasm caused by subarachnoid hemorrhage. Neurosurgery **36**(6): 1073–1081, 1995.

[58] Humphrey J.D., Baek S., Niklason L.E.: Biochemomechanics of cerebral vasospasm and its resolution: I. a new hypothesis and theoretical framework. Annals of Biomedical Engineering **35**(9): 1485–1497, 2007.

[59] Humphrey J.D., Rajagopal K.R.: A constrained mixture model for growth and remodeling of soft tissues. Mathematical Models and Methods in Applied Sciences **12**(3): 407–430, 2002.

[60] Humphrey J.D., Rajagopal K.R.: A constrained mixture model for arterial adaptations to a sustained step change in blood flow. Biomechanics And Modeling In Mechanobiology **2**(2): 109–126, 2003.

[61] Jiang C.F., Avolio A.P.: Characterisation of structural changes in the arterial elastic matrix by a new fractal feature: directional fractal curve. Medical & biological engineering & computing **35**(3): 246–252, 1997.

[62] Kachanov L.: Time of rupture process under creep conditions. IVZ Akad. Nauk, S.S.R., Otd Tech Nauk **8**: 26–31 (1958)

[63] Keeley F.W.: The synthesis of soluble and insoluble elastin in chicken aorta as a function of development and age. effect of a high cholesterol diet. Canadian journal of biochemistry **57**(11): 1273–1280 (1979)

[64] Kondo S., Hashimoto N., Kikuchi H., Hazama F., Nagata I., Kataoka H.: Cerebral aneurysms arising at nonbranching sites. an experimental study. Stroke **28**(2): 398–403; discussion 403–394, 1997.

[65] Krex D., Schackert H.K., Schackert G.: Genesis of cerebral aneurysms–an update. Acta Neurochirurgica **143**(5): 429–448; discussion 448–429, 2001.

[66] Lanir Y.: A structural theory for the homogeneous biaxial stress-strain relationships in flat collagenous tissues. Journal of biomechanics **12**(6): 423–436, 1979.

[67] Lanir Y.: Constitutive equations for fibrous connective tissues. Journal of biomechanics **16**(1): 1–12, 1983.

[68] Lee R.M.: Morphology of cerebral arteries. Pharmacology & therapeutics **66**(1): 149–173, 1995.

[69] Lematire J., Desmorat R.: Engineering damage mechanics: ductile, creep, fatigue and brittle failures. Springer, 2005.

[70] Li D.: Structural multi-mechanism model with anisotropic damage for cerebral arterial tissues and its finite element modeling. Ph.D. thesis, University of Pittsburgh, 2009.

[71] Li D., Robertson, A.: A structural multi-mechanism constitutive model for cerebral arterial tissue. Int. J. Solids Struct. **46**: 2920–2928, 2009.

[72] Li D., Robertson, A.M.: Finite element modeling of cerebral angioplasty using a multi-mechanism structural damage model. In: Proceedings of the ASME 2009 Summer Bioengineering Conference (SBC-2009), 2009.

[73] Li D., Robertson A.M.: A structural damage model for cerebral arterial tissue and angioplasty simulation. In: 10th US National Congress on Computational Mechanics (USNCCM X), 2009.

[74] Li D., Robertson A.M.: A structural multi-mechanism damage model for cerebral arterial tissue. J. Biomech. Eng. **131**: 8 pages, 2009. Doi: 10.1115/1.3202559

[75] Li D., Robertson A.M., Guoyu L.: Finite element modeling of cerebral angioplasty using a structural multi-mechanism anisotropic damage model. submitted for publication, 2011.

[76] Meng H., Swartz D.D., Wang Z., Hoi Y., Kolega J., Metaxa E.M., Szymanski M.P., Yamamoto J., Sauvageau E., Levy E.I.: A model system for mapping vascular responses to complex hemodynamics at arterial bifurcations in vivo. Neurosurgery **59**(5): 1094–100; discussion 1100–1, 2006.

[77] Meng H., Wang Z., Hoi Y., Gao L., Metaxa E., Swartz D.D., Kolega J.: Complex hemodynamics at the apex of an arterial bifurcation induces vascular remodeling resembling cerebral aneurysm initiation. Stroke **38**(6): 1924–1931, 2007.

[78] Miehe C.: Discontinuous and continuous damage evolution in ogden-type large-strain elastic materials. Eur. J. Mech. A/Solids **14**(5): 697–720, 1995.

[79] Mohan D., Melvin J.W.: Failure properties of passive human aortic tissue. i–uniaxial tension tests. Journal of biomechanics **15**(11): 887–902 (1982)

[80] Montes G.S.: Structural biology of the fibres of the collagenous and elastic systems. Cell Biol Int **20**(1): 15–27, 1996.

[81] Morimoto M., Miyamoto S., Mizoguchi A., Kume N., Kita T., Hashimoto N.: Mouse model of cerebral aneurysm: experimental induction by renal hypertension and local hemodynamic changes. Stroke; A Journal Of Cerebral Circulation **33**(7): 1911–1915, 2002.

[82] Mullins L.: Effect of stretching on the properties of rubber. Rubber Chem. Technol. **21**: 281–300, 1948.

[83] Mullins L.: Softening of rubber by deformation. Rubber Chemistry and Technology **42**: 339–362, 1969.

[84] Oktay H.: Continuum damage mechanics of balloon angioplasty. doctoral, University of Maryland, Baltimore County (1993)

[85] O'Rourke M.: Mechanical principles in arterial disease. Hypertension **26**: 2–9, 1995.

[86] O'Rourke M.F.: Vascular mechanics in the clinic. Journal of biomechanics **36**(5): 623–630, 2003.

[87] Peña E., Alastrué V., Laborda A., Martínez M.A., Doblaré M.: A constitutive formulation of vascular tissue mechanics including viscoelasticity and softening behaviour. Journal of biomechanics **43**(5): 984–989, 2010.

[88] Peña E., Peña J.A., Doblaré M.: On the mullins effect and hysteresis of fibreed biological materials: A comparison between continuous and discontinuous damage models. International Journal of Solids and Structures **46**(7–8): 1727–1735, 2009.

[89] Rachev A., Hayashi K.: Theoretical study of the effects of vascular smooth muscle contraction on strain and stress distributions in arteries. Annals of Biomedical Engineering 27(4): 459–468, 1999.

[90] Reuterwall O.: Über die Elästizität der Gefäßwände und die Methoden ihrer näheren Prüfung. Acta med. scand Suppl 2.: 1–175, 1921.

[91] Rhodin J.A.G.: Architecture of the vessel wall. In: R.M. Berne, N. Sperelakis (eds.) Vascular Smooth Muscle, *The Cardiovascular System*, vol. Vol 2 of Handbook of Physiology, Sect. 2: The Cardiovascular System., pp. 1–31. APS, Baltimore, 1979.

[92] Roach M.R., Burton A.C.: The reason for the shape of the distensibility curves of arteries. Canadian journal of biochemistry and physiology 35(8): 681–690, 1957.

[93] Roach M.R., Burton A.C.: The reason for the shape of the distensibility curves of arteries. Can. J. Biochem. Physiol. 35: 681–690, 1957.

[94] Rodriguez J., Goicolea J.M., Gabaldon F.: A volumetric model for growth of arterial walls with arbitrary geometry and loads. Journal of biomechanics 40(5): 961–971, 2007.

[95] Rodríguez J., Martufi G., Doblaré M., Finol E.: The effect of material model formulation in the stress analysis of abdominal aortic aneurysms. Annals of Biomedical Engineering 37(11): 2218–2221, 2009.

[96] Ronchetti I., Alessandrini A., Contri M., Fornieri C., Mori G., Quaglino D., Valdre U.: Study of elastic fibre organization by scanning force microscopy. Matrix Biol 17: 75–83, 1988.

[97] Sacks M.S.: Incorporation of experimentally-derived fibre orientation into a structural constitutive model for planar-collagenous tissues. Journal of Biomechanical Engineering-Transactions of the Asme 125(2): 280–287, 2003.

[98] Sacks M.S., Smith D.B., Hiester E.D.: A SALS device for planar connective tissue microstructural analysis. Ann. Biomed. Eng. 25: 678–689 (1997)

[99] Sacks M.S., Sun W.: Multiaxial mechanical behavior of biological materials. Annual Review of Biomedical Engineering 5: 251–284, 2003.

[100] Samila Z., Carter S.: The effect of age on the unfolding of elastin lamellae and collagen fibres with stretch in human carotid arteries. Can. J. Physiol. Pharmacol. 59: 1050–1057, 1981.

[101] Schievink W.I., Roiter V.: Epidemiology of cervical artery dissection. Frontiers of neurology and neuroscience 20: 12–15, 2005.

[102] Scott S., Ferguson G.G., Roach M.R.: Comparison of the elastic properties of human intracranial arteries and aneurysms. Canadian journal of physiology and pharmacology 50(4): 328–332, 1972.

[103] Shapiro S., Endicott S., Province M., Pierce J., Campbell E.: Marked longevity of human lung parenchymal elastic fibres deduced from prevalence of d-aspartate and nuclear weaponsrelated radiocarbon. J Clin Invest 87: 1828–1834, 1991.

[104] Sherratt M.: Tissue elasticity and the ageing elastic fibre. AGE 31: 305–325, 2009.

[105] Shifren A., Mecham R.P.: The stumbling block in lung repair of emphysema: elastic fibre assembly. Proceedings of the American Thoracic Society 3(5): 428–433 (2006)

[106] Sidorov S.: Finite element modeling of human artery tissue with a nonlinear multimechanism inelastic material. Ph.D. thesis, U. of Pittsburgh (2006)

[107] Simo J.C., Ju J.W.: Strain and stress-based continuum damage models- i. formulation. International Journal of Solids and Structures 23: 821–840, 1987.

[108] Spencer A.: Constitutive theory for strongly anisotropic solids. In: A. Spencer (ed.) Continuum Theory of the Mechanics of Fibre-Reinforced Composites, *CISM Courses and Lectures*, vol. 282. Springer (1984)

[109] Spencer A.J.M.: Theory of invariants. In: A.C. Eringen (ed.) Continuum Physics, vol. I, pp. 239–253. Academic Press, 1971.

[110] Valentin A., Cardamone L., Baek S., Humphrey J.: Complementary vasoactivity and matrix remodelling in arterial adaptations to altered flow and pressure. J. R. Soc. Interface 6: 293–306, 2009.

[111] Wagenseil J., Mecham R.: Vascular extracellular matrix and arterial mechanics. Physiological Reviews 89(3): 957–989, 2009.

[112] Watton P., Ventikos Y., Holzapfel G.: Modelling the mechanical response of elastin for arterial tissue. J. Biomech. **42**: 1320–1325, 2009.

[113] Weber T., Auer J., Eber B., O'Rourke M.F.: Relationship between reduced elasticity of extracardiac vessels and left main stem coronary artery disease. European heart journal **25**(21): 1966–1967, 2004.

[114] Weber T., Auer J., O'Rourke M.F. Kvas E., Lassnig E., Lamm G., Stark N., Rammer M., Eber B.: Increased arterial wave reflections predict severe cardiovascular events in patients undergoing percutaneous coronary interventions. European heart journal **26**(24): 2657–2663, 2005.

[115] Wiechert L., Metzke R., Wall W.A.: Modeling the mechanical behaviour of lung tissue at the micro-level. Mechanics of Biological and bioinspired materials in Journal of Engineering Mechanics **135**(5): 434–438, 2009. DOI 10.1061/(ASCE)0733-9399(2009)135:5(434).

[116] Wojak J.C., Dunlap D.C., Hargrave K.R., DeAlvare L.A., Culbertson H.S., Connors J. Jr.: Intracranial angioplasty and stenting: long-term results from a single center. AJNR Am. J. Neuroradiol. **27**(9): 1882–1892, 2006.

[117] Wolinsky H., Glagov S.: Structural basis for the static mechanical properties of the aortic media. Circulation research **14**: 400–413, 1964.

[118] Wulandana R., Robertson A.: Use of a multi-mechanism constitutive model for inflation of cerebral arteries. In: First Joint BMES/EMBS Conference, vol. 1, p. 235. Atlanta, GA, 1999.

[119] Wulandana R., Robertson A.M.: An inelastic multi-mechanism constitutive equation for cerebral arterial tissue. Biomech. Model. Mechanobiol. **4**(4): 235–248, 2005.

[120] Zeng Z., Chung B.J., Durka M., Robertson A.M.: An in vitro device for evaluation of cellular response to flows found at the apex of arterial bifurcations. In: R. Rannacher, A. Sequeira (eds.) Advances in Mathematical Fluid Mechanics: Dedicated to Giovanni Paolo Galdi on the Occasion of his 60th Birthday. Springer-Verlag, New York, 2010.

[121] Zipfel W.R., Williams R.M., Christie R., Nikitin A.Y., Hyman B.T., Webb W.W.: Live tissue intrinsic emission microscopy using multiphoton-excited native fluorescence and second harmonic generation. Proceedings of the National Academy of Sciences of the United States of America **100**(12): 7075–7080, 2003.

[122] Zollikofer C.L., Chain J., Salomonowitz E., Runge W., Bruehlmann W.F., Castaneda-Zuniga W.R., Amplatz K.: Percutaneous transluminal angioplasty of the aorta. light and electron microscopic observations in normal and atherosclerotic rabbits. Radiology **151**(2): 355–363, 1984.

[123] Zoumi A., Lu X., Kassab G., Tromberg B.: Imaging coronary artery microstructure using second harmonic and two-photon fluorescence microscopy. Biophys J **87**: 2778–2786 (2004)

[124] Zulliger M., Stergiopulos N.: Structural strain energy function applied to the ageing of the human aorta. Journal of biomechanics **40**(14): 3061–3069, 2007.

[125] Zulliger M.A., Rachev A., Stergiopulos N.: A constitutive formulation of arterial mechanics including vascular smooth muscle tone. Am J Physiol-Heart C **287**(3): H1335–H1343, 2004.

7

Arterial growth and remodelling is driven by hemodynamics

Luca Cardamone, and Jay D. Humphrey

Abstract. Experimental observations highlight the importance of altered hemo-dynamics on arterial function and adaptation [27, 28, 29]. We discuss a class of mechano-biological models for growth and remodelling (G&R) of the arterial wall that describe the intimate interaction between hemodynamics, cell activity, and arterial wall mechanics. For some applications the artery can be described as a thin walled structure: for example, basic adaptations to perturbed pressure and flow, cerebral aneurysms, and vasospasms have been successfully modelled treating the vascular wall as a membrane. A multiple-time scales membrane model is described and illustrative results discussed. Future patient-specific models of large arteries and pathologies as atherosclerosis and abdominal aortic aneurysms require a full 3D model of the interaction between the blood flow and the growing vessel. We discuss the extension of the model to thick walled vessels and some preliminary results.

7.1 Introduction

7.1.1 Arterial structure

The vasculature consists of a complex system of arteries, arterioles, capillaries, venules, and veins. Each vessel serves a unique function and exhibits unique behaviour. The microstructure of the normal arterial wall varies with location along the vascular tree, age, species, local adaptation and disease; thus, one must focus on the particular vessel and condition of interest. Nonetheless, arteries can be cate-

Luca Cardamone (✉)
Sector of Functional Analysis and Applications, SISSA–International School for Advanced Studies, Trieste, Italy
e-mail: cardamon@sissa.it

Jay D. Humphrey
Department of Biomedical Engineering, Yale University, New Haven, Connecticut, USA
e-mail: jay.humphrey@yale.edu

Ambrosi D., Quarteroni A., Rozza G. (Eds.): Modeling of Physiological Flows.
DOI 10.1007/978-88-470-1935-5_7, © Springer-Verlag Italia 2012

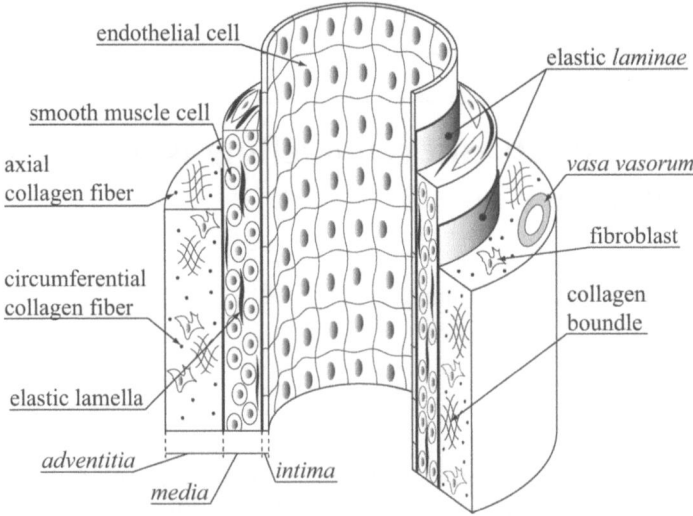

endothelial cell

smooth muscle cell

axial
collagen fiber

circumferential
collagen fiber

elastic lamella

adventitia

media

intima

elastic *laminae*

vasa vasorum

fibroblast

collagen
boundle

Fig. 7.1. Structural anatomy of an artery. See [20] for more detailed pictures underlining peculiar features of elastic and muscular arteries

gorized according to two general types: "elastic arteries", including the aorta, main pulmonary arteries, common carotids, and common iliacs, and "muscular arteries", which include the coronaries, cerebrals, femorals and renals. Elastic arteries tend to be larger-diameter vessels located closer to the heart, whereas muscular arteries are smaller-diameter vessels closer to the arterioles. Transitional arteries, such as the external carotids, exhibit some characteristics of the elastic and muscular types [20].

Regardless of the type, all arteries consist of three layers: the *tunica intima*, *tunica media*, and *tunica adventitia* (Fig. 7.1). The *intima* is similar in most elastic and muscular arteries, typically consisting of a monolayer of endothelial cells and an underlying thin (∼80 nm) basal *lamina*. Exceptions include the aorta and coronary arteries in which the *intima* may contain a subendothelial layer of connective tissue and axially oriented smooth muscle cells. Endothelial cells are usually flat and elongated in the direction of the blood flow; exceptions occur near bifurcations wherein the blood flow is complex and the cells are often polygonal in shape. Endothelial cells are interconnected and may communicate via in-plane junctions or with underlying smooth muscle cells via short processes that extend through the basal *lamina* and into the *media*. The basal *lamina* consists largely of net-like type IV collagen, the adhesion molecules laminin and fibronectin, and some proteoglycans; it provides some structural support to the arterial wall, but acts primarily as an adherent meshwork on which the endothelial cells can grow. The *media* contains smooth muscle cells that are embedded in an extracellular matrix of elastin and collagen (primarily types I, III, and V) as well as an aqueous ground substance matrix containing proteoglycans. Although the orientation and distribution of medial constituents varies with species and location along the vascular tree, vascular smooth muscle tends to be oriented helically, albeit nearly circumferentially in many vessels.

This preferential orientation is expected, in part, for the primary roles of the contraction of vascular smooth muscle are to modify the distensibility of large arteries or to regulate the luminal diameter in medium and small arteries. Finally, the *adventitia* consists primarily of a dense network of type I collagen fibres with admixed elastin, nerves, fibroblasts and *vasa vasorum*. The adventitial collagen fibres tend to have an axial orientation in most arteries and they are slightly undulated in the basal state. Although the *adventitia* comprises only $\sim 10\%$ and $\sim 50\%$ of the arterial wall in elastic and muscular constituents, respectively, it is thought to limit acute overdistension in all vessels. That is, the collagenous *adventitia* may serve primarily as a protective sheath, similar to the *epicardium* of the heart. Nonetheless, the presence of nerves within the *adventitia* also allows innervation of the outer *media* via the diffusion of neurotransmitters. The fibroblasts are responsible for regulating the connective tissue, particularly the type I collagen. In many vessels, the *adventitia* is contiguous with perivascular tissue, which often provides additional structural support. Exceptions, again, are the cerebral arteries, which are sometimes sorrounded mainly by cerebro-spinal fluid.

7.1.2 Arterial functions and the roles of hemodynamics

The aforementioned classification of arteries as elastic or muscular actually arises more from considerations of function than structure. To appreciate this, consider the following. The heart is a pulsatile pump that ejects blood into the vasculature only during its contractile phase (i.e., systole). Whereas contraction of the heart results in forward flow of most of the ejected blood, some of the blood is "stored" in the large arteries as they distend under high pressure. When the blood pressure drops during diastole, these distended vessels recoil elastically and thereby augment the flow of blood by supplying a second pressure pulse. Moreover, augmentation of blood flow via the recoil of large arteries helps convert it from a pulsatile to a nearly steady flow in the capillaries, which in turn aids the requisite transport processes. It is the large amount of intramural elastin and type III collagen in the *media* that gives rise to this important "elastic behaviour" of the large arteries, and hence the terminology. In contrast, the smaller arteries and arterioles play a key role in regulating local blood flow via vasoconstriction or vasodilation, that is, by muscular activity. Note, therefore, that arterial smooth muscle is partially contracted in its homeostatic or basal state – this is referred to as basal tone. Further contraction or alternatively relaxation of the abundant smooth muscle in these vessels thus results in a decreased or increased lumen, which in turn controls the resistance to the blood flow. It is in this way that blood can be routed to regions having an increased need for oxygen or nutrient exchanges (e.g., skeletal muscles during exercise or digestive system following eating) or away from regions where there is an injury to the vasculature (e.g., to minimize bleeding). Likewise, because of their ability to control blood flow, muscular arteries and arterioles play a key role in thermal regulation within the body by routing blood toward or away from the skin. Muscular arteries are thus those vessels that perform their primary function via smooth muscle activity.

As we have seen, the three primary cells within the arterial wall are the endothelial cells in the *intima*, the smooth muscle cells in the *media*, and the fibroblasts in the *adventitia*. The luminal surface of the endothelial has potent anticoagulant function. This property of the endothelium is in contrast to, for example, the highly thrombogenic character of the intramural type I and type III collagen. Clearly, then, an important role of the endothelium is to act as a smooth, nonthrombogenic lining that separates the wall contents from the flowing blood. The endothelium allows transport of substances to and from the bloodstream, however; hence, it must be considered as a selective barrier. Notwithstanding the importance of these functions of an intact endothelium, it was long thought that these were its only two roles. It is now known however that the endothelium is very active biologically: it can regulate coagulative processes, synthesize vasoactive substances (e.g., nitric oxide (NO) and endothelin-1 (ET-1)), produce growth-regulatory molecules (e.g., vascular endothelial growth factor (VEGF), platelet derived growth factor (PDGF), fibroblast growth factor (FGF)), and synthesize connective tissue (e.g., type IV collagen and proteoglycans). Moreover, many blood-borne vasoactive substances affect the arterial wall only in the presence of an intact endothelium, thus it also serves as a chemo-transducer. Conversely, many biological functions of the endothelium are signaled mechanically as, for example, via changes in local blood flow and the associated wall shear stress: increased flow downregulates the synthesis of the vasoconstrictor ET-1 and upregulates that of the vasodilators like NO, thus resulting in a flow-related vasodilation. Endothelial cells also tend to elongate and align themselves in the direction of the blood flow, which thereby reorients their actin microfilaments and alters cytoskeletal properties. Hence, although it is a small fraction of the arterial wall, the endothelium plays a major role in vascular mechanics.

Smooth muscle cells comprise about 25 % to 60 % of the arterial wall by dry weight. They can increas in size (i.e., hypertrophy) or increase in number via cell division (i.e., hyperplasia) depending on the stimulus. Structurally, smooth muscle cells contain a Ca^{2+} regulated actin-myosin contractile apparatus, albeit not arranged in sarcomeres as in striated muscles, and they can possess a well-developed Golgi apparatus which enables the synthesis of connective tissue, a capability that is very important during development but also plays a detrimental role in aging and different pathologies. Note, too, that various signals are capable of changing the phenotype of the mature smooth muscle cell from contractile to synthetic. Contraction of a vascular smooth muscle occurs much more slowly than that of a striated muscle, typically beginning 5 to 100 ms after the initial stimulus and taking on the order of 1 to 10 seconds to achieve its maximum level. Nevertheless, smooth muscles can maintain maximum contraction at a steady level for much longer periods than striated muscle, provided a stimulus persists, and with less energy expenditure (perhaps as little as 0.25 % to 5 % of the energy required for comparable contraction of skeletal muscle). Indeed, this ability is essential in the maintenance of the basal tone.

Despite some differences, the general relations between the force, length, and velocity of shortening are similar for smooth and striated muscle, both able to generate an active muscle stress of about 0.3 MPa. Like skeletal muscle, smooth muscle generates its maximum force at a unique length (often denoted with L_0). That is, there is a length at which the muscle prefers to act, it being able to generate less active force at lengths less than or greater than L_0. In contrast to striated muscle, however, which develop their maximal active force at a length where passive tension is absent, a significant passive force is required to achieve L_0 in smooth muscle prior to activation. On the molecular level, intracellular free calcium Ca^{2+} is the primary determinant of contractility. Increases in intracellular Ca^{2+} associate with contraction and decreases with relaxation. Finally, fibroblasts are important players in the maintenance of the *adventitia* via the synthesis of type I collagen. In case of tissue damage, the fibroblasts migrate quickly to the site of injury, proliferate and then synthesize new collagen. Such activity is regulated in part by growth factors. It has been observed that fibroblasts production of collagen filaments is strongly dependent on the stress level in the tissue and on the concentration of vasoactive molecules. Thus fibroblasts activity depends, ultimately, on the hemodynamic loads.

7.2 Models

7.2.1 Thin-wall models

Models proposed for the G&R of living tissues traditionally account for a single time scale at which the biological processes take place [2, 12, 13]. Nevertheless the influence of phenomena happening on different time scales might be relevant for G&R. Some examples of phenomena that influence long-term adaptation, although happening on shorter time scales, are the cardiac cycle, the respiratory cycle, and the circadian cycles. For vascular growth and adaptation, the dynamics on the cardiac cycle and the hemodynamic loads induced by the pumping action of the heart are certainly non negligible [15, 36]. The characteristic G&R time (days to weeks) is several orders of magnitude higher than the cardiac cycle time scale (seconds to fraction of a second depending on species). This difference in characteristic evolution times justifies the assumption of a multiple time scales approach to formulate the G&R problem under hemodynamic loads. In particular let us denote by s the G&R time variable and by t the cardiac cycle time and assume that negligible growth occurs during the cardiac cycle [17].

In several cases [3, 5, 37] the artery can be modelled as a thin-walled, circular cylinder of thickness h, internal radius a, and fixed length L subject to internal pressure P and wall shear stress τ_w induced by the blood flow Q. Even in this simple case, the description of blood flow rate, velocity and pressure result from the solution of a fluid-structure interaction problem [11, 39]. For the sake of simplicity let us assume that the blood flow rate and pressure are assigned as inputs into the model and that they are prescribed by a Fourier series on the cardiac cycle time scale t

$$P(s,t) = P_m(s) + \sum_i \left[A_i(s)\cos\left(i\frac{2\pi}{T(s)}t\right) + B_i(s)\sin\left(i\frac{2\pi}{T(s)}t\right) \right], \qquad (7.1)$$

$$Q(s,t) = Q_m(s) + \sum_i \left[C_i(s)\cos\left(i\frac{2\pi}{T(s)}t\right) + D_i(s)\sin\left(i\frac{2\pi}{T(s)}t\right) \right], \qquad (7.2)$$

where P_m and Q_m are the mean values of transmural pressure and flow rate, A_i, B_i, C_i, and D_i are the Fourier coefficients, and T is the cardiac period. These functions allow one to control arbitrarily the mean values, the wave form, the frequency of the transmural pressure and blood flow rate, and also their relative phase as a function of the G&R time variable s and therefore allow one to model separately the adaptation of the artery to the perturbation of each of these features of the hemodynamic loads.

Consistently with membrane theory [20] as well as prior work on aneurysms [4], vasospasm [21], and cerebral artery adaptation [37], the Cauchy membrane stress components, on the cardiac cycle time scale, are computed as

$$T_\theta(s,t) = \frac{1}{\lambda_z(s,t)} \left.\frac{\partial W}{\partial \lambda_\theta}\right|_{s,t} + h(s,t)\sigma_\theta^{\text{act}}(s,t), \tag{7.3}$$

$$T_z(s,t) = \frac{1}{\lambda_\theta(s,t)} \left.\frac{\partial W}{\partial \lambda_z}\right|_{s,t} + h(s,t)\sigma_z^{\text{act}}(s,t), \tag{7.4}$$

where W is the strain energy density of the wall tissue, $\sigma_\theta^{\text{act}}$ and σ_z^{act} are the smooth muscle active contributions to the Cauchy stress in the circumferential and axial directions, respectively, depending on the net constrictor concentration C (see Sect. 7.2.3). The same relationships hold on the G&R time scale with all the quantities depending only on s.

The mean values of pressure $P_m(s)$ and flow rate $Q_m(s)$ are assumed as reference values of pressure and flow, respectively, at each G&R time instant s hence, using Eqs. (7.3) and (7.4), the radial equilibrium on the G&R time scale is enforced by the equation

$$\frac{1}{\lambda_z(s)} \left.\frac{\partial W}{\partial \lambda_\theta}\right|_s + h(s)\sigma_\theta^{\text{act}}(s) = P_m(s)a(s). \tag{7.5}$$

On the cardiac cycle time scale we should, in principle, consider the inertial force; thus the radial equilibrium would assume the form

$$\frac{1}{\lambda_z(s,t)} \left.\frac{\partial W}{\partial \lambda_\theta}\right|_{s,t} + h(s,t)\sigma_\theta^{\text{act}}(s,t) = P(s,t)a(s,t) - \rho h(s,t)^2 \ddot{a}(s,t), \tag{7.6}$$

but inertial forces, as proved in [8] and [22], are largely negligible for the vast majority of species and vessel types and the quasi-static version of the equilibrium equations can be assumed also on the cardiac cycle time scale.

7.2.2 Modelling transmural inhomogeneity

Thin wall models are useful to capture the overall mechanical behaviour of the vessel and its evolution in response to perturbed hemodynamics, but cannot describe the local transmural differences of the biomechanical response. As described in the introduction, arterial layers are strongly inhomogeneous and populated by different types of cells that respond in very different ways to the external stimuli. For example, it is very common to observe adventitial hypertrophy in hypertension, due to

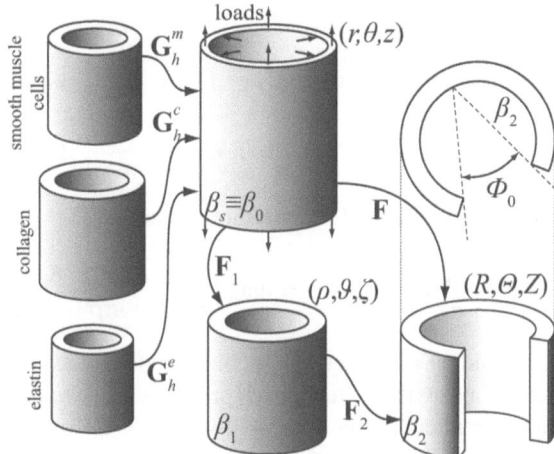

Fig. 7.2. Schema of the constrained mixture model of an arterial segment consisting of elastin, multiple families of collagen fibres, and smooth muscle. Shown, too, are the *in vivo* configuration, β_s, which is taken as a reference β_0 for the kinematics, a traction-free excised configuration, β_1, and a radially-cut, nearly stress-free configuration β_2. The deformation gradients are denoted by $\mathbf{F}_i$ $(i = 1, 2)$ and Φ_0 is the opening angle in β_2

the enhanced activity of the fibroblasts, while the *media* may not change its thickness or composition significantly [15]. Thus, a more detailed description of the wall mechanics is necessary to capture transmural differences in biomechanical response. Indeed the G&R model introduced in Sect. 7.2.3 can be easily applied to a thick-wall formulation. Let us introduce, then, a thick wall model of the artery to investigate the effects of transmural distributions of individual components and of their mechanical properties on the axial retraction, opening angle, and overall stress field in the vessel.

It proves convenient to study salient aspects of arterial wall mechanics using the semi-inverse approach of finite elasticity. In contrast to usual formulations, which use the "stress-free" configuration as a reference, we use the current, stressed configuration as a computational reference. Two reasons motivate this choice. First, most investigators prescribe the stress-free reference configuration (e.g., via an opening angle) based on empirical observations and then seek consequences of this reference on the *in vivo* state of stress. In contrast, we seek to determine consequences of material nonuniformity on both the *in vivo* state of stress and the residual stress related opening of a cut segment. Second, our work is motivated by the need to model growth and remodelling processes mathematically, which necessarily occur in the *in vivo*, stressed state. Hence, we prescribe the kinematics for an idealized axisymmetric artery via two successive motions (see Fig. 7.2): mappings of material points from a physiologically-relevant *in vivo* configuration β_s (with coordinates r, θ, z) associated with the finite extension and inflation of an intact cylindrical segment at time s to an intact but traction-free excised configuration β_1 (with coordinates ρ, ϑ, ζ) and then to a nearly stress-free, radially-cut configuration β_2 (R, Θ, Z). These mappings are given by

$$(r, \theta, z) \rightarrow (\rho, \vartheta, \zeta) : \rho = \rho(r), \ \vartheta = \theta, \ \zeta = \frac{z}{\lambda}, \tag{7.7}$$

$$(\rho, \vartheta, \zeta) \rightarrow (R, \Theta, Z) : R = R(\rho), \ \Theta = \frac{\pi - \Phi_0}{\pi}, \ Z = \frac{\zeta}{\Lambda}, \tag{7.8}$$

whereby the deformation gradients for these motions are

$$\mathbf{F}_1 = \text{diag} \left[\frac{\partial \rho}{\partial r}, \frac{\rho}{r}, \frac{1}{\lambda} \right], \ \mathbf{F}_2 = \text{diag} \left[\frac{\partial R}{\partial \rho}, \frac{(\pi - \Phi_0) R}{\pi \rho}, \frac{1}{\Lambda} \right], \tag{7.9}$$

with Φ_0 and Λ the residual stress related opening angle and axial stretch, respectively, and λ the additional axial stretch related primarily to the *in vivo* "prestretch"[1]. The total deformation gradient is thus computed via $\mathbf{F} = \mathbf{F}_2 \mathbf{F}_1$ and incompressibility is assumed to hold during transient motions, but not overall growth and remodelling, hence $\det \mathbf{F} = 1$ here. Assuming quasi-static motions [22], the Cauchy stress field $\boldsymbol{\sigma}$ associated with either the first ($\mathbf{F} = \mathbf{F}_1$ with $\mathbf{F}_2 = \mathbf{I}$) or the total ($\mathbf{F} = \mathbf{F}_2 \mathbf{F}_1$) motion can be computed via

$$\text{div} \boldsymbol{\sigma} = \mathbf{0}, \tag{7.10}$$

$$\boldsymbol{\sigma} = -p\mathbf{I} + \frac{\partial W}{\partial \mathbf{F}} \mathbf{F}^{\mathrm{T}} + \boldsymbol{\sigma}^{\text{act}}, \tag{7.11}$$

where p is a Lagrange multiplier that enforces incompressibility, W is the net strain energy function for the passive behaviour of the wall, and $\boldsymbol{\sigma}^{\text{act}}$ accounts for smooth muscle activity.

Recall the key fact that elastin is produced during the perinatal period and is normally stable thereafter [10, 26], thus it undergoes large multiaxial stretches as the artery grows to the adult configuration $\beta_0 \equiv \beta_s$. On the other hand, collagen and smooth muscle turn over continuously throughout life [26] and we assume that they are deposited at a preferred stretch during maturity. These assumptions result in higher "prestretches" in elastin than in collagen fibres and smooth muscle in maturity. That is, the different timing of deposition and values of deposition stretches for each constituent play particularly important roles because the associated prestretches can thereby vary transmurally, which together with the changing mass fractions of the constituents can strongly influence the unloaded length (i.e., net axial prestress) and the opening angle (i.e., net residual stress) of the vessel. To study this mathematically, consider the following.

Equilibrium of the *in vivo* configuration β_0, both local and global (integral) forms, requires

$$p(r) = P + \frac{\partial W}{\partial \lambda_r} \lambda_r \Big|_r - \int_{r_i}^{r} (\sigma_{\theta\theta} - \sigma_{rr}) \frac{dr}{r}, \tag{7.12}$$

---

[1] One can think of these deformation gradients as inverses of the usual tensors referred to the stress-free configuration [20].

and

$$\int_{r_i}^{r_a} (\sigma_{\theta\theta} - \sigma_{rr}) \frac{dr}{r} = P, \quad 2\pi \int_{r_i}^{r_a} \sigma_{zz} r dr = f, \tag{7.13}$$

where r_i and r_a denote inner (intimal) and outer (adventitial) radii in β_0 (associated with the deformation $\mathbf{F} = \mathbf{I}$) and P and f are the in vivo luminal pressure and axial force, respectively. Here, the in vivo configuration is assumed to correspond to mean transmural pressure and basal muscular tone in the basilar artery [37]. In order to ensure radial equilibrium, with the given in vivo luminal pressure P, Eq. (7.13)$_1$ can be solved for the adventitial radius r_a (or for the luminal radius in the case of a full G&R model) while the axial force f is computed explicitly from Eq. (7.13)$_2$. In this way, in all the simulations, the material parameters, the intimal radius, and the luminal pressure can be kept constant regardless of the distribution of constituent prestretches and mass fractions, while a slight variation in the outer radius preserves the radial equilibrium.

Equilibrium of the unloaded configuration β_1 similarly requires

$$p(\rho) = \frac{\partial W}{\partial \lambda_\rho} \lambda_\rho \Big|_\rho - \int_{\rho_i}^{\rho} (\sigma_{\vartheta\vartheta} - \sigma_{\rho\rho}) \frac{d\rho}{\rho}, \tag{7.14}$$

and

$$\int_{\rho_i}^{\rho_a} (\sigma_{\vartheta\vartheta} - \sigma_{\rho\rho}) \frac{d\rho}{\rho} = 0, \quad \int_{\rho_i}^{\rho_a} \sigma_{\zeta\zeta} \rho d\rho = 0, \tag{7.15}$$

where ρ_i and ρ_a denote inner (intimal) and outer (adventitial) radii in β_1 (associated with the deformation $\mathbf{F} = \mathbf{F}_1$). The two global equations can be solved to determine the inner radius ρ_i and the net in vivo axial prestretch λ for prescribed material properties and distribution of constituents prestretch and mass fraction.

Finally equilibrium of the excised, radially-cut configuration can be satisfied via the following [35]

$$p(R) = \frac{\partial W}{\partial \lambda_R} \lambda_R \Big|_R - \int_{R_i}^{R} (\sigma_{\Theta\Theta} - \sigma_{RR}) \frac{dR}{R}, \tag{7.16}$$

and

$$\int_{R_i}^{R_a} (\sigma_{\Theta\Theta} - \sigma_{RR}) \frac{dR}{R} = 0, \quad \int_{R_i}^{R_a} \sigma_{ZZ} R dR = 0, \quad \int_{R_i}^{R_a} \sigma_{\Theta\Theta} R dR = 0, \tag{7.17}$$

where R_i and R_a denote inner (intimal) and outer (adventitial) radii in β_2 (associated with the deformation $\mathbf{F} = \mathbf{F}_2\mathbf{F}_1$). Note that the additional global equilibrium equation enforces zero applied moments on the radially-cut section. The three global equations can be solved to determine the inner radius R_i, the net in vivo axial prestretch $\Lambda\lambda$, given λ from above, and the residual stress related opening angle Φ_0 for prescribed material properties and distribution of constituents prestretch and mass fraction.

7.2.3 Growth & remodelling: a multiple-time-scales approach

The main novelty of the proposed model is in the formulation of tissue growth and re-modelling and in the inclusion of muscle tone guided by hemodynamics. According to the theory of constrained mixtures proposed in [23], we assume the arterial wall to be composed of several structurally significant constituents that are deposited in the current configuration, on the G&R time scale, with a given prestretch $\mathbf{G}^k$ relative to their own natural (i.e., stress-free) configuration and that they follow the mixture deformation afterwards, with no relative motion. We usually assume that the deposition prestretches $\mathbf{G}^k$ do not depend on time. One can use different approximations for the collagen orientations, including continuous distributions about two diagonal directions [18]. Following [37], we assume a four collagen fibre families model (axial (z), circumferential (θ), and helical (h) inclined of $\pm\alpha^h$ degrees from the longitudinal axes) and denote elastin with e, collagen fibres with c, and smooth muscle with m. The structural constituents turnover continuously on the G&R time scale. Let us denote with $\mathbf{F}$ the deformation gradient from a given reference configuration β_0 to the current configuration on the G&R time β_s, and with $\mathbf{F}_s$ the deformation gradient from β_s to the current configuration on the cardiac cycle time β_s^t (see Fig. 7.3). Moreover we denote with τ a generic instant on the G&R time scale at which new constituent

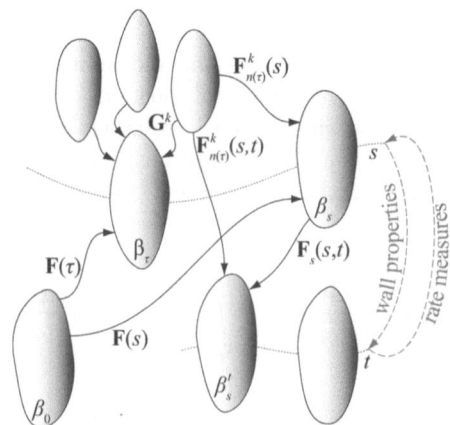

Fig. 7.3. Relevant configurations in the multiple-time scales G&R framework. At the current G&R time τ, the constituents are deposited in the current configuration on the G&R time scale β_τ at a given prestretch $\mathbf{G}^k$ independent of τ. The deformation gradient from a common reference configuration β_0 to the current configuration on the G&R time β_τ is denoted with $\mathbf{F}(\tau)$. The deformation gradient from the configuration β_s at G&R time s to the configuration β_s^t corresponding to the cardiac cycle time t (at G&R time s) is denoted by $\mathbf{F}_s(t)$. Grey, dashed arrows show the interconnections between the two time scales: the current wall geometries and material properties influence the solution on the cardiac cycle, in turn the mass production and removal rates as well as the constituents material properties depend on suitable rate measures on the cardiac cycle

fibres are incorporated in the tissue. Therefore, at G&R time s, the constituent k deposited at time τ experiences, according to [5] and [37], the deformation gradient

$$\mathbf{F}^k_{n(\tau)}(s) = \mathbf{F}(s)\mathbf{F}(\tau)^{-1}\mathbf{G}^k, \qquad (7.18)$$

while on the cardiac cycle time scale t, at a given G&R instant s, its deformation gradient is given by

$$\mathbf{F}^k_{n(\tau)}(s,t) = \mathbf{F}_s(t)\mathbf{F}(s)\mathbf{F}(\tau)^{-1}\mathbf{G}^k, \qquad (7.19)$$

were the further deformation gradient $\mathbf{F}_s(t)$ is superimposed to configuration β_s (corresponding to the mean pressure $P_m(s)$ and flow $Q_m(s)$) and brings to the current configuration on the cardiac cycle time scale β_s^t. It is worth noticing that, due to strong material non-linearity, the deformation gradient $\mathbf{F}^k_{n(\tau)}(s)$ does not correspond to the average of $\mathbf{F}^k_{n(\tau)}(s,t)$ over t. We assume that the constituents are deposited at the configuration of the vessel corresponding to the mean transmural pressure and flow rate on the cardiac cycle β_s (see Eqs. 7.1 and 7.2).

With the assumption of time scales separation, the mass density of the mixture, that is a function of the G&R time s only, can be computed as $M(s) = \sum_k M^k(s)$ where M^k is the mass density of the constituent k given by the convolution integral [4, 37]

$$M^k(s) = M^k(0)Q^k(s) + \int_0^s m^k(\tau)q^k(s,\tau)d\tau, \qquad (7.20)$$

where $Q^k(s) \in [0,1]$ accounts for the mass fraction present at time $s_0 = 0$ (assumed as the instant at which the perturbation of the system from the normal conditions occurs) that survives at time s, $m^k \geq 0$ is the true mass density production rate function, and $q^k(s,\tau) \in [0,1]$ is the mass fraction of material produced at time $\tau \in [0,s]$ that survives at time s. The mass fraction of the constituent k at G&R time s is therefore given by $\phi^k(s) = M^k(s)/M(s)$. Similarly the mixture strain energy density density is given by the sum $W(s,t) = \sum_k W^k(\mathbf{F}(s,t))$ where W^k is the contribution of the constituent k computed via the convolution integral

$$W^k(\mathbf{F}(s,t))$$
$$= \frac{M^k(0)}{\rho}Q^k(s)\widehat{W}^k\left(\mathbf{F}^k_{n(0)}(s,t)\right) + \int_0^s \frac{m^k(\tau)}{\rho}q^k(s,\tau)\widehat{W}^k\left(\mathbf{F}^k_{n(\tau)}(s,t)\right)d\tau, \quad (7.21)$$

where ρ is the mixture mass density per unit volume, assumed as constant herein (i.e., changes in mass parallel changes in volume), and $\widehat{W}^k$ is the strain energy density per reference unit volume for the constituent k. Typical assumptions for $\widehat{W}^k$ are the neo-Hookean model for elastin and Fung-type exponentials for collagen fibres and passive smooth muscle cells.

Given a quantity $y(s,t)$, defined on the cardiac cycle time scale, we denote by $\widetilde{y}(s)$ a suitable measure of its rate of variation on the cardiac cycle at G&R time s.

An example of such a measure can be [8]

$$\tilde{y}(s) = \left[\frac{1}{T} \int_0^T |\dot{y}(s,t)|^p \, dt \right]^{\frac{1}{p}},$$

(7.22)

where p is a parameter and the dot denotes the derivative with respect to the cardiac cycle time t. This quantity defines the influence of the time "microscale" t on the time "macroscale" s, and it is sensitive to perturbations in the frequency as well as to arbitrary variations of the wave form of the variable $y(s,t)$ on the cardiac cycle. Once the metric that bridges the cardiac cycle time scale to the G&R time scale has been chosen, the general constitutive equations for the G&R process can be formulated.

We need to model the active contractile property endowed by the smooth muscle cells. We chose to model the smooth muscle activation level with a scalar quantity C denoting the net concentration of vasoconstrictors. Experimental evidence reveals that the concentration of vasoactive molecules is modulated by molecular pathways regulated by the endothelial cells according to the hemodynamic forces they experience. It is then reasonable to assume that C depends on the wall shear stress τ_w and on its rate of variation on the cardiac cycle $\tilde{\tau}_w$

$$C(s) = C(\tau_w(s), \tilde{\tau}_w(s)).$$

(7.23)

This constitutive assumption asserts that the level of smooth muscle activation is driven directly by hemodynamic processes on the cardiac cycle.

It has been shown that fibroblast activity is governed largely by the stress in the tissue [34, 37], on the rate of variation of hemodynamic loads on the cardiac cycle [8], and on the concentration of vasoactive molecules [29, 31, 32]. We thus assume that mass production rates m^k depend on a measure of the stress in the component σ^k, on its rate of variation on the cardiac cycle $\tilde{\sigma}^k$, and on the constrictor concentration in the tissue C, namely

$$m^k(s) = m^k\left(\sigma^k(s), \tilde{\sigma}^k(s), C(s)\right).$$

(7.24)

Because the stress and constrictor concentration depend on the hemodynamic loads, also the mass production rate depends ultimately on the hemodynamics.

Cell apoptosis and matrix degradation are complex phenomena that are also influenced by the tensional state induced by the hemodynamic loads. Thus we assume that the survival function $q^k(s, \tau)$ (i.e., the fraction of the constituent k that was deposited at time τ and survives at time s) depends on the time course of the stress in the constituent and of its rate over the cardiac cycle time scale, that is

$$q^k(s, \tau) = q^k\left(\sigma^k_{n(\tau)}([\tau,s]), \tilde{\sigma}^k_{n(\tau)}([\tau,s])\right).$$

(7.25)

This constitutive assumption reflects the importance of hemodynamics also on the removal rate of structurally significant constituents.

As discussed in [19] and [15], the stress rate experienced by the constituents might also affect their mechanical behaviour. This effect can be modelled by letting the material parameters c_i^k, involved in the definition of the strain energy densities $\widehat{W}^k$, be either functions of instantaneous values of suitable stress-rate measures or functionals of the time history of such measures over the G&R time interval $[\tau, s]$ (assuming that the constituent was deposited at time τ), that is

$$c_i^k(s) = c_i^k\left(\widetilde{\sigma}^k(s)\right) \quad \text{or} \quad c_i^k(\tau, s) = c_i^k\left(\widetilde{\sigma}^k([\tau, s])\right). \tag{7.26}$$

These functional dependencies take into account, for example, the possibility that the stiffness of the newly deposited fibres (i.e., collagen) can depend on the current stress rate of the tissue and that some constituents (i.e., elastin) might manifest early aging or fatigue/damage induced by severe increase in the stress rate over the cardiac cycle. These might be regarded as remodelling phenomena driven by the stress rate in the vessel and thus, ultimately, by the hemodynamics.

Constitutive assumptions (7.22)-(7.26) introduce the dependence of G&R, on time scale s, on the rate of phenomena happening on the cardiac cycle time scale t, explaining the coupling effect depicted as an upward arrow on the far right side of Fig. 7.3. Conversely, the hemodynamics on the cardiac cycle time scale t depends on the wall geometry and material properties at time s and this explains the downward arrow in Fig. 7.3. Therefore the two time scales are fully coupled in this multiscale formulation of the G&R problem.

7.3 Results

Thin-walled arterial models of the class described in Sect. 7.2.1 have been used to describe arterial adaptation to several diseases or hemodynamic perturbations. In [5, 21] the time course of a cerebral vasospasm was successfully described by a thin-wall model including endothelial damage and recovery. The model suggests that chemical and mechanical mediators of cellular and extracellular matrix turnover can differentially dominate the progression and resolution of vasospasm. Arterial adaptation to hypertension and perturbed blood flow rate were modelled in [37]. The model is able to predict arterial wall stiffening in hypertensive human basilar arteries and highlighted the complementary effects of active contraction and matrix turnover in response to severe perturbations in the mean blood flow rate. Adaptation to changes in the axial stretch was considered in [38]; the model predicts a marked increase in fibrous constituent production, leading to a compensatory lengthening that restores original mechanical behaviour after a step increase in axial length.

Nevertheless, all of these models are based on a quasi-static version of the one presented here, that is G&R is only driven by mean values on the cardiac cycle and not by the rate measures. Such models may fail in the accurate description of clinical scenarios in which dynamical features of the hemodynamics, like the pulsatility of blood flow rate and pressure, undergo significant perturbations. The case of aortic arch banding model of hypertension is one of these scenarios [15]. Indeed the aortic arch banding procedure induces severe change in pulse pressure and flow rate in the common carotid arteries but only modest changes in their mean values. The adaptation of a mouse left common carotid artery to the perturbation induced by the aortic banding procedure has been modelled in [8]. The numerical model proves that the adaptation process in this case is driven mainly by dynamic stimuli (i.e., stress rates over the cardiac cycle) and that constituents degradation and remodelling as described in Sect. 7.2.3 play a fundamental role. The model, in agreement with experimental measurements, predicts a significant increase in arterial caliber and thickness together with a significant reduction of axial stress and circumferential stiffness.

Concerning thick-wall models, full G&R formulations based on the constrained mixture theory are under developement [25]. Nevertheless important results can be obtained by making suitable assumptions on the distribution of the quantities without modelling the entire G&R process [7]. Using the assumption of constrained mixture, we can assume a strain energy density of the form

$$W = \phi^e W^e (\mathbf{F}^e) + \sum_{c=1}^{4} \phi^c W^c (\lambda^c) + \phi^m W^m (\lambda^m), \qquad (7.27)$$

where ϕ^k are mass fractions for each structurally significant constituent and W^k are individual stored energy functions. In Eq. (7.27) $\mathbf{F}^e$, λ^c, and λ^m represent the elastin deformation gradient, the collagen fibre stretch, and the smooth muscle stretch, respectively. These quantities are defined with respect to individual stress-free configurations (see Fig. 7.2).

It has been shown in [7] that, for basilar arteries, the axial retraction and opening angle are extremely sensitive to the prestretches of the constituents. In particular the elastin prestretches play a fundamental role. Elastin is the most stable constituent in the wall and is deposited early during development in parallel layers starting from the inner wall and proceeding outward under vessel growth driven by an increased blood flow rate and luminal pressure; this biological peculiarity suggest that inner layers deposited earlier during development may experience larger prestretches than the outer layers. This biologically based assumption leads to a realistic prediction of the opening angle and axial retraction in a homogeneous wall and leads to a nearly uniform distribution of stress with the radial position. Moreover the model is able to reproduce the measured increase in the opening angle with muscle contraction [30] and the increase in opening angle with age due to a change in transmural distribution of collagen and elastin [16].

Several other models [1, 2, 35] have been employed to describe the transmural variation of vascular mechanical properties and its effect on the opening angle and axial retraction. Nevertheless a microstructurally motivated model, like the one introduced here, gives further insight in the biological origin of mechanical quantities and is able to reproduce several different sets of experimental measurements by simply making reasonable assumptions on the transmural distribution of constituents and mechanical properties.

7.4 Discussion

Experimental observations over the past few decades revealed that mechanical stimuli induced by the hemodynamics play a fundamental role in arterial development, adaptation, and disease progress [6, 9, 14, 15, 27, 28, 29, 33]. Starting from the histology of the arterial wall and cellular functions, we proposed a class of models to describe arterial growth, remodelling, and adaptation to hemodynamic perturbations. The model accounts for multiple structurally significant constituents, namely collagen, smooth muscle, and elastin, that are allowed to turnover continuously in time, within the framework of constrained mixture theory [23, 24]. Constituents turnover rates depend on certain measures of stress, stress rate of variation over the cardiac cycle, and the concentration of vasoactive molecules. These mechanical and chemical stimuli for tissue maintenance and adaptation are all directly governed by the hemodynamic forces, via mechanical equilibrium and wall shear stress affecting endothelial cell activity. Introduction of these metrics guiding tissue G&R are based on experimental observations and represent a reasonable extension of mechanics to the mechanobiology of living tissues.

Although the model formulation is suitable for three-dimensional analyses, the framework has been applied mostly to thin walled, cylindrical arteries to model their adaptations to changes in hemodynamics. Simple functional forms for the constitutive equations (7.23)–(7.26) were able to reproduce and give a microstructural interpretation of basic arterial adaptation to hypertension [8, 37], vasospasm [4], and flow perturbations [37]. In particular it has been shown [8] that dynamic features of the pressure and flow waves on the cardiac cycle should be accounted for in order to explain accurately some experimental findings, asserting that mean values of pressure and flow are not the only hemodynamical quantities guiding arterial adaptation and response to mechanical perturbations. Nevertheless, more modelling and experimental effort is needed in order to fully characterize cellular response to mechanical stimuli.

Thin wall models proved useful to describe homogenized properties of the vascular tissue, but they are not suitable to describe intramural inhomogeneity of tissue composition and functions. Arterial layers can show very different mechanical properties, due to the presence of different structurally significant constituents, and mechanobiological responses to mechanical stimuli, due to the different kind of cells present in the layers. In order to capture these radial gradients of properties, the models need to be extended to thick walled formulation. Axisymmetry can still be assumed in many cases for straight arteries. Preliminary simulations [8] show that distributing elastin prestretches according to its deposition process during development and biological stability explains several experimental results on axial retraction and opening angle.

Models can be used for the double purpose of guiding new experiments and testing new hypotheses. In turn, experiments can suggest appropriate values for model parameters and functional forms. Thus modelling represents a fundamental step of a cycle in the process of understanding the development and adaptations of living tissues to changes in mechanical environment. The models described here have several applications including further promising developments such as full 3D G&R models and coupled fluid-structure interaction for studying aneurysms and patient specific geometries.

References

[1] Alastrué V., Peña E., Martínez M.A., Doblaré M.: Assessing the use of the "opening angle method" to enforce resdidual stresses in patient–specific arteries. Ann. Biomed. Eng. **35**: 1821–1837, 2007.

[2] Alford P.W., Humphrey J.D., Taber L.A.; Growth and remodeling in a thick-walled artery model: effects of spatial variations in wall constituents. Biomech. Mod. Mechanobiol. **7**: 245–262, 2008.

[3] Baek S., Gleason R.L., Rajagopal K.R., Humphrey J.D.: Theory of small on large: Potential utility in computations of fluid–solid interactions in arteries. Comput. Methods Appl. Mech. Engrg. **196**: 3070–3078, 2007.

[4] Baek S., Rajagopal K.R., Humphrey J.D.: A theoretical model of enlarging intracranial fusiform aneurysms. J. Biomech. Eng. **128**(1): 142–9, 2006.

[5] Baek S., Valentín A., Humphrey J.D.: Biochemomechanics of cerebral vasospasm and its resolution: II. constitutive relations and model simulations. Ann. Biomed. Eng. **35**(9): 1498–1509, 2007.

[6] Burton A.C.: Relation of structure to function of the tissues of the wall of blood vessels. Physiol. Rev. **34**: 619–642, 1954.

[7] Cardamone L., Valentín A., Eberth J.F., Humphrey J.D.: Origin of axial prestretch and residual stress in arteries. Biomech. Mod. Mechanobiol. **8**: 431–446, 2009.

[8] Cardamone L., Valentín A., Eberth J.F., Humphrey J.D.: Modeling carotid artery adaptations to dynamic alterations in pressure and flow over the cardiac cycle. Math. Med. Biol. 2010.

[9] Dancu M.B., Berardi D.E., Vanden Heuvel J.P., Tarbell J.M.: Asynchronous shear stress and circumferential strain reduces endothelial NO synthase and cyclooxygenase-2 but induces endothelin-1 gene expression in endothelial cells. Arterioscler. Thromb. Vasc. Biol. **24**: 2088–2094, 2004.

[10] Davis E.C.: Elastic lamina growth in the developing mouse aorta. J. Histochem. Cytochem. **43**: 1115–1123, 1995.

[11] Demiray H.: Wave propagation through a viscous fluid contained in a prestressed thin elastic tube. Int. J. Eng. Sci. **30**(11): 1607–1620, 1992.

[12] Di Carlo A., Quiligotti S.: Growth and balance. Mech. Res. Comm. **29**: 449–456, 2002.

[13] Doblaré M. et al.: Anisotropic bone remodelling model based on a continuum damage-repair theory. Journal of Biomechanics **35**(1): 1–17, 2002.

[14] Dobrin P.B., Canfield T., Sinha S.: Development of longitudinal retraction of carotid arteries in neonatal dogs. Experientia **31**: 1295–1296, 1975.

[15] Eberth J.F., Gresham V.C., Reddy A.K., Popovic N., Wilson E., Humphrey J.D.: Importance of pulsatility in hypertensive carotid artery growth and remodeling. J. Hypertension **27**: 2010–2021, 2009.

[16] Feldman S.A., Glagov S.: Transmural collagen and elastin in human aortas: reversal with age. Atherosclerosis **13**: 385–394, 1971.

[17] Figueroa C.A., Baek S., Taylor C.A., Humphrey J.D.: A computational framework for fluid-solid-growth modeling in cardiovascular simulations. Computer Methods in Applied Mechanics and Engineering **198**(45–46): 3583–3602, 2009.

[18] Gasser T., Ogden R., Holzapfel G.: Hyperelastic modelling of arterial layers with distributed collagen fibre orientations. Journal of the Royal Society Interface **3**(6): 15, 2006.

[19] Greenwald S.: Ageing of the conduit arteries. The Journal of Pathology **211**(2): 157–172, 2007.

[20] Humphrey J.D.: Cardiovascular Solid Mechanics: Cells, Tissues, and Organs. Springer–Verlag, New York, 2002.

[21] Humphrey J.D., Baek S., Niklason L.E.: Biochemomechanics of cerebral vasospasm and its resolution: I. a new hypothesis and theoretical framework. Ann. Biomed. Eng. 35(9): 1485–1497, 2007.

[22] Humphrey J.D., Na S.: Elastodynamics and arterial wall stress. Ann. Biomed. Eng. 30(4): 509–23, 2002.

[23] Humphrey J.D., Rajagopal K.R.: A constrained mixture model for growth and remodeling of soft tissues. Math. Models. Methods. Appl. Sci. 128(3): 407–30, 2002.

[24] Humphrey J.D., Rajagopal K.R.: A constrained mixture model for arterial adaptations to a sustained step change in blood flow. Biomech. Mod. Mechanobiol. 22: 109–126, 2003.

[25] Karšaj I., Humphrey J.: A mathematical model of evolving mechanical properties of intraluminal thrombus. Biorheology 46(6): 509–527, 2009.

[26] Langille B.L.: Arterial remodeling: relation to hemodynamics. Can. J. Physiol. Pharmacol. 74(7): 834–41, 1996.

[27] Langille B.L., Bendeck M.P., Keeley F.W.: Adaptations of carotid arteries of young and mature rabbits to reduced carotid blood flow. Am. J. Physiol. 256(4 Pt 2): H931–9, 1989.

[28] Langille B.L., O'Donnell F.: Reductions in arterial diameter produced by chronic decreases in blood flow are endothelium-dependent. Science 231(4736): 405–407, 1986.

[29] Malek A., Izumo S.: Physiological fluid shear stress causes downregulation of endothelin-1 mRNA in bovine aortic endothelium. Am. J. Physiol. 263(2 Pt 1): C389–96, 1992.

[30] Matsumoto T., Hayashi K.: Stress and strain distribution in hypertensive and normotensive rat aorta considering residual strain. J. Biomech. Eng. 118(1): 62–73, 1996.

[31] Rizvi M.A.D., Katwa L., Spadone D.P., Myers P.R.: The effects of endothelin-1 on collagen type I and type III synthesis in cultured porcine coronary artery vascular smooth muscle cells. J. Mol. Cell Cardiol. 28(2): 243–252, 1996.

[32] Rizvi M.A.D., Myers P.R.: Nitric oxide modulates basal and endothelin-induced coronary artery vascular smooth muscle cell proliferation and collagen levels. J. Mol. Cell Cardiol. 27(7): 1779–1789, 1997.

[33] Rodbard S.: Vascular caliber. Cardiology 60(1): 4–49, 1975.

[34] Taber L.A.: A model for aortic growth based on fluid shear and fiber stresses. J. Biomech. Eng. 120(3): 348–54, 1998.

[35] Taber L.A., Humphrey J.D.: Stress-modulated growth, residual stress, and vascular heterogeneity. J. Biomech. Eng. 123: 528–535, 2001.

[36] Tada S., Tarbell J.M.: A computational study of flow in a compliant carotid bifurcation-stress phase angle correlation with shear stress. Ann. Biomed. Eng. 33(9): 1202–1212, 2005.

[37] Valentín A., Cardamone L., Baek S., Humphrey J.D.: Complementary vasoactivity and matrix remodeling in arterial adaptations to altered flow and pressure. J. Roy. Soc. Interface 6: 293–306, 2009.

[38] Valentín A., Humphrey J.: Modeling effects of axial extension on arterial growth and remodeling. Medical and Biological Engineering and Computing 47(9): 979–987, 2009.

[39] Zamir M.: The Physics of Pulsatile Flow. Springer, New York, 2000.

8

The VPH-Physiome Project: standards, tools and databases for multi-scale physiological modelling

Peter Hunter, Chris Bradley, Randall Britten, David Brooks, Luigi Carotenuto, Richard Christie, Alejandro Frangi, Alan Garny, David Ladd, Caton Little, David Nickerson, Poul Nielsen, Andrew Miller, Xavier Planes, Martin Steghoffer, Alistair Young, and Tommy Yu

Abstract. The VPH/Physiome project is developing tools and model databases for computational physiology based on three primary model encoding standards: CellML, SBML and FieldML. For the modelling community these standards are the equivalent of the DICOM standard for the clinical imaging community and it is im-

Peter Hunter (✉) and Chris Bradley
Auckland Bioengineering Institute (ABI), University of Auckland, New Zealand and Department of Physiology, Anatomy and Genetics (DPAG), University of Oxford, United Kingdom
e-mail: p.hunter@auckland.ac.nz

Randall Britten and David Brooks
Auckland Bioengineering Institute (ABI), University of Auckland, New Zealand

Luigi Carotenuto
Center for Computational Imaging and Simulation Technologies in Biomedicine (CISTIB), Universitat Pompeu Fabra, Barcelona, Spain

Richard Christie
Auckland Bioengineering Institute (ABI), University of Auckland, New Zealand

Alejandro Frangi
Center for Computational Imaging and Simulation Technologies in Biomedicine (CISTIB), Universitat Pompeu Fabra, Barcelona, Spain

Alan Garny
Department of Physiology, Anatomy and Genetics (DPAG), University of Oxford, United Kingdom

David Ladd, Caton Little, David Nickerson, Poul Nielsen and Andrew Miller
Auckland Bioengineering Institute (ABI), University of Auckland, New Zealand

Xavier Planes, Martin Steghoffer
Center for Computational Imaging and Simulation Technologies in Biomedicine (CISTIB), Universitat Pompeu Fabra, Barcelona, Spain

Alistair Young and Tommy Yu
Auckland Bioengineering Institute (ABI), University of Auckland, New Zealand

Ambrosi D., Quarteroni A., Rozza G. (Eds.): Modeling of Physiological Flows.
DOI 10.1007/978-88-470-1935-5_8, © Springer-Verlag Italia 2012

portant that the tools adhere to these standards to ensure that models from different groups can be curated, annotated, reused and combined. This chapter discusses the development and use of the VPH/Physiome standards, tools and databases, and also discusses the minimum information standards and ontology-based metadata standards that are complementary to the markup language standards. Data standards are not as well developed as the model encoding standards (with the DICOM standard for medical image encoding being the outstanding exception) but one new data standard being developed as part of the VPH/Physiome suite is BioSignalML and this is described here also. The PMR2 (Physiome Model Repository 2) database for CellML and FieldML files is also described, together with the Application Programming Interfaces (APIs) that facilitate access to the models from the visualisation (cmgui and GIMIAS) or computational (OpenCMISS, OpenCell/OpenCOR and other) software.

8.1 Introduction

Over the last 50 years biomedical science has yielded a wealth of reductive knowledge of the human body, most notably with the Human Genome Project, but has made relatively little effort to integrate this knowledge back into physiological understanding and therefore evidence-based medical treatment (Hunter and Viceconti, 2009). In almost every other area of human activity – communication, transport, entertainment, weather and climate prediction, to name a few – the physical and engineering sciences play a crucial role. Bioengineering has of course had a substantial impact on medical practice through medical devices and diagnostic imaging equipment, but the impact from 150 years of scientific progress in physics, engineering and mathematics on the biomedical sciences, and evidence-based medicine, has been relatively slight (Hunter, 2004; Hunter, 2006).

To help bring the quantitative and predictive power of the mathematical sciences into the biological sciences, and to take advantage of an engineering approach to biological materials and a more systematic approach to the knowledge management of multi-scale physiological data, the International Union of Physiological Sciences (IUPS) initiated the Physiome Project in 1997[1]. A related US initiative by the Inter-

---

[1] The concept of a "Physiome Project" was presented in a report from the Commission on Bioengineering in Physiology to the International Union of Physiological Sciences (IUPS) Council at the 32nd World Congress in Glasgow in 1993. The term "physiome" comes from "physio" (life) + "ome" (as a whole), and is intended to provide a "quantitative description of physiological dynamics and functional behaviour of the intact organism" [2,3,4]. A satellite workshop "On designing the Physiome Project", organized and chaired by the Chair of the IUPS 'Commission on Bioengineering in Physiology' (Prof Jim Bassingthwaighte), was held in Petrodvoretz, Russia, following the 33rd World Congress in St Petersburg in 1997. A symposium on the Physiome Project was held at the 34th World Congress of IUPS in Christchurch, New Zealand, in August 2001 and the Physiome Project was designated by the IUPS executive as a major focus for IUPS during the next decade.

agency Modelling and Analysis Group (IMAG) was begun in 2003[2]. Following a workshop in 2005 (Nørager et al., 2005) and the publication of the STEP[3] ('Strategy for a European Physiome') roadmap in 2006, the European Commission initiated the Virtual Physiological Human (VPH) project[4] in 2007. These projects and similar initiatives in Japan[5] and Korea are now, to some extent, coordinated and are collectively referred to here as the 'VPH-Physiome' Project.

The primary goal of the international VPH-Physiome Project is to develop, promote and facilitate the use of computational models, data and software tools (including web services) for understanding the human body as an integrated system. This will allow the development of strategies for disease prediction and medical treatment that are more subject-specific and scientifically based. Medical practice will benefit from technologies in which digital data enables predictable outcomes through quantitative models that integrate physical processes across multiple spatial scales. Computational tools are also being developed to link individual patient data with virtual population databases using mathematical models of biological processes.

We discuss some of the standards, model and data repositories, tools and workflows being developed by the VPH-Physiome project to support multi-scale modelling in clinical settings. The application of these tools is illustrated in the context of multi-scale modelling of the heart in which physical processes are coupled at both cellular and organ level scales.

8.2 Infrastructure for the VPH-Physiome Project

The VPH-Physiome Project aims to provide a systematic framework for understanding physiological processes in the human body in terms of anatomical structure and biophysical mechanisms at multiple length and time scales. The importance of establishing a solid foundation for the VPH by creating model and data standards, together with mechanisms for achieving model reproducibility and reuse, was recognized in the STEP Roadmap. The framework includes modelling languages for encoding systems of differential-algebraic equations (DAEs, using for example, CellML[6] and SBML[7]) and the spatially varying fields used with systems of partial differential equations (PDEs using FieldML[8]). In both cases the parameters and variables in the mathematical models are annotated with metadata that provides the biological

---

[2] This coordinates various US Governmental funding agencies involved in multi-scale bioengineering modelling research including NIH, NSF, NASA, the Dept of Energy (DoE), the Dept of Defense (DoD), the US Dept of Agriculture and the Dept of Veteran Affairs. See www.nibib.nih.gov/Research/MultiScaleModelling/IMAG.

[3] The STEP ("Strategy for a European Physiome") report is available at www.europhysiome.org/roadmap

[4] See www.vph-noe.eu for details on the 15 projects funded under the VPH calls.

[5] See www.physiome.jp for details on the Japanese Physiome Project.

[6] www.cellml.org

[7] www.sbml.org

[8] www.fieldml.org

or biophysical meaning. The languages encourage modularization and have import mechanisms for creating complex models from modular components. Model repositories have been established, together with freely available Open Source software tools to create, visualize and execute the models. The CellML model repository[9], for example (discussed below), now includes models for a wide variety of subcellular processes.

Many models of biological processes at the subcellular level ignore the detailed 3D structure of the cell and model the cell excitability, mechanics, calcium (or other second messenger) transients, motility, signalling, metabolism or gene regulation, etc, in a 'lumped parameter' system of DAEs. These models generally either include explicit discretisation of spatial variation, or make the approximating assumption that the model will remain spatially homogeneous (the well-stirred reactor assumption). Sometimes the models also include non-linear algebraic equations or need to solve constrained optimisation problems as part of the solution strategy, but they do not require the solution of PDEs. In some cases the models are 'systems physiology' models at the whole body level, but again without the need for solving PDEs. The markup language CellML (www.cellml.org) has been developed to provide an unambiguous definition of these lumped parameter models. The language is designed to support the definition and sharing of models of biological processes by including information about: model structure (how the parts of a model are organizationally related to one another); mathematics (equations describing the underlying biological processes); and metadata (additional information about the model that allows scientists to search for specific models or model components in a database or other repository, and which indicates the biological meaning associated with a model and its parts, allowing the model to be manipulated or interpreted correctly).

The development of a biophysically based mathematical model is, at present, a creative endeavor, often requiring a great deal of insight into the physical processes being modelled and personal judgment about the approximations needed. Once created, however, a model should stand independent of its creator and be reproducible and testable by others. The model and data files, that together can demonstrate reproducibility of a model on an automated basis, are called the *reference description* of the model. This includes protocols for running the model with appropriate parameter sets and comparing simulation results against suitably encoded experimental data (possibly multiple times to generate typical phenotypic outputs). The issue of robustness and reproducibility is particularly important when a model representing some small component of physiological function is incorporated into a more comprehensive model – and especially one that is to be used in a clinical setting. To be worthy of reuse in this fashion, each independently developed component should be demonstrably 'correct' for the function it claims to represent, in the sense both of *biological validity* – it matches some aspect of biological reality – and *mathematical validity* – for example, it has consistent units and does not violate physical principles such as conservation of mass or charge.

---

[9] www.cellml.org/models

The general strategy for developing the modelling standards is as follows:

- develop markup languages (MLs) for encoding models, including metadata, and data;
- develop application programming interfaces (APIs) for interacting with the MLs;
- develop libraries of open source tools that can read and write ML-encoded files;
- develop data and model repositories based on the MLs;
- develop reference descriptions to demonstrate model reproducibility;
- develop services/libraries for a variety of tasks including access to automated scripts to run the models and compare results against experimental data, optimize parameter values for new experimental data and provide sensitivity analyses for changes in model parameters.

Fig. 8.1 provides an overview of how the standards, tools and databases work together. Models at the cell, tissue or organ level are encoded in the CellML, SBML or FieldML standards (using Open Source tools) and deposited in model repositories. These models are curated (to ensure consistency of units and other checks on the validity of the models) and annotated (to provide controlled vocabulary annotations for the variables and parameters in the model). A model can then be loaded from a database into the simulation environment: e.g. cmgui, GIMIAS or OpenCMISS

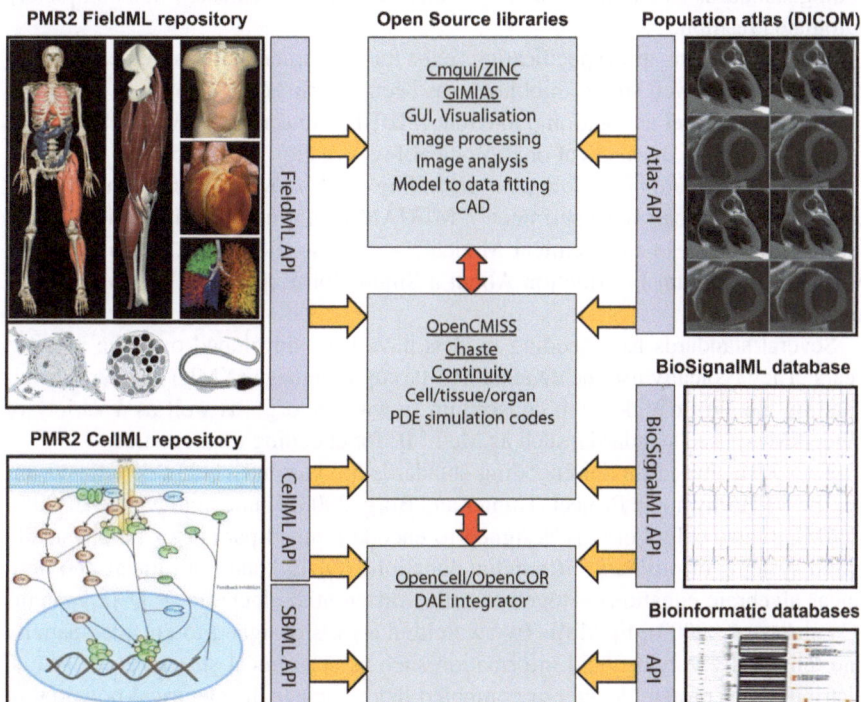

Fig. 8.1. The overall framework for the use of the software tools, together with the model and data repositories based on the VPH/Physiome standards

in the case of the FieldML model repository; OpenCMISS or OpenCell/OpenCOR in the case of the CellML model repository. Population data is stored in a separate atlas database along with the medical images (in DICOM format). Another XML standard called SED-ML is being developed to specify the numerical simulation itself, including the choice of numerical algorithm. Time varying signal data encoded in the BioSignalML format is stored in a separate database that can also be accessed from the computational software environments (OpenCMISS and OpenCell/OpenCOR) via an API. Finally, genomic and proteomic bioinformatic data is accessed from a number of databases via a separate API. Details on these standards, tools and databases are presented in the following sections (Hunter and Viceconti, 2009).

8.3 Syntax and semantics of VPH/Physiome models

Key steps that have been taken in the last 10 years to help improve the robustness and reproducibility of published models are the development of specifications for the "minimum information" needed for a model, and the development of model encoding standards so that a model can be checked and implemented in a completely automated fashion.

Minimum information specifications exist for experimental measurements in specific areas and a web site at mibbi.org has been established to provide a portal for these and for model annotation. Note that MIBBI is an acronym for "Minimum Information about a Biomedical or Biological Investigation". Examples of minimum information standards for biological modelling are MIAME ("Minimum Information about a Microarray Experiment"), MIRIAM ("Minimal Information Required In the Annotation of biochemical Models" – see also biomodels.net/miriam) and MIASE ("Minimum Information About a Simulation Experiment" – see biomodels.net/miase).

Several standards for encoding models have been developed over the past ten years. They typically use the eXtensible Markup Language (XML) standard developed by the WorldWide Web consortium (www.w3c.org), as well as a variety of other XML-based standards, such as MathML for encoding mathematics.

Two XML-based model encoding standards are currently being developed under the VPH/Physiome Project (Hunter and Borg, 2003; Hunter and Nielsen, 2005): CellML (www.cellml.org) is designed to encode lumped parameter biophysically based systems of ordinary differential equations (ODEs) and both linear and nonlinear algebraic equations – together called differential algebraic or DAE systems (Cuellar et al., 2003). FieldML (www.fieldml.org) is designed to encode spatially and temporally varying field information such as anatomical structure, the spatial distribution of protein density or computed fields such as the electrical potential or oxygen concentration throughout a tissue (Christie et al., 2009). A third markup language called the 'Systems Biology Markup Language' or SBML (sbml.org) has been developed by the systems biology community. This has similar expressiveness

Table 8.1. The minimum information standards, syntax and semantics being developed for data, models and simulation experiments

	Data	*Models*	*Simulation*
Minimal requirements	MIBBI[a]	MIRIAM[b]	MIASE[c]
Standard formats	PDB[d], DICOM[e], BioSignalML[f]	SBML[g], CellML[h], FieldML[i]	SED-ML[j]
Ontologies	GO[k], Biopax[l], FMA[m], SBO[n], OPB[o]GO	Biopax, FMA, SBO, OPB	KiSAO[p]

[a] www.mibbi.org
[b] biomodels.net/miriam
[c] biomodels.net/miase
[d] www.rcsb.org/pdb
[e] medical.nema.org
[f] www.embs.org/techcomm/tc-cbap/biosignal.html
[g] www.sbml.org
[h] www.cellml.org
[i] www.fieldml.org
[j] www.sed-ml.org
[k] www.geneontology.org
[l] www.biopax.org
[m] sig.biostr.washington.edu/projects/fm/AboutFM.html
[n] www.ebi.ac.uk/sbo
[o] sig.biostr.washington.edu/projects/biosim/opb-intro.html
[p] www.ebi.ac.uk/compneur-srv/kisao

to CellML but is targeted more specifically to representing models of biochemical reactions.

A useful way of viewing the development of standards is shown in Table 8.1, including specification of the minimal requirements for data, models and a simulation experiment, along with the standards for the syntax of the data, models or simulation experiments and the ontologies for annotating the semantic meaning of terms in the data, models or simulation experiments.

8.4 CellML

CellML separates the syntax of a model (e.g. the mathematical equations encoded in MathML) from the semantics (the biological and biophysical meanings of the model components and parameters) defined in the model metadata by reference to suitable ontologies. In addition, models can be broken down into components; this facilitates building complex models by importing modular components defined in libraries. SBML is more closely tied to the concepts of biochemical and genetic networks, and does not maintain such a clear separation between the mathematical representation and the biological semantics. FieldML deals with the encoding of anatomy at multiple spatial scales by allowing hierarchies of material coordinate systems that preserve anatomical relationships (e.g. coronary arteries embedded in a deforming myocardial tissue that is itself part of a heart contained within a torso). These three standards are recommended both by the US National Institutes of Health (NIH) and by the European Commission for the VPH project.

CellML has a simple structure based upon connected components. These components abstract concepts by providing well-defined interfaces to other components, and encapsulate concepts by hiding details from other components. Connections provide the means for sharing information by associating variables visible in the interface of one component with those in the interface of another component. Consistency is enforced by requiring that all variables be assigned appropriate physical units which must be dimensionally compatible when variables are connected. Public and private interfaces enable encapsulation hierarchies, providing further mechanisms for information hiding and abstraction. Model reuse is facilitated by the import element, enabling new models to be constructed by combining existing hierarchies of components into model hierarchies. The CellML 1.1 standard is available at www.cellml.org/specifications/cellml_1.1. An overview of CellML structure is shown in Fig. 8.2. An overview of CellML metadata standards is given in Beard et al., 2009.

Other markup languages include NeuroML (neuroml.org), used by the neuroscience modelling community and InSilicoML (physiome.jp/insilicoml), developed by Physiome Project groups in Japan.

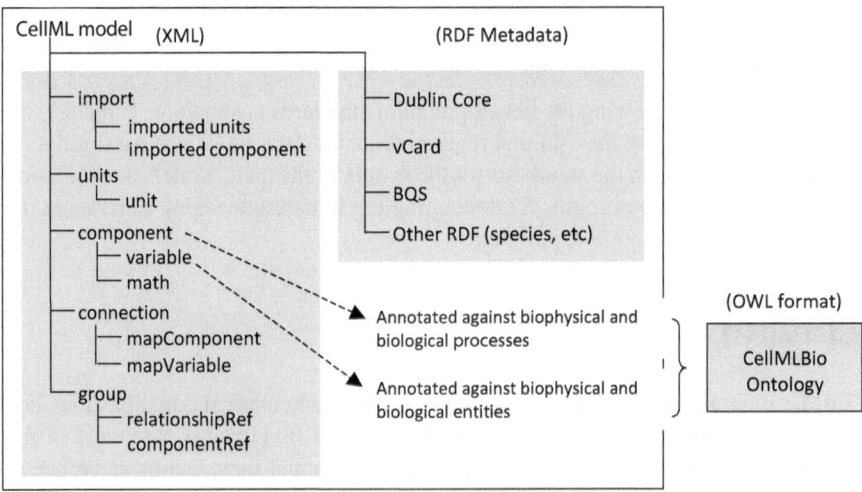

Fig. 8.2. Entities in a CellML model. The CellML model file on the left contains the base model with its imports, units, components, connections and groups described in XML format, and the metadata in RDF format. The annotation of CellML variables with biological and biophysical meaning is handled via cmeta:id links to terms stored in RDF format in a separate OWL file. Further metadata are needed to specify the simulation parameters (using the SED-ML standard, see sed-ml.org) and a graphical output standard. (Reproduced with permission from [30, 31])

The CellML API is the software used to read and write CellML models and to provide a number of services, listed below, that are associated with the use of CellML models. The API can be viewed as a collection of interfaces that describe objects from CellML models, together with operations and attributes on each object. It is formally described in IDL[10]. *Bindings* describe how the IDL maps to particular programming languages. *Bridges* describe how to access the objects, operations, and attributes of one binding from another one, and vice versa. Documentation of the API is provided via Doxygen style comments on each interface.

Along with the API, a reference implementation is provided. This is a full Open Source implementation of all API interfaces, written in C++ (the binding is called the Physiome C++ Mapping or PCM). Bindings have been written between PCM and Java, PCM and Javascript (via XPCOM), PCM and Python, and PCM and CORBA. The goal is to provide a common interface between implementations (CellML libraries) and applications, so that implementations are interchangeable and applications can efficiently communicate parsed models. This makes it easier for applications to process CellML properly, and aims to increase support for, and the correctness of, CellML processing. It also avoids unnecessary duplication of effort spent on CellML processing software.

The API is divided up into a 'core' part (providing the basic data model) and services. The core provides an object for every element type. MathML is accessed using the W3C MathML Document Object Model (DOM) specification and extension elements are accessed via the W3C DOM Specification. Optionally, implementations can provide access to the full W3C DOM alongside access via the API (the reference API does this).

The **API services** are as follows: The CellML Variable Association Service (**CeVAS**) makes dealing with variables connected to each other in multiple components a lot easier. The MathML Language Expression Service (**MaLaES**) allows MathML fragments to be converted into text based expressions. This is used as part of the CCGS service (discussed below). The language is described in a format called MAL. The CellML Units Simplification and Expansion Service (**CUSES**) allows CellML Units to be resolved into the base units they are made up of, thus facilitating unit comparison. The CellML Code Generation Service (**CCGS**) allows models to be translated into imperative code. The MaLaES transformer can be provided so that a wide range of languages are supported (some, like FORTRAN77 do, however, require substantial post-processing). The CellML Language Export Definition Service (**CeLEDS** *and* **CeLEDSExporter**) provides an XML wrapper around MAL. The CeLEDSExporter provides an XML description of all configurable parts of CCGS code generation, so that a single XML file can define an imperative language to generate code in. The CellML Integration Service (**CIS**) is used to run numerical simulations of models. Results (or reasons for failure) are provided back to the application by a call-back interface as they come in. Parameters can be perturbed via the API without recompiling the model, e.g. for sampling-based sensitivity analyses. The Validation Against CellML Specification Service (**VACSS**) allows models

---

[10] Interface Description Language. See www.OMG.org/cgi-bin/doc?formal/02-06-39

to be validated. Errors and warnings can be representational (invalid XML or elements in the wrong place) or semantic. Dimensional inconsistency on connections is an error; dimensional or units inconsistency in equations triggers warnings. The SED-ML processing Service (**SproS**) provides a way to access SED-ML descriptions of simulation experiments (see below) from the CellML API. Currently this is just the object model (SED-OM), but more services for fully processing SED-ML descriptions referencing CellML models are planned.

Note that the API provides parsing of RDF for query lookup and provides access to all RDF in a model. Planned future developments include better support for interpreted languages (e.g. dynamic querying of interfaces supported by an object), better SED-ML support, more language bindings and many other planned items – see tracker.physiomeproject.org. For the current draft of the CellML metadata specification, see cellml.org/specifications/metadata/mcdraft.

8.5 OpenCell/OpenCOR

OpenCell is a software environment for working with CellML models (authoring, visualizing, simulating). It is Open Source[11] and written in C++, hosted on Source-Forge, and uses the Mozilla XUL framework (www.opencell.org). An example of the graphical interface is illustrated in Fig. 8.3

A new version of OpenCell is being developed based on Qt/C++ and to distinguish this project from the previous XUL/C++ project, the software is currently called 'OpenCOR' (opencor.ws). OpenCOR is developed, built, tested and packaged on Windows 7, Ubuntu 11.4 (Natty Narwhal; both the 32-bit and 64-bit versions); and Mac OS X 10.6 (Snow Leopard).

OpenCOR will be used to organise, edit, simulate and analyse CellML and, in the future, SBML files. Some general features are:

- available both as a command line tool and through a graphical user interface;
- fully customisable and multi-lingual interface;
- support for automatic updates (both for development and release versions);
- access to the CellML Model Repository and BioModels Database through web services;
- support for ontologies through web services to RICORDO and support for SED-ML;
- two-dimensional (2D) and 3D representation of a model (using SVG and cmgui, respectively).

Some more specific features are as follows:

Organising files
- Access to the CellML Model Repository and BioModels Database to search, download, upload, etc. a model;

---

[11] Distributed under a tri-licence MPL, GPL or LGPL.

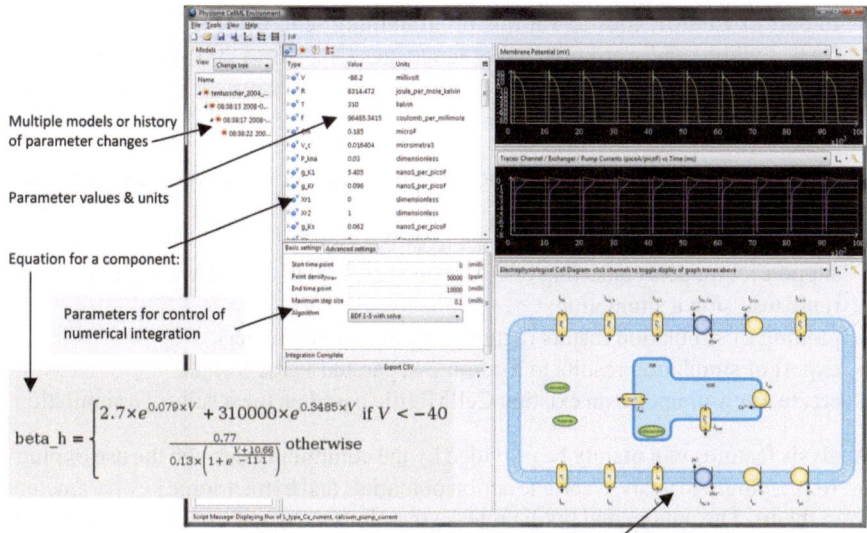

Multiple models or history
of parameter changes

Parameter values & units

Equation for a component:

Parameters for control of
numerical integration

$$\text{beta\_h} = \begin{cases} 2.7 \times e^{0.079 \times V} + 310000 \times e^{0.3485 \times V} & \text{if } V < -40 \\ \dfrac{0.77}{0.13 \times \left(1 + e^{\frac{V+10.66}{-11.1}}\right)} & \text{otherwise} \end{cases}$$

Graphical user interface based on SVG (a 2D XML graphics standard) for
displaying model topology and controlling visibility of displayed results

Fig. 8.3. A screen-shot of the Open Source software OpenCell. This is one of a number of software programs that read and write CellML files and run simulations (www.cellml.org/tools). The particular model displayed here is the human cardiac electrophysiology model [49] – see www.cellml.org/models/tentusscher_noble_noble_panfilov_2004_version05. (Reproduced with permission from [49])

- access to a local file browser;
- access to a file organiser to virtually arrange files (e.g. a virtual folder that contains a symbolic link to all the CellML files used in a given modelling study).

Editing files

- Editing of a CellML file is possible using one of several views:
  - a raw XML view and a COR-like view (i.e. a compact format, see cor.physiol.ox.ac.uk);
  - an improved COR-like view (i.e. a COR-like view without CellML-specific concepts such as public/private interfaces, encapsulation hierarchy and connections);
  - a tree-like view (with the ability to show/hide various CellML features);
  - a graphical view (using SVG or cmgui, and based on the metadata/ontology information present in a CellML file).
- Graphical rendering of unit/variable definitions, mathematical equations, etc.
- CellML validation (list of warnings/errors highlighting a particular problem).
- Basic metadata support (i.e. model authors/description/curators, etc.).
- Advanced metadata support so that, using domain-specific ontologies (through RICORDO), models can be comprehensively annotated.
- Visual difference of two CellML files (only for text-based views).

- Export of CellML files to various programming (e.g. C/C++, F77, Java MAT-LAB, Python) and word processing languages (MS Word 2007/2010 and T_EX).
- Import/export from/to SBML.

Simulating files

- Editing of simulation (e.g. start/end points) and numerical solver parameters (dependent on the chosen solver);
- editing of model parameters (using a tree-like view);
- support for models consisting of DAEs (using CVODE and IDA as default solvers);
- run/pause/stop a simulation;
- plotting of simulation results (against any model parameter);
- export of simulation results to a comma-separated value format;
- create a new or update an existing CellML file based on the results of a simulation.

Analysis features will mainly be provided by the community through the use of plug-ins (e.g. a plugin to analyse cardiac action potentials and extract some key parameters from them). The anticipated public release date for OpenCOR, at which point it will replace the existing OpenCell software, is December 2011.

8.6 FieldML

To cater for models that do include spatial information, another markup language called FieldML (fieldml.org) is being developed. FieldML files contain all the parameters and expressions needed to mathematically define fields over multidimensional manifolds representing space, time and other domains. The most common form of parameterization is a finite element mesh, where the representation is built out of nodal parameters, mesh topology and the element basis functions that interpolate the nodal parameters over the elements. However, FieldML is intended to handle more general parameterizations than just finite element fields and also includes dense data formats (e.g. for images embedded inside models) and arbitrary functions of existing fields.

FieldML is being developed as a standard for communicating computational field descriptions and data (Christie et al., 2009). The fundamental idea behind its design is to define the spaces that make up a body of interest (the domain – a manifold) from basic primitives, and to map the positions on those spaces to values (the fields) using explicit mathematical expressions.

Important principles in its design include:

- use of a minimal set of interoperable concepts;
- placing no limits on the generality of the domain (discrete, continuous or any combination; representing space, time, parameter-space, population, etc. or any combination) so that all quantities are 'field-like';
- having no 'low-level objects', only domains/sets with attributes mapped to them;
- extensibility to support arbitrarily complex field functions;
- supporting reuse of domain descriptions and functional expressions;

- supporting reuse of domain descriptions and functional expressions;
- efficiency in storage and performance;
- a preference for homogeneous data (fixed-dimension manifolds; N homogeneous arrays instead of one inhomogeneous collection)

The approach taken in FieldML contrasts with typical 'finite element' field interchange formats which build fields out of fixed data structures such as 'nodes' and 'elements' with fixed relationships between them which limit the description of more general fields. FieldML is equally able to represent these legacy formats, but the definition of sets of such 'objects' and the relationships between them, notably the mapping of parameters for interpolation, must be explicitly described. FieldML is essentially a meta-language for field interchange.

Reading and writing to the FieldML format is supported by the companion FieldML API which provides a C API for accessing and working with FieldML documents and objects plus support for bulk data input/output (I/O). See www.fieldml.org for further details.

8.7 Cmgui

Cmgui[12] is an advanced open source 3D visualisation software package for visualizing and manipulating models of fields defined over manifolds. Cmgui is primarily used to visualize the dynamic geometry and solution fields obtained from numerical simulations of multi-scale and multi-physics physiological models. It is coded in C/C++ and consists of data structures for describing hierarchical models built from mathematical fields, graphics conversion operations for making field visualisations, high-quality OpenGL rendering, and interactive manipulation tools. Fundamental field types supported include finite element interpolation with high order basis functions, 2D and 3D images and Computer Aided Design (CAD) geometry. Fields may also be derived and manipulated by standard mathematical expressions, image processing filters and other specialist operators.

Cmgui libraries are currently built containing all the components in Fig. 8.4 left of the C API. However work is progressing to separate it into smaller libraries such as the "core" Regions and Fields which clients may wish to use on their own. There are already facilities for configuring out blocks of specialized functionality such as ITK image processing filters and CAD fields. Most of the headers in the C API are already divided between the blocks shown.

The following sections summarize what each block delivers to a potential client of the cmgui library.

Regions and fields
The core part of the cmgui library is an interface to an in-memory representation of Regions and Fields, broadly following the FieldML data model (Christie et al., 2009). The implementation of these data structures is designed to support interactive

---

[12] Distributed under a tri-license MPL, GPL or LGPL. See also Christie et al., 2002.

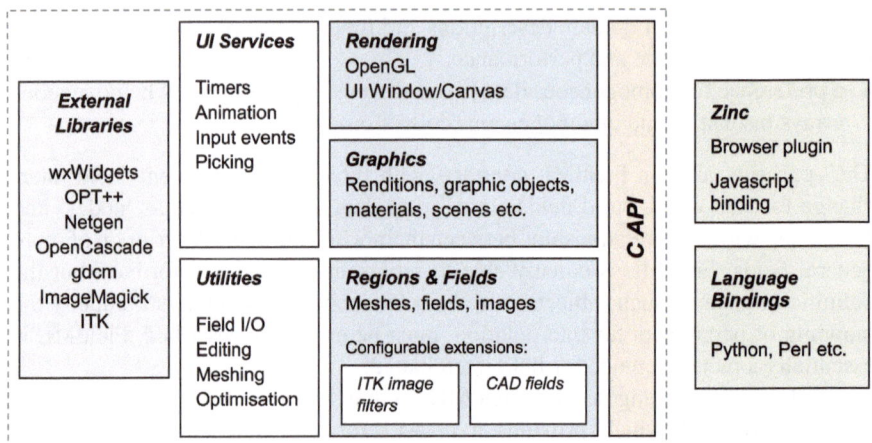

Fig. 8.4. Key components of cmgui, including some of the external Open Source libraries contained in the build. wxWidgets provides GUI widgets, OPT++ is a numerical optimisation library, Netgen is for tetrahedral meshing, OpenCascade is a CAD library, gdcm is for DICOM I/O, ImageMagick is for image I/O, and ITK is the Insight Toolkit for image processing and segmentation

graphical applications where client code needs to be notified of changes to fields, permitting for example graphics to be automatically updated to reflect the changes.

Regions are a container of fields, domains and their sub-objects, as well as child regions, permitting cmgui models to have an hierarchical structure. Regions provide separate namespaces for their sub-objects: except where explicitly imported, fields and other region sub-objects, such as nodes and elements, are independent of any same-named object in another region. Regions provide notification to clients for changes to the fields they contain.

Within regions, fields store model and other data, or compute values from other fields. Originally, cmgui fields consisted of nodal parameters and continuously interpolated element fields. This has been expanded to include other data such as images, and what we call "computed fields" which evaluate their values in terms of other field values, implementing anything from basic mathematical operations (add, subtract, cross product etc.) to invoking algorithms from libraries such as ITK image processing.

Cmgui field types implement basic field functionality for query, evaluation and value assignment, but may extend this with type-specific APIs. Field types offer software modularity: the main method of extending cmgui is to define new field types to implement the required function or to wrap one or more field-like features of an existing data structure. Cmgui doesn't yet have a plug-in architecture for adding new field types, but this is feasible.

Cmgui supports reading and writing of fields and models in its own EX format, and reading and writing of images in various formats via ImageMagick/gdcm. It supports import from the latest FieldML format, and export to FieldML is planned for implementation shortly. At this time FieldML data is translated on reading into internal cmgui mesh and field structures, which although fairly flexible do not support all the rich expressions possible in the FieldML language. Current plans are to migrate cmgui's internal data structures to be fully FieldML compatible so that FieldML can become the native file format for cmgui. Note that FieldML will eventually support serialization of computed fields.

Graphics

A proportion of the graphics data consists of some globally managed resources:
- graphical materials giving basic material colours and lighting parameters, with optional texturing with image fields plus OpenGL shader programs:
  - spectrums, which map field values to colours, modifying an underlying material;
  - static graphics objects or 'glyphs' which can be drawn at points in the scene;
  - Scenes which control what is visible in a rendering device by filtering the graphics for a model.

The remainder of the graphics code is essentially concerned with making graphical visualisations of the fields in a region, which is termed its "rendition". These consist of algorithms for making particular types of graphics from compatible field definitions:
- lines and cylinders from 1-D elements;
- surfaces from 2D elements:
  - points illustrated with glyphs from point data (e.g. nodes) and at discrete locations in n-D elements;
  - iso-surfaces/iso-lines from volume/surface elements;
  - streamlines.

Each of the factors affecting the appearance of these visualisations is controlled by a field, including the choice of coordinate field, data field for colouring with spectrums, texture coordinates (surfaces), orientation and scaling fields (for point glyphs), iso-scalar field (iso-surfaces/lines), stream vector field (streamlines), visibility field (where supported) etc. This degree of control combined with the arbitrary field functions, gives cmgui most of its flexibility and power.

The output of each of these algorithms is a graphics object which stores graphics primitives in a form ready to be efficiently sent to a rendering engine such as OpenGL.

Rendering

Rendering is a separate block in the architecture diagram; it is feasible to remove it from the cmgui library to have regions, fields and graphical conversion without any user-interface. Some rendering outputs such as VRML are also UI-less, but OpenGL output requires code to interface with the windowing system, for which cmgui supports the respective drawing canvas objects for wxWidgets, gtk, motif, win32 and carbon (Mac). Support for rendering to a Qt canvas is planned.

OpenGL rendering in cmgui is powerful and high quality, supporting advanced features such as controllable tessellation of curved surfaces, arbitrary shaders, order-independent transparency, anti-aliasing, per-pixel lighting, and off-screen rendering to large images.

Other Services
Cmgui includes core utilities for manipulating fields and field domains, including optimisation and mesh generation.

The cmgui library offers other services such as timers for synchronizing animation and events for receiving input from devices. Some of these services are currently required by the graphics layer but are potentially able to be decoupled – or can be left unused by a client of the library. To support wider use of the cmgui library from other applications, there is a current plan to make the library work with an external user interface main loop.

Cmgui supports selected export and import of graphics in STL, IGES, VRML, Wavefront OBJ and other formats.

C API
The cmgui library exposes a C API to outside clients, even for objects that are implemented using C++. Although it requires casting functions to access type-specific API, for example with specialised field types, the use of C greatly increases the simplicity of linking into other software (c.f. problems with different function name mangling produced by different C++ compilers).

The direct API is now fairly comprehensive and includes the "execute command" function to invoke any legacy command available from the standalone cmgui application. The API has reached a level of maturity where reasonable guarantees against breaks can be made.

Some work has been done to wrap the C API in an inline C++ 'shim' API to restore type polymorphism and provide an object method interface style. This has been used as input to SWIG to automate generation of Python language bindings. The completion of this work is a priority as use of cmgui via its API natively in scripting languages is demanded by many users and is one of the proposed replacements of the stand-alone cmgui application.

The cmgui library is integrated into GIMIAS (from version 1.3) where it offers rendering capabilities from the GIMIAS interface and access to cmgui functionality via the API.

Cmgui related tools
Cmgui is also offered in its original stand-alone application form which presents a basic wxWidgets user interface with generic dialogs for controlling command entry, graphic generation, materials, spectrums and graphics windows. It has an integrated Perl interpreter to support variables, program flow and other features with a small number of hand-written perl bindings to API functions

Zinc is a browser plugin enabling the cmgui visualisation engine to be used with accelerated 3D graphics in web pages with custom interfaces, and controlled via Javascript bindings to the API. A rewritten version incorporating the new cmgui API is at alpha release and nearing final release. It will run on standards-compliant browsers in linux, Windows and Mac OS X. It is currently tested on Firefox and Google Chrome with Opera and Safari also supporting the npruntime technology it uses. Implementation for Microsoft Internet Explorer is not yet planned but would require use of different technology to bind to the cmgui API and rendering code.

Zinx is a javascript library that provides easy methods for loading and rendering models through Zinc. This allows the rapid development of web pages for displaying and interacting with cmgui models. Rich interfaces can be developed with the use of other javascript libraries like jQuery for HTML interaction and jQuery UI for advanced effects and widgets. Zinx is currently based on the old version of Zinc, but once updated its users should be seamlessly migrated.

8.8 GIMIAS

GIMIAS (Graphical Interface for Medical Image Analysis and Simulation) is a software framework designed to be an integrative tool for fast prototyping of medical applications[13]. It is a workflow-oriented environment for advanced biomedical image computing and simulations, and it can be extended through the development of problem-specific plug-ins.

GIMIAS is particularly tailored to integrate tools from medical imaging, computational modelling (e.g. openCMISS) and computer graphics to provide scientific developers and researchers with a software framework for building a wide variety of tools. Multi-modal image processing, personalised model creation, numerical simulation and visualisation of simulation results are some of the possible applications for which GIMIAS has been designed. The aim of the framework is to combine tools from different areas of knowledge providing a framework for multi-disciplinary research and clinical study.

GIMIAS allows building medical prototypes for clinical evaluation, in the areas of medical image analysis, modelling and simulation. It simplifies the integration of tools needed to build new clinical workflows, and the use of common libraries for user interfaces, visualisation, image processing, DICOM access, etc, that are commonly used as standard in Virtual Physiological Human (VPH) research.

The prototypes developed on GIMIAS can be verified by end users in real scenarios and with real data at early development stages, thus reducing the time and effort required to get new concepts from research to the clinical environment.

GIMIAS provides a graphical user interface with all main data IO, visualisation and interaction functions for images, meshes and signals. It includes additional tools for image segmentation, mesh editing, signal navigation and specific visualisations. It also provides a simple Application Programming Interface (API) that allows developers to create their own tools as plug-ins that can be dynamically loaded and combined in different ways to create their own customized medical applications. The GIMIAS API provides developers with standard services such as rendering, data access and advanced visualisation of multi-modal imaging data, among others.

---

[13] GIMIAS is released under a BSD license.

GIMIAS is built over widely used open source C++ based libraries like the Visualisation Toolkit (VTK, supported by Kitware Inc.), the Insight Toolkit (ITK, also supported by Kitware Inc.), the DICOM Toolkit (DCMTK, supported by Offis in Germany), the Medical Imaging Interaction Toolkit (MITK, developed at the Division of Medical Informatics, Deutsches Krebsforschungszentrum (DKFZ), Germany) and other commonly used C++ libraries (Boost, wxWidgets, CXXTest, among others).

Clinician and researcher end-user features

- Direct integration with PACS systems. Conformity to DICOM Standard: Query/ Retrieve/Send.
- Multimodal 2D and 3D image interactive visualisation:
  - control of the visualisation window (Window Width/Level);
  - zoom and displacement;
  - image rotation (+90, -90) and vertical and horizontal views;
  - image annotations, including Landmarks, Measurements and Regions of interest (ROI);
  - movie control: play forward, play back, infinite loop, speed control, etc.
- Clinical workflow navigation to easily walk through the workflow steps. Usable interface for clinical environment.
- Data: 3D images, surface meshes, volumetric meshes, signals and landmarks.
- Visualisation: Multiplanar reformation, OrthoSlice, MultiSlice, Direct volume rendering, XRay, Maximum-intensity projection.
- IO formats: DICOM, vtk, vti, stl, Nifty, Analyze.
- Available on Windows platform, 32 and 64 bits. Beta version available on Linux platform and Mac.
- Tools include: a DICOM browser and PACS connection; a manual segmentation tool; Surface mesh manipulation tools: Smoothing, volume closing, mesh optimisation, tetra generation, ring cut, skeletonization; a Signal viewer; a Landmark selection tool; Image processing tools.

An example of the GIMIAS user interface is shown in Fig. 8.5.

Developer's features

- C++ language, Plug-in architecture, and Cross-platform using standard CMake tool.
- Easy integration of external libraries using CSnake tool.
- Based on common open source libraries (VTK, ITK, MITK, BOOST, wxWidgets, cmgui).
- Command line plugins to easily extend the framework.
- Automated GUI generation: Uses XML to describe the parameters of a filter (compatible with 3D Slicer). Use automated GUI generation to create the graphical interface.

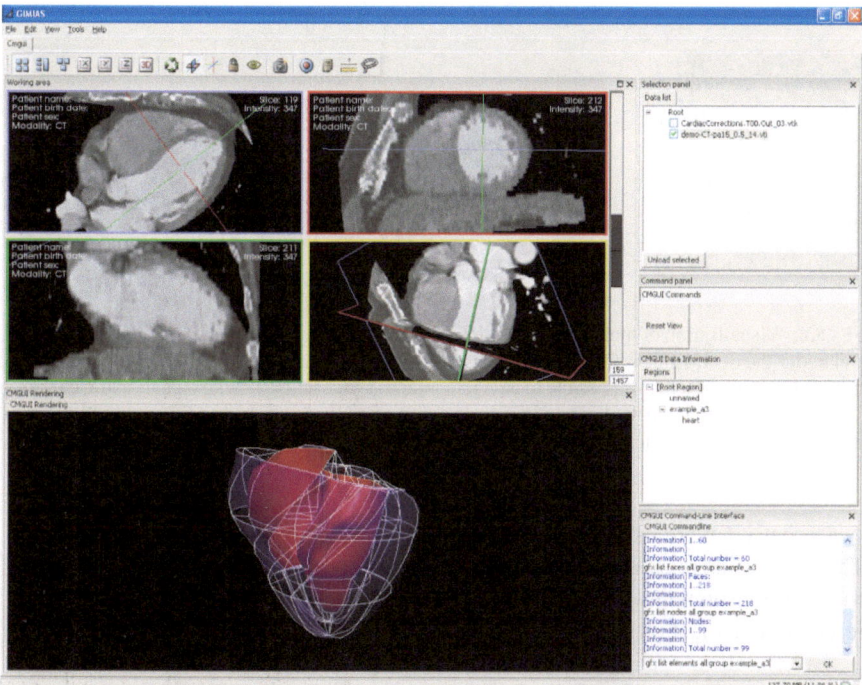

Fig. 8.5. Visualisation of CT image of the heart on GIMIAS rendering, based on VTK (on top) and the mechanical simulation model of the heart in cmgui rendering, based on OpenGL (on the bottom), visualized together under the same user interface

- Extensible Processing layer: supporting different processing data libraries integrated into the framework. For example, GIMIAS handles ITK image, VTK image, MITK image, cmgui image and automatic conversion between them.
- Extensible Rendering layer: support for different rendering libraries. Cmgui and MITK have been integrated as plugins that provide different rendering views. VTK image or a vtkPolyData can be rendered in cmgui or MITK view. The camera parameters of rendering windows using different rendering techniques can be synchronised.
- VTK, MITK and ITK execution module: Dynamic library module can execute filters (like vtkImageThreshold) directly using a DLL (like vtkImaging.dll). This reduces the effort needed to integrate processing toolkits.

GIMIAS can seamlessly integrate signals and images in the same workflow (Fig. 8.6). GIMIAS allows visualisation of multiple channels of physiological signals and their navigation maintaining temporal synchronization. Additionally, it is possible to simultaneously navigate through images and signals, for instance in cardiac imaging studies where simultaneous ECG and blood pressure are monitored.

It's also possible to add annotations to the signals, choosing from a predefined set of typical biomedical annotations or defining new custom annotations names. The user can also specify the duration of each annotation along the signal.

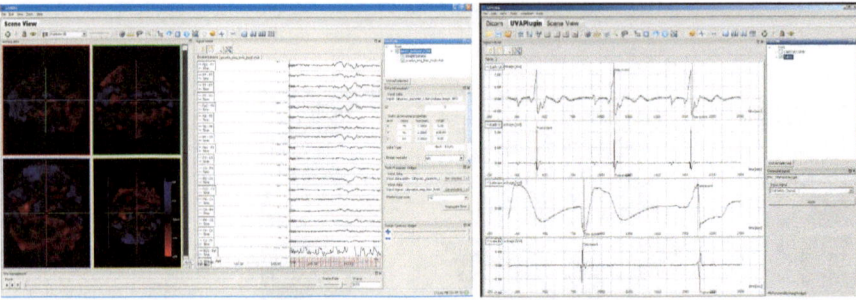

Fig. 8.6. Visualisation of multi-modal multichannel physiological signals with GIMIAS integrated and under the development framework where image analysis, image-based modelling and simulation tools are pipelined. It is possible to navigate images synchronously with signals and vice versa. The left picture shows an application in epilepsy with spatially synchronised EEG and SPECT data while the right picture shows multiple ECG signals simultaneously visualized

Recently Taverna's workflow manager[14] has been partially integrated in GIMIAS. This integration represents a first step towards the creation of visual workflows in distributed environments, given the ability of Taverna of running workflow steps on different remote machines and to graphically control the workflow execution (Fig. 8.8).

Taverna is an open source Workflow Management System written in Java. The integration of Taverna in GIMIAS allows the user to discover GIMIAS data processing services, create a workflow using a graphical designer, and execute the workflow locally or remotely using the SSH protocol.

All GIMIAS command line plugins are automatically available in Taverna. When a new command line plugin is added into GIMIAS, it will be automatically available in Taverna Workbench thanks to the GIMIAS plugin provider. The user can also configure a remote SSH plugin provider that will discover GIMIAS command line plugins in a remote computer.

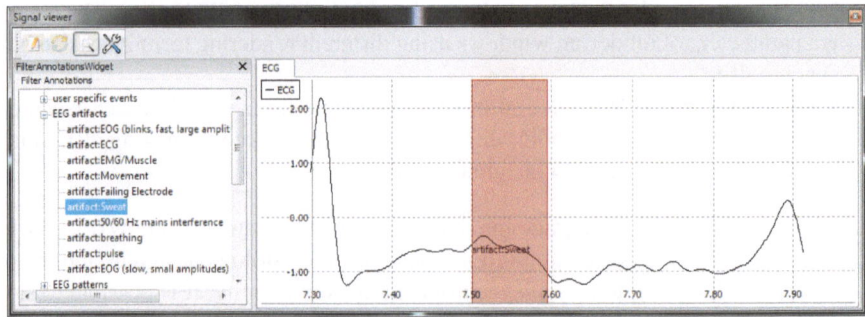

Fig. 8.7. Signals can be loaded or saved using several formats. The currently supported signal formats are BrainVision, GDF and CSV, but with support for BioSignalML planned

---

[14] www.taverna.org.uk

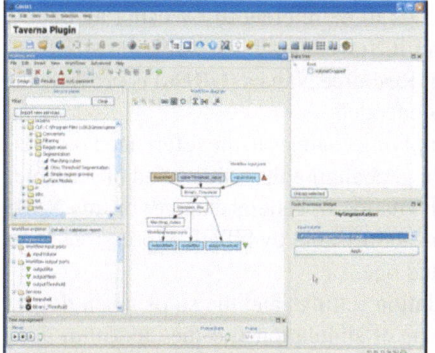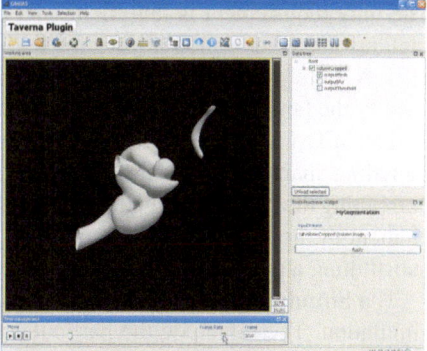

Fig. 8.8. Integration of the Taverna workflow manager within GIMIAS allowing visual design of complex workflows and the option of local or remote execution of each module. On the left the definition of a single segmentation workflow from cerebrovascular 3dRA images, on the right the resulting surface mesh visualized in GIMIAS

The user can drag and drop the available services to the workflow diagram area and connect the inputs and outputs of the used services. Parameter sweep and basic conditional flow control parameters can be used, specifying, for example, ranges for input parameters or the execution of a certain pipeline on a specified type of result.

When the workflow is executed, the input data for the workflow is automatically passed from GIMIAS to Taverna. When the service being executed is located remotely, the data is transferred using SSH protocol between local and remote computers. Once the workflow is finished, the output data will be automatically loaded in GIMIAS.

For example to create the simple segmentation workflow described in Fig. 8.8, the following processing services have been used: ITK Binary threshold, ITK Gaussian blur and VTK marching cubes. During execution, the input image specified in GIMIAS is processed and the resulting surface mesh is automatically rendered in the GIMIAS visualisation environment.

8.9 SED-ML[15]

The Simulation Experiment Description Markup Language (SED-ML) is an XML-based format for encoding simulation experiments, following the requirements defined in the MIASE guidelines (biomodels.net/miase). SED-ML specifies which model to use, the experimental task to run and which result to produce.

SED-ML is built of five main classes: the Model Class, the Simulation Class, the Task Class, the DataGenerator Class, and the Output Class.

The **Model class** is used to reference the models used in the simulation experiment. SED-ML itself is independent of the model encoding underlying the models.

---

[15] Comments here are modified versions of comments on the SED-ML website sed-ml.org

The only requirement is that the model needs to be referenced by using an unambiguous identifier which allows for finding it, for example by using an MIRIAM URI. A set of predefined language Uniform Resource Names (URNs) is provided to specify the language in which the model is encoded

The SED-ML Change class facilitates applying changes to the referenced models, including changes on the XML attributes, e.g. changing the value of an observable, computing the change of a value using mathematics, or general changes on any XML element of the model representation that is addressable by XPath expressions, e.g. substituting a piece of XML with an updated one.

The **Simulation class** defines the simulation settings and the steps taken during simulation. These include the particular type of simulation and the algorithm used for the execution of the simulation; preferably an unambiguous reference to such an algorithm should be given, using controlled vocabulary, or ontologies. One example for an ontology of simulation algorithms is the Kinetic Simulation Algorithm Ontology (KiSAO). Further information encodable in the Simulation class includes the step size, simulation duration, and other simulation-type dependent information.

SED-ML makes use of the notion of a **Task class** to combine a defined model (from the Model class) and a defined simulation setting (from the Simulation class). A task always holds one reference each.

The raw simulation result sometimes does not correspond to the desired output of the simulation (e.g. one might want to normalize a plot before output, or apply post-processing such as a mean-value calculation). The **DataGenerator class** allows for the encoding of post-processing to be applied to the simulation results before output. To define data generators, any addressable variable or parameter of any defined model (from instances of the Model class) may be referenced, and new entities might be specified using MathML definitions.

The **Output class** defines the output of the simulation, in the sense that it specifies what shall be plotted in the output. To do so, an output type is defined (e.g. 2D-plot, 3D-plot or data table), and each corresponding axis or column is assigned to one of the previously specified instances of the DataGenerator class.

8.10 BioSignalML

More than 100 different file formats exist to store and exchange biosignal data (Schlögl, 2008). Several are likely to be used in a given project, not by design, but because manufacturers, software vendors and in-house developers are unlikely to be using a common format, with archived recordings in yet more formats; the numbers compound when collaborating with other research teams. Although a number of conversion tools and standard libraries exist, e.g. libRASCH (Schneider, 2007) and the BioSig Project (Schlögl, 2009), they do not provide any standard view of metadata contained in a biosignal file. The lack of a standard, domain-neutral framework for working with biosignals has hampered their easy interchange between disciplines and integration with physiological modelling software.

In order to address problems arising from the multitude of formats and lack of standard metadata, the BioSignalML project (www.embs.org/TECHCOMM/tc-cbap/biosignal.html) is developing a framework for the exchange and storage of physiological time-series data (biosignals). This framework:

- encapsulates common features of biosignal file formats in an abstraction layer;
- uses ontologies to define biosignal terms and attributes;
- can be extended to incorporate domain-specific concepts and terms;
- provides software tools and libraries that allow the use of disparate signal file formats;
- includes a repository component that allows recordings to be accessed using standard web software as a Linked Data resource.

The project is initially working with biosignal recordings associated with sleep research. Polysomnograms contain a comprehensive range of physiological signals, including those measuring brain, heart, breath, eye and muscle activity, often at different sampling rates. Recordings are usually over several hours, with possibly several recordings for a single patient. A variety of signal storage formats are used, including proprietary formats, the European Data Format (EDF and EDF+; Kemp & Olivan, 2003), the Stanford Data Format (SDF; Sahul et al., 2003) and the WaveForm DataBase (WFDB; Moody, 2011). BioSignalML will facilitate the use of biosignals with physiological modelling languages such as CellML, with an HDF5-based file format (Folk et al., 1999) specified for biosignal exchange and storage and a streaming format used for real-time connections between simulation components.

BioSignal metadata

Some questions addressed are: What are general attributes of biosignals? What does a particular signal represent? When was it recorded? How and by whom? What processing has been applied? What is the purpose of a recording besides being a collection of signals? Do other people know our meaning of the terms we have used to describe properties? Current biosignal file formats usually have a limited number of fields for metadata and these are usually only pertinent to the domain the format was designed for. These metadata fields often contain free-format text with no controlled vocabulary or ontology to specify content, leading to possible future ambiguity.

BioSignalML specifies a core set of terms and relationships for metadata. Some of these are defined in the BioSignalML ontology; others are taken from existing standard ontologies; additional domain-specific concepts and terms can easily be added. Metadata statements about biosignals are made using RDF, with the Web Ontology Language (OWL) being used to specify ontologies. The use of both RDF and OWL allows for biosignal data to be integrated into the broader context of the Semantic Web without restricting future applications and extensions. Some of the ontologies applicable to biosignal annotation are well established international standards; others are at different stages of development. Both the OBO Foundry (www.obofoundry.org) and NCBO BioPortal (bioportal.bioontology.org) provide repositories of publicly available biological and biomedical ontologies. Ontologies used include:

- Dublin Core Terms (dublincore.org/documents/dcmi-terms).
- Timeline Ontology (purl.org/NET/c4dm/timeline.owl).
- Foundational Model of Anatomy (sig.biostr.washington.edu/projects/fm).
- Relation Ontology (www.obofoundry.org/ro).
- Cardiovascular Research Grid ECG Ontology (www.cvrgrid.org/?q= Ontologies).
- Sleep Domain Ontology (www.bioontology.org/sleep-ontology).

Concepts from the abstract model have been realised as a set of objects and methods in an Application Programming Interface (API) and software library. This library allows signals and their metadata to be created and accessed in a format independent way and forms the basis of a web-accessible biosignal repository.

BioSignalML repository

Biosignal recordings are usually stored as files on a computer system in whatever format they were recorded in. Exchanging recordings with colleagues will often involve file copying and possibly format conversion. The emphasis is usually on obtaining signal data for processing and analysis without accompanying metadata, this being kept and exchanged in the form of laboratory notes.

As an alternative to simply working with recordings as computer files, BioSignalML provides a repository application in which metadata is treated as a first-class component of a biosignal. The repository has been designed as an easy to use, extensible, cross-platform resource that integrates with existing signal processing workflows. The repository stores BioSignalML resources – **Recordings**, **Signals**, **Annotations** and **Events**. Each resource has its own URI; a request for a resource returns a representation of it. This could be actual signal data in some format; an HTML web page describing the object; or an RDF description of the resource complete with metadata links to other resources; the particular representation is determined by the request. The repository provides a HTTP interface and may be accessed using standard web-browsers.

Signal recordings submitted to the repository are kept in their original format. Metadata about BioSignalML objects contained in the recording are extracted into an RDF triplestore. Each recording has a Named Graph (Carroll et al., 2005) holding its metadata, so that all metadata from the recording can be referenced by a single resource allowing statements to be made for provenance and access control. Domain-specific metadata is mapped to ontological terms in a user extensible way, allowing future refinement as ontologies are developed.

Signal recordings are either exchanged with the repository as files or as a telemetry stream. File formats currently supported include EDF/EDF+, WFDB, SDF, an HDF5 based format, and a proprietary one; adding new formats requires a software module for conversion into the BioSignalML abstract model along with mapping statements for domain-specific metadata.

The use of URIs to identify biosignals is fundamental to the repository and its interfaces. In line with Linked Data guidelines (Berners-Lee, 2006), the BioSignalML repository uses the "http" scheme for URIs, except for local filesystem resources. Because full "http" URIs can be unwieldy for users, the API and user tools allow base prefixes to be given and used to construct relative URIs. Temporal segments of a Recording or Signal can be requested by using a comma separated list of intervals as the query component of the object's URI; an interval can be either in the form of "start-end" or "start:duration". Signal data returned in response will always span the requested interval(s); data outside of the request may also be returned, depending on the actual format requested.

8.11 PMR2

PMR2 (Physiome Model Repository 2) is the software project that provides the model repository for CellML and FieldML models. Note that SBML models are available from biomodels.net.

The key features of PMR2 are:

- facilitated model exchange directly between modellers, without reliance on a central repository;
- a detailed revision history for each mode;
- user access workflows to control privacy when required;
- embedded workspaces to enable model reuse and promote modular model development.

Workspaces & model revision history

All the files related to a particular model are stored together in PMR2 within a defined *workspace*, which can be regarded as a folder. The implementation of workspaces uses the Mercurial Distributed Version Control System (DVCS). By providing version tracking, this system ensures that users within a group cannot accidentally overwrite or purge changes of other users. Furthermore, each change made to a model, or its associated files, is recorded as a single changeset: a time-stamped, informative comment from an identifiable user, which describes the changes they have made. As model files are progressively altered, the changesets preserve the history of model development. Finally, even if two users simultaneously change the same model file, distinct changesets allow their work to be later merged in a controlled manner. Another advantage associated with using a DVCS is that it allows users to collaborate directly with each other, independent of a centralized online repository, as collaborators of any particular model will have a complete clone of the workspace with the model and related files they are working on. This allows each individual collaborator to work and commit changes to their local copy, creating new changesets which may then be shared between themselves, or pushed to a centralized repository when their work is ready to be reviewed or released.

Embedded workspaces

The previous version of the model repository software could only handle CellML 1.0 models. By introducing the concept of workspaces, PMR2 has enabled the storage of CellML 1.1 models: models that use the *import* feature can reference components and units from other existing models. PMR2 v0.2 has developed this concept further by including support for embedded workspaces, in which one or more workspaces can be included, by reference, into another workspace. Embedded workspaces are intended to manage the separation of the core model from its subcomponents, and thereby facilitate the sharing and reuse of model components independently from the source model. As this separation enables the development of the subcomponents to proceed independently of the main model, the version of the workspaces embedded is also tracked. Changes made to a workspace which has been embedded in another workspace will not affect the embedding workspace until the author explicitly chooses to update the version to use. This gives the author the opportunity to review the changes to the embedded workspace, and ensure that they won't adversely affect the models in the embedding workspace. Finally, embedded workspaces enable the import of components via relative URIs, thus promoting modular model development.

Model exposures

In addition to the concept of a workspace as a folder, the other main feature of PMR2 is the presentation view of the model and any associated data which may be within the workspace. As the contents of the workspace at the revision corresponding to any changeset are immutable, one can select a single changeset and create an *exposure* from it. Presently, creating an exposure leads to both the generation of the presentation view of the contents of a workspace as at the selected revision, and to the exposure's URL being included in the main repository category listings. Currently, a curator may annotate their assessment of the quality of the coded version of the model using *curation stars*. PMR2 has been designed to be extensible to enable the support of the range of presentation styles required for the different file types. This requires a system where plug-ins can be installed with ease onto an instance of PMR2; this system has been implemented and built on the Zope Component Architecture (Baiju, 2007), which makes use of the adapter pattern (Gamma et al., 1995) to register and activate plugins based on their names. This system allows software developers to construct specific plugins, to generate the presentation styles required for any particular model type, which can be installed and enabled on an instance of PMR2. This enables modellers to use them to render pages to describe models, or activate specific browser plug-ins to create a richer web interface for viewing models.

The access control and presentational layer of PMR2 is managed by Plone (Aspell, 2007), a Content Management System. The access control features of Plone allow authorized users to manage permissions for other users, such as allowing a user to view a private workspace, push changesets, create exposures and update workflow states such as expiring exposures.

The 600 models currently listed in the CellML component of the PMR2 repository are categorised under the following headings:

- Calcium Dynamics
- Cardiovascular Circulation
- Cell Cycle
- Cell Migration
- Circadian Rhythms
- Electrophysiology
- Endocrine
- Excitation-Contraction Coupling

- Gene Regulation
- Hepatology
- Immunology
- Ion Transport
- Mechanical Constitutive Laws
- Metabolism

- Myofilament Mechanics
- Neurobiology
- pH Regulation
- PKPD
- Signal Transduction
- Synthetic Biology

These categories change as new classes of models are developed. The 'PKPD' (PharmacoKinetic-PharmacoDynamic) classification, for example, has recently been added to accommodate the models being developed by this community and encoded in CellML.

8.12 OpenCMISS[16]

We discuss one of several software packages being developed for multiscale physiological simulation of partial differential equations (PDEs). Others are Chaste (comlab.ox.ac.uk/chaste) and Continuity (cmrg.ucsd.edu/Continuity).

Background to CMISS

CMISS is an acronym for 'Continuum Mechanics, Image analysis, System identification and Signal processing'. Code development for CMISS began in Auckland in 1980 with the goal of creating a bioengineering finite element code for solving multiple coupled partial differential equations. The initial applications were for the heart physiome project, primarily for electro-mechanics (Hunter & Smaill, 1989; Hunter, Pullan and Smaill, 2003), but it rapidly expanded to support applications in other organs such as the lungs, with coupling between soft tissue mechanics, heat and moisture transport and airway fluid mechanics (Tawhai et al., 2000, 2004). While finite element methods provided the primary numerical technique, the code was further developed in the 1980s to include boundary element methods, for electrical current flow in the thorax (Bradley et al., 1997), and finite difference equations based on curvilinear grids.

Some of the unique features of CMISS (in comparison to commercial finite element codes) that have been carried over into OpenCMISS are: (i) the use of cubic Hermite basis functions to preserve C^1 or G^1 continuity across element boundaries where this is important for efficient field representation (Bradley et al., 1997); (ii) the use of variable order basis functions (e.g. bicubic Hermite by linear Lagrange); and (iii) the close relationship that is preserved between the representation of tissue structure and the organ anatomy, through the use of material structure fields defined with respect to embedded material coordinates.

The CMISS code (including cmgui) has been at the heart of many multi-scale physiome projects over the past 20 years, but as the VPH/Physiome modelling standards evolved it became clear that a redevelopment of both programs was needed to

---

[16] A more extensive overview of OpenCMISS, on which this section is based, is given in Bradley et al., 2011.

take advantage of both the new modelling standards and repositories and the increasing move (in the academic world at least) to freely available Open Source software. The OpenCMISS project was therefore begun in 2005 as an Open Source collaboration between the University of Auckland in New Zealand and the University of Oxford in the UK. This collaboration was extended last year to include King's College London, Universitat Pompeu Fabra, Barcelona, Spain, the Norwegian University of Life Sciences and the University of Stuttgart, Germany. This year, the Shenzhen Institute of Advanced Technology, in Shenzhen, China, has also joined.

The OpenCMISS code has been developed over the last six years as a distributed-memory code to replace the shared-memory CMISS code that has supported a number of organ system physiome projects, including the cardiac physiome project and a number of similar projects on the lungs, digestive system and musculo-skeletal system.

The design requirements for the OpenCMISS code are that it encompasses multiple sets of physical equations (such as finite elasticity coupled with fluid mechanics and reaction-diffusion) and that it links subcellular and tissue level biophysical processes into organ level processes. In the heart physiome project, for example, the large deformation mechanics of the myocardial wall need to be coupled to both ventricular flow and embedded coronary flow, and the reaction-diffusion equations that govern the propagation of electrical waves through myocardial tissue need to be coupled with equations that describe the ion channel currents that flow through the cardiac cell membranes.

Design goals
The first goal is that OpenCMISS should be a flexible library rather than a large monolithic application. A library-based code means that it is considerably easier to incorporate physiome and bioengineering models into clinical or commercial applications as a library that can be wrapped by a customized interface. The library should be modular, extensible and programmable. This allows for the library itself to be customized and/or extended in whatever way is appropriate for the end application.

The second design goal is generality. Previous experience with the CMISS modelling environment indicated the importance of developing code in as general a way as possible. Generalized data structures, in which the data for diverse modelling problems are expressed in a common format, allow for easier coupling between different problems. This is especially true for future, unforeseen, coupled problems that may arise from future applications. The goal of generality does, however, often mean that there is some trade-off with the computational performance of code. As the computational size of physiome and bioengineering models can be very large it is extremely important that computational performance is carefully considered. But it is our view that it is better to optimize a more general code armed with the knowledge of exactly what the problem is than to prematurely optimize a specific code which could then limit the applicability of that code.

The third design goal is that OpenCMISS should be an inherently parallel code and that the parallel environment should be as general as possible. As the computational demands of solving models increases due to increased resolution or complexity of the models, parallel processing is needed. However, optimal parallel processing strategies depend on the particular problem being solved. Also the lifetime of modelling codes is often an order of magnitude greater than the lifetime of the computer hardware and it is notable that the architecture of parallel machines has changed over the last few decades from vector processors, to symmetric multiple processors (SMPs), to clusters of processors, to clusters of multiple core processors, through to using General Purpose Graphical Processing Units (GPGPUs). Code that assumes a particular parallel algorithm or a particular parallel architecture may not be appropriate for a future problem or future parallel hardware. For these reasons a design goal of OpenCMISS is that the code uses a general $n \times p(n) \times e(p)$ hierarchical parallel environment where n is the number of computational nodes, $p(n)$ is the number of processing systems on the computational node and e(p) is the number of processing elements for each processing system. Examples of this hierarchy are:

- multi-core or SMP $n = 1, p(n) > 1, e(p) = 1;$
- pure cluster $n > 1, p(n) = 1, e(p) = 1;$
- multi-core cluster $n > 1, p(n) > 1, e(p) = 1;$
- multi-core cluster with GPUs $n > 1, p(n) > 1, e(p) > 1.$

The fourth design goal is that OpenCMISS should be able to be used, understood and developed by both novices and experts. Modern bioengineering and physiome science requires a team of scientists, graduate students and post-doctoral researchers from varied backgrounds, each with a different skill set. It is unrealistic to expect that each member of the team will become an expert in every area of modelling and computation. The design of OpenCMISS thus abstracts and encapsulates model details in a number of objects of hierarchical complexity. The hierarchy of these objects allows complex details to be hidden from the users, if required, and the object interface allows an expert to manipulate object parameters whilst the novice user makes use of sensible default parameter values for common cases.

The final design goal, as mentioned earlier, is to incorporate Application Programming Interfaces (APIs) for the physiome markup languages CellML and FieldML.

Software systems

OpenCMISS is written in Fortran 95/2003 with an object-based approach for high level objects. It has bindings for Fortran and C and uses SWIG (swig.org) interfaces for C++ and python. It uses the Mozilla trilicense (mozilla.org/MPL) that is being used for other open source Physiome projects. Standard software engineering practices are followed, including the use of a source code repository on Source-Forge (sourceforge.net/projects/opencmiss), Doxygen for documentation, testing via a Buildbot system, validation against analytic test cases, and a tracker system – all described further below.

Although the main source code revision control system for OpenCMISS is currently hosted on SourceForge using Subversion, distributed version control systems (DVCSs) such as Git or Mercurial, are currently being evaluated as an alternative.

One perceived advantage of distributed version control systems is the fit with a research oriented development model, allowing for code to be written for research that is destined to eventually be released as open source, but kept closed-source initially until the corresponding research articles have been published. DVCSs allow for version control and collaboration during the closed phase, and when released as open source, the original version history can be kept intact, or possibly summarised.

Other SourceForge services used are:

- a wiki
  (sourceforge.net/apps/mediawiki/opencmiss/index.php?title=Main_Page);
- mailing lists
  (sourceforge.net/mail/?group_id=201176);
- a web-based source code revision viewer using Trac
  (sourceforge.net/apps/trac/opencmiss).

Note that the source code revision viewer allows views of the code revision history via a web browser, and also provides an RSS feed of code changes. A mirror of the Trac revision viewer is also available, and is hosted at the ABI (svnviewer.bioeng. auckland.ac.nz/projects/opencmiss).

The OpenCMISS project uses the Physiome project issue tracker (tracker.physio meproject.org) for project planning. The Bugzilla tracker system is used for managing both feature planning and bug reporting. A draft version of the OpenCMISS development plan is stored in the issue tracker database. The tracker allows tracker items to have dependencies on other tracker items, and these can be viewed by means of interactive web based tree views and graph views. This allows for dynamic planning, where the views of the plan are updated instantly as each status is updated. Modifications to plan are also instantly visible. Stakeholders have the option of receiving status updates via customizable RSS feeds or by means of customizable e-mail alerts.

The OpenCMISS project makes extensive use of example programs. These examples serve a number of purposes including demonstrating the capability of OpenCMISS, documenting how to solve certain equations and as a means for testing and validating the code. Although examples are currently stored in the OpenCMISS source repository, the plan is that example models and their field descriptions are hosted as part of a physiome model repository. OpenCMISS examples undergo an extensive validation process in which example solutions are compared with an analytic solution to the problem (if available). The example solutions are checked for convergence and tested in parallel. Once the example has demonstrated that it is solving correctly it is then tested nightly against the current code using a Build-bot testing system. The BuildBot automated testing system for OpenCMISS (autotest.bioeng.auckland.ac.nz) is a web based system used for automated daily building, testing and documentation generation. BuildBot provides a web based configuration facility, and views of current and historical test results. These results are also available via e-mail alerts and an RSS feed.

The generated documentation is updated daily to match the source code head revision. The Doxygen documentation system is used to extract specially formatted comments from the source code and also parses the source code structure and generates documentation targeted at developers using the OpenCMISS library, as well as developers contributing to the OpenCMISS project (available from cmiss.bioeng. auckland.ac.nz/OpenCMISS/doc/programmer).

A second Subversion system svn.physiomeproject.org/svn/opencmissextras hosted by the ABI is used to assist developers in setting up a development environment for OpenCMISS, mainly by facilitating retrieval of compatible versions of the third-party libraries and tools on which OpenCMISS depends.

Finally, the main OpenCMISS website (opencmiss.org) uses the ModX content management system.

Open source libraries

OpenCMISS builds on a number of successful software projects used in the modelling community. In accordance with an open source design philosophy, OpenCMISS aims to use software libraries that are, where at all possible, themselves open source. In cases where a library is not open source that library and its functionality are considered optional so that it is possible to build a completely open-sourced product.

For parallel computations in a heterogeneous multiprocessing environment OpenCMISS uses the MPI standard (mpi-forum.org) for distributed parallelisation and the OpenMP (openmp.org) standard for shared memory parallelisation. OpenCMISS has been tested with the MPICH2 (mcs.anl.gov/research/projects/ mpich2), Open MPI (open-mpi.org), MVAPICH (mvapich.cse.ohio-state.edu) open source MPI libraries as well as vendor specific MPI libraries from Intel and IBM. Recently, OpenCMISS developers have been investigating parallel computations on GPGPUs using CUDA (nvidia.com/object/cuda_home_new.html). However, as CUDA is proprietary technology developers have started to consider OpenCL (khronos.org/opencl) for programming GPGPUs.

In order to calculate optimal mesh partitions OpenCMISS uses the parallel graph partitioning package ParMETIS (glaros.dtc.umn.edu/gkhome/metis/parmetis/ overview). The licensing options for ParMETIS allow it to be used freely only for educational and research purposes by non-profit institutions. For situations where this copyright is too restrictive OpenCMISS also uses Scotch (gforge.inria.fr/projects/scotch) for graph partitioning.

For numerical solvers OpenCMISS makes use of a number of third-party libraries. Particular use is made of libraries developed through the US Department of Energy SciDAC (scidac.gov), TOPS (scidac.gov/math/TOPS.html) and ACTS (acts.nersc.gov) projects. For linear and non-linear system solvers OpenCMISS uses PETSc (mcs.anl.gov/petsc) for iterative Krylov sub-space linear system solvers. In addition a number of direct linear solvers such as MUMPS (graal.ens-lyon.fr/MUMPS), SuperLU_DIST (crd.lbl.gov/~xiaoye/SuperLU) and PaStiX (gforge.inria.fr/projects/pastix) are available in OpenCMISS through a PETSc interface. OpenCMISS also uses

- SUNDIALS (computation.llnl.gov/casc/sundials/main.html) for differential-algebraic equations;
- Hypre (computation.llnl.gov/casc/hypre/software.html) for preconditioners;
- TAO (mcs.anl.gov/research/projects/tao) for optimisation;
- SLEPc (grycap.upv.es/slepc) for eigenproblems;
- BLACS (netlib.org/blacs) and ScaLAPACK (netlib.org/scalapack) for linear algebra.

For input and output OpenCMISS uses the CellML (cellml.org) and FieldML (fieldml.org) APIs (Miller et al., 2010). FieldML will ultimately use standard libraries such HDF5 (hdfgroup.org/HDF5) and/or NetCDF (unidata.ucar.edu/software/netcdf) for parallel I/O of large data sets.

Multi-physics modelling

The major features in OpenCMISS for dealing with coupled multi-physics models include a flexible system for describing multiple physical models and complex problem work flows, methods for coupling different physical systems together and the ability to handle different spatial and temporal scales via FieldML concepts and CellML models. In order to provide as general a modelling environment as possible OpenCMISS separates the equations and data that describe the physics from the numerical and computational operations on those equations that make up the workflow for the problem. As the computational operations are then independent of the physics, coupling different physical systems is easier as the overall solution process can be formed by doing a sequence of numerical operations on a common system of equations.

There are methods for coupling different physical systems of equations both in the same region of interest and across different regions of interest. Coupled physics in the same region uses a consistent FieldML description of each individual physical problem and allows for coupled equations through the sharing of common variables. Since each problem's associated data is stored using the same data structures, the individual equations for each problem can be formulated using variables from different problems as easily as they can be formulated using variables from their problem.

Coupling across regions can be strong or weak. For strong coupling, different equation sets from different physical systems are coupled together by allowing the degrees-of-freedom (DOFs) in one set of equations to be linearly related to the DOFs in another set of equations. This then allows for a strongly coupled set of equations that govern the combined physics to be formed through row and column manipulations of the individual equation sets. As inter-region coupling involves different regions, the strong coupling is defined by using explicit interfaces. These interfaces ensure that information is passed between regions in a controlled manner.

For weak coupling an intermediate object is used to relate the equation data in one region with the equation data in a neighbouring region. The intermediate object can then store the information that is required to couple the DOFs in a weak, or integral, sense. Examples of weak methods supported by OpenCMISS include Lagrange, augmented Lagrange and penalty methods. For the weak coupling methods the weak relationships are formed using a number of general interface conditions that allow for reuse of the coupling method in a number of different physical situations.

The final key feature for multi-scale, multi-physics problems in OpenCMISS is the use of CellML and FieldML. CellML models allow for the physics at a small spatial scale to be abstracted and viewed as a single point model within another model at a larger spatial scale. OpenCMISS also allows for CellML models to be evaluated and integrated at a different temporal scale than other models. Further spatial and temporal scales can be incorporated into OpenCMISS problems using hierarchical field concepts from FieldML which allow data to be viewed at different levels.

8.13 Physiome standards based multiscale modelling

In an illustrative example demonstrating the utility and capabilities of the Physiome-style modelling paradigm described in the previous sections, we present here a renal nephron modelling portal (Nickerson et al., 2011). This portal (Fig. 8.9; and available at www.abi.auckland.ac.nz/nephron) provides an interface for browsing a collection of related models of the renal nephron spanning the spatial scales from individual transport pathways through to the whole nephron. The renal nephron is the primary functional unit of the kidney, with approximately one million nephrons in a human kidney.

The renal nephron portal uses CellML to encode cellular and subcellular models and the models are available from the CellML model repository (see models.cellml.org/exposure/4f9 for an example collection). The renal nephron is a 1-dimensional model in which we have described the distribution of cell types. This allows the portal to present the user with a graphical rendering of the nephron from which they are able to select the different anatomical segments and be presented with information related to that segment, such as the various cell models in the database for that segment. From a given cell model, the user is then able to investigate the various transport proteins expressed in that cell type and their representation in the various cellular models. These relationships are currently captured statically in the portal using unique integer identifiers. We plan to replace these static relationships with dynamic database queries using more advanced ontological annotation. In this manner we are able to extend the portal to dynamically incorporate new models and simulation results added to the various model repositories and data sources. Similarly, by simply providing a different base model with associated annotations the portal can be utilized by any Physiome standards based modelling project or organ system. For example, an annotated heart model and web service access to related data sources would enable the portal presented in Fig. 8.9 to be dynamically presented to cardiac Physiome users.

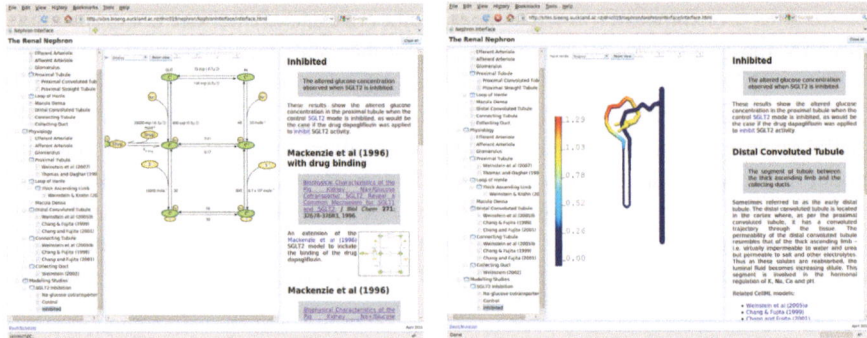

Fig. 8.9. Screenshots from the renal nephron modelling portal available at www.abi.auckland.ac.nz/nephron. The two images illustrate the current extremes of the spatial scales encompassed by the portal, with the individual transporter on the left and simulation results from the whole nephron on the right

8.14 The cardiac physiome project[17]

The following example is intended to illustrate the use of these VPH-Physiome standards and tools in a clinically focused multi-scale heart modelling project, since cardiac modelling provides a good example of the use of multi-physics and multi-scale physiome modelling standards and tools. At the organ level the models include the coupled electro-mechanics of the myocardium, with detailed models of tissue structure, coupled to blood flow in the ventricles and coronaries. At the cell level the models include membrane ion channel electrophysiology, calcium transport, myofilament mechanics, signalling pathways and metabolic pathways. We first briefly review the technologies used for imaging cardiac structure at multiple length scales and how these images are segmented and used to construct anatomically based models on which the equations representing physical laws can be solved.

Cardiac Imaging technologies
The structure of biological tissues can be imaged at spatial scales from protein and subcellular levels to tissue, organ and organ system levels. Some of these imaging techniques are illustrated for the heart in Fig. 8.10. Starting at the smallest scale, X-ray diffraction techniques yield protein structure at about 0.5nm resolution (Fig. 8.10a). Electron tomography has the capacity to image 3D sub-cellular structure at the 5nm level (Fig. 8.10b). With image registration across multiple scanned tissue samples, confocal imaging can provide 3D reconstruction of complete transmural cross-sections of, for example, the ventricular wall (see Fig. 8.10c). Multiphoton and confocal imaging routinely gives 500nm resolution on soft tissues, although with the use of single molecule fluorescent markers such as GFPs coupled to quenching techniques that eliminate background illumination (Soeller et al., 2009), optical imaging techniques are now pushing down to 50nm resolution. Micro-CT, in

---

[17] This section is based on Hunter & Viceconti, 2009.

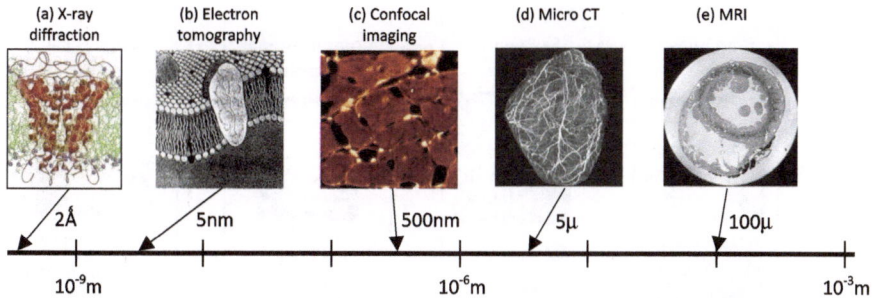

(a) X-ray diffraction (b) Electron tomography (c) Confocal imaging (d) Micro CT (e) MRI

2Å   5nm   500nm   5µ   100µ

10^{-9}m 10^{-6}m 10^{-3}m

Fig. 8.10. Imaging techniques used to measure cardiac structure and function at multiple spatial scales

which a specimen is imaged with a collimated X-ray source from multiple incident angles to build a 3D tomographic reconstruction, provides about 5nm resolution on suitable contrast enhanced tissue (Fig- 8.10d). Magnetic resonance imaging (MRI) has the capacity to image intact organs at about 0.1mm resolution (see Fig. 8.10e) and when used with tagging can provide direct information on tissue movement in a clinical setting. A number of other imaging technologies, such as helical scan CT, positron emission tomography (PET) and ultrasound (US), are used clinically for non-invasive imaging of structure and function at about 1mm resolution.

Image segmentation and modelling at different spatial scales

The process of creating anatomical models from these images and then solving the equations representing biophysical processes is illustrated at various spatial scales in Fig. 8.11, from organ system (top) to subcellular (bottom). In each case the process of creating the anatomical model is similar: The raw image data in the first column on the left is processed and segmented to provide a clearly defined boundary (second column) from which a computational mesh can be generated (third column). Finally, the computational mesh (right hand column) is used to solve systems of equations representing the governing physical principles – such as conservation of mass, conservation of momentum and conservation of charge.

The first row (labeled 'organ system') in Fig. 8.11 shows an image of the torso, followed by the segmented boundaries of the structures within the torso (heart, lungs, thoracic muscles and skin). The 3rd and 4th images are the computational mesh and the potential distribution on the skin surface generated by currents flowing out of the heart and into the thoracic cavity. The second ('organ') row shows an MRI image of the heart in the torso, from which the endocardial and epicardial ventricular surfaces (2nd image) are segmented to produce the computational heart mesh (3rd image). The final image shows the predictions of the electrical activation and mechanical contraction of the ventricular myocardium. The third ('tissue') row shows first the raw data from confocal imaging, followed by the segmented collagen structure, then a computational mesh based on the myocardial sheet structure and finally a prediction of the activation wavefront moving through the tissue structure. The fourth ('cell') row shows an electron microscopy image of the cardiac myocyte fol-

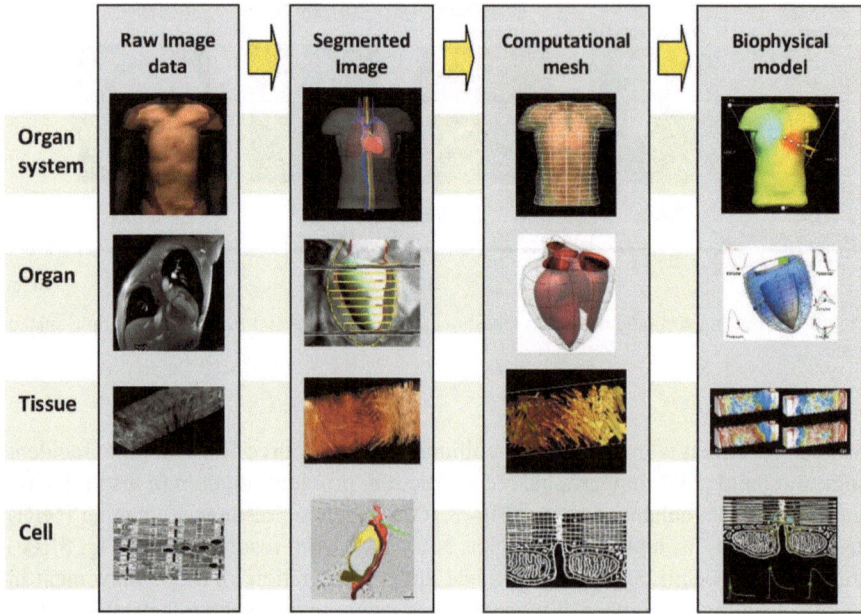

Fig. 8.11. Illustration of the process (left to right) of segmenting image data to create computational meshes on which the equations representing biophysical processes can be solved for each of the four spatial scales characteristic of the physiome project – organ system, organ, tissue and cell. A similar process, not shown, occurs at the protein level, where X-ray diffraction data gives rise to 3D atomic structure on which molecular modelling allows protein phenotype to be predicted. In fact a model at any one scale needs to merge different physical processes that operate at each of these spatial scales. The biophysical processes being modelled in the right hand column are, from bottom to top, calcium diffusion from ryanodine receptor (RyR) release sites to troponin-C (TnC) binding sites in the cardiac myocyte, electrical activity in a transmural block of myocardial tissue, the spread of electrical activation and the subsequent mechanical contraction around the left ventricle, and the spread of cardiac generated electrical activity over the torso together with the computed vector cardiogram (Reproduced with permission from [21])

lowed by segmented surfaces of the transverse tubules and surrounding junctional endings of the sarcoplasmic reticulum. This is used to create the two dimensional mesh that is then used to examine calcium fluxes released from ryanodine receptors and taken up first by binding to troponin C (to activate myofilament contraction) and then the SERCA pumps that recycle calcium back to the release sites. Not shown in Fig. 8.11 is the multiplicity of biophysical processes that operate at each of these levels.

Linking spatial scales

The hierarchy of spatial scales for heart modelling is illustrated in Fig. 8.12. Note that a number of physical processes must be integrated at each scale. For example, at the organ level, the large deformation mechanics of the ventricular wall is coupled to the computational fluid dynamics of blood flow within the ventricles and in

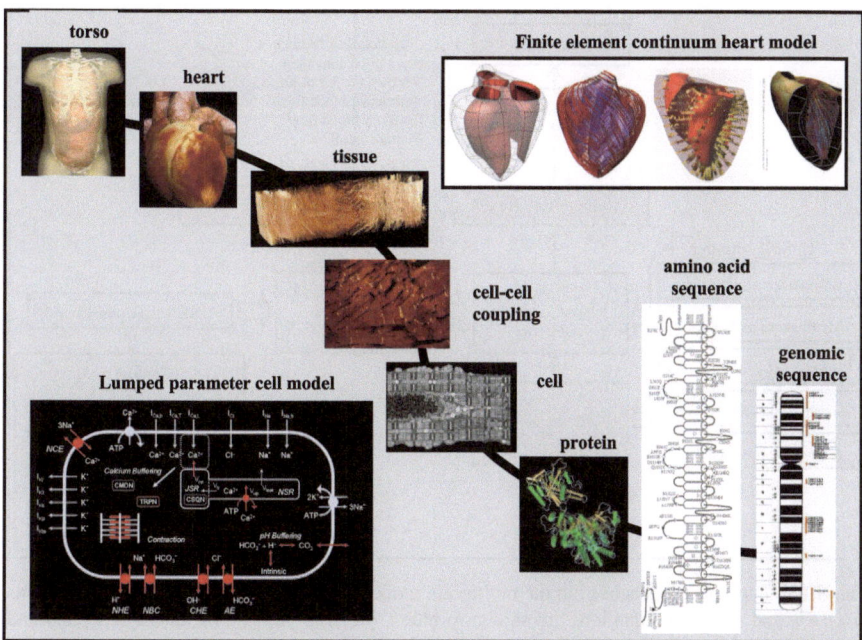

Fig. 8.12. The multi-scale cardiac physiome modelling hierarchy from genes to the whole organism. Parameters used in a model at one scale can often be derived from a more detailed model at a lower spatial scale

the coronary vasculature, and provides the moving domain for solving the reaction-diffusion equations that determine the propagating myocardial activation wavefront. The current that drives the activation processes, however, is generated by membrane ion channels and these in turn are linked at the sub-cellular level to many biophysical processes, including calcium transport, myofilament mechanics, metabolic pathways, signal transduction pathways, proton and bicarbonate control and gene regulatory pathways.

The process of linking multiple biophysical processes across these scales is illustrated in Fig. 8.13. At the organ level the processes are myocardial activation (Hooks et al., 2002), ventricular wall mechanics (Nash & Hunter, 2001), ventricular blood flow and heart valve mechanics (Nordsletten et al., 2007), coronary blood flow (Smith et al., 2002; 2004) and neural control. All of these processes must be modelled within an integrated framework to capture the substantial interactions, and all have the same requirement to link down to tissue (LeGrice et al., 1995; 1997) and cell properties. At the cell level the various cellular processes such as electrophysiology, calcium transport, myofilament mechanics, metabolic pathways and signalling networks, also need to be modelled together and these are described using the CellML framework. In some cases the cellular level models, such as ion channel electrophysiology, have parameters that are derived from lower level processes, such as Markov models and these in turn will at some future stage be formed by coarse

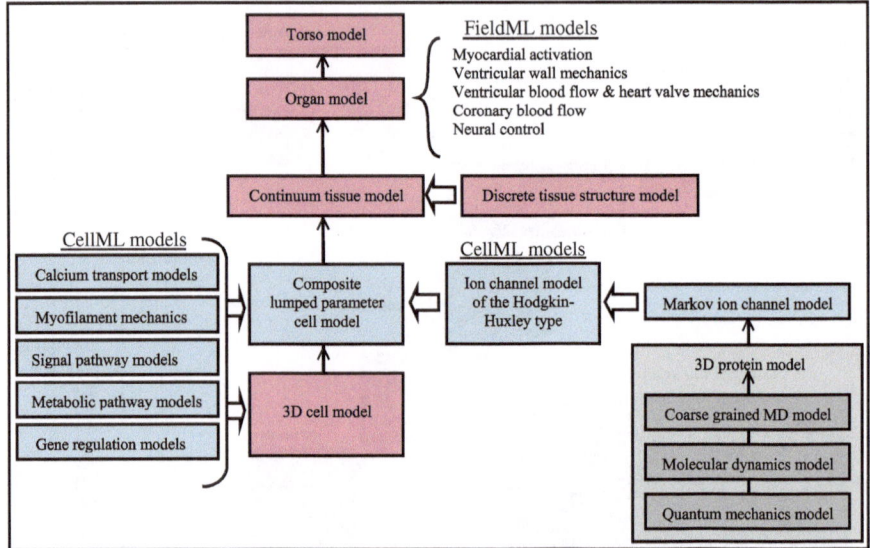

Fig. 8.13. Types of models used in the multiscale modelling hierarchy. Models based on systems of ODEs and algebraic equations are shown in blue (so called 'lumped parameter' models) and these are encoded in CellML. Models that require the solution of partial differential equations are shown in pink and are encoded in FieldML. The FieldML models link to CellML models at material points in the tissue. The arrows above are shown as unidirectional but, in fact, information flows both ways. The models shown in gray will be linked into the cardiac modelling hierarchy in the future.

grained approximations of molecular dynamics models, as shown on the lower right in Fig. 8.13.

We next discuss building composite models by combining cellular models of different physical processes. The tool used to build the composite myocyte model is OpenCell (see Fig. 8.3).

Building a composite cell model

Along with cardiac excitability (included here via Pandit et al., 2001), several other cell level processes need to be modelled since they influence the electrical behaviour of the cardiomyocyte. Most importantly, cardiac cells contract. The wave of electrical propagation passing over the cell membrane and down into the invaginations of this membrane, called 'transverse-' or 'T-tubules', releases Ca from internal stores located at points where the internal reticular network called the 'sarcoplasmic' reticulum (responsible for soaking up calcium from the cytoplasm) is adjacent to the T-tubules. Release of Ca from these stores through ryanodine receptors is initiated by voltage-activated Ca channels in the external membrane (included here via Hinch et al., 2004). The released Ca diffuses to troponin-C binding sites on the contractile myofilaments and initiates force production (included here via Niederer et al., 2006). It is necessary to consider Ca transients and mechanical contraction along with the excitability of cardiac cells because the action potential is heavily influenced by Ca movements across the membrane, and these in turn are influenced by the mechanical shortening of the cell. Further coupling occurs through stretch-dependent ion channels (Kohl et al., 2006).

The CellML versions of each of these models are available as follows:

- *Pandit electrophysiology model*
  www.cellml.org/models/pandit_clark_giles_demir_2001_version07;

- *Hinch calcium dynamics model*
  www.cellml.org/models/hinch_greenstein_tanskanen_xu_winslow_2004_version01;

- *Niederer active contraction model*
  www.cellml.org/models/niederer_hunter_smith_2006_version01.

The process of coupling these models in an integrated cell model using the CellML 1.1 import mechanism is shown in Fig. 8.14, where A includes only the Pandit electrophysiology model, B includes the Hinch calcium dynamics and C includes the Niederer active contraction model. D is the resulting composite model (Terkildsen et al., 2008). The mathematical components of these separately developed models are identified with standard biological terms from an ontology such as GO (www.geneontology.org).

The composite model shown in Fig. 8.14 includes the basic cellular processes needed to support coupled electro-mechanics in the heart. There are, however, many other cellular processes that influence these ones and are needed in more comprehensive studies of whole heart function. For example, the aerobic metabolic pathway that controls ATP production has been modelled (Beard, 2005) and is available in the CellML model repository. The control of key proteins involved in excitatory, calcium and mechanical function in cardiac cells via β-adrenergic (Saucerman & McCulloch, 2000; Saucerman et al., 2003), CaM-kinase (Livshitz & Rudy, 2007) and IP$_3$ (Cooling et al., 2007) pathways have also been modelled and are available in CellML. The CellML models for these processes are:

- *metabolic pathways*
  www.cellml.org/models/beard_2005_version01;

- *β-adrenergic signalling*
  www.cellml.org/models/saucerman_mcculloch_2004_version01;

- *CaM-kinase*
  www.cellml.org/models/livshitz_rudy_2007_version01;

- *IP$_3$ signalling*
  www.cellml.org/models/cooling_hunter_crampin_2007_version01.

Clinical applications of heart modelling
An application of the ML-encoded models and data to clinical diagnostics and treatment planning is illustrated in Fig. 8.15. A 3D stack of DICOM-encoded images of the patient's heart are loaded into a software environment where they are seg-

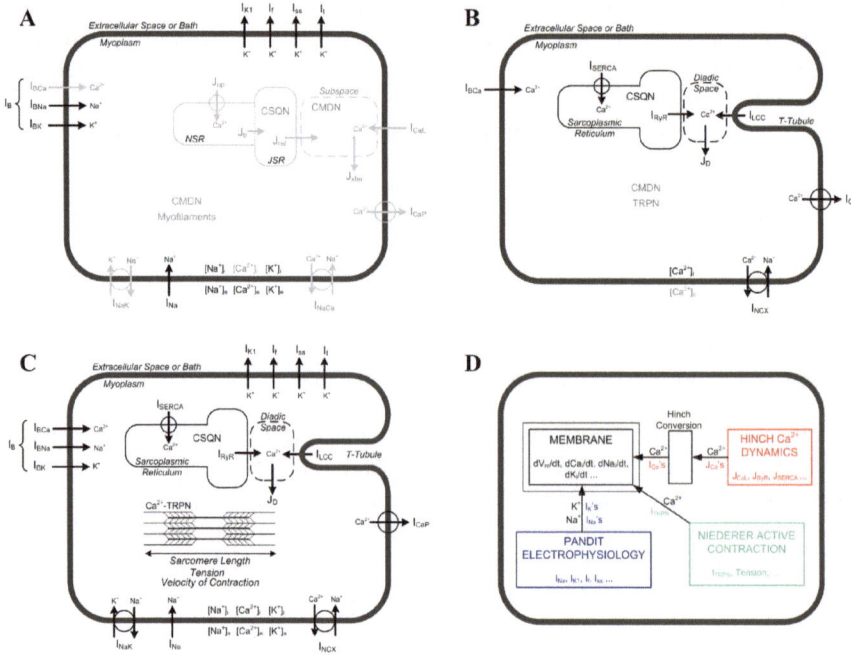

Fig. 8.14. Building a composite coupled electrophysiology-calcium-mechanics cell model from three separately developed cell models (see text)

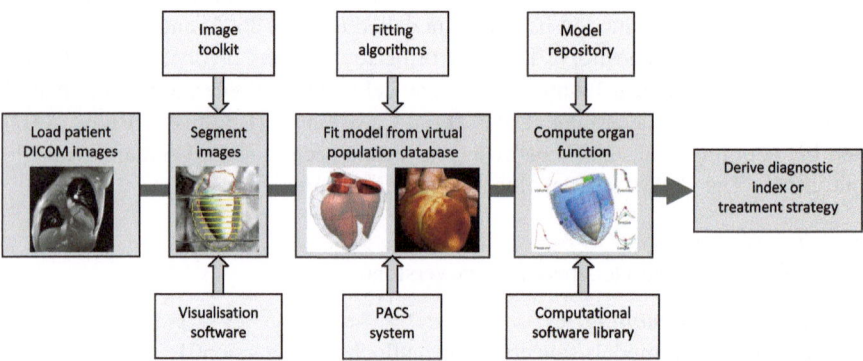

Fig. 8.15. The sequence of steps used in evaluating clinical indices or designing a treatment strategy by fitting models to patient images and then computing physiological function using biophysically based models. Illustrations here are for cardiac images and models from the euHeart project (www.euheart.eu), but the workflow is equally applicable to all organ systems

mented (using an image processing toolkit) and fitted with generic finite element models from a virtual population database (stored in a PACS system). The patient's anatomically-specific model is then combined with pre-existing models (from the model repository) and solved subject to patient-specific boundary conditions (e.g.

blood pressure or flow profiles). This stage uses computational modelling software available from an open source library (Garny et al., 2008). The outcome of the patient-specific computation is a diagnostic index or treatment strategy.

8.15 Population atlases

The Cardiac Atlas Project (www.cardiacatlas.org) is establishing a structural and functional atlas of the heart (Fonseca et al., 2011)[18] by combining cardiac modelling and biophysical analysis methods with a structural database for the comprehensive mapping of heart structure and function. Cardiac MRI examinations provide detailed, quantitative data on heart structure and function, and standardized protocols are now routinely used in a number of studies. Comprising cardiac magnetic resonance imaging (MRI) examinations, together with derived functional analyses and associated clinical variables, the Cardiac Atlas Project is developing a database, which will be extendible to allow inclusion of data from a variety of imaging and other sources. The initial goals of the project are to facilitate statistical analysis of regional heart shape and wall motion characteristics, across population groups, via the application of parametric mathematical modelling tools. Modelling tools and analysis methods developed by the University of Auckland are being combined with a database for neuroimaging and related clinical data and probabilistic mapping infrastructure developed by the UCLA Center for Computational Biology (CCB). The project is part of the National Centers for Biological Computing (NCBC) collaboration program and is funded by the US National Institutes of Health. The Cardiac Atlas Project aims to significantly improve the evaluation of cardiac performance and disease processes, establish characteristic parameters of cardiac structure and function on a regional basis, and enable the evaluation of clinical cases in relation to the statistical distributions within patient subgroups.

Three main open-source software components have been developed: i) a database with a web-interface, ii) a modelling client for 3D+time visualisation and parametric description of shape and motion, and iii) open data formats for semantic characterization of models and annotations. The database was implemented using a three-tier architecture utilizing MySQL, JBoss and Dcm4chee, in compliance with the DICOM standard to provide compatibility with existing clinical networks and devices. Parts of Dcm4chee were extended to access image specific attributes as search parameters. To date, approximately 3000 cardiac imaging examinations have been de-identified in a HIPAA compliant manner and uploaded to the database.

---

[18] All software components developed by the CAP are open source and are freely available under the Mozilla Public License Version 1.1 (mozilla.org/MPL/MPL-1.1.txt).

8.16 Discussion and future directions

The goals of the VPH/Physiome project are to improve the quantitative and integrative understanding of biological processes through anatomically and biophysically based modelling, and the application of these models to medical practice in areas such as diagnostics, therapy, device design, surgical planning and drug discovery. Achieving these goals requires the development of model and data encoding standards, together with the establishment of web-accessible databases and freely available open source software libraries that are based on the standards. The Physiome project has been underway for 15 years and the VPH project for 5 years. Through these projects there is now a robust standards framework in place, including the CellML, SBML and FieldML standards and their associated databases, as well as a number of open source codes that incorporate the APIs for the model and data repositories. However, much remains to be done and the comments below identify some of the gaps and necessary future developments.

The CellML and FieldML standards provide a means of verifying that models are properly curated (have consistent units, obey appropriate physical laws, etc) and are annotated (including the use of metadata to assign controlled vocabulary terms from biological and biophysical ontologies to the variables and parameters of the model to ensure, for example, that the models can be retrieved in electronic searches and imported into more complex composite models). Checks on the validity of the models occurs at the time of reviewing for publication in a peer-reviewed scientific journal, but much more needs to be done to clarify the provenance of model parameters (i.e. which species, tissue type, cell type, etc, were used in the experimental determination of a parameter and under what experimental conditions?) and the limitations of the model for application to a new setting. Other important aspects are the robustness of the model: how sensitive are the model predictions to small changes in the model parameters and how well does the the model reflect biological robustness (e.g. parameter values obtained for one cell may differ substantially from those measured in a neighbouring cell). The CellML standard is being developed under the VHP-Share project to include the stochastic variability of parameters and to provide for multiple instantiation of the same model in adjacent cell types using stochastically generated parameter values consistent with the specified distribution function for that parameter.

Note that the simulation codes themselves need to be demonstrably verified and databases of benchmark problems are being developed to help establish this. Competitions are also now being run to encourage these developments e.g. the STACOM challenge (cardiacatlas.org/web/guest/stacom2011).

Finally a third planned markup language, called ModelML, is being developed to encode the workflows needed to set up the governing physical equations, boundary conditions and initial conditions for a modelling problem in computational physiology. Currently this is done with Perl or Python scripts, but a marked up form of workflow will facilitate dealing with the solution of complex problems involving many different types of physical equation systems. ModelML may turn out to be an extension of or merge with SED-ML.

Note that one of the major goals of the VPH/Physiome project is to link clinical imaging data (and associated models of tissue/organ structure and function) with genomics and proteomics data and models for an individual patient using more subject-specific and scientifically based models. To achieve this goal the VPH/Physiome project has to integrate much more with experimental data and models at the level of molecular systems biology – there are currently good links with the systems biology community dealing with biochemical networks (who have adopted the SBML standard), but much stronger connection is needed with the molecular biophysics community for whom biophysically based models take account of 3D spatial information. This will be a major goal for the next few years.

Finally, establishing a robust infrastructure for ensuring the availability of verified, validated, annotated and reproducible models is a necessary first step in providing a quantitative modelling framework for medical applications, but applying these models in a clinical setting brings a whole new set of challenges. Two new European projects in particular are now addressing these clinical connection issues. One is the RICORDO project (ricordo.eu) led by EBI, which is developing a multiscale ontological framework in support of the VPH community to improve interoperability amongst its data and modelling resources. The other is the VPH-Share project which will provide the essential services, building on the VPH/Physiome computational infrastructure, for the sharing of clinical and research data and tools, facilitating the construction and operation of new VPH workflows, and collaborations between the members of the VPH community.

The biomedical community, working with the bioengineering community, now has the opportunity to assemble the molecular pieces, from 50 years of reductionist science, to understand genotype-phenotype relationships by linking databases of genetic and proteomic data to anatomy and function at the cell, tissue and organ levels. Biophysically based computational modelling of the human body, applied to human physiology and the diagnosis and treatment of disease, could revolutionise 21st century bio-sciences and medicine. We have discussed the infrastructure being built for the VPH-Physiome project and illustrated its use with a clinically oriented project on diagnosis and treatment planning for heart disease. The success of this and many other similar exciting opportunities is highly dependent on the development, adoption and integration of ICT and eHealth infrastructures using the tools, model repositories and workflows being developed for the VPH-Physiome project.

Acknowledgements. The development of standards, tools and databases for the VPH/Physiome project is being funded by many public good funding agencies in Europe (e.g. the EU ICT VPH 2, 4 & 6 calls and particularly the NoE and euHeart projects), the US (the MSM Physiome RFPs) and many other countries including the UK (especially the Wellcome Trust), Japan and New Zealand. The authors thank the many people from many different groups around the globe who have contributed to the infrastructure described here – for details see the websites given for the various software projects described in the document. Funding from the Wellcome Trust for the Heart Physiome Project and the European Union for the VPH Network of Excellence (VPH NoE FP7-ICT2008-223920) and the euHeart project (VPH euHeart FP7-ICT2008-224495) is gratefully acknowledged.

References

1. Aspell M.: Professional Plone Development. Packt Publishing, 2007.
2. Beard D.A.: A biophysical model of the mitochondrial respiratory system and oxidative phosphorylation. PLoS. Comp. Biol. **1**(4), 2005.
3. Beard D.A., Britten R., Cooling M.T, Garny A., Halstead M.D.B., Hunter P.J., Lawson J, Lloyd C.M., Marsh J., Miller A., Nickerson D.P., Nielsen P.M.F., Nomura T., Subramanium S., Wimalaratne S.M., Yu T.: CellML metadata: Standards, associated tools and repositories. Philosophical Transactions of the Royal Society A**367**(1895): 1845–1867, 2009.
4. Bradley C.P., Pullan A.J., Hunter P.J.: Geometric modelling of the human torso using cubic Hermite elements. Annals of Biomedical Engineering **25**(1): 96–111, 1997.
5. Bradley et al.: OpenCMISS: A multi-physics & multi-scale computational infrastructure for the VPH/Physiome project. Progress in Biophysics and Molecular Biology, In press 2011.
6. Carroll J.J., Bizer C., Hayes P., Stickler P.: Named graphs, provenance and trust, In: Proceedings of the 14th International Conference on World Wide Web, ser. WWW '05. New York, NY, USA: ACM, pp. 613—622, 2005.
7. Christie G.R., Blackett S.A., Hunter P.J., Bullivant D.P.: Modelling and visualising the heart. Computing and Visualisation in Science **4**: 227–235, 2002.
8. Christie G.R., Nielsen P.M.F., Blackett S.A., Bradley C.P., Hunter P.J.: FieldML: Standards, tools and repositories. Phil. Trans. Roy. Soc. A **367**: 1869–1884, 2009.
9. Cooling M., Hunter P.J., Crampin E.J.: Modelling hypertrophic IP_3 transients in the cardiac myocyte. Biophys. J. **93**: 3421–3433, 2007.
10. Cuellar A.A., Lloyd C.M., Nielsen P.F., Bullivant D.P., Nickerson D.P., Hunter P.J.: An overview of CellML 1.1, a biological model description language. SIMULATION: Transactions of The Society for Modelling and Simulation International **79**(12): 740–747, 2003.
11. Folk M., Cheng A., Yates K.: HDF5: A file format and I/O library for high performance computing applications. In Proceedings of Supercomputing'99 (CD-ROM), 1999.
12. Fonseca C.G., Backhaus M., Bluemke D.A., Britten R.D., Chung J.D., Cowan B.R., Dinov I.D., Finn J.P., Hunter P.J., Kadish A.H., Lee D.C., Lima J.A.C., Medrano-Gracia P., Shivkumar K., Tao W., Young A.A.: The Cardiac Atlas Project – An imaging database for computational modelling and statistical atlases of the heart. Bioinformatics, in press 2011.
13. Gamma E., Helm R., Johnson R., Vlissides J.: Design patterns: Elements of reusable object-oriented software. Addison-Wesley, Reading, MA, 1995.
14. Garny A., Nickerson D., Cooper J., Weber dos Santos R., McKeever S., Nielsen P., Hunter P.: CellML and associated tools and techniques. Phil. Trans. Roy. Soc. A **366**: 3017–3043, 2008.
15. Hinch R., Greenstein J.R., Tanskanen A.J. et al.: A simplified local control model of calcium-induced calcium release in cardiac ventricular myocytes. Biophys. J. **87**: 3723–3736, 2004.
16. Hooks D.A., Tomlinson K.A., Marsden S.G. et al.: Cardiac microstructure: Implications for electrical propagation and defibrillation in the heart. Circ. Res. **9**: 331–338, 2002.
17. Hunter P.J.: Modelling living systems: The IUPS/EMBS Physiome Project. Proceedings of the IEEE **94**: 678–691, 2006.
18. Hunter P.J.: The IUPS Physiome Project: a framework for computational physiology. Progress in Biophysics and Molecular Biology **85**(2–3): 551–569, 2004.
19. Hunter P.J., Borg T.K.: Integration from proteins to organs: The Physiome Project. Nature Reviews Molecular and Cell Biology **4**, 237–243, 2003.
20. Hunter P.J., Nielsen P.M.F.: A strategy for integrative computational physiology. Physiology **20**: 316–325, 2005.
21. Hunter P.J., Pullan A.J., Smaill B.H.: Modelling total heart function. Ann Review of Biomedical Engineering **5**: 147–177, 2003.
22. Hunter P.J., Smaill B.H.: The analysis of cardiac function: a continuum approach. Prog. Biophys. Molec. Biol. **52**: 101–164, 1989.
23. Hunter P.J., Viceconti M.: The VPH-Physiome Project: Standards and tools for multi-scale modelling in clinical applications. IEEE Reviews in Biomedical Engineering **2**: 40–53, 2009.

24. Kemp B., Olivan J.: European data format 'plus'(EDF+), an EDF alike standard format for the exchange of physiological data. Clinical Neurophysiology **114**(9): 1755–1761, 2003.
25. Kereiakes D.J.: Interpreting the COURAGE trial. PCI is no better than medical therapy for stable angina? Seeing is not believing. Cleve. Clin. J. Med. **74**(9): 637–8, 640–2, 2007.
26. Kohl P.: Bollensdorff C.: Garny A.: Effects of mechanosensitive ion channels on ventricular electrophysiology: experimental and theoretical models. Exp. Physiol. **91**(2): 307–321, 2006.
27. LeGrice I.J., Hunter P.J., Smaill B.H.: Laminar structure of the heart: a mathematical model. Am. J. Physiol. **272**: H2466–H2476, 1997.
28. LeGrice I.J., Smaill B.H., Chai L.Z. et al.: Laminar structure of the heart: ventricular myocyte arrangement and connective tissue architecture in the dog. Am. J. Physiol. **269**: H571–H582, 1995.
29. Livshitz L.M., Rudy Y.: Regulation of Ca^{2+} and electrical alternans in cardiac myocytes: role of CAMKII and repolarizing currents. AJP: Heart and Circulatory Physiology **292**: H2854–H2866, 2007.
30. Lloyd C.M., Halstead M.D.B., Nielsen P.M.F.: CellML: its future, present and past. Prog. Biophys. Mol. Biol. **85**: 433–450, 2004.
31. Lloyd C.M., Lawson J.R., Hunter P.J., Nielsen P.F.: The CellML Model Repository. Bioinformatics **24**(18): 2122–2123, 2008.
32. Miller A.K., Marsh J., Reeve A., Garny A., Britten R., Halstead M., Cooper J., Nickerson D.P., Nielsen P.M.F.: An overview of the CellML API and its implementation. BMC Bioinformatics **11**: 178, 2010.
33. Moody G.B.: WFDB Applications Guide, 10th edition, physionet.org/physiotools/wag/wag.htm, 2011.
34. Nash M.P., Hunter P.J.: Computational mechanics of the heart. J. Elasticity **61**(1–3): 113–141, 2001.
35. Nickerson D.P., Hunter P.J.: The Noble cardiac ventricular electrophysiology models in CellML. Prog. Biophys. Molec. Biol. **90**: 346–359, 2006.
36. Nickerson D.P., Terkildsen J., Hamilton K.L., Hunter P.J.: A tool for multi-scale modelling of the renal nephron. Interface Focus **1**: 417–425, 2011.
37. Niederer S.A., Hunter P.J., Smith N.P.: A quantitative analysis of cardiac myocyte relaxation: a simulation study. Biophys. J. **90**(5): 1697–722, 2006.
38. Nordsletten D.A., Hunter P.J., Smith N.P.: Conservative arbitrary lagrangian-eulerian forms for boundary driven and ventricular flows. Int. J. Num. Meth. Fluids **56**(8): 1457–1463, 2007.
39. Nørager S., Iakovidis I., Cabrera M., Özcivelek R. (eds.): Towards virtual physiological human: multilevel modelling and simulation of the human anatomy and physiology – White Paper Edited by DG INFSO & DG JRC, Nov 2005. http://ec.europa.eu/information_society/activities/health/docs/events/barcelona2005/ec-vph-white-paper2005nov.pdf.
40. Pandit S.V., Clark R.B., Giles W.R. et al.: A mathematical model of action potential heterogeneity in adult rat left ventricular myocytes. Biophys. J. **81**(6): 3029–3051, 2001.
41. Saucerman J.J., McCulloch A.D.: Mechanistic systems models of cell signalling networks: a case study of myocyte adrenergic regulation. Prog. Biophys. and Mol. Biol. **11**: 369–391, 2000.
42. Saucerman J.J., Brunton L.L., Michailova A.P., McCulloch A.D.: Modeling beta-adrenergic control of cardiac myocyte contractility in silico. Journal of Biological Chemistry **48**: 47997–48003, 2003.
43. Savio-Galimberti E., Frank J., Inoue M., Goldhaber J.I., Cannell M.B., Bridge J.H., Sachse F.B.: Novel features of the rabbit transverse tubular system revealed by quantitative analysis of three-dimensional reconstructions from confocal images. Biophys. J. **95**(4): 2053–2062, 2008.
44. Schlögl A.: Dataformats supported by BioSig biosig.sf.net, pub.ist.ac.at/?schloegl/biosig/ TESTED 2008.
45. Schlögl A.: An overview on data formats for biomedical signals. In Image Processing, Biosignal Processing, Modelling and Simulation, Biomechanics, ser. IFMBE Proceedings,

Dössel O. and Schlegel A. (eds.), World Congress on Medical Physics and Biomedical Engineering, Springer, 25/4: 1557–1560, 2009.

46. Schneider R.: About libRASCH. www.librasch.org/librasch/, 2007.
47. Smith N.P., Nickerson D.P., Crampin E.J., Hunter P.J.: Multiscale computational modelling of the heart. Acta Numerica **13**: 371–431, 2004.
48. Smith N.P., Pullan A.J., Hunter P.J.: An anatomically based model of coronary blood flow and myocardial mechanics. SIAM J. Appl. Maths. **62**: 990–1018, 2002.
49. ten Tusscher K.H.W.J., Noble D., Noble P.J., Panfilov A.V.: A model for human ventricular tissue, American Journal of Physiology **286**: H1573–H1589, 2004.
50. Terkildsen J.R., Niederer S., Crampin E.J., Hunter P.J. Smith N.P.: Using Physiome standards to couple cellular functions for cardiac excitationcontraction. Exp. Physiol. **93**: 919–929, 2008.

9

The role of the variational formulation in the dimensionally-heterogeneous modelling of the human cardiovascular system

Pablo J. Blanco, and Raúl A. Feijóo

Abstract. The modelling of the cardiovascular system entails dealing with different phenomena pertaining to different time, constitutive and geometrical scales. Specifically, the problem of integrating various geometrical scales can be understood from a kinematical point of view, which means to integrate models with different kinematics, and in particular different dimensionality. In this context, all the variational machinery can be employed to derive consistent variational formulations according to the underlying kinematical hypotheses that rule over the corresponding models. In this work we discuss the application of variational formulations to model the blood flow in the cardiovascular system making use of heterogeneous representations. Two examples of applications are used to show the capabilities and potentialities of the present approach.

9.1 Introduction

The cardiovascular system (CVS) can be viewed as the integration of different parts, at different geometric/time scales and different physiological mechanisms, into a *whole* with rather complex structural and functional behaviour. These different components, scales and functionalities are inseparably coupled by virtue of the interactions between local and global phenomena. In such context, the interplay of such

Pablo J. Blanco
LNCC – National Laboratory for Scientific Computing and INCT-MACC – National Institute of Science and Technology in Medicine Assisted by Scientific Computing, Av. Getúlio Vargas 333, Quitandinha, 25651-075, Petrópolis, RJ, Brazil
e-mail: pablo.j.blanco@gmail.com

Raúl A. Feijóo (✉)
LNCC – National Laboratory for Scientific Computing and INCT-MACC - National Institute of Science and Technology in Medicine Assisted by Scientific Computing, Av. Getúlio Vargas 333, Quitandinha, 25651-075, Petrópolis, RJ, Brazil
e-mail: feijooraul@gmail.com

Ambrosi D., Quarteroni A., Rozza G. (Eds.): Modeling of Physiological Flows.
DOI 10.1007/978-88-470-1935-5_9, © Springer-Verlag Italia 2012

different levels of integration in the CVS will be the focus of developments in computational modelling in the forthcoming years. As a matter of fact, this kind of integrative models will be capable of providing a better understanding of the processes involved in the onset and development of vascular diseases [12, 20, 32, 41, 51, 59] such as stenotic plaques, aneurism growth and rupture, elevated arterial pressure, atherosclerosis, among others. Moreover, the capabilities of such models will go beyond these applications, thus providing guidelines to assist and plan surgical procedures.

Regarding the fluid dynamics aspects of the CVS we identify what we call levels of integration: (i) the overall systemic behaviour, (ii) the hemodynamics of large arteries and (iii) the local circulation in specific districts. Several models have been proposed to take into account the relevant phenomena at each level of integration. For instance, specific vessels can be modelled using a full 3D representation, whereas the systemic arteries can be accounted for using simplified 1D representations and peripheral beds may be seen as lumped vascular entities, that is 0D models. In this sense, it is possible to select the level of complexity of each component based on the type of data available about a certain vascular entity and the information to be retrieved from the model. Particularly, modelling the blood flow in deformable vessels comprises several challenging issues. Indeed, the propagatory nature of the pulse wave, which is related to the compliant properties of the arterial walls, poses the problem of defining boundary conditions once a specific district has been artificially isolated from the rest of the system.

Thus, the simulation of blood flow in specific vessels can be carried out through several approaches. The simplest one is based just on the imposition of boundary conditions obtained from patient records (see for instance [47, 48]). A more complex approach, presented in [28, 62, 63], is based on accounting for the phenomena occurring downstream the 3D region through the computation of the downstream vascular impedance. Finally, the most sophisticated approach consists in coupling dimensionally-heterogeneous models (3D-1D-0D models). With this approach a high level of detail can be attained in specific zones (3D models), while the physical phenomena corresponding to the remaining part of the system is modelled using dimensionally-reduced (1D-0D) models. This kind of formulation ensures proper systemic functioning from which physiological regimes for 3D simulations can be readily achieved. This line of research was pioneered in [17] and continued in [6, 8, 9, 18, 19, 34, 43, 60].

In [17, 18, 60] the authors make use of *a priori* assumptions which are directly incorporated into the set of partial differential equations with the aim of coupling full 3D model based on the Navier-Stokes equations with simplified distributed and/or lumped parameter equations. Specifically, this is accomplished by prescribing the continuity of some of the involved quantities (for instance flow rate and mean pressure) at the so-called coupling interfaces (points to which the heterogeneous models converge). In turn, in [19, 23] the concept of defective boundary conditions is introduced so as to write down a well-posed problem by means of further assumptions, in view of the lack of information available at the communicating interfaces between a 3D model and the rest of the system.

Nevertheless, notice that the problem is always three-dimensional, and that hypotheses are introduced in certain parts of the system to simplify the problem. When performing such reduction to a 1D or 0D model we are introducing, at the coupling interfaces, incompatibilities between the corresponding fields. The present work proposes a different perspective: *to explore the role of the variational formulation in providing an appropriate extended variational formulation from which the coupling conditions are naturally derived*. Then, we will see a *N*D (*N* = 0; 1; 2) simplified model as the reduction of a complete 3D model through the introduction of suitable kinematical restrictions. As mentioned earlier, due to the dissimilarity between the underlying kinematics, we generated discontinuities in the involved fields. In this new context, the original variational formulation, valid for fields that are continuous in the sense of the trace of the functions over such a coupling interface, needs to be reformulated in order to accommodate such discontinuities.

The goal of the present work is to give an account of the state-of-the-art in the realm of extended variational formulations for problems where fields can become discontinuous as the result of kinematical considerations introduced at some artificial internal boundary (coupling interface). This follows the ideas introduced in [3, 6]. This formulation is applied in the modelling of fluid flow and more specifically in the hemodynamics field. However, notice that these ideas can be extended straightforwardly to a great variety of problems. A theoretical account including the extended variational formulation for such models was introduced by the authors with regard to solid mechanics, [3, 7].

This work is organized as follows. In Sect. 2 an abstract extended variational principle is formulated so as to highlight the basic ideas and mechanical concepts which are behind the extended variational formulation for coupling dimensionally-heterogeneous models. In Sect. 3 we apply the extended formulation to the problem of the flow of an incompressible fluid and discuss several situations of potential interest. In Sect. 4 we construct a heterogeneous model of the CVS for which we may consider two different topological descriptions: an open-loop systemic network or a closed-loop model of the entire system. In these two alternative global configurations we embed 3D models of arterial vessels to model blood flow in specific districts. Finally, in Sect. 5 numerical experiments are presented, illustrating the capabilities of this approach.

9.2 Abstract Extended Variational Formulation

As pointed out in the introduction, the coupling of dimensionally-heterogeneous models for modelling and simulation of the CVS will be dealt with an extended variational formulation from which the coupling conditions are naturally derived. The equivalent approach from the differential viewpoint is known in the literature as *geometrical multi-scale modelling* [16, 17, 18, 50].

In fact, as it will be seen, the coupling conditions between the models correspond to the so-called *Weierstrass-Erdmann Corner Conditions (WECCs)* associated to this extended variational formulation at the coupling interfaces. Furthermore, since these WECCs are *natural boundary conditions* that are satisfied automatically by the solution of the extended variational formulation, the desired coupling conditions are derived in a consistent way with respect to the kinematical restrictions adopted for each dimensionally-reduced model.

Therefore our aim in this section is to develop this extended variational formulation. At first this will be done in a compact (abstract) form. In doing this we believe that the basic ideas and mechanical concepts behind this new variational principle will emerge more clearly without being obscured by excessive mathematical details. The application of such ideas to the specific problem of our concern is presented in the forthcoming section.

Let Ω be an open and bounded domain in $\mathbb{R}^3$. Then, the compact form of a Classical Primal (kinematical) Variational Formulation of a generalized problem defined in Ω governed by a kinematics whose regularity is characterized by the space $\mathscr{Q}$ can be cast as follows.

Problem 1. Virtual Powers Principle. Find $\mathbf{u} \in \mathscr{U}$ such that

$$(\mathscr{R}(\mathbf{u}), \mathbf{v})_{\mathscr{Q}} = 0 \qquad \forall \mathbf{v} \in \mathscr{V}. \tag{9.1}$$

The above expression corresponds to the well-known Virtual Power Principle (VPP) in which $\mathscr{U} \subset \mathscr{Q}$ represents the set of kinematically admissible fields $\mathbf{u}$. In general, this set is a linear manifold given by a translation in $\mathbf{u}_o$ (an arbitrary element of $\mathscr{U}$) of the subspace $\mathscr{V}$ of $\mathscr{Q}$ defining the kinematically admissible virtual variation fields $\mathbf{v}$. Taking into account essential boundary conditions prescribed on $\Gamma_D \subset \partial\Omega$, this means that

$$\begin{aligned} \mathscr{U} &= \{\mathbf{u} \in \mathscr{Q}; \mathbf{u}_{|\Gamma_D} = \bar{\mathbf{u}}\}, \\ \mathscr{V} &= \{\mathbf{v} \in \mathscr{Q}; \mathbf{v}_{|\Gamma_D} = 0\}. \end{aligned} \tag{9.2}$$

In Eq. (9.1), $\mathscr{R}(\cdot)$ is the *equilibrium* operator from $\mathscr{Q}$ into $\mathscr{Q}'$, generally non-linear and time-dependent, and $(\cdot, \cdot)_{\mathscr{Q}}$, which is linear with respect to $\mathbf{v}$, stands for the virtual power between the kinematical virtual variation field $\mathbf{v}$ and elements belonging to $\mathscr{Q}'$, the corresponding dual space of $\mathscr{Q}$.

Problem (9.1) is well-posed provided that (see [44]) $\mathscr{Q}$ is a separable, reflexive Banach space and $\mathscr{R}$ satisfies the following properties: (i) $\mathscr{R}$ is weakly sequentially continuous, (ii) the restriction of $\mathscr{R}$ to finite-dimensional suspaces of $\mathscr{Q}$ is continuous, (iii) $\mathscr{R}$ is coercive and (iv) $\mathscr{R}$ is bounded.

For the problems we are aiming at in this work it is important to remark that we choose the space $\mathscr{Q}$ to guarantee the continuity of the fields $\mathbf{u}$ through any (sufficiently smooth) internal interface Γ_a in the sense defined by the space of traces of elements in $\mathscr{Q}$ over that boundary, $\mathscr{T}_{\Gamma_a}(\mathscr{Q})$. We will denote this continuity property by

$$[\![\mathbf{u}]\!]_{|\Gamma_a} = 0 \quad \text{in the sense given by } \mathscr{T}_{\Gamma_a}(\mathscr{Q}), \forall \Gamma_a. \tag{9.3}$$

Furthermore, in the sense of the duality product associated to $\mathscr{T}_{\Gamma_a}(\mathscr{Q})$, the variational Problem 1 gives, as a natural boundary condition (known in the literature as *jump equation* or WECC), the continuity of the traction on any sufficiently smooth

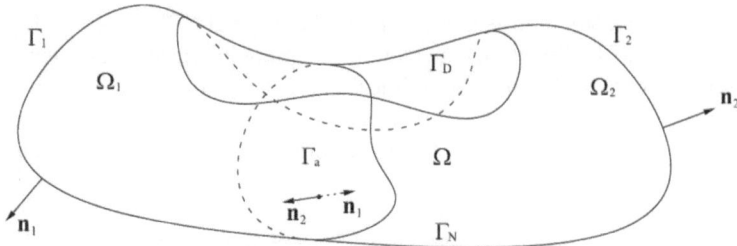

Fig. 9.1. Partitioned domain for the definition of the abstract problem

internal boundary Γ_a over which no external distributed action (force) is assumed to be applied (see Fig. 9.1), that is

$$\mathscr{R}(\mathbf{u})_{|\Gamma_a^-} = -\mathscr{R}(\mathbf{u})_{|\Gamma_a^+}, \tag{9.4}$$

where the minus sign in the right-hand side appears because we have adopted as positive orientation the direction given by the unit normal vector $\mathbf{n}_1$, and where Γ_a^- and Γ_a^+ correspond to the quantitites associated to the left and right partitions, respectively, of the domain Ω in Fig. 9.1.

9.2.1 A partition with equal kinematics

Suppose now that we split the domain Ω into two parts, Ω_1 and Ω_2, by a sufficiently smooth (arbitrary) internal boundary Γ_a (see again Fig. 9.1) such that $\Omega = \left(\Omega_1 \cup \Omega_2\right)^\circ$. Moreover, instead of using kinematically admissible fields $\mathbf{u}$ satisfying the continuity given by Eq. (9.3), we want to work with a pair of fields $(\mathbf{u}_1, \mathbf{u}_2)$ ($\mathbf{u}_i$ defined in $\Omega_i, i = 1, 2$) that maintain the same regularity of the original problem on each partition but are discontinuous over Γ_a. In other words we have

$$\mathscr{U}_i = \{\mathbf{u}_i \in \mathscr{Q}_{|\Omega_i}; \mathbf{u}_{i|\Gamma_{D_i}} = \bar{\mathbf{u}}_i\}, \qquad i = 1, 2, \tag{9.5}$$

where Γ_{D_i} is the part of Γ_D belonging to the boundary of $\Omega_i, i = 1, 2$, and where $\mathscr{Q}_{|\Omega_i}$ denotes the restriction of the kinematics $\mathscr{Q}$ to the subdomain Ω_i which is defined as follows

$$\mathscr{Q}_{|\Omega_i} = \{\mathbf{z}; \mathbf{z} = \tilde{\mathbf{z}}_{|\Omega_i}, \tilde{\mathbf{z}} \in \mathscr{Q}\} \qquad i = 1, 2. \tag{9.6}$$

Due to the discontinuity over Γ_a, the variational formulation given by Eq. (9.1) is no longer valid. Hence we need to reformulate the problem in order to find the same solution characterized by this equation in the case of sufficiently regular data. In this new situation we observe that the continuity condition

$$\mathbf{u}_{1|\Gamma_a} = \mathbf{u}_{2|\Gamma_a} \quad \text{in the sense given by } \mathscr{T}_{\Gamma_a}(\mathscr{Q}), \tag{9.7}$$

has to be obtained as *natural boundary condition* of the new variational problem. The way to obtain this new extended variational formulation is very well known (see

[11, 58]) and, basically, consists in adding to the original VPP, given by Eq. (9.1), the virtual powers (duality products), over the coupling interface Γ_a, between the discontinuity and admissible variations of the traction, and between the traction and admissible variations of the discontinuity (see Problem 2 below). Following the above considerations, the new extended variational formulation takes the following form.

Problem 2. *Classical* Extended Variational Formulation. For $\gamma \in [0,1]$ find $(\mathbf{u}_1, \mathbf{u}_2, \mathbf{t}_1, \mathbf{t}_2) \in \mathscr{U}_1 \times \mathscr{U}_2 \times \mathscr{T}_{\Gamma_a}(\mathscr{Q})' \times \mathscr{T}_{\Gamma_a}(\mathscr{Q})'$ such that

$$
\begin{aligned}
&(\!(\mathscr{R}(\mathbf{u}_1), \mathbf{v}_1)\!)_{\mathscr{Q}_{|\Omega_1}} + (\!(\mathscr{R}(\mathbf{u}_2), \mathbf{v}_2)\!)_{\mathscr{Q}_{|\Omega_2}} \\
&\quad + \gamma \langle \mathbf{t}_1, (\mathbf{v}_1 - \mathbf{v}_2)\rangle_{\mathscr{T}_{\Gamma_a}(\mathscr{Q})' \times \mathscr{T}_{\Gamma_a}(\mathscr{Q})} + (1-\gamma)\langle \mathbf{t}_2, (\mathbf{v}_1 - \mathbf{v}_2)\rangle_{\mathscr{T}_{\Gamma_a}(\mathscr{Q})' \times \mathscr{T}_{\Gamma_a}(\mathscr{Q})} \\
&\quad + \gamma \langle \mathbf{s}_1, (\mathbf{u}_1 - \mathbf{u}_2)\rangle_{\mathscr{T}_{\Gamma_a}(\mathscr{Q})' \times \mathscr{T}_{\Gamma_a}(\mathscr{Q})} + (1-\gamma)\langle \mathbf{s}_2, (\mathbf{u}_1 - \mathbf{u}_2)\rangle_{\mathscr{T}_{\Gamma_a}(\mathscr{Q})' \times \mathscr{T}_{\Gamma_a}(\mathscr{Q})} \\
&\qquad\qquad \forall (\mathbf{v}_1, \mathbf{v}_2, \mathbf{s}_1, \mathbf{s}_2) \in \mathscr{V}_1 \times \mathscr{V}_2 \times \mathscr{T}_{\Gamma_a}(\mathscr{Q})' \times \mathscr{T}_{\Gamma_a}(\mathscr{Q})', \quad (9.8)
\end{aligned}
$$

where $\mathscr{V}_i$ is the kinematical virtual variation space associated to $\mathscr{U}_i, i = 1,2$.

It is worth noting that the above problem is consistent with Problem 1 in the sense that for any $\gamma \in [0,1]$ the solutions of both problems are exactly the same provided the data are regular enough. In fact, for any $\gamma \in [0,1]$ and from Eq. (9.8) we obtain that

$$
\langle \mathbf{s}_\gamma, (\mathbf{u}_1 - \mathbf{u}_2)\rangle_{\mathscr{T}_{\Gamma_a}(\mathscr{Q})' \times \mathscr{T}_{\Gamma_a}(\mathscr{Q})} = 0 \qquad \forall \mathbf{s}_\gamma = \gamma \mathbf{s}_1 + (1-\gamma)\mathbf{s}_2 \in \mathscr{T}_{\Gamma_a}(\mathscr{Q})'. \quad (9.9)
$$

In other words, the above equation guarantees the continuity of the solution as required by Problem 1 (see Eq. (9.3)). However, in this classical extended variational formulation the continuity is **not imposed** but is **automatically** satisfied by the solution $(\mathbf{u}_1, \mathbf{u}_2)$ of Problem 2.

On the other hand, denoting $\mathbf{t}_\gamma = \gamma \mathbf{t}_1 + (1-\gamma)\mathbf{t}_2, \gamma \in [0,1]$, and from Eq. (9.8) we also obtain

$$
\langle \mathbf{t}_\gamma + \mathscr{R}(\mathbf{u}_1)_{|\Gamma_a^-}, \mathbf{v}_1 \rangle_{\mathscr{T}_{\Gamma_a}(\mathscr{Q})' \times \mathscr{T}_{\Gamma_a}(\mathscr{Q})} = 0 \qquad \forall \mathbf{v}_1 \in \mathscr{V}_1, \qquad (9.10)
$$

$$
\langle \mathbf{t}_\gamma - \mathscr{R}(\mathbf{u}_2)_{|\Gamma_a^+}, \mathbf{v}_2 \rangle_{\mathscr{T}_{\Gamma_a}(\mathscr{Q})' \times \mathscr{T}_{\Gamma_a}(\mathscr{Q})} = 0 \qquad \forall \mathbf{v}_2 \in \mathscr{V}_2. \qquad (9.11)
$$

The above equations together with (9.9) ensure that the jump equation required by Problem 1 (see Eq. (9.4)) is satisfied, and works as natural boundary condition. Furthermore, the above expression gives the mechanical meaning of the dual variable $\mathbf{t}_\gamma$ (a generalized force over Γ_a).

Here we want to highlight that: *since the kinematics adopted on both parts of the domain are the same, the solution of* Problem 2 (i) *does not depend on the value of the parameter* γ *and* (ii) *is exactly the same as the solution of* Problem 1 *provided the data are regular enough*. Despite the fact that this is well known in the literature, it is important to note that Problem 2 gives us a new insight for the case when **different kinematics** are taken on each partition of the domain, as is the case when coupling dimensionally-heterogeneous models. In what follows we are going to develop this approach, which, to the best of our knowledge, is new.

9.2.2 A partition with different kinematics

Now let us turn our attention to the case in which a different kinematics is adopted in a part of the domain, for example, in Ω_1. This kinematics is characterized by the space $\mathcal{Q}_1$. Since this kinematics is obtained by introducing known restrictions on the behaviour and/or by increasing the regularity of the original kinematics (3D model) characterized by the space $\mathcal{Q}$, it follows that $\mathcal{Q}_1 \subset \mathcal{Q}_{|\Omega_1}$ and therefore $\mathcal{T}_{\Gamma_a}(\mathcal{Q}_1) \subset \mathcal{T}_{\Gamma_a}(\mathcal{Q}_{|\Omega_1})$. For example, if the domain Ω_1 has a symmetry of revolution and the boundary conditions guarantee that the fields have the same property, then we can adopt a kinematics with symmetry of revolution, and the associated model will be defined in a 2-dimensional domain (see Fig. 9.2). On the other hand, if Ω_1 has a characteristic length L that in one dimension is much greater than the length l_{orth} in any orthogonal (transversal) direction ($L \gg l_{orth}$) and the boundary conditions allow us to consider a known behaviour over the transversal direction, then we can adopt a kinematics with this property, and the associated model will be defined in a 1-dimensional domain (see Fig. 9.3).

Given $\mathcal{Q}_1$ we have that $\mathcal{T}_{\Gamma_a}(\mathcal{Q}_1)$ is defined, and then its dual space $\mathcal{T}_{\Gamma_a}(\mathcal{Q}_1)'$ is automatically defined as well, as is the duality pairing between elements of this two spaces, which is denoted by

$$\langle s_1, v_1 \rangle_{\mathcal{T}_{\Gamma_a}(\mathcal{Q}_1)' \times \mathcal{T}_{\Gamma_a}(\mathcal{Q}_1)}. \tag{9.12}$$

Moreover, and for the sake of simplicity, let us introduce similar notations for the partition Ω_2 where the original kinematics has been maintained: (i) $\mathcal{Q}_2$ ($\mathcal{Q}_2 = \mathcal{Q}_{|\Omega_2}$); (ii) $\mathcal{T}_{\Gamma_a}(\mathcal{Q}_2) = \mathcal{T}_{\Gamma_a}(\mathcal{Q}_{|\Omega_2})$; (iii) $\mathcal{T}_{\Gamma_a}(\mathcal{Q}_2)'$ and the corresponding duality pairing is denoted by

$$\langle s_2, v_2 \rangle_{\mathcal{T}_{\Gamma_a}(\mathcal{Q}_2)' \times \mathcal{T}_{\Gamma_a}(\mathcal{Q}_2)}. \tag{9.13}$$

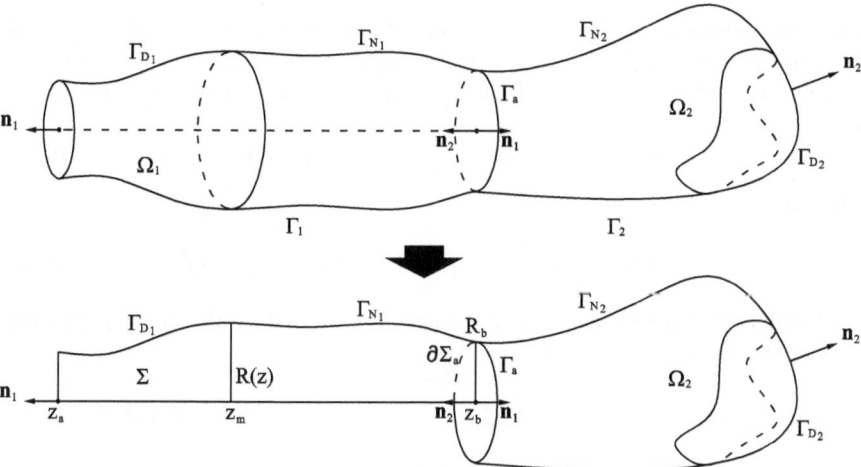

Fig. 9.2. Two-dimensional kinematics on Ω_1 (symmetry of revolution), yielding 3D-2D models

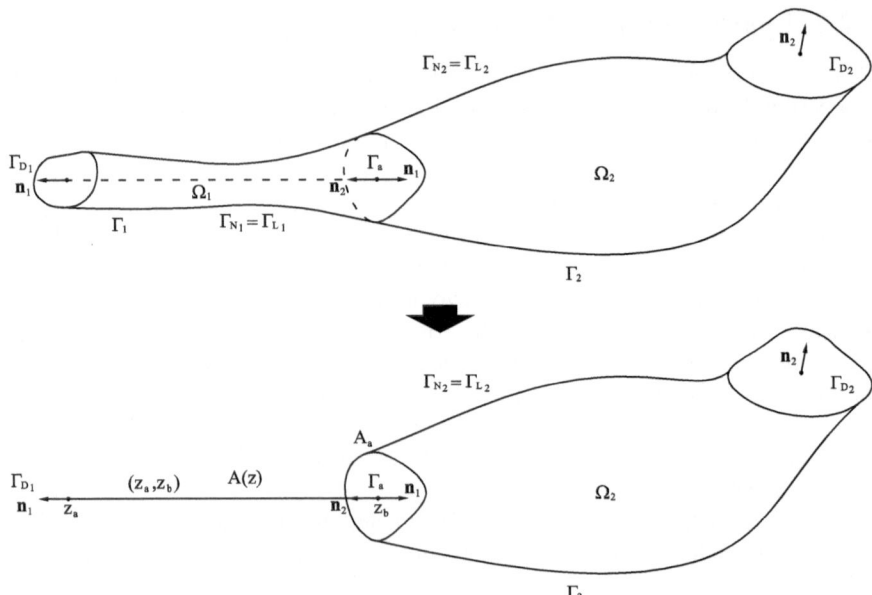

Fig. 9.3. One-dimensional kinematics on Ω_1, yielding 3D-1D models

Since $\mathscr{Q}_1 \subset \mathscr{Q}_{|\Omega_1}$ then $\mathscr{T}_{\Gamma_a}(\mathscr{Q}_1) \subset \mathscr{T}_{\Gamma_a}(\mathscr{Q}_{|\Omega_1}) = \mathscr{T}_{\Gamma_a}(\mathscr{Q}_{|\Omega_2}) = \mathscr{T}_{\Gamma_a}(\mathscr{Q}_2)$, and it follows that $\mathscr{T}_{\Gamma_a}(\mathscr{Q}_2)$ can be described by the following direct decomposition

$$\mathscr{T}_{\Gamma_a}(\mathscr{Q}_2) = \mathscr{T}_{\Gamma_a}(\mathscr{Q}_1) \oplus (\mathscr{T}_{\Gamma_a}(\mathscr{Q}_1)')^{\perp}, \qquad (9.14)$$

where the elements of $(\mathscr{T}_{\Gamma_a}(\mathscr{Q}_1)')^{\perp}$ are orthogonal, in the sense of the duality product defined by Eq. (9.12), to any element belonging to $\mathscr{T}_{\Gamma_a}(\mathscr{Q}_1)'$. In other words, for all fields $\mathbf{u}_2 \in \mathscr{Q}_2$ the trace over Γ_a, $\mathbf{u}_{2|\Gamma_a} \in \mathscr{T}_{\Gamma_a}(\mathscr{Q}_2)$, can be uniquely decomposed as

$$\mathbf{u}_{2|\Gamma_a} \in \mathscr{T}_{\Gamma_a}(\mathscr{Q}_2) \quad \Leftrightarrow \quad \mathbf{u}_{2|\Gamma_a} = \mathbf{u}_{21} + \mathbf{u}_{2f}, \qquad (9.15)$$

where $\mathbf{u}_{21} \in \mathscr{T}_{\Gamma_a}(\mathscr{Q}_1)$ and

$$\langle \mathbf{s}_1, \mathbf{u}_{2f} \rangle_{\mathscr{T}_{\Gamma_a}(\mathscr{Q}_1)' \times \mathscr{T}_{\Gamma_a}(\mathscr{Q}_1)} = 0 \qquad \forall \mathbf{s}_1 \in \mathscr{T}_{\Gamma_a}(\mathscr{Q}_1)'. \qquad (9.16)$$

It is worth noting that by using the above decomposition the following expression is well defined for $\mathbf{u}_{2|\Gamma_a} \in \mathscr{T}_{\Gamma_a}(\mathscr{Q}_2)$

$$\langle \mathbf{s}_1, \mathbf{u}_2 \rangle_{\mathscr{T}_{\Gamma_a}(\mathscr{Q}_1)' \times \mathscr{T}_{\Gamma_a}(\mathscr{Q}_1)} = \langle \mathbf{s}_1, \mathbf{u}_{21} \rangle_{\mathscr{T}_{\Gamma_a}(\mathscr{Q}_1)' \times \mathscr{T}_{\Gamma_a}(\mathscr{Q}_1)} \qquad \forall \mathbf{s}_1 \in \mathscr{T}_{\Gamma_a}(\mathscr{Q}_1)'. \quad (9.17)$$

Analogously to (9.14), we introduce the following decomposition

$$\mathscr{T}_{\Gamma_a}(\mathscr{Q}_2)' = \mathscr{T}_{\Gamma_a}(\mathscr{Q}_1)' \oplus (\mathscr{T}_{\Gamma_a}(\mathscr{Q}_1))^{\perp}, \qquad (9.18)$$

where $(\mathscr{T}_{\Gamma_a}(\mathscr{Q}_1))^{\perp}$ is the orthogonal space to $\mathscr{T}_{\Gamma_a}(\mathscr{Q}_1)$ in the sense of the duality product defined by (9.13). That is, for all $s_2 \in \mathscr{T}_{\Gamma_a}(\mathscr{Q}_2)'$ we can write

$$s_2 \in \mathscr{T}_{\Gamma_a}(\mathscr{Q}_2)' \quad \Leftrightarrow \quad s_2 = s_{21} + s_{2f}, \tag{9.19}$$

where $s_{21} \in \mathscr{T}_{\Gamma_a}(\mathscr{Q}_1)'$ and

$$\langle s_{2f}, u_1 \rangle_{\mathscr{T}_{\Gamma_a}(\mathscr{Q}_2)' \times \mathscr{T}_{\Gamma_a}(\mathscr{Q}_2)} = 0 \qquad \forall u_1 \in \mathscr{T}_{\Gamma_a}(\mathscr{Q}_1). \tag{9.20}$$

This yields

$$\langle s_2, u_1 \rangle_{\mathscr{T}_{\Gamma_a}(\mathscr{Q}_2)' \times \mathscr{T}_{\Gamma_a}(\mathscr{Q}_2)} = \langle s_{21}, u_1 \rangle_{\mathscr{T}_{\Gamma_a}(\mathscr{Q}_1)' \times \mathscr{T}_{\Gamma_a}(\mathscr{Q}_1)} \qquad \forall u_1 \in \mathscr{T}_{\Gamma_a}(\mathscr{Q}_1). \tag{9.21}$$

For two arbitrary elements $u_2 \in \mathscr{T}_{\Gamma_a}(\mathscr{Q}_2)$ and $s_2 \in \mathscr{T}_{\Gamma_a}(\mathscr{Q}_2)'$, decompositions (9.14) and (9.18) lead to

$$\langle s_2, u_2 \rangle_{\mathscr{T}_{\Gamma_a}(\mathscr{Q}_2)' \times \mathscr{T}_{\Gamma_a}(\mathscr{Q}_2)} = $$
$$\langle s_{21}, u_{21} \rangle_{\mathscr{T}_{\Gamma_a}(\mathscr{Q}_1)' \times \mathscr{T}_{\Gamma_a}(\mathscr{Q}_1)} + \langle s_{2f}, u_{2f} \rangle_{\mathscr{T}_{\Gamma_a}(\mathscr{Q}_2)' \times \mathscr{T}_{\Gamma_a}(\mathscr{Q}_2)}. \tag{9.22}$$

On the other hand, since now the kinematics at both sides of Γ_a are different, it is necessary to redefine the sets $\mathscr{U}_i$ of kinematically admissible fields u_i, $i = 1, 2$

$$\mathscr{U}_i = \{u_i \in \mathscr{Q}_i; u_{i|\Gamma_{D_i}} = \bar{u}_i\}. \tag{9.23}$$

From the above considerations, we can formulate a **new** extended variational statement which we named **Non-Classical Extended Variational Formulation**.

Problem 3. *Non-Classical* Extended Variational Formulation. For $\gamma \in [0,1]$ find $(u_1, u_2, t_1, t_2) \in \mathscr{U}_1 \times \mathscr{U}_2 \times \mathscr{T}_{\Gamma_a}(\mathscr{Q}_1)' \times \mathscr{T}_{\Gamma_a}(\mathscr{Q}_2)'$ such that

$$(\!(\mathscr{R}_1(u_1), v_1)\!)_{\mathscr{Q}_1} + (\!(\mathscr{R}_2(u_2), v_2)\!)_{\mathscr{Q}_2}$$
$$+ \gamma \langle t_1, (v_1 - v_2) \rangle_{\mathscr{T}_{\Gamma_a}(\mathscr{Q}_1)' \times \mathscr{T}_{\Gamma_a}(\mathscr{Q}_1)} + (1 - \gamma) \langle t_2, (v_1 - v_2) \rangle_{\mathscr{T}_{\Gamma_a}(\mathscr{Q}_2)' \times \mathscr{T}_{\Gamma_a}(\mathscr{Q}_2)}$$
$$+ \gamma \langle s_1, (u_1 - u_2) \rangle_{\mathscr{T}_{\Gamma_a}(\mathscr{Q}_1)' \times \mathscr{T}_{\Gamma_a}(\mathscr{Q}_1)} + (1 - \gamma) \langle s_2, (u_1 - u_2) \rangle_{\mathscr{T}_{\Gamma_a}(\mathscr{Q}_2)' \times \mathscr{T}_{\Gamma_a}(\mathscr{Q}_2)}$$
$$\forall (v_1, v_2, s_1, s_2) \in \mathscr{V}_1 \times \mathscr{V}_2 \times \mathscr{T}_{\Gamma_a}(\mathscr{Q}_1)' \times \mathscr{T}_{\Gamma_a}(\mathscr{Q}_2)'. \tag{9.24}$$

In the above formulation, $\mathscr{R}_i$ is the *equilibrium* operator associated to the kinematics characterized by $\mathscr{Q}_i$, $i = 1, 2$ (recalling that $\mathscr{Q}_2 = \mathscr{Q}_{|\Omega_2}$, so $\mathscr{R}_2 = \mathscr{R}$).

As we will show, Problem 3 is γ-dependent with γ now playing a fundamental role from the mechanical point of view. In fact, this parameter governs how the continuity over Γ_a of the primal variables (u_1 and u_2) and of the dual ones (t_1 and t_2) is established. Here it is important to notice the difference between Problem 2 and Problem 3 with respect to γ. In Problem 3 the parameter γ defines the kinematics that will prevail on the coupling interface. This is not the case for Problem 2 since the kinematics are the same on both sub-domains.

Let us adopt $\gamma = 1$. All terms associated with the factor $(1 - \gamma)$ in Eq. (9.24) drop. From the mechanical point of view this means that over Γ_a all contributions to the virtual power in the sense of the kinematics characterized by $\mathscr{Q}_2$ are not considered. With the introduction of decomposition (9.15), and using (9.19) to write $\mathscr{R}_2(\mathbf{u}_2)_{|\Gamma_a} = \mathscr{R}_{21}(\mathbf{u}_2)_{|\Gamma_a} + \mathscr{R}_{2f}(\mathbf{u}_2)_{|\Gamma_a}$, Eq. (9.24) delivers the following natural conditions over Γ_a

$$\langle \mathbf{s}_1, (\mathbf{u}_1 - \mathbf{u}_{21}) \rangle_{\mathscr{T}_{\Gamma_a}(\mathscr{Q}_1)' \times \mathscr{T}_{\Gamma_a}(\mathscr{Q}_1)} = 0 \qquad \forall \mathbf{s}_1 \in \mathscr{T}_{\Gamma_a}(\mathscr{Q}_1)', \qquad (9.25)$$

$$\langle \mathbf{t}_1 + \mathscr{R}_1(\mathbf{u}_1)_{|\Gamma_a}, \mathbf{v}_1 \rangle_{\mathscr{T}_{\Gamma_a}(\mathscr{Q}_1)' \times \mathscr{T}_{\Gamma_a}(\mathscr{Q}_1)} = 0 \qquad \forall \mathbf{v}_1 \in \mathscr{T}_{\Gamma_a}(\mathscr{Q}_1), \qquad (9.26)$$

$$\langle \mathbf{t}_1 - \mathscr{R}_{21}(\mathbf{u}_2)_{|\Gamma_a}, \mathbf{v}_{21} \rangle_{\mathscr{T}_{\Gamma_a}(\mathscr{Q}_1)' \times \mathscr{T}_{\Gamma_a}(\mathscr{Q}_1)} = 0 \qquad \forall \mathbf{v}_{21} \in \mathscr{T}_{\Gamma_a}(\mathscr{Q}_1), \qquad (9.27)$$

$$\langle \mathscr{R}_{2f}(\mathbf{u}_2)_{|\Gamma_a}, \mathbf{v}_{2f} \rangle_{\mathscr{T}_{\Gamma_a}(\mathscr{Q}_2)' \times \mathscr{T}_{\Gamma_a}(\mathscr{Q}_2)} = 0 \qquad \forall \mathbf{v}_{2f} \in (\mathscr{T}_{\Gamma_a}(\mathscr{Q}_1)')^{\perp}. \quad (9.28)$$

From Eq. (9.25) we have

$$\begin{cases} \mathbf{u}_1 = \mathbf{u}_{21} & \text{in the sense of } \mathscr{T}_{\Gamma_a}(\mathscr{Q}_1), \\ \mathbf{u}_{2f} \neq 0 & \text{obtained after solving (9.24),} \end{cases} \qquad (9.29)$$

that, in short, means

$$\mathbf{u}_1 = \mathbf{u}_2 \qquad \text{in the sense of } \mathscr{T}_{\Gamma_a}(\mathscr{Q}_1). \qquad (9.30)$$

It is important to remark that Eq. (9.29) does not characterize the component $\mathbf{u}_{2f}$, but it just states that its value is different from zero, and that it can be determined after solving (9.24).

Hence, due to the different underlying kinematics, the continuity of the primal variables is not satisfied in the sense required by the original problem (Problem 9.1). In fact, the fluctuation $\mathbf{u}_{2f}$ is not necessarily null (see (9.29)$_2$) and is obtained after solving the variational equation (9.24). On the other hand from (9.26)–(9.28) we obtain

$$\begin{cases} \mathbf{t}_1 = -\mathscr{R}_1(\mathbf{u}_1)_{|\Gamma_a} = \mathscr{R}_{21}(\mathbf{u}_2)_{|\Gamma_a} & \text{in the sense of } \mathscr{T}_{\Gamma_a}(\mathscr{Q}_1)', \\ \mathscr{R}_{2f}(\mathbf{u}_2)_{|\Gamma_a} = 0 & \text{in the sense of } \mathscr{T}_{\Gamma_a}(\mathscr{Q}_2)', \end{cases} \qquad (9.31)$$

which is

$$-\mathscr{R}_1(\mathbf{u}_1)_{|\Gamma_a} = \mathscr{R}_2(\mathbf{u}_2)_{|\Gamma_a} \qquad \text{in the sense of } \mathscr{T}_{\Gamma_a}(\mathscr{Q}_2)'. \qquad (9.32)$$

For simplicity, let us take now $\gamma = 0$. Then the natural boundary conditions over Γ_a are given by (we also make use of decompositions (9.15) and (9.19))

$$\langle \mathbf{s}_2, (\mathbf{u}_1 - \mathbf{u}_2) \rangle_{\mathscr{T}_{\Gamma_a}(\mathscr{Q}_2)' \times \mathscr{T}_{\Gamma_a}(\mathscr{Q}_2)} = 0 \qquad \forall \mathbf{s}_2 \in \mathscr{T}_{\Gamma_a}(\mathscr{Q}_2)', \qquad (9.33)$$

$$\langle \mathbf{t}_{21} + \mathscr{R}_1(\mathbf{u}_1)_{|\Gamma_a}, \mathbf{v}_1 \rangle_{\mathscr{T}_{\Gamma_a}(\mathscr{Q}_1)' \times \mathscr{T}_{\Gamma_a}(\mathscr{Q}_1)} = 0 \qquad \forall \mathbf{v}_1 \in \mathscr{T}_{\Gamma_a}(\mathscr{Q}_1), \qquad (9.34)$$

$$\langle \mathbf{t}_2 - \mathscr{R}_2(\mathbf{u}_2)_{|\Gamma_a}, \mathbf{v}_2 \rangle_{\mathscr{T}_{\Gamma_a}(\mathscr{Q}_2)' \times \mathscr{T}_{\Gamma_a}(\mathscr{Q}_2)} = 0 \qquad \forall \mathbf{v}_2 \in \mathscr{T}_{\Gamma_a}(\mathscr{Q}_2). \qquad (9.35)$$

From Eq. (9.33) we have

$$\begin{cases} \mathbf{u}_1 = \mathbf{u}_{21} & \text{in the sense of } \mathscr{T}_{\Gamma_a}(\mathscr{Q}_1), \\ \mathbf{u}_{2f} = 0 & \text{in the sense of } (\mathscr{T}_{\Gamma_a}(\mathscr{Q}_1)')^\perp, \end{cases} \tag{9.36}$$

that is

$$\mathbf{u}_1 = \mathbf{u}_2 \qquad \text{in the sense of } \mathscr{T}_{\Gamma_a}(\mathscr{Q}_2). \tag{9.37}$$

In this case the continuity condition of the primal variables is satisfied in exactly the same sense as in the original problem, noting that now $\mathbf{u}_{2f}$ is identically null over Γ_a. From the mechanical point of view this is a stronger restriction over the behaviour of $\mathbf{u}_2$ on Γ_a that certainly impacts on the behaviour of the solution in Ω_2. On the other hand, from (9.34)-(9.35) we obtain

$$\begin{cases} \mathbf{t}_{21} = -\mathscr{R}_1(\mathbf{u}_1)_{|\Gamma_a} = \mathscr{R}_{21}(\mathbf{u}_2)_{|\Gamma_a} & \text{in the sense of } \mathscr{T}_{\Gamma_a}(\mathscr{Q}_1)', \\ \mathbf{t}_{2f} = \mathscr{R}_{2f}(\mathbf{u}_2)_{|\Gamma_a} \neq 0 & \text{obtained after solving (9.24),} \end{cases} \tag{9.38}$$

implying that

$$-\mathscr{R}_1(\mathbf{u}_1)_{|\Gamma_a} = \mathscr{R}_2(\mathbf{u}_2)_{|\Gamma_a} \qquad \text{in the sense of } \mathscr{T}_{\Gamma_a}(\mathscr{Q}_1)'. \tag{9.39}$$

Again, due to the difference in the governing kinematics, the continuity of the dual variables is not satisfied as in the original Problem 9.1 (see Eq. (9.4)). In fact there is a fluctuation of the traction coming from Ω_2, and defined by the variational equation (9.24), which is not identically null as expressed by (9.38)$_2$.

Before closing this section it is worth summarizing the results already obtained with the non-classical extended variational formulation given by Problem 3.

1. The non-classical extended variational formulation given by Problem 3 gives an appropriate framework to deal with the coupling of dimensionally-heterogeneous models through the concept of kinematically incompatible models.
2. With this variational formulation the coupling conditions to be prescribed on the coupling interface are already accounted for, and are satisfied by the solution as natural boundary conditions.
3. The selection criterion for choosing the value of γ is dictated by purely mechanical considerations, since it defines the sense in which the continuity of the primal and dual variables is addressed by the formulation at the continuous level.
4. The continuity equations for $\gamma = 1$ are (see (9.30) and (9.32))

$$\begin{cases} \mathbf{u}_1 = \mathbf{u}_2 & \text{in the sense of } \mathscr{T}_{\Gamma_a}(\mathscr{Q}_1), \\ -\mathscr{R}_1(\mathbf{u}_1)_{|\Gamma_a} = \mathscr{R}_2(\mathbf{u}_2)_{|\Gamma_a} & \text{in the sense of } \mathscr{T}_{\Gamma_a}(\mathscr{Q}_2)', \end{cases} \tag{9.40}$$

and for $\gamma = 0$ are (see (9.37) and (9.39))

$$\begin{cases} \mathbf{u}_1 = \mathbf{u}_2 & \text{in the sense of } \mathscr{T}_{\Gamma_a}(\mathscr{Q}_2), \\ -\mathscr{R}_1(\mathbf{u}_1)_{|\Gamma_a} = \mathscr{R}_2(\mathbf{u}_2)_{|\Gamma_a} & \text{in the sense of } \mathscr{T}_{\Gamma_a}(\mathscr{Q}_1)'. \end{cases} \tag{9.41}$$

5. The stiffness of the coupled system exhibited in the case $\gamma = 1$ is lower than the case $\gamma = 0$. Indeed, the constraint in the former case is far weaker than in the latter. For instance, in the example of Sect. 9.3.2 below, the constraint (9.51a) (obtained for $\gamma = 1$) is weaker than (9.52a) (obtained for $\gamma = 0$).

6. From the computational point of view, the parameter γ also has an interesting role. In this sense, when $\gamma = 1$, less unknowns are required as a result of the weaker condition to be imposed, reducing the computational cost necessary to obtain the solution. In fact, for the same example of Sect. 9.3.2, it is easy to see that for $\gamma = 1$ in the discrete problem we have to obtain an approximation of $\mathbf{t}_1$ which is a real number, while for $\gamma = 0$ we have to compute an approximation of $\mathbf{t}_2$ which is a field in $\mathbf{H}^{-1/2}(\Gamma_a)$.

A last comment regarding the definition of γ is in order. In the formulation of Problem 3 γ ranges over the closed interval $[0, 1]$. The purpose of this was to introduce a convex combination of the two duality products involved in the problem, namely $\langle \cdot, \cdot \rangle_{\mathscr{T}_{\Gamma_a}(\mathscr{Q}_1)' \times \mathscr{T}_{\Gamma_a}(\mathscr{Q}_1)}$ and $\langle \cdot, \cdot \rangle_{\mathscr{T}_{\Gamma_a}(\mathscr{Q}_2)' \times \mathscr{T}_{\Gamma_a}(\mathscr{Q}_2)}$. Nonetheless, it can be shown that the result achieved with $\gamma \in (0, 1)$ is the same as with $\gamma = 0$, that is, its value is actually immaterial in that case. Moreover, it could have been sufficient to define Problem 3 for $\gamma \in \{0, 1\}$.

9.3 Application to the coupling of fluid flow models

This section describes some common situations encountered when trying to couple incompatible models in fluid dynamics. In Sect. 9.2 we discussed the coupling between two models with different kinematics. The differences in the kinematics arise from differences in the nature of the models. As a result of the two underlying kinematics it was possible to identify two different senses in which the coupling equations, naturally derived from the extended variational Problem 3, are satisfied. Let us give some examples of scenarios of interest in which formulation 3 provides coupling equations. For the sake of brevity we will skip the presentation of the entire flow models, assuming that the reader is aware of the physical elements present in each case.

In the dimensional reduction process of a mathematical model we can find two different situations concerning the kinematical restrictions: (i) a simple dimensional reduction and (ii) a dimensional reduction plus augmented regularity. The two examples presented in the forthcoming sections for fluid flow problems correspond to situation (i). In such cases the finite-dimensional problem does not entail further complications from the point of view of the mathematical problem. On the other hand, situation (ii) demands a careful manipulation of the trace spaces over the coupling interfaces due to the mismatch in the regularity conditions. Examples of this class can be found in the analysis of beams, plates and shells in solid mechanics [7]. In such cases, the finite-dimensional analysis should be carried out *ad hoc*.

9.3.1 Coupling 3D and 2D models

Suppose we have a flow model with symmetry of revolution set up in Ω_1, that is a 2D model, and a full flow model in Ω_2, with a circular coupling interface Γ_a. In this case, at any time t, the velocity fields are described by functions of the form

$$\mathbf{u}_1 \to \mathbf{u}_1(r,z) \in \mathscr{Q}_1 \qquad \mathbf{u}_2 \to \mathbf{u}_2(\mathbf{x}) \in \mathscr{Q}_2, \qquad (9.42)$$

where r and z are the radial and axial coordinates in Ω_1 (denoted now by Σ), respectively, and $\mathbf{x}$ is the three-dimensional position vector in Ω_2 (see Fig. 9.2). Note that the above form of $\mathbf{u}_1$ is equivalent to writing $\mathbf{u}_1(r,z) = u_{1,r}(r,z)\mathbf{e}_r + u_{1,z}(r,z)\mathbf{e}_z$, $\mathbf{e}_r, \mathbf{e}_z$ being the unit vectors in the cylindrical system of coordinates. In this case the spaces $\mathscr{Q}_i$ and $\mathscr{T}_{\Gamma_a}(\mathscr{Q}_i)$, $i=1,2$ are (the dual spaces follow directly)

$$\mathscr{Q}_1 = \mathbf{H}_r^1(\Sigma) \qquad\qquad \mathscr{T}_{\Gamma_a}(\mathscr{Q}_1) = \mathbf{H}_r^{1/2}(\partial\Sigma_a), \qquad (9.43)$$

$$\mathscr{Q}_2 = \mathbf{H}^1(\Omega_2) \qquad\qquad \mathscr{T}_{\Gamma_a}(\mathscr{Q}_2) = \mathbf{H}^{1/2}(\Gamma_a). \qquad (9.44)$$

Here, $\partial\Sigma_a$ and Γ_a correspond to the same coupling interface viewed from the domains Σ and Ω_2 (see again Fig. 9.2). The space $\mathbf{H}_r^1(\Sigma)$ is the r-weighted Sobolev space defined as the set of measurable functions v with norm (see [2])

$$\|v\|_{H_r^1(\Sigma)}^2 = \sum_{\ell=0}^1 \sum_{k=0}^\ell \|\partial_r^k \partial_z^{\ell-k} v\|_{L_r^2(\Sigma)}^2 \quad \text{and} \quad \|v\|_{L_r^2(\Sigma)}^2 = \int_\Sigma v^2(r,z)\, r\, dr dz. \qquad (9.45)$$

We will express the vector $\mathbf{x}$ in terms of cylindrical coordinates over Γ_a, that is $\mathbf{x}_{|\Gamma_a} \to (r,\phi,z_{|\Gamma_a})$. In this situation, the choice $\gamma = 1$ produces the following coupling equations

$$\mathbf{u}_1(r,z_{|\partial\Sigma_a}) = \frac{1}{2\pi} \int_0^{2\pi} \mathbf{u}_2(r,\phi,z_{|\Gamma_a})\, d\phi \qquad \text{for } r \in \partial\Sigma_a, \qquad (9.46a)$$

$$-p_1(r,z_{|\partial\Sigma_a})\mathbf{n} + 2\mu\boldsymbol{\varepsilon}_{r,z}(\mathbf{u}_1(r,z_{|\partial\Sigma_a}))\mathbf{n} = $$
$$-p_2(r,\phi,z_{|\Gamma_a})\mathbf{n} + 2\mu\boldsymbol{\varepsilon}(\mathbf{u}_2(r,\phi,z_{|\Gamma_a}))\mathbf{n} \qquad \text{for } (r,\phi) \in \Gamma_a, \qquad (9.46b)$$

where $\varepsilon(\mathbf{u}_2) = (\nabla\mathbf{u}_2)^s$ is the symmetric part of the gradient of $\mathbf{u}_2$, and $\boldsymbol{\varepsilon}_{r,z}(\mathbf{u}_1)$ is the corresponding symmetric part of the gradient of $\mathbf{u}_1$ in cylindrical coordinates (r,z). Eq. (9.46a) condenses the dependence upon ϕ and establishes the continuity in a mean sense from the point of view of the angular coordinate, whereas Eq. (9.46b) extends the value of the dual quantity from the 2D model, defined $\forall r$, to all values of the pair (r,z) in the 3D model. That is, for a fixed value $r = r^*$, we have $-p_2(r^*,\phi,z_{|\Gamma_a})\mathbf{n} + 2\mu\boldsymbol{\varepsilon}(\mathbf{u}_2(r^*,\phi,z_{|\Gamma_a}))\mathbf{n}$ constant $\forall\phi$. In other words, Eq. (9.46a) stands for Eq. (9.30), that is the continuity of the primal variable in the sense of $\mathbf{H}_r^{1/2}(\partial\Sigma_a)$, whereas Eq. (9.46b) expresses the continuity of the dual variable like in (9.32), in the sense of $\mathscr{T}_{\Gamma_a}(\mathscr{Q}_2)' = \mathbf{H}^{-1/2}(\Gamma_a)$.

Analogously, for $\gamma = 0$ we obtain the following coupling equations

$$\mathbf{u}_1(r,z_{|\partial\Sigma_a}) = \mathbf{u}_2(r,\phi,z_{|\Gamma_a}) \qquad \text{for } (r,\phi) \in \Gamma_a, \qquad (9.47a)$$

$$-p_1(r,z_{|\partial\Sigma_a})\mathbf{n}+2\mu\boldsymbol{\varepsilon}_{r,z}(\mathbf{u}_1(r,z_{|\partial\Sigma_a}))\mathbf{n}=$$

$$\frac{1}{2\pi}\int_0^{2\pi}\left[-p_2(r,\phi,z_{|\Gamma_a})\mathbf{n}+2\mu\boldsymbol{\varepsilon}(\mathbf{u}_2(r,\phi,z_{|\Gamma_a}))\mathbf{n}\right]d\phi \quad \text{for } r\in\partial\Sigma_a. \quad (9.47\text{b})$$

In this case the opposite situation occurs, as expressed by Eqs. (9.37) and (9.39), respectively.

9.3.2 Coupling 3D and 1D models

In this case let us assume that over Ω_1 a 1D model is set up, so we have that $\Omega_1 \equiv I$ (I an interval), and we keep a full 3D model in Ω_2 (see Fig. 9.3). Specifically, for all t we consider the following kinematics

$$\mathbf{u}_1 \to u_{1,z}(z) \in \mathscr{Q}_1 \qquad \mathbf{u}_2 \to \mathbf{u}_2(\mathbf{x}) \in \mathscr{Q}_2, \qquad (9.48)$$

where $u_{1,z}$ is the component of $\mathbf{u}_1$ in the axial direction $\mathbf{e}_z$, so it would be equivalent to define $\mathbf{u}_1$ as $\mathbf{u}_1(z) = u_{1,z}(z)\mathbf{e}_z$. In this case, the spaces $\mathscr{Q}_i$ and $\mathscr{T}_{\Gamma_a}(\mathscr{Q}_i)$, $i = 1,2$ are

$$\mathscr{Q}_1 = H^1(I) \qquad\qquad \mathscr{T}_{\Gamma_a}(\mathscr{Q}_1) = \mathbb{R}, \qquad (9.49)$$

$$\mathscr{Q}_2 = \mathbf{H}^1(\Omega_2) \qquad\qquad \mathscr{T}_{\Gamma_a}(\mathscr{Q}_2) = \mathbf{H}^{1/2}(\Gamma_a). \qquad (9.50)$$

Here the coupling interface corresponds to Γ_a when viewed from Ω_2, and to the point ζ_a when viewed from I, as seen in Fig. 9.3. As in the previous case, we write $\mathbf{x}_{|\Gamma_a} \to (x,y,z_{|\Gamma_a})$. For this case, noting that $\mathbf{n}$ coincides with $\mathbf{e}_z$ over Γ_a, $\gamma = 1$ yields

$$u_{1,z}(z_{|\zeta_a}) = \frac{1}{|\Gamma_a|}\int_{\Gamma_a}\mathbf{u}_2(x,y,z_{|\Gamma_a})\cdot\mathbf{n}\,d\Gamma_a \quad \text{at } \zeta_a, \qquad (9.51\text{a})$$

$$-p_1(z_{|\zeta_a})+2\mu\frac{\partial u_{1,z}}{\partial z}(z_{|\zeta_a}) =$$
$$-p_2(x,y,z_{|\Gamma_a})+2\mu\boldsymbol{\varepsilon}(\mathbf{u}_2(x,y,z_{|\Gamma_a}))\mathbf{n}\cdot\mathbf{n} \quad \text{for } (x,y)\in\Gamma_a, \quad (9.51\text{b})$$

$$0 = 2\mu\boldsymbol{\varepsilon}(\mathbf{u}_2(x,y,z_{|\Gamma_a}))\mathbf{n}\cdot\mathbf{t} \quad \text{for } (x,y)\in\Gamma_a, \qquad (9.51\text{c})$$

where $\mathbf{t}$ is any vector tangent to the coupling interface Γ_a. In Eq. (9.51a) the dependence on (x,y) is eliminated, establishing the continuity in a mean sense from the point of view of the entire area Γ_a, whereas in Eq. (9.51b) an extension is made such that the value of the quantity from the 1D model, a real value, is mapped to all the points (x,y) over Γ_a from the 3D model. Finally, Eq. (9.51c) establishes that the traction over Γ_a, when viewed from the 3D model, has no component tangent to the coupling interface. Here the continuity of the velocity field is given in the sense of $\mathbb{R}$ (mean value) while the dual variable is continuous in a strong sense, that is the sense of $\mathbf{H}^{-1/2}(\Gamma_a)$.

In the same manner, for $\gamma = 0$ we obtain the following coupling equations

$$u_1(z_{|\zeta_a})\mathbf{n} = \mathbf{u}_2(x,y,z_{|\Gamma_a}) \quad \text{for } (x,y)\in\Gamma_a, \qquad (9.52\text{a})$$

$$-p_1(z_{|\zeta_a}) + 2\mu \frac{\partial u_{1,z}}{\partial z}(z_{|\zeta_a}) =$$

$$\frac{1}{|\Gamma_a|} \int_{\Gamma_a} \left[-p_2(x,y,z_{|\Gamma_a}) + 2\mu \boldsymbol{\varepsilon}(\mathbf{u}_2(x,y,z_{|\Gamma_a}))\mathbf{n} \cdot \mathbf{n} \right] d\Gamma_a \quad \text{at } \zeta_a. \quad (9.52b)$$

Conversely, for this case we have the strong continuity (pointwise) of the velocity field, in the sense of $\mathbf{H}^{1/2}(\Gamma_a)$ (it is considered to be constant over the section) and a weak continuity (mean value) of the dual field, in the sense of $\mathbb{R}$.

When looking at (9.52a), the constant velocity profile may suggest that inconsistencies arise in the representation of no-slip boundary conditions over the lateral boundary. For instance, this profile may lead us to think that no viscous effects should be considered in the 3D model, while it is obvious that we should take into account such viscous effects. In this respect, this model is not consistent in the sense of reproducing the same profile. Nevertheless, this inconsistency can be removed if we further refine the 1D model, for example, by modifying the velocity profile and introducing a parabolic, or Womersley-like, profile characterized by a function φ. That is, a more complex situation can be considered assuming that the velocity profile in the 1D model is given by

$$\mathbf{u}_1 \to u_{1,z}(x,y,z)\mathbf{e}_z = \varphi(x,y)\bar{u}_{1,z}(z)\mathbf{e}_z, \quad (9.53)$$

where $\frac{1}{|\Gamma_a|} \int_{\Gamma_a} \varphi(x,y) d\Gamma_a = 1$ is such that $\frac{1}{|\Gamma_a|} \int_{\Gamma_a} \mathbf{u}_1 \cdot \mathbf{e}_z d\Gamma_a = \bar{u}_{1,z}$, and $\varphi_{|\Gamma_{L1}} = 0$ (see Fig. 9.3). The spaces are similar to those previously introduced. Eventually, the viscous contribution to the dual variable in the 1D domain will change, according to the specific form of the function φ. Nevertheless, the way in which the problem is formulated is completely equivalent.

9.3.3 About fluid-structure interaction

In our present case, the fluid problem we consider both above and in the numerical examples below is the blood flow problem in deforming domains. Such deformation is induced by lateral wall displacements in the direction of the normal to the wall as a function of the value of the pressure (in a pointwise sense). Therefore, we have an independent ring model for the 3D wall (an algebraic equation -elasticity- or an ordinary differential equation -viscoelasticity-) which does not introduce any difficulty in the definition of the operator $\mathscr{R}_2$ for the 3D model, nor incompatibilities with the structural model of the 1D representation (an independent ring model as well). This is why we refer to our problem just as a "fluid problem", although we clearly consider deformations of the domains. However, and as explicitly mentioned before, we focus our presentation in the role of the variational formulation of the fluid problem only.

In order to consider the fluid-structure interaction using our variational problem for the structure, the operator $\mathscr{R}_2$ must be properly extended. In such case, not only the fluid, but also the solid may undergo kinematical incompatibilities at coupling interfaces. Hence, the same treatment used for the fluid formulation would be required by the solid problem for the incompatibilities in the solid displacement between the 1D model and the 3D model. In this respect, the abstract formulation presented here is valid when the incompatibilities arise in just one physical quantity (the velocity field in our case). A more general framework is needed when considering coupled phenomena with multiple kinematical incompatibilities at coupling interfaces, such as the case of the classical fluid-structure interaction problem. For the interested reader, examples of incompatibilities in solid mechanics using this same framework have been addressed in [7].

9.4 3D-1D-0D heterogeneous model of the cardiovascular system

Firstly we discuss the use of the variational formulation (9.24) in this context, and then we succintly describe each of the models used in setting up either a 3D-1D-0D open-loop model of the arterial network or a 3D-1D-0D closed-loop model of the entire cardiovascular system.

9.4.1 On the use of the non-classical extended variational formulation

In the present work 3D models of arterial vessels will always be embedded in the 1D arterial network. Therefore, we have two different situations: (a) coupling 1D and 0D models and (b) coupling 3D and 1D models. Case (a) does not involve any difficulty because the 0D model and the 1D model share the same kinematics (characterized simply by a real number) at the coupling interfaces, which are denoted by Γ_{01}. Case (b) is the one that raises more questions. Although the machinery presented in Sect. 9.2 applies to both cases, the latter brings a richer theoretical treatment as a result of the lack of compatibility in the kinematics. The interfaces between 1D and 3D models will be denoted by Γ_{13}.

The value of γ is relevant for the kinematical incompatibility emerging at Γ_{13}, while it is immaterial for the coupling at Γ_{01} (since both kinematics for the 0D and the 1D models are governed by a single real number over the coupling interface). We consider in the 1D model the kinematical hypothesis that led to the coupling equations seen in Sect. 9.3.2. In particular, we take $\gamma = 1$. As discussed in the previous section, this imposes the continuity of the velocity field in the sense of the mean value (in the sense of $\mathbb{R}$) over the coupling interface Γ_{13}. Otherwise, note that $\gamma = 0$ implies the imposition of the velocity profile at the coupling interfaces in the 3D model, as expressed by Eq. (9.52a), which is too strong in a fluid flow problem.

Hence with this choice the continuity of the traction force (dual variable) on Γ_{13} is satisfied pointwise (indeed, from the variational formulation we obtained a constant force in the 3D model as given by (9.51b)–(9.51c)) and the velocity profile is allowed to take quite arbitrary shapes in the 3D solution. This provides a richer framework in which, for example, it is possible to capture Womersley-like velocity profiles, that is, situations in which the velocities normal to the cross-section area of the vessel can be positive and negative at the same time.

It must be said that other coupling techniques are based on the pointwise imposition of the velocity profile (see [13, 27]), restricting the solution at the entrances and/or the exits of the 3D domain. This approach is embraced by formulation (9.24) when taking $\gamma = 0$, for which we would obtain equations of the form of (9.52a)–(9.52b) with a given function $\varphi(x,y)$ characterizing the velocity profile, as in (9.53).

Note that while the velocity profiles can exhibit large variations over the cross-sectional areas of the vessels, the pressure remains, in most cases, almost uniform (as long as the coupling interfaces are somewhat normal to the axial direction of the vessel). In this sense the choice $\gamma = 1$ considered here appears more adequate for the type of problems faced in the application examples. Moreover, instead of the simple 1D model adopted here (derived from a flat velocity profile), more sophisticated 1D models could be considered. For all these models, the proposed non-classical extended variational formulation will yield automatically the appropriate coupling conditions satisfied by the primal (kinematic) and dual (traction) variables over the coupling interfaces.

Specifically, in the construction of our 1D model we neglect the viscous effects associated to the axial gradients of the velocity field, that is, we neglect terms of the form $\frac{\partial u_{1,z}}{\partial z}$. A simple dimensional analysis justifies this consideration [25]. Therefore, in the coupling conditions no viscous terms will emerge from the 1D domain to the coupling interface (left-hand side of (9.51b)). In such a case, the traction over the 3D domain is exclusively due to the pressure in the 1D model (see Sect. 9.4.8 below).

Finally, the reader interested in more details and applications of this extended variational formulation to others physical problems, like heat conduction and solid mechanics, should refer to [3, 4, 5, 7].

9.4.2 Open/closed-loop representations of the cardiovascular system

As mentioned previously, the dimensionally-heterogeneous approach to model the systemic circulation is able to couple several elements of interest in the analysis: (i) the complexity of blood flow circulation in specific arterial districts such as bifurcations, aneurisms and other geometrical singularities in general (by using 3D models with geometry coming from patient-specific medical image data); (ii) the complex systemic behaviour that leads to the conformation of the cardiac pulse (by using 1D models) and (iii) the influence of the remaining (complementary) part of the circulatory system like arterioles, capillaries, venules, veins, pulmonary circulation, heart chambers and valves, among others (by using 0D models).

In the literature several approaches have been adopted that integrate different levels of circulation in the sense introduced in the previous paragraphs. Mostly, models based on lumped representations were employed to accomplish this task, incorporating 0D models to simulate flow in the larger arteries, veins and cardiac circulation [31, 35, 38, 52, 54, 55]. Distributed models to simulate the blood flow in compliant vessels have represented an exhaustive area of research over the last decades [1, 33, 46, 53, 56, 57, 64]. More recently, 1D models of the arterial circulation have been coupled to 0D models of the venous-cardiac-pulmonary circulation to study the influence of arterial stenoses on wave propagation [40]. In particular, the 1D model employed in Liang and Takagi, [40], which was taken from Stergiopulos et al., [56], comprises 55 arterial segments and a 0D lumped representation for the peripheral/venous/pulmonary and cardiac circulations. In this work, valves are modelled using an ideal model of a diode, not allowing for backflow to occur. This last point was addressed in [31], in which phenomenological models of the cardiac valves were proposed in order to model more accurately the opening and closing phases of the valves accounting for certain pathological conditions like valve regurgitation and stenosis. In the field of modelling blood flow in specific vessels, several works dealt with the use of heterogeneous representations in order to couple local and global hemodynamics phenomena. This has been mostly carried out using 3D and 1D (or 0D) models to couple blood flow in complex arterial geometries with either full or partial models for the systemic dynamics [6, 8, 9, 18, 21, 28, 43, 60, 62].

In the context of the previous paragraphs, the model introduced here for the entire cardiovascular system borrows the most important features of the different models available in the literature, and their integration leads to a model with more descriptive and predictive capabilities.

Evidently, the dynamics captured by the model depends on whether the model is open or closed. This can be understood by looking at the constituent vascular entities in each model, as described below.

- Open-loop model of the CVS composed by:
  - a 1D model of the arterial network;
  - 0D Windkessel models for the arterioles and capillaries;
  - 3D models for specific vessels.

- Closed-loop model of the CVS composed by:
  - a 1D model of the arterial network;
  - 0D Windkessel models for the arterioles and capillaries;
  - 3D models for specific vessels;
  - 0D models for venules, veins and cavas;
  - 0D models for the four cardiac chambers;
  - 0D models for the four cardiac valves.

Fig. 9.4 shows a schematic representation of the closed-loop CVS. Notice that the venous circulation is coupled to the arterial network through three points: the upper and lower body connections and the aortic root, closing the loop. The open-loop model is obtained by removing the elements from the capillary level to the aortic root. In the latter case boundary conditions are imposed at the Windkessel terminals (constant pressure), and at the aortic root (a given flow rate).

The glossary of the terms used in Fig. 9.4 is given in Table 9.1. In this table N_{sa} denotes the total number of systemic arteries and N_{wlb} and N_{wub} are the number of Windkessel models pertaining to the lower and upper parts of the body, respectively, with $N_w = N_{wlb} + N_{wub}$ the total number of Windkessel elements in the arterial side.

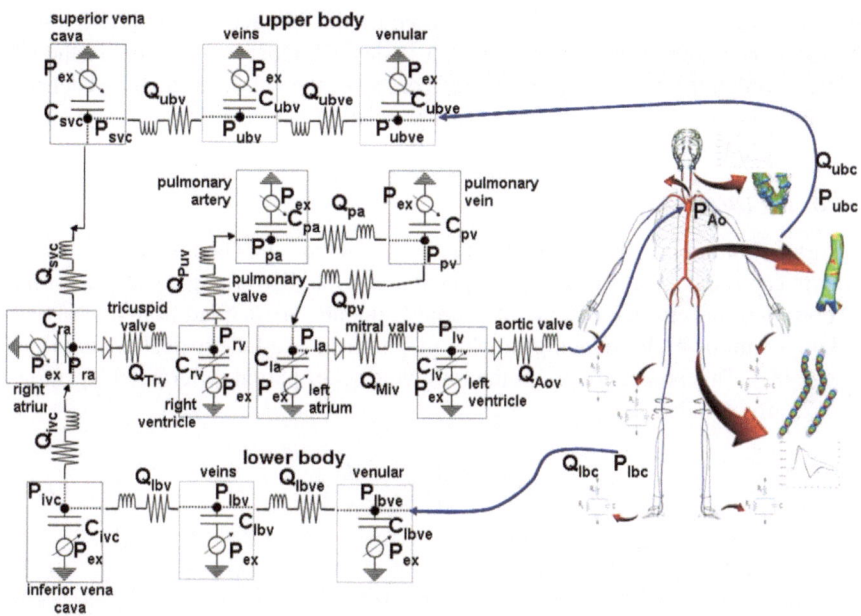

Fig. 9.4. Closed-loop model of the cardiovascular system via 3D-1D-0D models

Table 9.1. Glossary of the terms used in Fig. 9.4

sa,m:	systemic arteries, $m = 1, \ldots, N_{sa}$	w,k:	Windkessel models, $m = 1, \ldots, N_w$
wlb,k:	lower body Windkessel models, $k = 1, \ldots, N_{wlb}$	wub,k:	upper body Windkessel models, $k = 1, \ldots, N_{wub}$
lbc:	lower body capillaries	lbve:	lower body venule
lbv:	lower body veins	ubc:	upper body capillaries
ubve:	upper body venule	ubv:	upper body veins
ivc:	inferior vena cava	svc:	superior vena cava
pa:	pulmonary artery	pv:	pulmonary vein
ra:	right atrium	rv:	right ventricle
la:	left atrium	lv:	left ventricle
Trv:	tricuspid valve	Puv:	pulmonary valve
Miv:	mitral valve	Aov:	aortic valve
ao:	aorta artery	3D:	three-dimensional model

In the forthcoming sections each model is described more in detail, and the parameters that define each model are also given. In Tables 9.2–9.7 we collect all the data used in setting the 1D-0D closed-loop model of the cardiovascular system. In particular, the parameters used in the 1D-0D model have been assigned or estimated on the basis of the data reported in [1, 22, 24, 38, 39, 40, 45, 49, 61]. The values of the parameters used in the model for the peripheral circulation, the heart model and the model of the cardiac valves are based on the data reported in [22, 24, 40]. The mechanical parameters of arterial walls in the 3D models match those of the 1D model to which the specific vessel is coupled. As for the values of the blood density and viscosity, we take $\rho = 1.04\,\text{g/cm}^3$ and $\mu = 0.04\,\text{dyn s/cm}^2$, respectively.

9.4.3 1D models

The arterial blood flow is modelled as the flow of a fluid in 1D compliant vessels in order to capture the wave propagation phenomena. The governing equations for the 1D portion of the arterial system (all arterial segments) are derived from the Navier-Stokes equations by introducing suitable geometrical and kinematical assumptions (see (9.48)). This procedure yields the following set of partial differential equations (see [25] for details):

$$\frac{\partial Q_{sa,m}}{\partial t} + \frac{\partial}{\partial x}\left(\beta \frac{Q_{sa,m}^2}{A_{sa,m}}\right) = -\frac{A_{sa,m}}{\rho}\frac{\partial P_{sa,m}}{\partial x} - \frac{\pi D}{\rho}\tau_{osa,m} \quad m = 1,\ldots,N_{sa}, \quad (9.54)$$

$$\frac{\partial A_{sa,m}}{\partial t} + \frac{\partial Q_{sa,m}}{\partial x} = 0 \quad m = 1,\ldots,N_{sa}, \quad (9.55)$$

with

$$\tau_{osa,m} = f_r \frac{\rho \tilde{u}_{sa,m}|\tilde{u}_{sa,m}|}{8} \qquad Q_{sa,m} = \tilde{u}_{sa,m}A_{sa,m}, \qquad (9.56)$$

where $Q_{sa,m}$ is the flow rate, $A_{sa,m}$ is the cross-sectional area of the artery (D its diameter), $\tilde{u}_{sa,m}$ the mean value of the axial velocity, x the axial coordinate, $P_{sa,m}$ the mean pressure, ρ the blood density, $\tau_{osa,m}$ the viscous shear stress acting on the arterial wall, f_r a Darcy friction factor (in this work a fully developed parabolic velocity profile is considered) and β is the momentum correction factor ($\beta = 1$ is considered here).

The system is closed by introducing a constitutive law for the arterial wall. Here the following visco-elastic model is used [29, 30]

$$P_{sa,m} = P_0 + \frac{Eh_0}{R_0}\left(\sqrt{\frac{A_{sa,m}}{A_0}} - 1\right)$$
$$+ \frac{Kh_0}{R_0}\frac{1}{2\sqrt{A_0 A_{sa,m}}}\frac{\partial A_{sa,m}}{\partial t} \quad m = 1,\ldots,N_{sa}, \quad (9.57)$$

R being the radius of the artery, E an effective Young modulus, K the viscosity of the wall, h the thickness of the arterial wall. The subscript 'o' denotes quantities evaluated at the reference pressure P_0.

The arterial tree is described by 128 arterial segments (see [1]) in which continuity of mass and continuity of pressure are imposed at arterial junctions. Calling N_{jn} the total number of junctions with $N_{cg,j}$ arterial segments converging to junction j

we have

$$\sum_{n=1}^{N_{cg,j}} Q_{sa|j,n} = 0 \quad j = 1,\ldots,N_{jn}, \tag{9.58}$$

$$P_{sa|j,n} = P_{sa|j,1} \quad n = 2,\ldots,N_{cg,j} \quad j = 1,\ldots,N_{jn}, \tag{9.59}$$

where the pair $(Q_{sa|j}, P_{sa|j})$ indicates the restriction of the flow rate and pressure of a given systemic artery at the junction j.

The geometrical and mechanical parameters used in the 1D model of the arterial tree agree with the model proposed in [1] and are presented in Table 9.2.

9.4.4 0D model for the arterioles and capillaries

The peripheral circulation is represented through Windkessel models [54, 56] characterized by three elements R_c, R_a and C_a to model the capillaries and arterioles, respectively (see Fig. 9.5).

In Fig. 9.5 (Q_i, P_i) and (Q_o, P_o) denote the quantities at the input and output, respectively. For a single Windkessel element k these quantities are $(Q_{sa|w,k}, P_{sa|w,k})$ and $(Q_{w,k}, P_{w,k})$. The former pair represents the quantities at the arterial segment converging to the corresponding Windkessel segment k, whereas in the latter pair we have the quantities from the venous side. Then the balance equations for these models are the following

$$\frac{dQ_{sa|w,k}}{dt} = \frac{1}{R_c R_a C_a} \left[R_a C_a \frac{d}{dt}(P_{sa|w,k} - P_{w,k}) \right. \tag{9.60}$$

$$\left. + (P_{sa|w,k} - P_{w,k}) - (R_c + R_a)Q_{sa|w,k} \right] \quad k = 1,\ldots,N_w, \tag{9.61}$$

$$Q_{sa|w,k} = Q_{w,k} \quad k = 1,\ldots,N_w. \tag{9.62}$$

The Windkessel element is a model for the arterial-venous interface. Thus the $P_{sa|w,k}$ is the pressure from the arterial side (at the input of the compartment) while $P_{w,k}$ is the pressure from the venous side (at the output of the compartment), more precisely

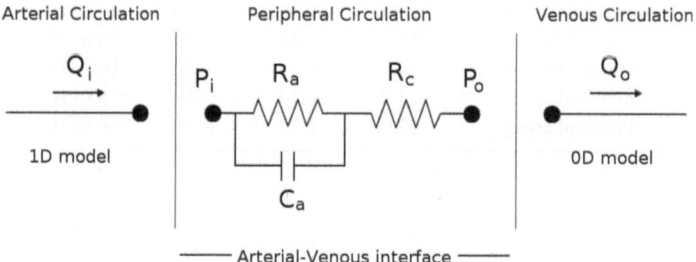

Fig. 9.5. The 0D Windkessel model for the peripheral circulation

Table 9.2. Part A. Geometric and mechanical parameters of the arterial segments

Segment	Description	L [cm]	R_0 [cm]	h_0 [cm]	$E \left[\frac{dyn}{cm^2}\right]$	$k \left[\frac{dyn\ s}{cm^2}\right]$
1	Ascending aorta	4.0	1.45	0.163	$4.0 \cdot 10^6$	$4.4 \cdot 10^4$
2	Aortic arch	2.0	1.12	0.132	$4.0 \cdot 10^6$	$4.4 \cdot 10^4$
5	Aortic arch	3.9	1.07	0.127	$4.0 \cdot 10^6$	$4.4 \cdot 10^4$
11	Thoracic aorta	5.2	1.00	0.120	$4.0 \cdot 10^6$	$4.4 \cdot 10^4$
21	Thoracic aorta	5.2	0.95	0.116	$4.0 \cdot 10^6$	$4.4 \cdot 10^4$
34	Thoracic aorta	5.2	0.95	0.116	$4.0 \cdot 10^6$	$4.4 \cdot 10^4$
50	Abdominal aorta	5.3	0.87	0.108	$4.0 \cdot 10^6$	$4.4 \cdot 10^4$
65	Abdominal aorta	5.3	0.57	0.080	$4.0 \cdot 10^6$	$4.4 \cdot 10^4$
75	Abdominal aorta	5.3	0.57	0.080	$4.0 \cdot 10^6$	$4.4 \cdot 10^4$
49	Celiac artery	1.0	0.39	0.064	$4.0 \cdot 10^6$	$4.4 \cdot 10^4$
61	Gastric artery	7.1	0.18	0.045	$4.0 \cdot 10^6$	$4.4 \cdot 10^4$
62	Splenic artery	6.3	0.28	0.054	$4.0 \cdot 10^6$	$4.4 \cdot 10^4$
63	Hepatic artery	6.6	0.22	0.049	$4.0 \cdot 10^6$	$4.4 \cdot 10^4$
64	Renal artery	3.2	0.26	0.053	$4.0 \cdot 10^6$	$4.4 \cdot 10^4$
66	Superior mesenteric	5.9	0.43	0.069	$4.0 \cdot 10^6$	$4.4 \cdot 10^4$
67	Gastric artery	3.2	0.26	0.053	$4.0 \cdot 10^6$	$4.4 \cdot 10^4$
83	Inferior mesenteric	5.0	0.16	0.043	$4.0 \cdot 10^6$	$4.4 \cdot 10^4$
4	L. common carotid	8.9	0.37	0.063	$4.0 \cdot 10^6$	$4.4 \cdot 10^4$
10	L. common carotid	8.9	0.37	0.063	$4.0 \cdot 10^6$	$4.4 \cdot 10^4$
20	L. common carotid	3.1	0.37	0.063	$4.0 \cdot 10^6$	$4.4 \cdot 10^4$
12	R. common carotid	8.9	0.37	0.063	$4.0 \cdot 10^6$	$4.4 \cdot 10^4$
22	R. common carotid	8.9	0.37	0.063	$4.0 \cdot 10^6$	$4.4 \cdot 10^4$
3	L. subclavian artery	3.4	0.42	0.067	$4.0 \cdot 10^6$	$4.4 \cdot 10^4$
6	Brachiocephalic artery	3.4	0.62	0.086	$4.0 \cdot 10^6$	$4.4 \cdot 10^4$
82,84	Common iliac	5.8	0.52	0.076	$4.0 \cdot 10^6$	$4.4 \cdot 10^4$
89,92	External iliac	8.3	0.29	0.055	$4.0 \cdot 10^6$	$4.4 \cdot 10^4$
90,91	Internal iliac	5.0	0.20	0.040	$16.0 \cdot 10^6$	$1.78 \cdot 10^5$
98,99	External iliac	6.1	0.27	0.053	$4.0 \cdot 10^6$	$4.4 \cdot 10^4$
104,107	Femoral artery	12.7	0.24	0.050	$8.0 \cdot 10^6$	$8.9 \cdot 10^4$
105,106	Profundis artery	12.6	0.23	0.049	$16.0 \cdot 10^6$	$1.78 \cdot 10^5$
109,110	Femoral artery	12.7	0.24	0.050	$8.0 \cdot 10^6$	$8.9 \cdot 10^4$
111,112	Popliteal artery	9.4	0.20	0.047	$8.0 \cdot 10^6$	$8.9 \cdot 10^4$
113,114	Popliteal artery	9.4	0.20	0.050	$4.0 \cdot 10^6$	$4.4 \cdot 10^4$
115,118	Anterior tibial artery	2.5	0.13	0.039	$16.0 \cdot 10^6$	$1.78 \cdot 10^5$
119,124	Anterior tibial artery	15.0	0.10	0.020	$16.0 \cdot 10^6$	$1.78 \cdot 10^5$
125,128	Anterior tibial artery	15.0	0.10	0.020	$16.0 \cdot 10^6$	$1.78 \cdot 10^5$
116,117	Posterior tibial artery	16.1	0.18	0.045	$16.0 \cdot 10^6$	$1.78 \cdot 10^5$
121,122	Posterior tibial artery	16.1	0.18	0.045	$16.0 \cdot 10^6$	$1.78 \cdot 10^5$
120,123	Peroneal artery	15.9	0.13	0.039	$16.0 \cdot 10^6$	$1.78 \cdot 10^5$
126,127	Peroneal artery	15.9	0.13	0.019	$16.0 \cdot 10^6$	$1.78 \cdot 10^5$

Table 9.2. Part B. Geometric and mechanical parameters of the arterial segments

Segment	Description	L [cm]	R_0 [cm]	h_0 [cm]	$E \left[\frac{dyn}{cm^2} \right]$	$k \left[\frac{dyn\ s}{cm^2} \right]$
31,37	Carotid (internal)	5.9	0.18	0.045	$8.0 \cdot 10^6$	$8.9 \cdot 10^4$
32,36	External carotid	11.8	0.15	0.042	$8.0 \cdot 10^6$	$8.9 \cdot 10^4$
33,35	Superior thyroid artery	4.0	0.07	0.020	$8.0 \cdot 10^6$	$8.9 \cdot 10^4$
43,56	Lingual artery	3.0	0.10	0.030	$8.0 \cdot 10^6$	$8.9 \cdot 10^4$
44,55	Internal carotid	5.9	0.13	0.039	$8.0 \cdot 10^6$	$8.9 \cdot 10^4$
45,54	Facial artery	4.0	0.10	0.030	$16.0 \cdot 10^6$	$1.78 \cdot 10^5$
46,53	Middle cerebral	3.0	0.06	0.020	$16.0 \cdot 10^6$	$1.78 \cdot 10^5$
47,52	Cerebral artery	5.9	0.08	0.026	$16.0 \cdot 10^6$	$1.78 \cdot 10^5$
48,51	Opthalmic artery	3.0	0.07	0.020	$16.0 \cdot 10^6$	$1.78 \cdot 10^5$
60,68	Internal carotid	5.9	0.08	0.026	$16.0 \cdot 10^6$	$1.78 \cdot 10^5$
73,77	Superficial temporal	4.0	0.06	0.020	$16.0 \cdot 10^6$	$1.78 \cdot 10^5$
74,76	Maxilliary artery	5.0	0.07	0.020	$16.0 \cdot 10^6$	$1.78 \cdot 10^5$
7,15	Internal mammary	15.0	0.10	0.030	$8.0 \cdot 10^6$	$8.9 \cdot 10^4$
8,14	Subclavian artery	6.8	0.40	0.066	$4.0 \cdot 10^6$	$4.4 \cdot 10^4$
9,13	Vertebral artery	14.8	0.19	0.045	$8.0 \cdot 10^6$	$8.9 \cdot 10^4$
16,26	Costo-cervical artery	5.0	0.10	0.030	$8.0 \cdot 10^6$	$8.9 \cdot 10^4$
17,25	Axilliary artery	6.1	0.36	0.062	$4.0 \cdot 10^6$	$4.4 \cdot 10^4$
18,24	Suprascapular	10.0	0.20	0.052	$8.0 \cdot 10^6$	$8.9 \cdot 10^4$
19,23	Thyrocervical	5.0	0.10	0.030	$8.0 \cdot 10^6$	$8.9 \cdot 10^4$
27,41	Thoraco-acromial	3.0	0.15	0.035	$16.0 \cdot 10^6$	$1.78 \cdot 10^5$
28,40	Axilliary artery	5.6	0.31	0.057	$4.0 \cdot 10^6$	$4.4 \cdot 10^4$
29,39	Circumflex scapular	5.0	0.10	0.030	$16.0 \cdot 10^6$	$1.78 \cdot 10^5$
30,38	Subscapular	8.0	0.15	0.035	$16.0 \cdot 10^6$	$1.78 \cdot 10^5$
42,57	Brachial artery	6.3	0.28	0.055	$4.0 \cdot 10^6$	$4.4 \cdot 10^4$
58,70	Profunda brachi	15.0	0.15	0.035	$8.0 \cdot 10^6$	$8.9 \cdot 10^4$
59,69	Brachial artery	6.3	0.26	0.053	$4.0 \cdot 10^6$	$4.4 \cdot 10^4$
71,79	Brachial artery	6.3	0.25	0.052	$4.0 \cdot 10^6$	$4.4 \cdot 10^4$
72,78	Superior ulnar collateral	5.0	0.07	0.020	$16.0 \cdot 10^6$	$1.78 \cdot 10^5$
80,86	Inferior ulnar collateral	5.0	0.06	0.020	$16.0 \cdot 10^6$	$1.78 \cdot 10^5$
81,85	Brachial artery	4.6	0.24	0.050	$4.0 \cdot 10^6$	$4.4 \cdot 10^4$
87,94	Ulnar artery	6.7	0.21	0.049	$8.0 \cdot 10^6$	$8.9 \cdot 10^4$
88,93	Radial artery	11.7	0.16	0.043	$8.0 \cdot 10^6$	$8.9 \cdot 10^4$
95,102	Ulnar artery	8.5	0.19	0.046	$8.0 \cdot 10^6$	$8.9 \cdot 10^4$
96,101	Interossea artery	7.9	0.09	0.028	$16.0 \cdot 10^6$	$1.78 \cdot 10^5$
97,100	Radial artery	11.7	0.16	0.043	$8.0 \cdot 10^6$	$8.9 \cdot 10^4$
103,108	Ulnar artery	8.5	0.19	0.046	$8.0 \cdot 10^6$	$8.9 \cdot 10^4$

Table 9.3. Windkessel terminals corresponding to each arterial segment (for numbers see Table 9.2)

Terminal	R_c $\left[\frac{\text{dyn s}}{\text{cm}^2\,\text{ml}}\right]$	R_a $\left[\frac{\text{ml cm}^2}{\text{dyn}}\right]$	C_a	Terminal	R_c $\left[\frac{\text{dyn s}}{\text{cm}^2\,\text{ml}}\right]$	R_a $\left[\frac{\text{ml cm}^2}{\text{dyn}}\right]$	C_a
125,128	62781.6	251356	1.00E-006	30,38	84015.4	198562	1.00E-006
126,127	31792.6	127400	1.00E-006	29,39	126252	506156	0.00E+000
121,122	21693	86769.2	2.00E-006	27,41	84015.4	198562	1.00E-006
105,106	12280.8	49008.4	4.00E-006	58,70	35120.4	140280	1.00E-006
90,91	15724.8	62781.6	3.00E-006	72,78	252504	1007720	0.00E+000
83	31103.8	125104	2.00E-006	80,86	370720	1480640	0.00E+000
64	3971.8	15839.6	1.20E-005	97,100	33055.4	131990.6	1.00E-006
66	3329.2	13314	1.40E-005	96,101	159530	635852	0.00E+000
67	3971.8	15839.6	1.20E-005	103,108	22265.6	89065.2	2.00E-006
61	23758	94918.6	2.00E-006	33,35	44303	176750	1.00E-006
62	8608.6	34433	5.00E-006	48,51	62666.8	250208	1.00E-006
63	6851.6	27431.6	7.00E-006	47,52	51189.6	204302	1.00E-006
9,13	22037.4	88032	2.00E-006	46,53	92163.4	368424	1.00E-006
7,15	89409.6	358092	1.00E-006	43,56	22265.6	88950.4	2.00E-006
19,23	89409.6	358092	1.00E-006	45,54	31448.2	126252	1.00E-006
18,24	20774.6	83326.6	2.00E-006	74,76	62666.8	250208	1.00E-006
16,26	89409.6	358092	1.00E-006	73,77	92163.4	368424	1.00E-006

at the venules. Therefore, it is

$$P_{w,k} = \begin{cases} P_{wub,k} & k = 1,\dots,N_{wub}, \\ P_{wlb,k} & k = 1,\dots,N_{wlb}, \end{cases} \qquad (9.63)$$

where $P_{wub,k}$ and $P_{wlb,k}$ are the pressures at the venules in the upper and lower body parts, respectively.

The values adopted for the parameters R_c, R_a and C_a used in the Windkessel terminals are given in Table 9.3.

9.4.5 0D model of the venous and pulmonary circulation.

The models to simulate the blood flow through the venules, veins, superior and inferior vena cava and also the pulmonary arteries and pulmonary veins, are mathematically formulated in terms of an electric analog model. A single compartment is shown in Fig. 9.6 in which R and L denote the resistance and inertance of the circuit, respectively; C is the compliance of the compartment, P_i and P_o are the pressures at the input and output of the compartment, P_{ex} is the external pressure which can also be a function of time and, finally, Q_i and Q_o are the blood inflow and outflow of the compartment.

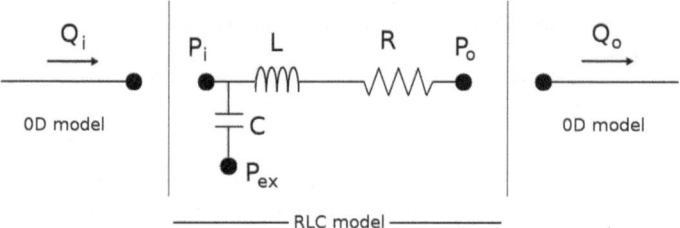

Fig. 9.6. Single-compartment circuit representation

The governing equations for this generic single compartment are

$$L\frac{dQ_o}{dt} + RQ_o = P_i - P_o, \tag{9.64}$$

$$C\frac{d}{dt}(P_i - P_{ex}) = Q_i - Q_o. \tag{9.65}$$

In particular, the coupling between the venules and the Windkessel models is carried out using the following equations

$$\sum_{k=1}^{N_{wub}} Q_{wub,k} = Q_{ubc} \qquad P_{wub,k} = P_{ubc} \qquad k = 1,\dots,N_{wub}, \tag{9.66}$$

$$\sum_{k=1}^{N_{wlb}} Q_{wlb,k} = Q_{lbc} \qquad P_{wlb,k} = P_{lbc} \qquad k = 1,\dots,N_{wlb}. \tag{9.67}$$

In this case the upper and lower venules compartments receive $(Q_i, P_i) = (Q_{ubc}, P_{ubc})$ and $(Q_i, P_i) = (Q_{lbc}, P_{lbc})$, respectively. The equations for the remaining compartments are easily obtained from (9.64)–(9.65) by accommodating the notation introduced in Table 9.1.

Table 9.4 and Table 9.5 present the data used in the 0D model of the venous and pulmonary circulation, respectively.

9.4.6 Elastance model for the cardiac chambers

Elastance-based modelling of the heart has been adopted in this study to describe each of the four cardiac chambers. The blood pressure in each cardiac chamber, denoted by P_i, is given by

$$P_i - P_{ex} = (E_A e(t) + E_B)(V_{ch} - V_{0,ch}) + \alpha_{ch}|P_i|\frac{dV_{ch}}{dt}, \tag{9.68}$$

where V_{ch} is the cardiac volume of the chamber and $V_{0,ch}$ refers to the dead volume of such chamber, α_{ch} is a coefficient that incorporates viscoelastic behaviour, E_A is the amplitude of elastance, E_B is the baseline value of elastance, and $e(t)$ is a normalized

Table 9.4. Data used in the 0D model of the venous circulation system

| | Lower body | | | Upper body | | |
	Venules	Veins	Inferior cava	Venules	Veins	Superior cava
R [dyn cm^{-2} s ml^{-1}]	53.32	11.997	0.6665	186.62	39.99	0.6665
L [dyn cm^{-2} s^2 ml^{-1}]	1.333	0.6665	0.6665	1.333	0.6665	0.6665
C [ml dyn^{-1} cm^2]	0.00112528	0.05626407	0.01125281	0.00037509	0.01125281	0.00375094

Table 9.5. Data used in the 0D models for the compartments in the pulmonary circulation

	Pulmonary arteries	Pulmonary veins
R [dyn cm^{-2} s ml^{-1}]	106.64	13.33
L [dyn cm^{-2} s^2 ml^{-1}]	0.0	0.0
C [ml dyn^{-1} cm^2]	0.00309077	0.060015

time-varying function of the elastance which for ventricles is (see [40])

$$e_v(t) = \begin{cases} \frac{1}{2}\left[1 - \cos\left(\pi \frac{t}{T_{vc}}\right)\right] & 0 \leq t \leq T_{vc}, \\ \frac{1}{2}\left[1 + \cos\left(\pi \frac{(t - T_{vc})}{T_{vr}}\right)\right] & T_{vc} < t \leq T_{vc} + T_{vr}, \\ 0 & T_{vc} + T_{vr} < t \leq T, \end{cases} \tag{9.69}$$

and for atria is

$$e_a(t) = \begin{cases} \frac{1}{2}\left[1 + \cos\left(\pi \frac{(t + T - t_{ar})}{T_{ar}}\right)\right] & 0 \leq t \leq t_{ar} + T_{ar} - T, \\ 0 & t_{ar} + T_{arp} - T < t \leq t_{ac}, \\ \frac{1}{2}\left[1 - \cos\left(\pi \frac{(t - t_{ac})}{T_{ac}}\right)\right] & t_{ac} < t \leq t_{ac} + T_{ac}, \\ \frac{1}{2}\left[1 + \cos\left(\pi \frac{(t - t_{ar})}{T_{ar}}\right)\right] & t_{ac} + T_{ac} < t \leq T. \end{cases} \tag{9.70}$$

Here, the subscript v denotes the ventricles and a the atria, T is the duration of a cardiac cycle, T_{vc}, T_{ac}, T_{vr} and T_{ar} refer to the times of ventricular/atrial contraction/relaxation, and t_{ac}, t_{ar} are the instants at which the atria begin to contract and relax, respectively.

Furthermore, the volume is related to the inflow and outflow as usual

$$\frac{dV_{ch}}{dt} = Q_i - Q_o. \tag{9.71}$$

Table 9.6 shows the data used in the computational implementation of the elastance model of the right and left halves of the heart.

Table 9.6. Data used in the elastance model of the right and left halves of the heart

Chambers	Right atrium	Right ventricle	Left atrium	Left ventricle
E_A [dyn cm^{-2} ml^{-1}]	79.98	733.15	93.31	3665.75
E_B [dyn cm^{-2} ml^{-1}]	93.31	66.65	119.97	106.64
T_c [s]	0.17	0.34	0.17	0.34
T_r [s]	0.17	0.15	0.17	0.15
t_c [s]	0.80	–	0.80	–
t_r [s]	0.97	–	0.97	–
V_0 [ml]	4.0	10.0	4.0	5.0
α	0.0005	0.0005	0.0005	0.0005

9.4.7 Non-ideal diode model for the heart valves

The momentum balance in each heart valve is such that we take into account the non-linear behaviour by which the flow can be inverted when the valve closes. The model employed is shown in Fig. 9.7 and has been partially inspired by the valve model presented in [31].

The governing equation for this valve model is the following

$$L\frac{dQ_o}{dt} + RQ_o + B|Q_o|Q_o = \Xi(P_i - P_o), \qquad (9.72)$$

where L is the inertance of the fluid, R is the viscous resistance, B accounts for the flow separation phenomenon and P_i and P_o are the input and output pressure values in the compartment. The non-binary state of the valve depends upon the presence of the coefficient Ξ. This coefficient simulates the behaviour of the orifice of the valve, and is a function of the opening angle of the valve, denoted by θ, as follows

$$\Xi = \frac{(1 - \cos\theta)^4}{(1 - \cos\theta_{max})^4}, \qquad (9.73)$$

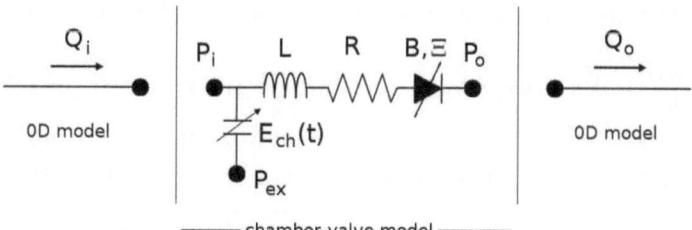

Fig. 9.7. Generic chamber-valve element with a non-ideal diode

Table 9.7. Data used in the non-ideal diode models of the cardiac valves

Valves	Tricuspid	Pulmonar	Mitral	Aortic
R [dyn cm^{-2} s ml^{-1}]	0.006	0.006	0.006	0.006
L [dyn cm^{-2} s^2 ml^{-1}]	0.005	0.005	0.005	0.005
B [dyn cm^{-2} s^2 ml^{-2}]	0.0064	0.00756	0.0064	0.00756
θ_{max} [°]	75.0	75.0	75.0	75.0
θ_{min} [°]	5.0	5.0	5.0	5.0
k_P/I [rad s^{-2} dyn^{-1} cm^2]	4.126032	4.126032	4.126032	4.126032
k_F/I [s^{-1}]	50.0	50.0	50.0	50.0
k_Q/I [rad s^{-1} ml^{-1}]	2.0	2.0	2.0	2.0
k_V/I [rad s^{-1} ml^{-1}]	3.5	3.5	3.5	7.0

where θ_{max} is the maximum angle of opening. A heuristic angular momentum balance equation for θ is used here (see [31] for the derivation and details) as follows:

$$I\frac{d^2\theta}{dt^2} + k_F\frac{d\theta}{dt} = k_P(P_i - P_o) + k_Q Q_o \cos\theta + \begin{cases} k_V Q_o \sin(2\theta) & \text{if } P_i \geq P_o \\ 0 & \text{if } P_i < P_o \end{cases}, \quad (9.74)$$

where I is the momentum of inertia of the valve. The solution of this balance equation is constrained to

$$\theta = \begin{cases} \theta_{min} & \text{if } \theta < \theta_{min}, \\ \theta_{max} & \text{if } \theta > \theta_{max}. \end{cases} \quad (9.75)$$

In this way, a valve can undergo malfunctioning in two distinct ways (or combination of them): by *stenosis*, if the valve is narrowed (reduced value of θ_{max}), and by *incompetence* or *insufficiency*, when the valve is leaky and fails to prevent prominent backward flow (increased value of θ_{min}). The data used in the computational implementation is presented in Table 9.7

9.4.8 3D models

As already stated, we consider the existence of 3D models accounting for all the complexity of three-dimensional blood flow in specific vessels of interest. Then we make use of the Navier-Stokes equations in moving domains (ALE formulation), that is

$$\rho\frac{\partial \mathbf{u}_{3D}}{\partial t} + \rho\nabla\mathbf{u}_{3D}(\mathbf{u}_{3D} - \mathbf{v}_{FR}) - \mu\triangle\mathbf{u}_{3D} + \nabla P_{3D} = \mathbf{f} \qquad \text{in } \Omega, \quad (9.76)$$

$$\text{div}\,\mathbf{u}_{3D} = 0 \qquad \text{in } \Omega, \quad (9.77)$$

plus proper coupling conditions (see Sect. 9.3) at Γ_i $\qquad i = 1,\ldots,N_{cf},$ $\quad (9.78)$

where $\mathbf{u}_{3D}$ is the fluid velocity, $\mathbf{v}_{FR}$ the velocity of the frame of reference consistent with the ALE formulation, P_{3D} the pressure field, $\mathbf{f}$ the volume body force, ρ and μ

are the density and the viscosity, respectively, and Γ_i, $i = 1,\ldots,N_{cf}$, the interfaces in the 3D model to be coupled with the 1D model of the arterial tree. This set of equations must be provided with proper coupling boundary conditions (naturally given by the proposed non-classical extended variational formulation) and a proper constitutive relation relating the displacement of the arterial wall with the pressure (structural model). The coupling boundary conditions for $\gamma = 1$ are given by (see (9.51))

$$Q_{sa|3D,i} = \int_{\Gamma_i} \mathbf{u}_{3D} \cdot \mathbf{n} \, d\Gamma_i \qquad\qquad i = 1,\ldots,N_{cf}, \qquad (9.79)$$

$$P_{sa|3D,i}\mathbf{n} = \left[(P_{3D}\mathbf{I} - 2\mu\boldsymbol{\varepsilon}(\mathbf{u}_{3D}))\mathbf{n}\right]_{|\Gamma_i} \qquad i = 1,\ldots,N_{cf}, \qquad (9.80)$$

where the pairs $(Q_{sa|3D,i}, P_{sa|3D,i})$, $i = 1,\ldots,N_{cf}$, denote the restriction of the flow rate and pressure in the systemic arteries which converge to the N_{cf} coupling interfaces of the 3D model. As said in Sect. 9.4.1, here we observe the consequences of having neglected the viscous effects due to axial gradients in the 1D model. On the other hand, for the constitutive equation for the arterial wall, denoted by $\partial\Omega_w$, we choose an independent rings wall model consistent to that used for the 1D model, yielding

$$P_{3D} - P_0 = \frac{Eh_0}{R_0^2}\delta_{3D} + \frac{kh_0}{R_0^2}\frac{\partial\delta_{3D}}{\partial t}, \qquad (9.81)$$

$$\mathbf{u}_{3D} = \frac{\partial\delta_{3D}}{\partial t}\mathbf{n} = \mathbf{v}_{FR}, \qquad (9.82)$$

where $\delta_{3D}\mathbf{n} = \mathbf{w}_{FR}$ is the displacement of the surface $\partial\Omega_w$ in the direction of its normal $\mathbf{n}$. The displacement $\mathbf{w}_{FR}$ of the frame of reference is extended to the interior of the domain by considering its harmonic extension, obtained by solving a Laplace problem in the reference fluid domain.

Other constitutive behaviours for the blood can be incorporated in the model, giving rise to an equation similar to (9.76) but now valid for non-Newtonian fluids, like the one obtained when the regularized Casson model is used. This fact does not affect the coupling equations derived from Problem 3.

9.5 Numerical applications

In this section we present two study cases (one open-loop, one closed-loop) to show the capabilities of the heterogeneous approach for modelling the cardiovascular system. These cases address two pathological situations in which aneurisms have developed in the abdominal aorta and in the cerebral artery, respectively. The blood flow is modelled in both cases as a Newtonian fluid.

We approximate the problem using the finite-element method for both the 3D and the 1D models. For the 3D problem the *mini* element is used with SUPG stabilization, and a Crank-Nicolson finite-difference discretization in time is performed. In particular, in view of our independent ring model for the structure, our fluid-structure

problem is solved monolithically. The non-linearities are treated using Picard iterations. For the 1D model the finite element is applied to the equations written along
the characteristic lines with stabilization provided by least-square terms. The interested reader is referred to [6, 60] for a more detailed account of these well-known
numerical techniques that, not being the main focus of our investigation, will not
be discussed any further. To deal with the heterogeneous coupling we perform a
partition of the problem into two subproblems: (i) the stand-alone 3D model under
analysis and (ii) the complementary 1D-0D model (open loop or closed loop, depending on the case). For the coupling between these two submodels we employ
different techniques that will be commented in due course.

9.5.1 Blood flow in an abdominal aneurism

In this case we restrict the model to the open-loop representation of the CVS (see
Sect. 9.4.2). The geometry was obtained from medical images through standard segmentation procedures. Fig. 9.8 displays the final geometry and its posititiong within
the arterial network. Also in that figure the flow rate boundary condition at the aortic
root is shown.

    The finite-element mesh of the 3D model consists of 0.8M nodes and 4.7M tetra-
hedra. The time step used in the simulation was $6.25 \cdot 10^{-4}$ s. Two cardiac cycles,
of period $T = 0.8$ s, were run from at-rest initial conditions. In this case the coupling scheme for the interactions between the 3D and the open-loop 1D-0D models
is explicit, that is, we solved at each time step the 1D model and the 3D model just
once. Based on past experience, in fact, we have found that for the flow regimes
encountered in the abdominal aorta, together with the geometrical features of the
1D segment that is being substituted by a 3D geometry, makes the explicit approach
feasible, reliable and, evidently, computationally cheaper. The results presented correspond to the second heart beat.

    In Fig. 9.8 the flow rate and pressure at the coupling interfaces are presented. No-
tice the increased pressure drop in the left common iliac as a result of the reduction
in the arterial lumen right after the aneurism in that branch. Another remarkable fact
is the change in the flow rate at proximal and distal coupling interfaces. Indeed, the
aneurism acts as a large capacitor, changing the way in which the blood flow is released to the iliacs. Specifically, notice that the solutions obtained at these coupling
interfaces differ significantly from what would be expected in an aneurysm-free geometry like the one analyzed in [8].

    Concerning the 3D blood flow, the sequence shown in Fig. 9.9 presents the blood
flow patterns in the aneurismal region for several time instants throughout the cardiac cycle. Note that the blood flow is actually a jet flow impiging on the posterior
arterial wall and producing a quite complex secondary circulation in the anterior
region of the aneurism.

    We also carried out the calculation of the WSS and OSI hemodynamics indices.
Fig. 9.10 presents these two fields. Notice that the complex flow observed in Fig. 9.9
is manifest in the structure of the OSI field, which features larger values on the lateral zones of the arterial wall as a result of the vortical pattern induced by the jet

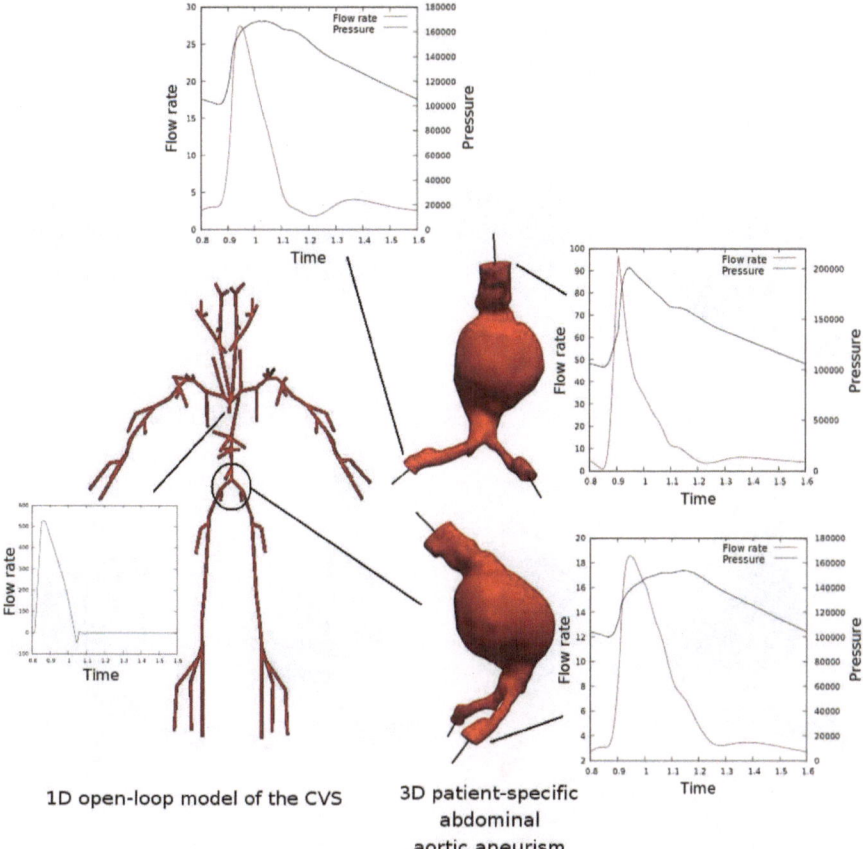

Fig. 9.8. Placement of the abdominal aneurism in the open-loop model of the CVS and results at coupling interfaces

flow entering the aneurism. In the aneurismal region, the WSS remains bounded in the posterior wall by 25 dyn/cm², whereas in the anterior wall this value reduces to 10 dyn/cm². These two indices are recognized to play a role in aneurism growth and rupture [15, 65] as well as in other pathologies [26, 32, 37, 42]. Here we point out another aspect of heterogeneous models, i.e. that they provide a physiologically realistic context to carry out 3D hemodynamics simulations.

9.5.2 Blood flow in a cerebral aneurism

In this application we employ the entire closed-loop representation of the CVS (see Sect. 9.4.2). As in the previous case, the geometry was obtained using standard segmentation procedures. In Fig. 9.11 we present the final geometry and its placement within the arterial network. Notice the bleb present in the aneurism.

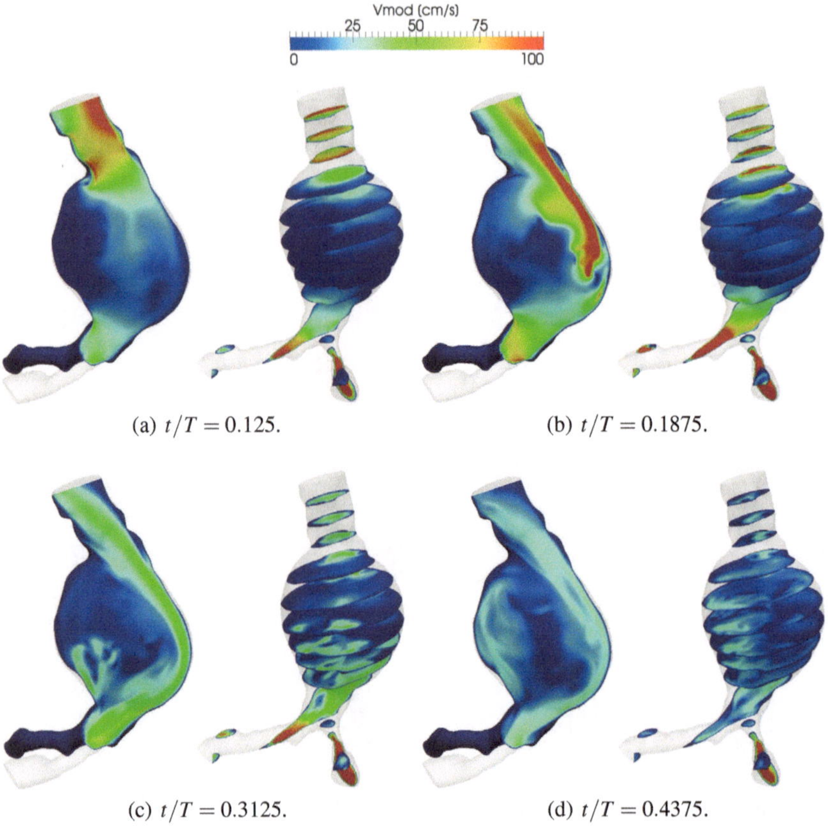

Vmod (cm/s)

(a) $t/T = 0.125$.

(b) $t/T = 0.1875$.

(c) $t/T = 0.3125$.

(d) $t/T = 0.4375$.

Fig. 9.9. Blood flow pattern in the aneurismal region

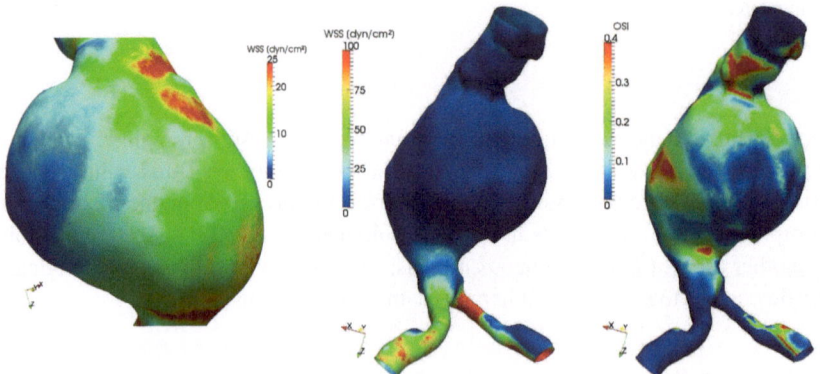

Fig. 9.10. WSS and OSI hemodynamics indexes in the abdominal aortic aneurism

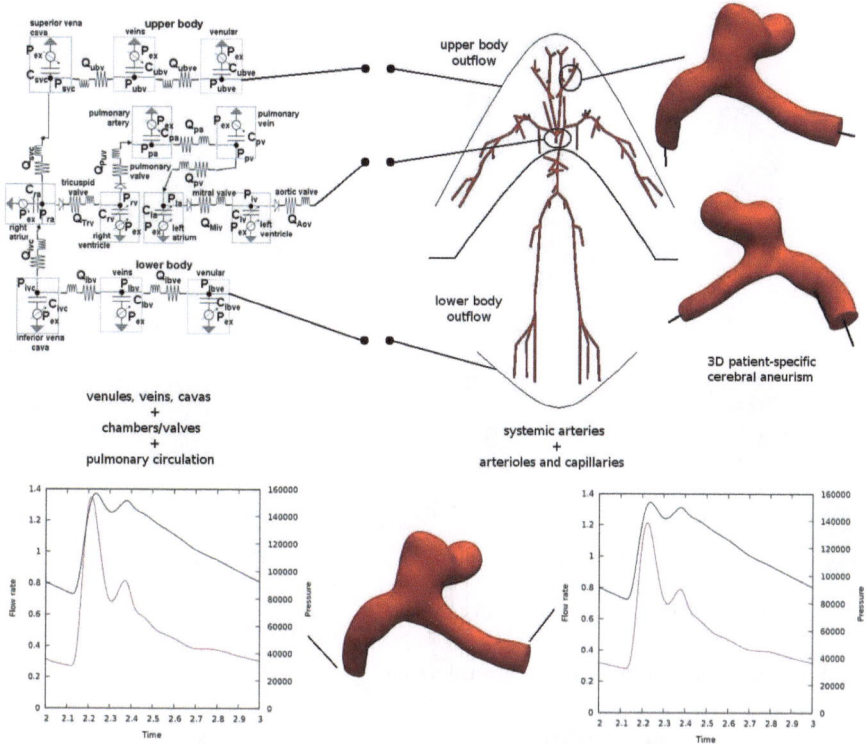

Fig. 9.11. Placement of the cerebral aneurism in the closed-loop model of the CVS and results at coupling interfaces

In this example, the finite-element mesh of the 3D model consists of 20K nodes and 120K tetrahedra, and the time step used was $1.25 \cdot 10^{-3}$ s. Three cardiac cycles, of period $T = 1.0$ s, were run from at-rest conditions. To perform the coupling between the 3D and the 1D-0D closed-loop model we used an iterative method based on a Broyden algorithm such that the equilibrium between the models at coupling interfaces is achieved at each time step. In this case, the explicit approach to couple the heterogeneous models results in an unstable coupling scheme even for time steps 30 times smaller than the one used here. In contrast to the previous example, in this case the geometrical features of the 1D cerebral artery, replaced by a 3D model, are significantly different. This motivated the use of strong coupling schemes. The interested reader is referred to [10, 36] for a detailed account of such coupling strategies.

The results at the coupling interfaces are presented in Fig. 9.11. In this case the pressure pulse is almost the same at proximal and distal locations, while the flow rate exhibits a decrease in the peak, which is the result of the presence of the enlarged capacitance of the arterial segment due to the aneurism (same effect that in the previous example).

Concerning the local blood flow, Fig. 9.12 presents a set of streamlines computed at different time instants. A stagnation zone can be observed in the deepest part of

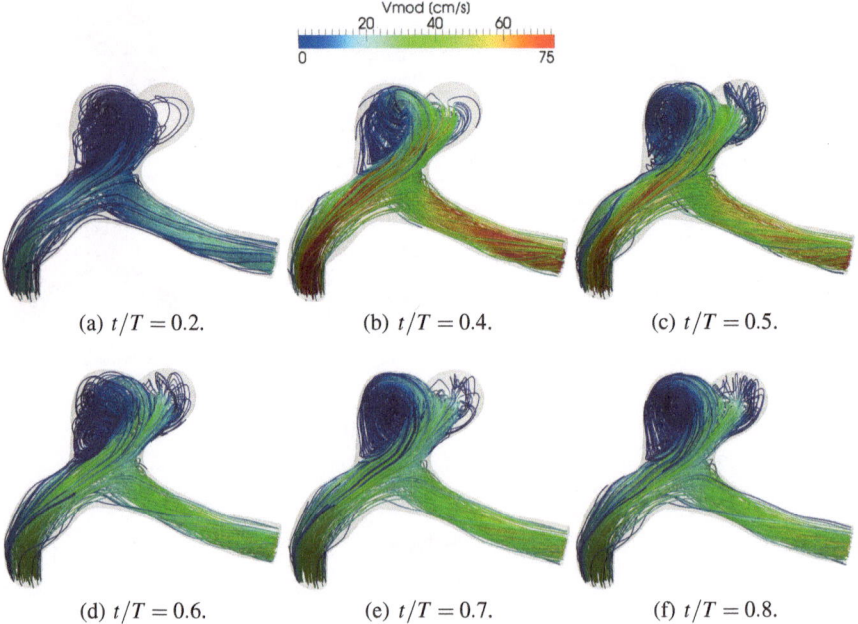

(a) $t/T = 0.2$.   (b) $t/T = 0.4$.   (c) $t/T = 0.5$.

(d) $t/T = 0.6$.   (e) $t/T = 0.7$.   (f) $t/T = 0.8$.

Fig. 9.12. Blood flow pattern in the aneurismal region

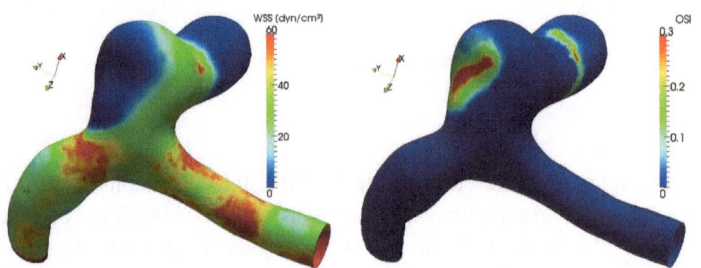

Fig. 9.13. WSS and OSI hemodynamics indices in the cerebral aneurism

the aneurism, inside the bleb, which leads to an oscillatory behaviour of the shear stresses as seen in Fig. 9.13, as well as to low values of shear stresses. The main vortical structure observed inside the aneurism in Fig. 9.12 produces also large values of the OSI index on the lateral part of the arterial wall.

The positioning of the aneurism is such that there is a jet inflow induced in the intra-aneurismal region, producing a marked main stream inside the aneurism and generating a large WSS region in the impaction zone (see Fig. 9.13). The bleb is adjacent to the impaction zone, near the region of large WSS over the aneurismal wall. This may suggest that the bleb occured in an impaction region with a former large value of WSS (see [14] and references therein for a complete discussion on this topic).

9.6 Conclusions

In this work a variational foundation for the problem of coupling dimensionally-heterogeneous models has been explored. It turned out that such an approach can be understood in the context of kinematical incompatibilities between models, where the heterogeneities arise as the result of kinematical hypotheses. The consequences of different considerations regarding the sense in which the models are coupled were discussed. Further attention was given to the application of these models in the hemodynamics field, which were employed to account for the modelling of either an open-loop model and a closed-loop model of the CVS with embedded 3D patient-specific geometries.

Two examples of pathophysiological scenarios were also studied that involved an abdominal aneurism and a cerebral aneurism, respectively. It is worth remarking that such an approach made it possible to set up 3D numerical simulations under quite realistic physiological conditions posed by the surrounding cardiovascular network, under the assumption that our model of the CVS is correctly calibrated. Particularly, in the case of the abdominal aortic aneurism, it was observed that the solutions at the coupling interfaces were substantially different from a hypothetical pure 1D solution, sustaining the need for this kind of heterogeneous modelling to correctly consider geometrical singularities present in the vasculature.

Therefore this class of formulations is able to deliver realistic cardiovascular simulations as well as versatility in the setting of different physiological and pathophysiological scenarios of interest. For instance, a whole area of research may be readily accounted for with this model, ranging from the study of the sensitivity of local blood flow features, and related hemodynamics variables, to alterations in global cardiovascular parameters and viceversa.

Acknowledgements. This work was partially supported by the Brazilian agencies CNPq and FAPERJ. The support of these agencies is gratefully acknowledged.

References

[1] Avolio A.P.: Multi-branched model of the human arterial system. Med. Biol. Engrg. Comp. **18**(6): 709–718, 1980.
[2] Bernardi C., Dauge M., Maday Y.: *Spectral Methods for Axisymmetric Domains*, volume Series in Applied Mathematics 3. Gauthier-Villars, Editions Scientifiques et Médicales Elsevier, Paris, 1999.
[3] Blanco P.J.: Kinematical incompatibility, immersed domains and multiscale constitutive modeling: Nexus with the modeling of the cardiovascular system (in portuguese). Ph.D. thesis, Laboratório Nacional de Computação Científica, Petrópolis, Brasil, 2008.
[4] Blanco P.J., Feijóo R.A.: Sensitivity analysis in kinematically incompatible models. Computer Methods in Applied Mechanics and Engineering **198**(41–44): 3287–3298, 2009.
[5] Blanco P.J., Discacciati M., Quarteroni A.: Modeling dimensionally-heterogeneous problems: Analysis, approximation and applications. Submitted to Numer. Math., 2010.
[6] Blanco P.J., Feijóo R.A., Urquiza S.A.: A unified variational approach for coupling 3D-1D models and its blood flow applications. Comp. Meth. Appl. Mech. Engrg. **196**(41–44): 4391–4410, 2007.

[7] Blanco P.J., Feijóo R.A., Urquiza S.A.: A variational approach for coupling kinematically incompatible structural models. Comp. Meth. Appl. Mech. Engrg. **197**(17-18): 1577–1602, 2008.

[8] Blanco P.J., Urquiza S.A., Feijóo R.A.: Assessing the influence of heart rate in local hemodynamics through coupled 3D-1D-0D models. International Journal for Numerical Methods in Biomedical Engineering **26**(7): 890–903, 2010.

[9] Blanco P.J., Pivello M.R., Urquiza, S.A., Feijóo R.A.: On the potentialities of 3D-1D coupled models in hemodynamics simulations. Journal of Biomechanics **42**(7): 919–930, 2009.

[10] Blanco P.J., Leiva J.S., Buscaglia G.C.: Black-box decomposition approach for computational hemodynamics: One-dimensional models. Comp. Meth. Appl. Mech. Engrg. **200**(13-16): 1389–1405, 2011.

[11] Brezzi F., Fortin M.: Mixed and Hybrid Finite Element Methods. Springer-Verlag, New York, 1991.

[12] Caro C.G., Fitz-Gerald J.M., Schroter R.C.: Atheroma and arterial wall shear dependent mass transfer mechanism for atherogenesis. Proc. Roy. Soc. London Biol. **B177**: 109–159, 1971.

[13] Cebral J., Castro M., Appanaboyina S., Putman C., Millan D., Frangi A.: Efficient pipeline for image-based patient-specific analysis of cerebral aneurysm hemodynamics: Technique and sensitivity. IEEE Trans. Med. Imaging **24**: 457–467, 2005.

[14] Cebral J.R., Sheridan M., Putman C.M.: Hemodynamics and bleb formation in intracranial aneurysms. American Journal of Neuroradiology **31**: 304–310, 2010.

[15] Dhar S., Tremmel M., Mocco J., Kim M., Yamamoto J., Siddiqui A.H., Hopkins L.N., Meng H.: Morphology parameters for intracranial aneurysm rupture risk assessment. Neurosurgery **63**: 185–197, 2008.

[16] Fernández M.A., Milišić V., Quarteroni A.: Analisys of a geometrical multiscale blood flow model based on the coupling of ODEs and hyperbolic PDEs. SIAM J. on Multiscale Model Simul. **4**(1),: 215–236, 2005.

[17] Formaggia L., Nobile F., Quarteroni A., Veneziani A.: Multiscale modelling of the vascular system: A preliminary analysis. Comp. Vis. Sci. **2**: 75–84, 1999.

[18] Formaggia L., Gerbeau J.F., Nobile F., Quarteroni A.: On the coupling of 3D and 1D Navier-Stokes equations for flow problems in compliant vessels, Comput. Methods Appl. Mech. Engrg. **191**(6–7): 561–582, 2001.

[19] Formaggia L., Gerbeau J.F. Nobile F., Quarteroni A.: Numerical treatment of defective boundary conditions for the Navier-Stokes equations. SIAM J. Numer. Anal. **40**(1): 376–401, 2002.

[20] Giddens D.P., Zarins C.K., Glagov S.: The role of fluid mechanics in the localization and detection of atherosclerosis. J. Biomech. Engrg. **115**: 588–594, (1993).

[21] Grinberg L., Anor T., Madsen J., Yakhot A., Karniadakis G.: Large-scale simulation of the human arterial tree. Clinical and Experimental Pharmacology and Physiology **36**: 194–205, 2009.

[22] Heldt T., Shim E., Kamm R., Mark R.: Computational modeling of cardiovascular response to orthostatic stress. J. Appl. Physiol. **92**: 1239–1254, 2002.

[23] Heywood J.G., Rannacher R., Turek S.: Artificial boundaries and flux and pressure conditions for the incompressible Navier Stokes equations. Int. J. Numer. Methods Fluids **22**: 325–352, 1996.

[24] Hoppensteadt F., Peskin C.: Modeling and Simulation in Medicine and the Life Sciences. Texts in Applied Mathematics, Springer, 2002.

[25] Hughes T.J.R., Lubliner J.: On the one-dimensional theory of blood flow in the larger vessels. Math. Biosci. **18**(1–2): 161–170, 1973.

[26] Keynton R.S., Evancho M.M., Sims R.L., Rodway N.V., Gobin A., Rittgers S.E.: Intimal hyperplasia and wall shear in arterial bypass graft distal anastomoses: an in vivo model study. J. Biomech. Engng. **123**: 464-473, 2001.

[27] Kim H.J., Figueroa C.A., Hughes T.J.R., Jansen K.E., Taylor C.A.: Augmented lagrangian method for constraining the shape of velocity profiles at outlet boundaries for three-dimensional finite element simulations of blood flow. Comp. Meth. Appl. Mech. Engrg. **198**(45–46): 3551–3566, 2009.

[28] Kim H.J., Vignon-Clementel I.E., Figueroa C.A., LaDisa J.F., Jansen, K.E. Feinstein J.A., Taylor C.A.: On coupling a lumped parameter heart model and a three-dimensional finite element aorta model. Ann. Biomed. Engng. **37**(11): 2153–2169, 2009.

[29] Kivity Y., Collins R.: Nonlinear fluid-shell interactions: application to blood flow in large arteries. In: Proceedings of the International Symposium on Discrete Methods Engineering: 476–488, 1974.

[30] Kivity Y., Collins R.: Nonlinear wave propagation in viscoelastic tubes: application to aortic rupture. J. Biomech. **7**(1): 67–76, 1974.

[31] Korakianitis T., Shi Y.: Numerical simulation of cardiovascular dynamics with healthy and diseased heart valves. J. Biomech. **39**(11): 1964–1982, 2006.

[32] Ku D.N., Giddens D.P., Zarins C.K., Glagov S.: Pulsatile flow and atherosclerosis in the human carotid bifurcation. Arteriosclerosis **5**(3): 293–302, 1985.

[33] Kufahl R.H., Clark M.E.: A circle of willis simulation using distensible vessels and pulsatile flow. J. Biomech. Engrg. **107**(2): 112–122, 1985.

[34] Laganà K., Dubini G., Migliavacca F., Pietrabissa R., Pennati G., Veneziani A., Quarteroni A.: Multiscale modelling as a tool to prescribe realistic boundary conditions for the study of surgical procedures. Biorheology **39**(3–4): 359–364, 2002.

[35] Lanzarone E., Liani P., Baselli G., Costantino M.L.: Model of arterial tree and peripheral control for the study of physiological and assisted circulation. Medical Engineering and Physics **29**(5): 542–555, 2007.

[36] Leiva J.S., Blanco P.J., Buscaglia G.C.: Iterative strong coupling of dimensionally–heterogeneous models. Int. J. Num. Meth. Engrg. **81**: 1558–1580, (2010).

[37] Li X.M., Rittgers S.E.: Hemodynamic factors at the distal end-to-side anastomosis of a bypass graft with different POS:DOS flow ratios. J. Biomech. Engng. **123**: 270–276, (2001).

[38] Liang F., Liu H.: A closed-loop lumped parameter computational model for human cardiovascular system. JSME International Journal Series C **48**(4): 484–493, 2005.

[39] Liang F., Liu H.: Simulation of hemodynamic responses to the valsalva maneuver: An integrative computational model of the cardiovascular system and the autonomic nervous system. J. Physiol. Sci. **56**(1): 45–65, 2006.

[40] Liang F., Takagi S., Himeno R., Liu H.: Multi-scale modeling of the human cardiovascular system with applications to aortic valvular and arterial stenoses. Med. Biol. Eng. Comput. **47**: 743–755, 2009.

[41] Löhner R., Cebral J., Soto O., Yim P., Burgess, J.E.: Applications of patient-specific CFD in medicine and life sciences, Int. J. Numer. Meth. Fluids **43**(6–7): 637–650, 2003.

[42] Murphy J.B., Boyle F.J.: A numerical methodology to fully elucidate the altered wall shear stress in a stented coronary artery. Cardiovascular Engineering and Technology, **1**: 256–268 2010.

[43] Migliavacca F., Balossino R., Pennati G., Dubini G., Hsia T-Y., de Leval M.R., Bove E.L.: Multiscale modelling in biofluidynamics: Application to reconstructive paediatric cardiac surgery. J. Biomech. **39**(6): 1010–1020, 2006.

[44] Oden J.T.: Applied Functional Analysis. Prentice-Hall, New Jersey, 1979.

[45] Olufsen M.S., Ottesen J.T., Tran H.T., Ellwein L.M., Lipsitz L.A., Novak V.: Blood pressure and blood flow variation during postural change from sitting to standing: model development and validation. J. Appl. Physiol. **99**(4): 1523–1537, 2005.

[46] Olufsen, M.S., Peskin, C.S., Kim, W.Y., Pedersen, E.M., Nadım, A., Larsen, J.: Numerical simulation and experimental validation of blood flow in arteries with structured-tree outflow conditions. Ann. Biomed. Engng. **28**(11): 1281–1299, 2000.

[47] Oshima, M., Torii, R., Kobayashi, T., Taniguchi, N., Takagi, K.: Finite element simulation of blood flow in the cerebral artery. Comput. Methods Appl. Mech. Engrg. **191**(6–7): 661–671, 2001.

[48] Perktold K., Rappitsch G.: Computer simulation of local blood flow and vessel mechanics in a compliant carotid artery bifurcation model. J. Biomech. **28**(7): 845–856, 1995.

[49] Pontrelli G.: A multiscale approach for modelling wave propagation in an arterial segment. Comp. Meth. Biomech. Biom. Engrg. **7**(2): 79–89, 2004.

[50] Quarteroni A., Veneziani A.: Analysis of a geometrical multiscale model based on the coupling of ODEs and PDEs for blood flow simulations. SIAM J. on Multiscale Model. Simul. **1**(2): 173–195, 2003.

[51] Quarteroni A., Tuveri M., Veneziani A.: Computational vascular fluid dynamics: problems, models and methods. Computing and Visualization in Science **2**(4): 163–197, 2000.

[52] Reichold J., Stampanoni M., Lena Keller A., Buck A., Jenny P., Weber B.: Vascular graph model to simulate the cerebral blood flow in realistic vascular networks. J. Cereb. Blood Flow Metab. **29**(8): 1429–1443, 2009.

[53] Reymond P., Merenda F., Perren F., Rüfenacht D., Stergiopulos N.: Validation of a one-dimensional model of the systemic arterial tree. Am. J. Physiol. Heart Circ. Physiol. **297**(1): H208–H222, 2009.

[54] Schaaf B.W., Abbrecht P.H.: Digital computer simulation of human systemic arterial pulse wave transmission: A nonlinear model. J. Biomech. **5**(4): 345–364, 1972.

[55] Spencer M.P., Deninson A.B.: The square-wave electro-magnetic flowmeter. Theory of operation and design of magnetic probes for clinical and experimental applications. I.R.E. Trans. Med. Elect. **6**: 220–228, 1959.

[56] Stergiopulos N., Young D.F., Rogge T.R.: Computer simulation of arterial flow with applications to arterial and aortic stenoses. J. Biomech. **25**(12): 1477–1488, 1992.

[57] Stettler, J.C., Niederer, P., Anliker, M.: Theoretical analysis of arterial hemodynamics including the influence of bifurcations. Part I: mathematical models and prediction of normal pulse patterns. Ann. Biomed. Engrg. **9**(2): 145–164, (1981).

[58] Strang G., Fix G.J.: An Analysis of the Finite Element Method. Prentice-Hall, New York, 1973.

[59] Taylor C.A., Hughes T.J.R., Zarins C.K.: Finite element modeling of three-dimensional pulsatile flow in the abdominal aorta: relevance to atherosclerosis. Ann. Biomed. Engrg. **26**(6): 975–987, 1998.

[60] Urquiza S.A., Blanco P.J., Vénere M.J., Feijóo R.A.: Multidimensional modelling for the carotid artery blood flow. Comput. Methods Appl. Mech. Engrg. **195**(33–36): 4002–4017, 2006.

[61] van Heusden K., Gisolf J., Stok W.J., Dijkstra S., Karemaker J.M.: Mathematical modeling of gravitational effects on the circulation: importance of the time course of venous pooling and blood volume changes in the lungs. Am J. Physiol. Heart Circ. Physiol. **291**(5), H2152–H2165, 2006.

[62] Vignon-Clementel I.E., Figueroa C.A., Jansen K.E., Taylor C.A.: Outflow boundary conditions for three-dimensional finite element modeling of blood flow and pressure in arteries. Comput. Methods Appl. Mech. Engrg. **195**(29–32): 3776–3796, 2006.

[63] Vignon-Clementel I.E., Figueroa C.A., Jansen K.E., Taylor C.A.: Outflow boundary conditions for 3D simulations of non-periodic blood flow and pressure fields in deformable arteries. Comput. Methods Biomech. Biomed. Engrg. **13**: 625–640, 2010.

[64] Wang J.J., Parker K.H.: Wave propagation in a model of the arterial circulation. J. Biomech. **37**(4): 457–470, 2004.

[65] Xiang J., Natarajan S.K., Tremmel M., Ma D., Mocco J., Hopkins L.N., Siddiqui A.H., Levy E.I., Meng H.: Hemodynamic-morphologic discriminants for intracranial aneurysm rupture. Stroke **42**: 144–152, 2011.

10

Multiscale modelling of hematologic disorders

Dmitry Fedosov, Igor Pivkin, Wenxiao Pan, Ming Dao, Bruce Caswell, and
George E. Karniadakis

Abstract. Parasitic infectious diseases and other hereditary hematologic disorders
are often associated with major changes in the shape and viscoelastic properties of
red blood cells (RBCs). Such changes can disrupt blood flow and even brain perfusion, as in the case of cerebral malaria. Modelling of these hematologic disorders
requires a seamless multiscale approach, where blood cells and blood flow in the entire arterial tree are represented accurately using physiologically consistent parameters. In this chapter, we present a computational methodology based on dissipative
particle dynamics (DPD) which models RBCs as well as whole blood in health and
disease. DPD is a Lagrangian method that can be derived from systematic coarse-
graining of molecular dynamics but can scale efficiently *up* to small arteries and can
also be used to model RBCs *down* to spectrin level. To this end, we present two com-

Dmitry Fedosov
Forschungszentrum Jülich, 52425 Jülich, Germany
e-mail: d.fedosov@fz-juelich.de

Igor Pivkin
Massachusetts Institute of Technology, Cambridge, MA 02139, e-mail: piv@mit.edu, currently at
University of Lugano, Via Giuseppe Buffi 13, CH-6904, Lugano, Switzerland
e-mail: igor.pivkin@usi.ch

Wenxiao Pan
Pacific Northwest National Laboratory, Richland, WA 99352, USA
e-mail: Wenxiao.Pan@pnl.gov

Ming Dao
Massachusetts Institute of Technology, Cambridge, MA 02139, USA
e-mail: mingdao@mit.edu

Bruce Caswell
Brown University, Providence, RI 02912, USA
e-mail: caswell@dam.brown.edu

George E. Karniadakis (✉)
Brown University, Providence, RI 02912, USA
e-mail: George_Karniadakis@brown.edu

Ambrosi D., Quarteroni A., Rozza G. (Eds.): Modeling of Physiological Flows.
DOI 10.1007/978-88-470-1935-5_10, © Springer-Verlag Italia 2012

plementary mathematical models for RBCs and describe a systematic procedure on extracting the relevant input parameters from optical tweezers and microfluidic experiments for single RBCs. We then use these validated RBC models to predict the behaviour of whole healthy blood and compare with experimental results. The same procedure is applied to modelling malaria, and results for infected single RBCs and whole blood are presented.

10.1 Introduction

The healthy human red blood cells (RBCs) are discocytes when not subjected to any external stresses and they are approximately 7.5 to 8.7 μm in diameter and 1.7 to 2.2 μm in thickness [1]. The membrane of the RBC is made up of a phospholipid bilayer and a network of spectrin molecules (cytoskeleton), with the latter largely responsible for the shear elastic properties of the RBC. The spectrin network is connected to bilayer via transmembrane proteins and together with the spectrin filaments and the cytosol inside the membrane determine the morphological structure of RBCs. This critical binding between the spectrin network and the lipid bilayer is actively controlled by ATP [2]. Parasitic infections or genetic factors can drastically change the viscoelastic properties and even the shape of RBCs [3]. For example, the parasite *Plasmodium falciparum* that invades the RBCs (Pf-RBCs) of most malaria patients affects drastically the RBC membrane properties resulting in a ten-fold increase of its shear modulus and a spherical shape at the later stages of the intra-cell parasite development [3]. In addition, Pf-RBCs develop knobs on their surface that serve as adhesion sites for the binding to other Pf-RBCs as well as healthy RBCs. This enhanced cytoadherence of Pf-RBCs in combination with their reduced deformability may cause blood flow obstruction especially through the smaller arterioles and capillaries. Sickle cell anemia is another blood disorder that affects the hemoglobin inside the RBCs causing dramatic changes in their shape and deformability. These changes combined with the increased internal viscosity affects the flow of sickled RBCs through the capillaries leading to flow occlusion [3, 4]. Other hereditary diseases with similar effects are spherocytosis and elliptocytosis [5]. In the former, RBCs become spherical with reduced diameter and carry much more hemoglobin than healthy RBCs. In the latter, RBCs are elliptical or oval in shape and of reduced deformability.

The common problem in the aforementioned hematologic disorders is the remodelling of the cytoskeleton and correspondingly a change in the structure and viscoelastic properties of individual RBCs, so studying their mechanical and rheological properties in vitro can aid greatly in the understanding and possible discovery of new treatments for such diseases. To this end, new advanced experimental tools are very valuable in obtaining the basic properties of single RBCs in health and disease, which are required in formulating multiscale methods for modelling blood flow in vitro and in vivo. Specifically, advances in experimental techniques now allow measurements down to the nanometer scale, and include micropipette aspiration [6, 7], RBC deformation by optical tweezers [8, 9, 10], optical magnetic twisting cytometry [11], three-dimensional measurement of membrane thermal fluctuations [12, 13], and observations of RBCs immersed in both shear and in pressure-driven flows [14, 15, 16, 17, 18]. Micropipette aspiration and optical tweezers techniques tend to deform the whole RBC membrane directly, while optical magnetic twisting cytometry and measurements of membrane thermal fluctuations probe the membrane properties locally. The macroscopic shear modulus of healthy cells is reported in the range of 2–12 μN/m from the two former techniques, while the two latter ones allow measurements of local rheological properties (e.g., the complex modulus).

These experiments provide sufficient evidence for a complex membrane mechanical response including its unique viscoelastic properties. In addition, Li et al. [19] suggest that metabolic activity or large strains may induce a continuous rearrangement of the erythrocyte cytoskeleton. Consequently, in their numerical model the RBC membrane may exhibit strain hardening or softening depending on certain conditions. Moreover, the cytoskeleton attachments can diffuse within the lipid bilayer, but such behaviour can be neglected at short time scales. Gov [20] proposed an active elastic network model, where the metabolic activity may affect the stiffness of the cell through the consumption of ATP. The activity induced by ATP would also greatly affect membrane undulations [2, 21] resulting in fluctuations comparable to an effective temperature increase by a factor of three. For parasitic infectious diseases, powerful imaging techniques have been developed in recent years, which allow to observe details of parasite development inside the RBC and also to gain information about the properties of the cell components [12, 22]. Fig. 10.1(a) shows the parasite *P. falciparum* inside an infected RBC during the ring stage of parasite development, which was obtained using soft x-ray imaging technique. The parasite and some elaborate structure, which extends from the parasite into the cell cytosol, can be clearly seen in the image.

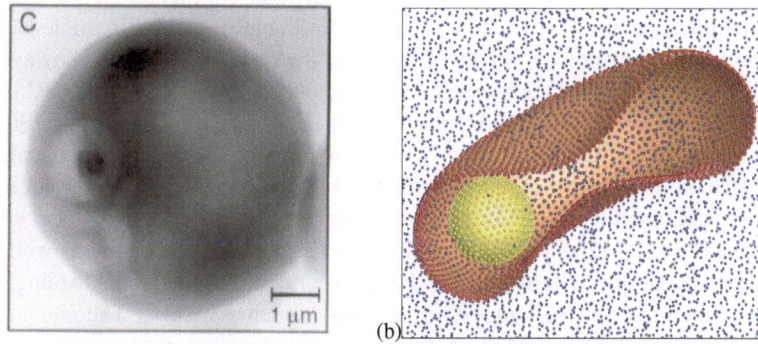

(a)   (b)

Fig. 10.1. (a) Soft x-ray micrograph of intra-erythrocytic ring stage *P. falciparum* malaria parasite imaged in RBC (Reproduced from [22]). (b) The computational RBC model consists of particles connected with links. The model is immersed into DPD fluid and fully interacts with it through pairwise forces. The internal DPD fluid has a higher viscosity to match the viscosity of RBC cytosol. The *P. falciparum parasite* is modelled as a rigid sphere of two microns in diameter

A number of numerical models have been developed recently including a continuum description [1, 23, 24, 25] and a discrete approximation on the spectrin molecular level [26, 27] as well as on the mesoscopic scale [28, 29, 30, 31]. Some of the models suffer from the assumption of purely elastic membrane, and are able to capture only the RBC mechanical response, but cannot quantitatively represent realistic RBC rheology and dynamics. Fully continuum (fluid and solid) models often suffer from non-trivial coupling between nonlinear solid deformations and fluid flow with consequential computational expense. Therefore, "semi-continuum" models [23, 25] of deformable particles which use immersed boundary or front-tracking techniques are developing rapidly. In these, a membrane is represented by a set of points which are tracked in Lagrangian fashion and are coupled to an Eulerian discretization of fluid domain. These models employ the same external and internal fluids and do not take into account the existing viscosity contrast between them. In addition, continuum models omit some mesoscopic and microscopic scale phenomena such as membrane thermal fluctuations which affect RBC rheology and dynamics [32]. On the microscopic scale, detailed spectrin molecular models of RBCs are much limited by the demanding computational expense. Therefore, we will focus here on an accurate mesoscopic modelling of red blood cells.

There exist several mesoscopic methods [28, 29, 30, 31] for modelling deformable particles such as RBCs. Dzwinel et al. [29] model RBCs as a volume of elastic material having an inner skeleton. This model does not take into account the main structural concept of red blood cell, namely a membrane filled with a fluid, and therefore it cannot capture properly the dynamics of RBCs, for example, the observed tumbling and tank-treading behaviour in shear flow [14, 33]. Three other aforementioned methods [28, 30, 31] employ a very similar approach to the method we will present here, where the RBC is represented by a network of springs in combination with bending rigidity and constraints for surface-area and volume conservation. Dupin et al. [28] couple the discrete RBC to a fluid described by the Lattice Boltzmann method [34]. They obtained promising results, however the model does not consider external and internal fluids separation, membrane viscosity, and thermal fluctuations. Noguchi and Gompper [30] employed Multiparticle Collision Dynamics [35] and present encouraging results on vesicles and RBCs, however they do not use realistic RBC properties and probe only a single aspect of RBC dynamics. Pivkin and Karniadakis [31] used Dissipative Particle Dynamics (DPD) [36] for a multiscale RBC model which will be the basis of the general multiscale RBC (MS-RBC) model we will present here. The MS-RBC model is able to successfully capture RBC mechanics, rheology, and dynamics; this very accurate model was first published in [37]. Potential membrane strain hardening or softening as well as the effects of metabolic activity can also be incorporated into the model leading to predictive capabilities on the progression of diseases such as malaria. Theoretical analysis of the hexagonal network yields its linear mechanical properties, and completely eliminates adjustment of the model parameters. Such models can be used to represent seamlessly the RBC membrane, cytoskeleton, cytosol, the surrounding plasma and even the parasite, e.g. in malaria-infected RBC, see Fig. 10.1. However, it is quite expensive computationally, and to this end, we also present a *low-dimensional* red blood cell model (LD-RBC), also based on DPD, that is more appropriate for blood flow simulations in large arterioles [38].

This chapter is organized as follows: in Sect. 10.2 we review the basic DPD theory, the two RBC models, as well as aspects of the aggregation and adhesion models that are especially important in modelling hematologic disorders. In Sect. 10.3 we present some details on how we can use diverse single-cell static and dynamic measurements to estimate key macroscopic parameters, which upon mapping to the network (microscopic) parameters serve as input to the models. In Sect. 10.4 we first present validation tests based on single-cell experiments. Subsequently, we present validation tests for whole blood, demonstrating that both models can predict the human blood viscosity in a wide range of shear rate values, including the low shear rate regime, where aggregation and rouleaux formation are responsible for the strong non-Newtonian blood behaviour. In Sect. 10.5, we apply to malaria the framework we developed, i.e. from single-cell-measurements parameter estimation to predicting the mechanical and rheological behaviour of infected blood in malaria. We conclude in Sect. 10.6 with a brief summary and a discussion on the potential of multiscale modelling to predicting the state and evolution of hematologic disorders.

10.2 Methods and models

We first review two formulations of the dissipative particle dynamics (DPD) method that we employ in modelling RBCs and blood flow. We then provide specific details on the multiscale RBC model (MS-RBC) and subsequently on the low-dimensional RBC model (LD-RBC), including the aggregation and adhesion models. Finally, we present details on the scaling from DPD units to physical units.

10.2.1 Dissipative particle dynamics: original method

Dissipative Particle Dynamics (DPD) [36, 39] is a mesoscopic particle method, where each particle represents a *molecular cluster* rather than an individual atom, and can be thought of as a soft lump of fluid. A first-principles derivation of the DPD method from the Liouville equation is presented in [40]. The DPD system consists of N point particles of mass m_i, position $\mathbf{r}_i$ and velocity $\mathbf{v}_i$. DPD particles interact through three forces: conservative ($\mathbf{F}_{ij}^C$), dissipative ($\mathbf{F}_{ij}^D$), and random ($\mathbf{F}_{ij}^R$) forces given by

$$
\begin{aligned}
\mathbf{F}_{ij}^C &= F_{ij}^C(r_{ij})\hat{\mathbf{r}}_{ij}, \\
\mathbf{F}_{ij}^D &= -\gamma\omega^D(r_{ij})(\mathbf{v}_{ij}\cdot\hat{\mathbf{r}}_{ij})\hat{\mathbf{r}}_{ij}, \\
\mathbf{F}_{ij}^R &= \sigma\omega^R(r_{ij})\frac{\xi_{ij}}{\sqrt{dt}}\hat{\mathbf{r}}_{ij},
\end{aligned}
\tag{10.1}
$$

where $\hat{\mathbf{r}}_{ij} = \mathbf{r}_{ij}/r_{ij}$, and $\mathbf{v}_{ij} = \mathbf{v}_i - \mathbf{v}_j$. The coefficients γ and σ define the strength of dissipative and random forces, respectively. In addition, ω^D and ω^R are weight functions, and ξ_{ij} is a normally distributed random variable with zero mean, unit variance, and $\xi_{ij} = \xi_{ji}$. All forces are truncated beyond the cutoff radius r_c, which defines the length scale in the DPD system. The conservative force is given by

$$F_{ij}^C(r_{ij}) = \begin{cases} a_{ij}(1 - r_{ij}/r_c) \ for \ r_{ij} \leq r_c, \\ 0 \ for \ r_{ij} > r_c, \end{cases} \tag{10.2}$$

where a_{ij} is the conservative force coefficient between particles i and j.

The random and dissipative forces form a thermostat and must satisfy the fluctuation-dissipation theorem in order for the DPD system to maintain equilibrium temperature T [41]. This leads to:

$$\omega^D(r_{ij}) = \left[\omega^R(r_{ij})\right]^2, \qquad \sigma^2 = 2\gamma k_B T, \tag{10.3}$$

where k_B is the Boltzmann constant. The choice for the weight functions is as follows

$$\omega^R(r_{ij}) = \begin{cases} (1 - r_{ij}/r_c)^k \ for \ r_{ij} \leq r_c, \\ 0 \ for \ r_{ij} > r_c, \end{cases} \tag{10.4}$$

where $k = 1$ for the original DPD method. However, other choices (e.g., $k = 0.25$) for these envelopes have been used [42, 43] in order to increase the viscosity of the DPD fluid.

The time evolution of velocities and positions of particles is determined by the Newton's second law of motion

$$d\mathbf{r}_i = \mathbf{v}_i dt, \tag{10.5}$$

$$d\mathbf{v}_i = \frac{1}{m_i} \sum_{j \neq i} \left(\mathbf{F}_{ij}^C + \mathbf{F}_{ij}^D + \mathbf{F}_{ij}^R\right) dt. \tag{10.6}$$

The above stochastic equations of motion can be integrated using a modified velocity-Verlet algorithm [39]; for systems governed by mixed hard-soft potentials sub-cycling techniques similar to the ones presented in [44] can be employed.

10.2.2 DPD method for colloidal particles

To simulate colloidal particles by single DPD particles, we use a new formulation of DPD, in which the dissipative forces acting on a particle are explicitly divided into two separate components: *central* and *shear* (non-central) components. This allows us to redistribute and hence, balance the dissipative forces acting on a single particle to obtain the correct hydrodynamics. The resulting method was shown to yield the quantitatively correct hydrodynamic forces and torques on a single DPD particle [45], and thereby produce the correct hydrodynamics for colloidal particles [46]. This formulation is reviewed below.

We consider a collection of particles with positions $\mathbf{r}_i$ and angular velocities Ω_i. We define $\mathbf{r}_{ij} = \mathbf{r}_i - \mathbf{r}_j$, $r_{ij} = |\mathbf{r}_{ij}|$, $\mathbf{e}_{ij} = \mathbf{r}_{ij}/r_{ij}$, $\mathbf{v}_{ij} = \mathbf{v}_i - \mathbf{v}_j$. The force and torque on particle i are given by

$$\mathbf{F}_i = \sum_j \mathbf{F}_{ij},$$

$$\mathbf{T}_i = -\sum_j \lambda_{ij} \mathbf{r}_{ij} \times \mathbf{F}_{ij}. \tag{10.7}$$

Here, the factor λ_{ij} (introduced in [47]) is included as a weight to account for the different contributions from the particles in different species (solvent or colloid) differentiated in sizes while still conserving the angular momentum. It is defined as

$$\lambda_{ij} = \frac{R_i}{R_i + R_j}, \tag{10.8}$$

where R_i and R_j denote the radii of the particles i and j, respectively. The force exerted by particle j on particle i is given by

$$\mathbf{F}_{ij} = \mathbf{F}_{ij}^U + \mathbf{F}_{ij}^T + \mathbf{F}_{ij}^R + \widehat{\mathbf{F}}_{ij}. \tag{10.9}$$

The radial conservative force $\mathbf{F}_{ij}^U$ can be that of standard DPD and is given in Eq. (10.2). The *translational force* is given by

$$\mathbf{F}_{ij}^T = -\left[\gamma_{ij}^\perp f^2(r)\mathbf{1} + (\gamma_{ij}^\parallel - \gamma_{ij}^\perp)f^2(r)\mathbf{e}_{ij}\mathbf{e}_{ij}\right] \cdot \mathbf{v}_{ij}$$
$$= -\gamma_{ij}^\parallel f^2(r_{ij})(\mathbf{v}_{ij} \cdot \mathbf{e}_{ij})\mathbf{e}_{ij} - \gamma_{ij}^\perp f^2(r_{ij})\left[\mathbf{v}_{ij} - (\mathbf{v}_{ij} \cdot \mathbf{e}_{ij})\mathbf{e}_{ij}\right]. \tag{10.10}$$

It accounts for the drag due to the relative translational velocity $\mathbf{v}_{ij}$ of particles i and j. This force is decomposed into two components: one along and the other perpendicular to the lines connecting the centers of the particles. Correspondingly, the drag coefficients are denoted by $\gamma_{ij}^\parallel$ and $\gamma_{ij}^\perp$ for a *central* and a *shear* components, respectively. We note that the central component of the force is identical to the dissipative force of standard DPD (Eq. (10.1)).

The *rotational force* is defined by

$$\mathbf{F}_{ij}^R = -\gamma_{ij}^\perp f^2(r_{ij})\left[\mathbf{r}_{ij} \times (\lambda_{ij}\Omega_i + \lambda_{ji}\Omega_j)\right], \tag{10.11}$$

while the *random force* is given by

$$\widehat{\mathbf{F}}_{ij}dt = f(r_{ij})\left[\frac{1}{\sqrt{3}}\sigma_{ij}^\parallel \mathrm{tr}[d\mathbf{W}_{ij}]\mathbf{1} + \sqrt{2}\sigma_{ij}^\perp d\mathbf{W}_{ij}^A\right] \cdot \mathbf{e}_{ij}, \tag{10.12}$$

where $\sigma_{ij}^\parallel = \sqrt{2k_B T \gamma_{ij}^\parallel}$ and $\sigma_{ij}^\perp = \sqrt{2k_B T \gamma_{ij}^\perp}$ are chosen to satisfy the fluctuation-dissipation theorem, $d\mathbf{W}_{ij}$ is a matrix of independent Wiener increments, and $d\mathbf{W}_{ij}^A$ is defined as $d\mathbf{W}_{ij}^{A\mu\nu} = \frac{1}{2}(d\mathbf{W}_{ij}^{\mu\nu} - d\mathbf{W}_{ij}^{\nu\mu})$. We can also use the generalized weight function $f(r) = (1 - \frac{r}{r_c})^k$ as in the previous section with $k = 0.25$ [48] in equations (10.10)- (10.12). The numerical results in previous studies [45, 49] showed higher accuracy with $k = 0.25$ compared to the usual choice $k = 1$. The standard DPD is recovered when $\gamma_{ij}^{\perp} \equiv 0$, i.e., when the *shear* components of the forces are ignored.

Colloidal particles are simulated as single DPD particles, similarly to the solvent particles but of larger size. The particle size can be adjusted with the coefficient a_{ij} of the conservative force (see Eq. (10.2)). However, the standard linear force in DPD defined in Eq. (10.2) is too soft to model any hard-sphere type of particles. To resolve this problem, we adopt an exponential conservative force for the colloid-colloid and colloid-solvent interactions, but keep the conventional DPD linear force for the solvent-solvent interactions. We have found that these hybrid conservative interactions produced colloidal particles dispersed in solvent without overlap, which was quantified by calculating the radial distribution function of colloidal particles [46]. Moreover, the timestep is not significantly decreased, in contrast to the small timesteps required for the Lennard-Jones potential [47]. The radial exponential conservative force is defined as

$$F_{ij}^{U} = \frac{a_{ij}}{1 - e^{b_{ij}}} \left(e^{b_{ij} r_{ij}/r_c^e} - e^{b_{ij}} \right), \tag{10.13}$$

where a_{ij} and b_{ij} are adjustable parameters, and r_c^e is its cutoff radius. The size of a colloidal particle can thus be controlled by adjusting the value of a_{ij} in Eq. (10.13).

10.2.3 Multiscale Red Blood Cell (MS-RBC) model

Here, we will use the DPD formulation described in Sect. 10.2.1. The average equilibrium shape of a RBC is biconcave as measured experimentally [24], and is represented by

$$z = \pm D_0 \sqrt{1 - \frac{4(x^2 + y^2)}{D_0^2}} \left[a_0 + a_1 \frac{x^2 + y^2}{D_0^2} + a_2 \frac{(x^2 + y^2)^2}{D_0^4} \right], \tag{10.14}$$

where $D_0 = 7.82\,\mu m$ is the average diameter, $a_0 = 0.0518$, $a_1 = 2.0026$, and $a_2 = -4.491$. The surface area and volume of this RBC are equal to $135\,\mu m^2$ and $94\,\mu m^3$, respectively.

In simulations, the membrane network structure is generated by triangulating the unstressed equilibrium shape described by (10.14). The cell shape is first imported into a grid generator to produce an initial triangulation based on the advancing-front method. Subsequently, free-energy relaxation is performed by flipping the diagonals of quadrilateral elements formed by two adjacent triangles, while the vertices are constrained to move on the prescribed surface. The relaxation procedure includes only elastic in-plane and bending energy components described below.

Fig. 10.2 shows the membrane model represented by a set of points $\{\mathbf{x}_i\}, i \in 1...N_v$ that are the vertices of a two-dimensional triangulated network on the RBC surface described by Eq. (10.14). The vertices are connected by N_s edges which form N_t triangles. The potential energy of the system is defined as follows

$$V(\{\mathbf{x}_i\}) = V_{in-plane} + V_{bending} + V_{area} + V_{volume}. \tag{10.15}$$

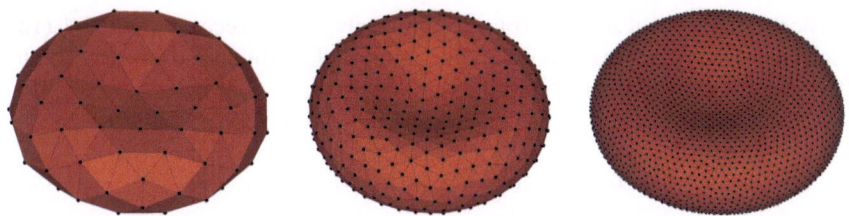

Fig. 10.2. MS-RBC membrane model with $N_v = 100, 500$, and 3000 from left to right, respectively

The in-plane elastic energy mimics the elastic spectrin network, and is given by

$$V_{in-plane} = \sum_{j \in 1...N_s} \left[\frac{k_B T l_m (3x_j^2 - 2x_j^3)}{4p(1-x_j)} + \frac{k_p}{(n-1)l_j^{n-1}} \right], \tag{10.16}$$

where l_j is the length of the spring j, l_m is the maximum spring extension, $x_j = l_j/l_m$, p is the persistence length, $k_B T$ is the energy unit, k_p is the spring constant, and n is a power. Note that the spring forces in membrane are a combination of conservative elastic forces, that may be expressed in terms of the energy potential above, and dissipative forces to be defined below. The first term in (10.16) corresponds to the attractive wormlike chain (WLC) potential, and the second term defines a repulsive force for $n > 0$ to be called the power force (POW), so that we abbreviate this spring model as WLC-POW. Note that if $n = 1$ the power force energy should be defined as $-k_p \log(l_j)$. A non-zero equilibrium spring length is defined by the balance of these two forces.

The bending energy represents the bending resistance of the lipid bilayer and is defined as

$$V_{bending} = \sum_{j \in 1...N_s} k_b [1 - cos(\theta_j - \theta_0)], \tag{10.17}$$

where k_b is the bending constant, θ_j is the instantaneous angle between two adjacent triangles having the common edge j, and θ_0 is the spontaneous angle. The above bending energy is a discretization [88] of the macroscopic Helfrich model [52].

The area and volume conservation constraints which account for area-incompressibility of the lipid bilayer and incompressibility of the inner cytosol, respectively, are expressed as

$$V_{area} = \frac{k_a (A - A_0^{tot})^2}{2A_0^{tot}} + \sum_{j \in 1...N_t} \frac{k_d (A_j - A_0)^2}{2A_0}, \tag{18a}$$

$$V_{volume} = \frac{k_v (V - V_0^{tot})^2}{2V_0^{tot}}, \tag{18b}$$

where k_a, k_d and k_v are the global area, local area and volume constraint coefficients, respectively. The terms A and V are the total area and volume of RBC, while A_0^{tot} and V_0^{tot} are the specified total area and volume, respectively. Note that the above expressions define global area and volume constraints, and the second term in Eq. (18a) incorporates the local dilatation constraint. Detailed description and discussion of the RBC model can be found in [37, 50].

Particle forces are derived from the above energies as follows

$$\mathbf{f}_i = -\partial V(\{\mathbf{x}_i\})/\partial \mathbf{x}_i, \qquad i \in 1...N_v. \tag{10.19}$$

Exact force expressions can be found in [51].

10.2.3.1 Mechanical properties

Linear analysis of the regular hexagonal network having the above energies yields a relationship between macroscopic elastic properties (shear, area-compression, and Young's moduli) of the network and model parameters [37, 50]. The membrane shear modulus is thus given by

$$\mu_0 = \frac{\sqrt{3}k_B T}{4p l_m x_0} \left(\frac{x_0}{2(1-x_0)^3} - \frac{1}{4(1-x_0)^2} + \frac{1}{4} \right) + \frac{\sqrt{3}k_p(n+1)}{4l_0^{n+1}}, \tag{10.20}$$

where l_0 is the equilibrium spring length and $x_0 = l_0/l_m$. The corresponding area-compression and Young's moduli are found as follows

$$K_0 = 2\mu_0 + k_a + k_d, \qquad Y_0 = \frac{4K_0\mu_0}{K_0 + \mu_0}. \tag{10.21}$$

The bending coefficient k_b of Eq. (10.17) can be expressed in terms of the macroscopic bending rigidity k_c of the Helfrich model [52] as $k_b = 2k_c/\sqrt{3}$.

10.2.3.2 Membrane viscoelasticity

The above model defines a purely elastic membrane, however the RBC membrane is known to be viscoelastic. To incorporate viscosity into the model, the spring definition is modified by adding viscous contribution through dissipative and random forces. Such a term fits naturally in the DPD method [36], where inter-particle dissipative interactions are an intrinsic part of the method. Straightforward implementation of the dissipative interactions as $\mathbf{F}_{ij}^D = -\gamma(\mathbf{v}_{ij} \cdot \mathbf{e}_{ij})\mathbf{e}_{ij}$ (γ is the dissipative parameter, $\mathbf{v}_{ij} = \mathbf{v}_i - \mathbf{v}_j$ is the relative velocity of vertices i and j connected by a spring, and $\mathbf{e}_{ij}$ is the direction along the spring with unit length) appears to be insufficient. Experience shows that small γ results in a negligible viscous contribution since $\mathbf{v}_{ij} \cdot \mathbf{e}_{ij} \sim 0$, while large values of γ require considerably smaller time steps to overcome the numerical instability. Better performance is achieved with a viscous spring dissipation term $-\gamma\mathbf{v}_{ij}$, which is similar to a "dashpot", and in combination with a spring force represents the Kelvin-Voigt model of a viscoelastic spring. For this term the fluctuation-dissipation balance needs to be imposed to ensure the maintenance of the equilibrium membrane temperature $k_B T$. We follow the general framework of the fluid particle model [53], and define $\mathbf{F}_{ij}^D = -\mathbf{T}_{ij} \cdot \mathbf{v}_{ij}$ and $\mathbf{T}_{ij} = \gamma^T \mathbf{1} + \gamma^C \mathbf{e}_{ij}\mathbf{e}_{ij}$, where γ^T and γ^C are the dissipative coefficients. This definition results in the dissipative interaction term of the kind

$$\mathbf{F}_{ij}^D = -\left[\gamma^T \mathbf{1} + \gamma^C \mathbf{e}_{ij}\mathbf{e}_{ij}\right] \cdot \mathbf{v}_{ij} = -\gamma^T \mathbf{v}_{ij} - \gamma^C (\mathbf{v}_{ij} \cdot \mathbf{e}_{ij})\mathbf{e}_{ij}, \qquad (10.22)$$

where the second term is analogous to the dissipative force in DPD. From the fluctuation-dissipation theorem, random interactions are given by

$$\mathbf{F}_{ij}^R dt = \sqrt{2k_BT} \left(\sqrt{2\gamma^T} \, \overline{d\mathbf{W}_{ij}^S} + \sqrt{3\gamma^C - \gamma^T} \frac{tr[d\mathbf{W}_{ij}]}{3} \mathbf{1} \right) \cdot \mathbf{e}_{ij}, \qquad (10.23)$$

where $tr[d\mathbf{W}_{ij}]$ is the trace of a random matrix of independent Wiener increments $d\mathbf{W}_{ij}$, and $\overline{d\mathbf{W}_{ij}^S} = d\mathbf{W}_{ij}^S - tr[d\mathbf{W}_{ij}^S]\mathbf{1}/3$ is the traceless symmetric part, while $d\mathbf{W}_{ij}^S = [d\mathbf{W}_{ij} + d\mathbf{W}_{ij}^T]/2$ is the symmetric part. Note, that the last equation imposes the condition $3\gamma^C > \gamma^T$. The defined dissipative and random forces in combination with an elastic spring constitute a viscoelastic spring whose equilibrium temperature k_BT is constant. To relate the membrane shear viscosity η_m and the dissipative parameters γ^T, γ^C we employ the idea used for the derivation of membrane elastic properties (see [37, 51] for details) and obtain the following relation

$$\eta_m = \sqrt{3}\gamma^T + \frac{\sqrt{3}\gamma^C}{4}. \qquad (10.24)$$

Clearly, γ^T accounts for a large portion of viscous contribution, and therefore γ^C is set to $\gamma^T/3$ in all simulations.

10.2.3.3 RBC-solvent boundary conditions

The RBC membrane encloses a volume of fluid and is itself suspended in a solvent. In particle methods, such as DPD, fluids are represented as a collection of interacting particles. Thus, in order to impose appropriate boundary conditions (BCs) between the membrane and the external/internal fluids two matters need to be addressed:

i) enforcement of membrane impenetrability to prevent mixing of the inner and the outer fluids;
ii) no-slip BCs imposed through pairwise point interactions between the fluid particles and the membrane vertices.

Membrane impenetrability is enforced by imposing bounce-back reflection of fluid particles at the moving membrane triangular plaquettes. The bounce-back reflection enhances the no-slip boundary conditions at the membrane surface as compared to specular reflection; however, it does not guarantee no-slip. Additional dissipation enhancement between the fluid and the membrane is required to achieve no-slip at the membrane boundary. For this purpose, the DPD dissipative force between fluid particles and membrane vertices needs to be properly set based on the idealized case of linear shear flow over a flat plate. In continuum, the total shear force exerted by the fluid on the area A is equal to $A\eta\dot{\gamma}$, where η is the fluid's viscosity and $\dot{\gamma}$ is the local wall shear-rate. In DPD, we distribute a number of particles on the wall to mimic the membrane vertices. The force on a single wall particle exerted by the sheared fluid can be found as follows

$$F_v = \int_{V_h} ng(r)F^D dV, \tag{10.25}$$

where F^D is the DPD dissipative force [53] between fluid particles and membrane vertices, n is the fluid number density, $g(r)$ is the radial distribution function of fluid particles with respect to the wall particles, and V_h is the half sphere volume of fluid above the wall. Here, the total shear force on the area A is equal to $N_A F_v$, where N_A is the number of wall particles enclosed by A. The equality of $N_A F_v = A\eta\dot{\gamma}$ results in an expression of the dissipative force coefficient in terms of the fluid density and viscosity, and the wall density N_A/A, while under the assumption of linear shear flow the shear rate $\dot{\gamma}$ cancels out. This formulation results in satisfaction of the no-slip BCs for the linear shear flow over a flat plate. It also serves as an excellent approximation for no-slip at the membrane surface in spite of the assumptions made. Note that in the absence of conservative interactions between fluid and wall particles $g(r) = 1$.

10.2.3.4 RBC aggregation interactions

For a blood suspension the attractive cell-cell interactions are crucial for simulation of aggregation into rouleaux. These forces are approximated phenomenologically with the Morse potential given by

$$\phi(r) = D_e \left[e^{2\beta(r_0-r)} - 2e^{\beta(r_0-r)} \right], \tag{10.26}$$

where r is the separation distance, r_0 is the zero force distance, D_e is the well depth of the potential, and β characterizes the interaction range. For the MS-RBC model the Morse potential interactions are implemented between every two vertices of separate RBCs if they are within a defined potential cutoff radius r_M as shown in Fig. 10.3. The Morse interactions consist of a short-range repulsive force when $r < r_0$ and of a long-range attractive force for $r > r_0$. However, such repulsive interactions cannot prevent two RBCs from an overlap. To guarantee no overlap among RBCs we employ a short range Lennard-Jones potential and specular reflections of RBC vertices

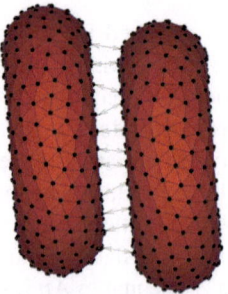

vertex-vertex interactions

Fig. 10.3. Aggregation interactions for the MS-RBC model

on membranes of other RBCs. The Lennard-Jones potential is defined as

$$U_{LJ}(r) = 4\varepsilon \left[\left(\frac{\sigma_{LJ}}{r} \right)^{12} - \left(\frac{\sigma_{LJ}}{r} \right)^{6} \right],$$ (10.27)

where ε and σ_{LJ} are energy and length characteristic parameters, respectively. These interactions are repulsive and vanish beyond $r > 2^{1/6}\sigma_{LJ}$. In addition, specular reflections of RBC vertices on surfaces of other RBCs are necessary due to coarseness of the triangular network, which represents the RBC membrane.

10.2.3.5 RBC adhesion interactions

Adhesion of Pf-RBCs to coated surfaces is mediated by the interactions between receptors and ligands which are the adhesion sites distributed on a cell and a surface, respectively. A potential bond between a receptor and a ligand may be formed only if the receptor is close enough to the free ligand, which is characterized by the reactive distance d_{on}. A ligand is called free if it is not bound to any receptors. During the time a receptor is within the distance d_{on} to a free ligand, a bond can be formed with on-rate k_{on}. Reversely, existing bonds are ruptured with off-rate k_{off} or if their length exceeds the rupture distance d_{off}. The rates k_{on} and k_{off} are defined as follows

$$k_{on} = k_{on}^0 \exp\left(-\frac{\sigma_{on}(l - l_0)^2}{2k_B T} \right), \qquad k_{off} = k_{off}^0 \exp\left(\frac{\sigma_{off}(l - l_0)^2}{2k_B T} \right),$$ (10.28)

where k_{on}^0 and k_{off}^0 are the reaction rates at the distance $l = l_0$ between a receptor and a ligand with the equilibrium spring length l_0 defined below. The effective on and off strengths σ_{on} and σ_{off} define a decrease or an increase of the corresponding rates within the interaction lengths d_{on} and d_{off}, and $k_B T$ is the unit of energy. The force exerted on the receptors and ligands by an existing bond is given by

$$F(l) = k_s(l - l_0),$$ (10.29)

where k_s is the spring constant. The probabilities of bond formation and dissociation are defined as $P_{on} = 1 - \exp(-k_{on}\Delta t)$ and $P_{off} = 1 - \exp(-k_{off}\Delta t)$, where Δt is the time step in simulations. This adhesion model is a slight modification of the well-known adhesive dynamics model developed by Hammer and Apte [54] primarily for leukocytes. In their model $\sigma_{on} = \sigma_{ts}$ and $\sigma_{off} = k_s - \sigma_{ts}$, where σ_{ts} is the transition state spring constant.

During the course of a simulation the receptor/ligand interactions are considered every time step. First, all existing bonds between receptors and ligands are checked for a potential dissociation according to the probability P_{off}. A bond is ruptured if $\xi < P_{off}$ and left unchanged otherwise, where ξ is a random variable uniformly distributed on $[0, 1]$. If a bond is ruptured the corresponding ligand is available for new binding. Second, all free ligands are examined for possible bond formations. For each free ligand we loop over the receptors within the distance d_{on}, and bond formation is attempted for each found receptor according to the probability P_{on}. This loop is terminated when a bond is formed. Finally, the forces of all remaining bonds are calculated and applied.

Note that this algorithm permits only a single bond per ligand, while receptors may establish several bonds if several ligands are free within their reaction radius. This provides an additional capability for the adhesive dynamics model compared with that employing one-to-one interactions between receptors and ligands. Also, this assumption appears to furnish a more realistic representation of adhesive interactions of Pf-RBCs with a coated surface. Pf-RBCs display a number of parasitic nanometer-size protrusions or knobs on the membrane surface [55, 56, 57], where receptors that mediate RBC adherence are clustered.

10.2.4 Low-Dimensional RBC (LD-RBC) model

Here, we will employ the DPD formulation presented in Sect. 10.2.2. The LD-RBC is modelled as a ring of 10 colloidal DPD particles connected by wormlike chain (WLC) springs. The intrinsic size of colloidal particle is determined by the radius of the sphere effectively occupied by a single DPD particle [46], which is defined by the distribution of its surrounding solvent particles.

To construct the cell model, however, we allow particles in the same RBC to overlap, i.e., the colloidal particles in the same cell still interact with each other through the soft standard DPD linear force (see Eq. (10.2)). The radius, a, of each colloidal particle is chosen to be equal to the radius of the ring, and hence the configuration of RBC is approximately a closed-torus as shown in Fig. 10.4.

The WLC spring force interconnecting all cell particles in each RBC is given by

$$F_{WLC}^U = \frac{k_B T}{p} \left[\frac{1}{4(1 - \frac{r_{ij}}{l_m})^2} - \frac{1}{4} + \frac{r_{ij}}{l_m} \right], \tag{10.30}$$

where r_{ij} is the distance between two neighboring beads, p is the persistence length, and l_m is the maximum allowed length for each spring. Since the cell has also bend-

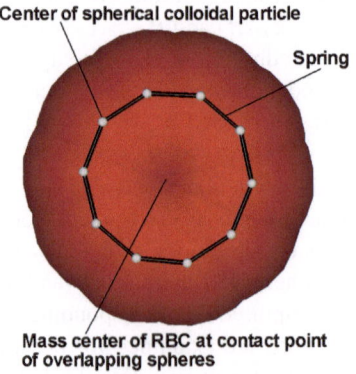

Center of spherical colloidal particle

Spring

Mass center of RBC at contact point
of overlapping spheres

Fig. 10.4. LD-RBC: A sketch of the low-dimensional closed-torus like RBC model

ing resistance, we incorporate into the ring model bending resistance in the form of "angle" bending forces dependent on the angle between two consecutive springs. The bending forces are derived from the cosine bending potential given by

$$U_{ijk}^{COS} = k_b[1 - cos\theta_{ijk}],$$ (10.31)

where k_b is the bending stiffness, and θ_{ijk} is the angle between two consecutive springs.

Here, p determines the Young's modulus, and along with l_m and a give the right size of RBC. To match both axial and transverse RBC deformations with the experimental data [9], k_b is adjusted to reach a good agreement, which also gives some contribution to the Young's modulus. The LD-RBC model does not have the membrane shear modulus.

Since the thickness of LD-RBC model is constant, we estimate the variations of the RBC volume and surface area under stretching by calculating the relative change of the area formed by the ring under stretching. For healthy RBCs we find that it varies within only 8 % in the range of all stretching forces [38]. Therefore, the surface-area and hence the volume of RBCs remain approximately constant in the LD-RBC model.

10.2.4.1 Number of particles in LD-RBC model

We examine the effect of coarse-graining on stretching response by varying the number of particles (N_c) to model the LD-RBC. Fig. 10.5 shows the RBC shape evolution from equilibrium (0 pN force) to 100 pN stretching force at different N_c. Note that an increase of the number of particles making up the RBC results in a smoother RBC surface. However, this feature seems to be less pronounced for higher N_c. Also, when we stretch the RBCs with different N_c, we find that an increase of N_c results in better agreement with the experimental data [9], but after $N_c = 10$, the change becomes very small [38]. To gain sufficiently good agreement and keep the computation cost low, we choose $N_c = 10$ for all the simulations shown herein; this is the accurate minimalistic model that we employ in our studies.

10.2.4.2 Aggregation model

For LD-RBC model, we also employ the Morse potential, see Eq. (10.26), to model the total intercellular attractive interaction energy. The interaction between RBCs derived from the Morse potential includes two parts: a short-ranged repulsive force and a weak long-ranged attractive force. The repulsive force is in effect when the distance between two RBC surfaces is $r < r_0$, where r_0 is usually in nanometer scale [58, 59, 60]. In our simulations, r_0 is chosen to be 200 nm.

Here, r is calculated based on the center of mass of RBCs, i.e., r is equal to the distance between the center of mass of two RBCs minus the thickness of a RBC. We also calculate the normal vector of each RBC ($\vec{n}_c$), which is used to determine if the aggregation occurs between two RBCs according to the angles formed by the normal

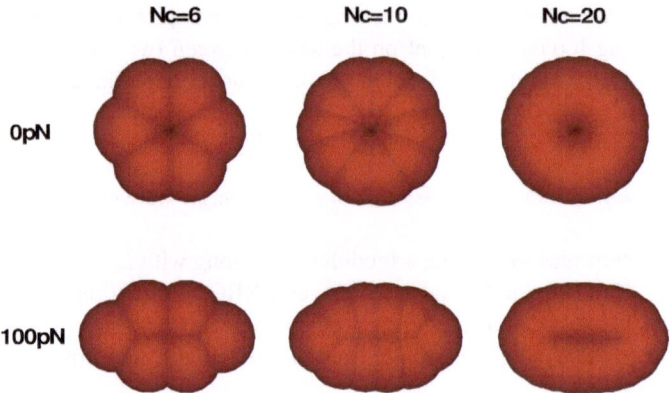

Fig. 10.5. LD-RBC shape evolution at different N_c (number of particles in LD-RBC model) and stretching forces

vectors of two RBCs with their center line. The RBC normal vector is defined as

$$\vec{n}_c = \frac{\sum \vec{v}_k \times \vec{v}_{k+1}}{N_c}, \vec{v}_k = \mathbf{x}_k - \mathbf{x}_c. \tag{10.32}$$

Here, $\mathbf{x}_k$ is the position of the kth particle in each RBC, $\mathbf{x}_c$ is the position of the center of mass, and N_c is the number of particles in each RBC. The center line $\vec{v}_{cij}$ of two RBCs (cell i and cell j) is defined as $\mathbf{x}_{ci} - \mathbf{x}_{cj}$. The angle formed by the normal vector of one cell with the center line is determined by their dot product

$$d_i = \frac{\vec{n}_{ci}}{\|\vec{n}_{ci}\|} \cdot \frac{\vec{v}_{cij}}{\|\vec{v}_{cij}\|}. \tag{10.33}$$

The Morse interaction is turned on if $d_i > d_c$ and $d_j > d_c$, otherwise, it is kept off. The critical value, d_c, is chosen to be equal to $cos(\pi/4)$, i.e., the critical angle (θ_c) to turn on/off the aggregation interaction is $\pi/4$. This value is found to be suitable to induce rouleaux formation, but exclude the disordered aggregation. The proposed aggregation algorithm can be further illustrated by a sketch in Fig. 10.6, where the aggregation between two neighbor RBCs is decided to be on/off according to their relative orientation.

10.2.5 Scaling of model and physical units

The dimensionless constants and variables in the DPD model must be scaled with physical units. The superscript M denotes that a quantity is in "model" units, while P identifies physical units (SI units). We define the length scale as follows

$$r^M = \frac{D_0^P}{D_0^M} \ m, \tag{10.34}$$

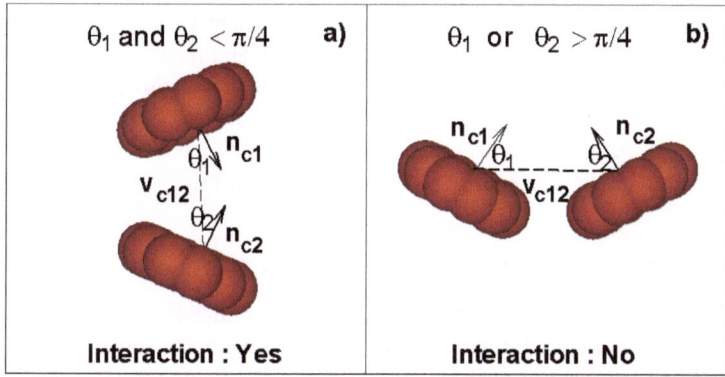

Fig. 10.6. Schematic of the aggregation algorithm. Here, the two neighbor RBCs (1 and 2) are to aggregate or not if the angles, θ_1 and θ_2, are smaller or greater than $\pi/4$

where r^M is the model unit of length, D_0 is the cell diameter, and m stands for meters. The energy per unit mass ($k_B T$) and the force unit ("N" denotes Newton) scales are given by

$$(k_B T)^M = \frac{Y^P}{Y^M} \left(\frac{D_0^P}{D_0^M}\right)^2 (k_B T)^P, \qquad N^M = \frac{Y^P}{Y^M} \frac{D_0^P}{D_0^M} N^P, \qquad (10.35)$$

where Y is the membrane Young's modulus. The time scale is defined as

$$\tau = \frac{D_0^P}{D_0^M} \frac{\eta^P}{\eta^M} \frac{Y^M}{Y^P} s, \qquad (10.36)$$

where η is a characteristic viscosity (e.g., solvent or membrane).

10.3 Parameter estimation

The models described in the previous section require as inputs "microscopic" parameters, e.g. the persistence length p for the WLC potential, but also other parameters, e.g. values of the membrane viscosity. These parameters may not be readily available in the literature and certainly they vary according to the RBC state, i.e a healthy or infected RBC. To this end, we aim to estimate most of the required parameters from single-cell measurements of macroscopic quantities, e.g. shear modulus, which can then be mapped to "microscopic" (network) parameters using analytical expressions, such as the one in Eq. (10.20). Specifically, the RBC model is compared against several available experiments which examine cell mechanics, rheology, and dynamics for healthy and diseased RBCs. First, we obtain the shear modulus using optical tweezers measurements of a stretched RBC. We then estimate the membrane rheological parameters using measurements from optical magnetic twisting cytometry and from the response of single RBC in shear flow. In all cases we run corresponding DPD simulations in order to compare and match the experimentally observed responses. In the following, we describe details of this procedure and we also demonstrate that while the parameters can be estimated in a relatively narrow regime, we can then predict accurately the single RBC mechanics, dynamics and rheology over a much wider range of operating conditions.

10.3.1 Shear modulus using optical tweezers

To mimic the optical tweezers experiments of [9] a modelled RBC undergoes stretching by applying a stretching force on both ends of the cell. The total stretching force f^M is applied to N_- and N_+ vertices ($N_- = N_+ = \varepsilon N_v$) along the negative and the positive directions, respectively. These vertices cover a near-spherical area on the RBC surface with $\varepsilon = 0.02$ which corresponds to the contact diameter of the attached silica bead with diameter $2\,\mu m$ used in experiments [9]. Note that the viscous properties of the membrane and of the suspending medium do not affect final stretching since the RBC response is measured after convergence to the equilibrium stretched state is achieved for given force.

Fig. 10.7 (left) compares the simulated axial and transverse RBC diameters with their experimental counterparts [9] for different coarse-graining levels starting from the spectrin-level ($N_v = 27344$) to the highly coarse-grained network of $N_v = 500$. Excellent correspondence between simulations and experiments is achieved for $\mu_0 = 6.3\,\mu N/m$ and $Y = 18.9\,\mu N/m$ independently of the level of coarse-graining. The small discrepancy between simulated and experimental transverse diameters is probably a consequence of the optical measurements being performed from only a single observation angle. Numerical simulations showed that RBCs subjected to stretching tend to rotate in y-z plane, and therefore measurements from a single ob-

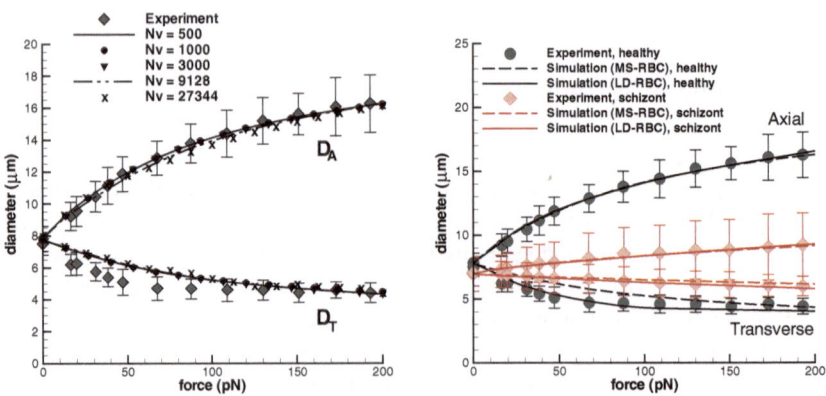

Fig. 10.7. MS-RBC (left): Stretching response of a healthy RBC for different coarse-graining levels. MS-RBC vs. LD-RBC (right): Stretching response of a healthy RBC and a Pf-RBC (schizont stage) with the experiments of [9]. D_A and D_T refer to the axial and transverse diameters (from [51, 61])

servation angle may result in underprediction of the maximum transverse diameter. However, the simulation results remain within the experimental error bars.

Next, we compare the MS-RBC versus the LD-RBC models; Fig. 10.7 (right) presents the axial and transverse RBC deformations for a healthy RBC and for a RBC at the latest stage (schizont) of intra-erythrocytic parasite development in malaria disease in comparison with experiments [9]. Simulation results are in excellent agreement with the experiments for both RBC models. The Young's modulus of a RBC is found to be 18.9 and 180.0 μN/m for healthy RBC and at the schizont stage, respectively, in case of the MS-RBC model, while the LD-RBC model yields the values of 20.0 and 199.5 μN/m for the RBC Young's modulus. Note that the low-dimensional RBC model is able to capture linear as well as non-linear RBC elastic response.

10.3.2 Membrane rheology from Twisting Torque Cytometry

Twisting torque cytometry (TTC) is the numerical analog of the optical magnetic twisting cytometry (OMTC) used in the experiments [11, 62], where a ferrimagnetic microbead is attached to RBC top and is subjected to an oscillating magnetic field. In simulations a microbead is attached to the modelled membrane, and is subjected to an oscillating torque as shown in Fig. 10.8 (left). In analogy with the experiments, the modelled RBC is attached to a solid surface, where the wall-adhesion is modelled by keeping stationary fifteen percent of vertices on the RBC bottom, while other vertices are free to move. The adhered RBC is filled and surrounded by fluids having viscosities much smaller than the membrane viscosity, and therefore, only the membrane viscous contribution is measured. The microbead is simulated by a set of vertices on the corresponding sphere subject to a rigid body motion. The bead attachment is modelled by including several RBC vertices next to the microbead bottom into the rigid motion.

A typical bead response to an oscillating torque measured in simulations is given in Fig. 10.8 (right). The bead displacement has the same oscillating frequency as the

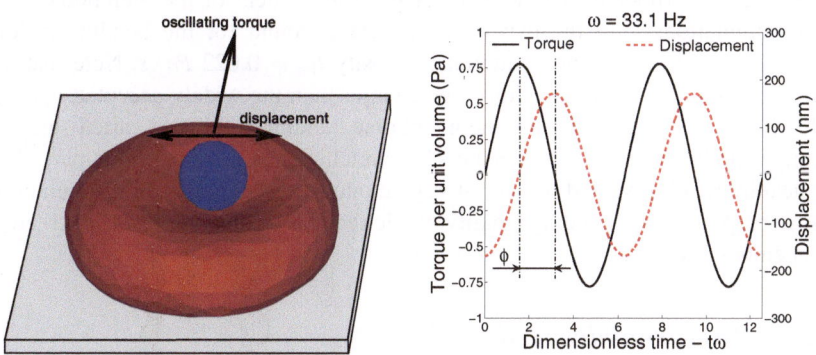

Fig. 10.8. Setup of the TTC (left) and the characteristic response of a microbead subjected to an oscillating torque (right)

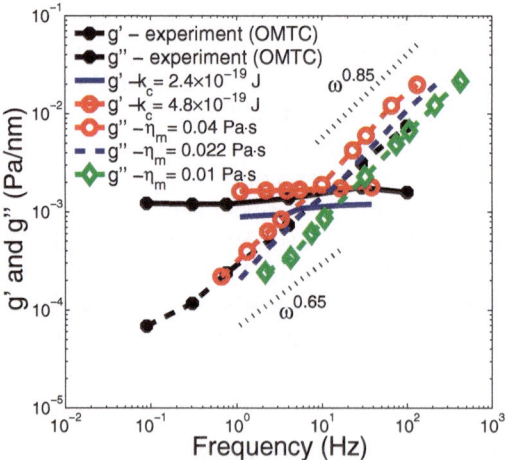

Fig. 10.9. Two-dimensional storage and loss (g' and g'') moduli of the RBC membrane obtained from simulations for different membrane viscosities and bending rigidities in comparison with the experiments [11] (from [51])

applied torque per unit volume, but it is shifted by a phase angle ϕ depending on the frequency. The phase angle can be used to derive components of the complex modulus according to linear rheology as follows

$$g'(\omega) = \frac{\Delta T}{\Delta d} \cos(\phi), \qquad g''(\omega) = \frac{\Delta T}{\Delta d} \sin(\phi), \qquad (10.37)$$

where $g'(\omega)$ and $g''(\omega)$ are the *two-dimensional* storage and loss moduli (G' and G'' in 3D), and ΔT and Δd are the torque and bead displacement amplitudes. Note that under the assumption of no inertial effects, the phase angle satisfies the condition $0 \leq \phi < \pi/2$.

Fig. 10.9 presents components of the complex modulus for healthy RBCs compared with experimental data of [11]. A good agreement of the membrane moduli in simulations with the experimental data is found for the bending rigidity $k_c = 4.8 \times 10^{-19}$ J and the membrane viscosity $\eta_m = 0.022$ $Pa \cdot s$. Note that this corresponds to the bending rigidity twice larger than the widely accepted value of 2.4×10^{-19} J. In Fig. 10.9 only the membrane bending rigidity is varied since the Young's modulus was obtained in the RBC stretching tests above. In summary, TTC for healthy RBCs revealed that the storage modulus (g') depends on the membrane elastic properties and bending rigidity, while the loss modulus (g'') is governed by the membrane viscosity.

10.3.3 RBC dynamics in shear flow

Experimental observations [14, 15, 63, 64] of RBC dynamics in shear flow show RBC tumbling at low shear rates and tank-treading at high shear rates. This be-

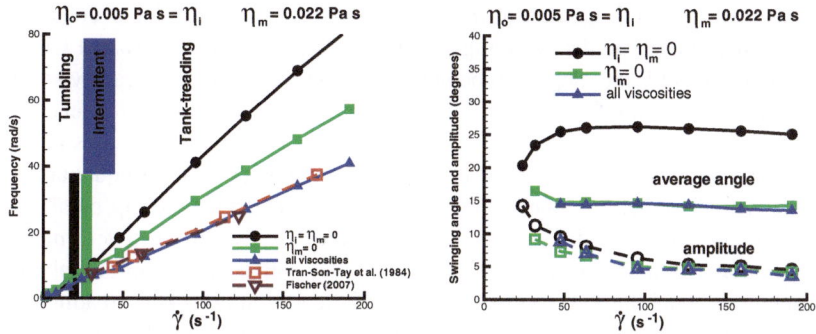

Fig. 10.10. Tumbling and tank-treading frequency (left) of a RBC in shear flow and swinging average angle and amplitude (right) for different cases: 1) $\eta_o = 5 \times 10^{-3}\ Pa \cdot s$, $\eta_i = \eta_m = 0$ (circles); 2) $\eta_o = \eta_i = 5 \times 10^{-3}\ Pa \cdot s$, $\eta_m = 0$ (squares); 3) $\eta_o = \eta_i = 5 \times 10^{-3}\ Pa \cdot s$, $\eta_m = 22 \times 10^{-3}\ Pa \cdot s$ (triangles) (from [51])

haviour is related to existence of a RBC minimum energy state shown in the experiments by Fischer [15], where a RBC relaxed to its original state marked by several attached microbeads after some time of tank-treading motion. Hence, the RBC has to exceed a certain energy barrier in order to transit into a tank-treading motion in shear flow.

Theoretical predictions [14, 33] attempt to capture RBC dynamics in shear flow depending on the shear rate and the viscosity contrast defined as $\lambda = (\eta_i + \eta_m)/\eta_o$. According to the theories, for a small $\lambda < 3$ a RBC tumbles at low shear rates and tank-treads at high shear rates. Near the tumbling-to-tank-treading transition there exists a narrow intermittent region where theories predict an instability such that RBC tumbling can be followed by tank-treading and vise versa. However, in case of a large viscosity contrast ($\lambda > 3$) the theories predict a well-defined tumbling regime followed by an intermittent region, while stable tank-treading may not be present. In addition, the tank-treading state is also characterized by RBC swinging around the tank-treading axes with certain frequency and amplitude.

A simulated RBC is suspended into a solvent placed between two parallel walls moving with constant velocities in opposite directions. Fig. 10.10 (left) shows tumbling and tank-treading frequencies with respect to shear rates in comparison with experiments [63, 64]. Comparison of the simulated dynamics with experiments showed that a purely elastic RBC with or without inner solvent (circles and squares) results in an overprediction of the tank-treading frequencies, because the membrane assumes no viscous dissipation. Addition of the membrane viscosity (triangles) reduces the values of the tank-treading frequencies and provides a good agreement with experiments for the membrane viscosity $\eta_m = 22 \times 10^{-3}\ Pa \cdot s$. Note that for all cases a finite intermittent region is observed and it becomes wider for a non-zero membrane viscosity. This result is consistent with the experiments, but it disagrees with the theoretical predictions. Similar results for the intermittent region were reported in simulations of viscoelastic vesicles [65]. Moreover, an increase in the internal fluid or membrane viscosities results in a shift of the tumbling-to-tank-treading transition to higher shear rates. Fig. 10.10 (right) shows the average RBC tank-treading angle and the swinging amplitude. The values are consistent with experimental data [14] and appear to be not very sensitive to the membrane viscosity. Note that the swinging frequency is equal to twice the tank-treading frequency.

In conclusion, the RBC model accurately captures membrane dynamics in shear flow, while the theoretical models can predict RBC dynamics *at most* qualitatively. The theoretical models assume ellipsoidal RBC shape and a fixed (ellipsoidal) RBC tank-treading path. Our simulations showed that a RBC is subject to deformations along the tank-treading axis. In addition, modelled RBCs show substantial shape deformations (buckling) in a wide range around the tumbling-to-tank-treading transition. A degree of these deformations depends on the Föppl-von Kármán number κ defined as YR_0^2/k_c, where $R_0 = \sqrt{A_0^{tot}/(4\pi)}$. As an example, if the RBC bending rigidity is increased by a factor of five, the aforementioned shape deformations become considerably smaller, while if the RBC bending rigidity is increased by a factor of ten, the shape deformations practically subside. The theoretical models do not take the bending rigidity into consideration, while experimental data are not conclusive on this issue. This again raises the question about the magnitude of bending rigidity of healthy RBCs since our simulations (TTC and RBC dynamics in shear flow) indicate that the RBC bending rigidity may be several times higher than the widely used value of $k_c = 2.4 \times 10^{-19} \, J$.

10.4 Validation

In the previous section we demonstrated how we can use experimental data from single-cell measurements to extract the input parameters for the models, but also, to partially validate the simulated biophysical behaviour of single RBCs. In this section, we extend this validation further by comparing simulation results based on the MS-RBC model as well as on the LD-RBC model with different experiments. First, we consider data from microfluidic experiments in channels with very small cross-sections, i.e., comparable to the smallest capillaries. We also compare with experimental results from the dynamic response of RBCs going though properly microfabricated geometric constrictions. Subsequently, we present simulation results for whole blood in terms of the flow resistance in tubes and compare against well known experimental results. Finally, we demonstrate how these multiscale simulations can be used as a "virtual rheometer" to obtain the human blood viscosity over a wide range of shear rate values. This includes the low shear rate regime, where the formation of rouleaux is shown to determine the strong non-Newtonian behaviour of blood.

10.4.1 Single RBC: comparison with microfluidic experiments

Microfabrication techniques allow manufacturing of channels with dimensions comparable to the smallest blood vessels. In recent years, microfluidic experiments have become popular in measuring properties of RBCs and other cells. Even though, at

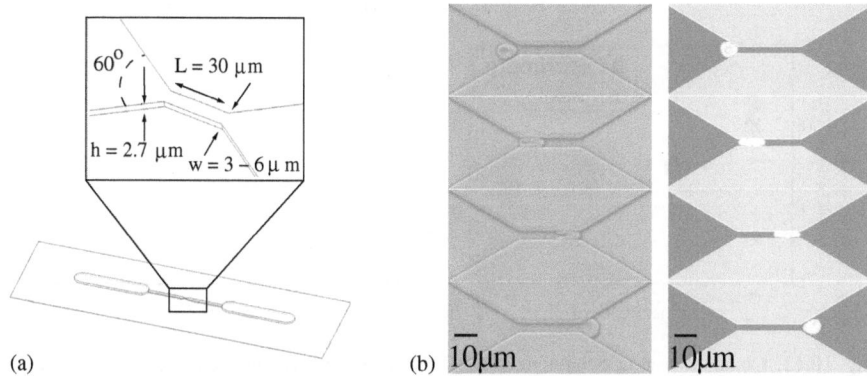

Fig. 10.11. (a) Schematic view of microfluidic channels used in experiments. (b) Shape characteristics of RBC traversal across microfluidic channels: Experimental (left) and simulated (right) images of erythrocyte traversal across 4 μm wide, 30 μm long, 2.7 μm high channel at room temperature and an applied pressure difference of 0.085 *kPa* (from [67])

present time, these experiments typically do not include biochemistry, they can provide quantitative information about the motion of a single RBC through the channels at controlled conditions. This information can be used to validate computational models. The two examples of RBC model validation using microfluidic devices described in this section are taken from refs.[66] and [67], where detailed description of experiments and simulations can be found.

The first set of experiments was performed in the S. Suresh lab at MIT. The channel structures used in these experiments are illustrated in Fig. ü10.11(a). At their narrowest point, these sharply converging/diverging channels are 30 μm long, 2.7 μm high and have widths ranging from 3 to 6 μm. The experiments were carried out at temperature 37°C and 41°C. High-speed imaging was used to measure and quantify the temperature-dependent flow characteristics and shape transitions of RBCs as they traversed microfluidic channels of varying size.

The fluid domain in DPD simulations corresponds to the middle part of the microfluidic device. The width of the flow domain is 60 μm, the length is 200 μm, and the height is 2.7 μm. The central part of the simulation domain is the same as in the experiment. Specifically, the flow is constricted to rectangular cross-section of 4, 5 or 6 μm in width and 2.7 μm in height. The walls are modelled by freezing DPD particles in combination with bounce-back reflection, similar to those in [68]. Periodic inlet/outlet boundary conditions are employed, and the flow is sustained by applying an external body force. The RBC model consists of 500 points. Bounce-back reflection is employed at the membrane surface. The internal RBC fluid is 9, 8.5 and 7.6 times more viscous than the external fluid in simulations corresponding to temperature of 22°C, 37°C and 41°C, respectively [69]. The effect of temperature in the experiment on the viscosity of the suspending medium is modelled by changing the viscosity of the DPD fluid surrounding the RBC. Specifically, the viscosity of the external fluid at 37°C and 41°C is decreased by 22 % and 28 % compared to the

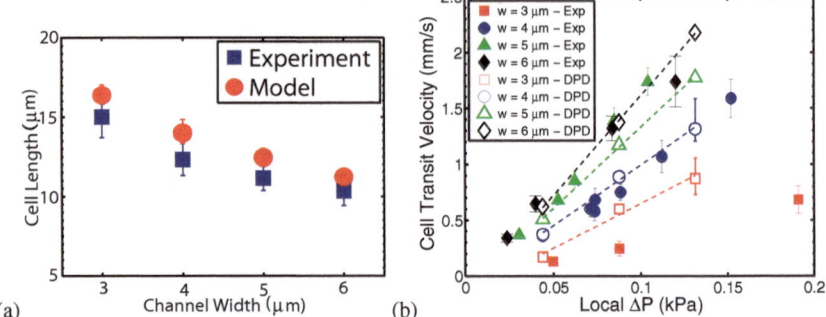

Fig. 10.12. Quantitative flow behaviour of RBC traversal of microfluidic channels. (a) Measured and simulated cell lengths at the center of the microfluidic channel for varying channel widths. (b) Comparison of DPD simulation results (open markers) with experimentally measured mean velocities (filled markers) of RBC traversal as a function of measured local pressure differences for 3, 4, 5 and 6 o µm channel widths (height = 2.7 µm, length = 30 µm). Error bars on experimental data points represent an average $+/-$ one standard deviation of a minimum of 18 cells. Error bars on modelling data points indicate minimum and maximum variations resulting from a case study exploring the sensitivity of the RBC traversal to channel geometry and cell volume (from [66])

viscosity at 22 °C, while the membrane viscosity is decreased by 50 % and 63.5 %, respectively, to match the experimentally measured RBC relaxation times at these temperatures.

Fig. 10.11(b) presents a qualitative comparison of experiment with the DPD model for RBC traversal across a 4 µm wide channel. Here, the cell undergoes a severe shape transition from its normal biconcave shape to an ellipsoidal shape with a longitudinal axis up to 200 % of the average undeformed diameter. Fig. 10.12(a) illustrates how the longitudinal axis of the cell, measured at the center of the channel, changes with different channel widths. Experimental and simulated longitudinal RBC axes typically differ no more than 10–15 %. Fig. 10.12(b) presents pressure–velocity relationships for RBC flow across channels of different cross-sectional dimensions. Average cell velocity measurements were taken between the point just prior to the channel entrance (the first frame in Fig. 10.11(b)) and the point at which the cell exits the channel (the final frame in Fig. 10.11(b)). The DPD model adequately captures the scaling of flow velocity with average pressure difference for 4–6 µm wide channels. The significant overlap in the experimental data for 5–6 µm wide channels can be attributed largely to variations in cell size and small variations in channel geometry introduced during their microfabrication. For the smallest channel width of 3 µm, the experimentally measured velocities are as much as half those predicted by the model. This may be attributed to several factors, including non-specific adhesive interactions between the cell membrane and the channel wall due to increased contact. Furthermore, this 3 µm × 2.7 µm (8.1 µm²) cross-section approaches the theoretical 2.8 µm diameter (6.16 µm²) limit for RBC transit of axisymmetric pores [70]. Therefore, very small variations in channel height (due, for example, to channel swelling/shrinking due to small variations in temperature and humidity) can have significant effects.

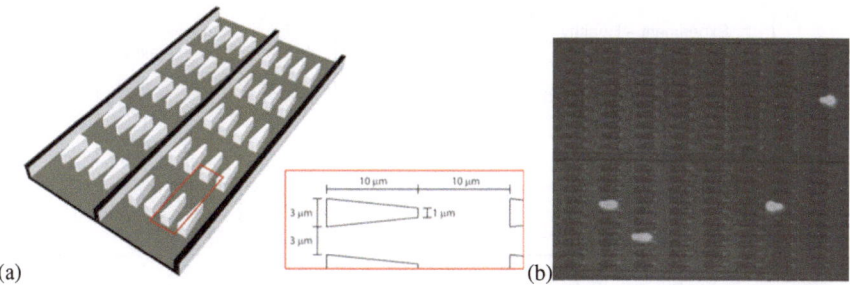

Fig. 10.13. (a) Two parallel channels, one with opening geometries that are the reverse of the other, are connected to common inlet and outlet reservoirs. The height of the device is 4.2 μm, the distance between the pillars is 3 μm, and the distance between rows of pillars is 10 μm. (b) Snapshot from video of labeled cells moving through the device. Liquid flow is from right to left (from [67])

For the effect of temperature on the flow dynamics of the RBC we refer the reader to [66] where the ratio of the local pressure gradient and average cell velocity ($\Delta P / V$) as a function of temperature is examined. The main finding is that there exists a threshold cross-section below which the RBC viscous components begin to play a significant role in its dynamic flow behaviour; this effect is less profound at higher temperatures. Since the energy dissipation in the membrane is typically higher than in the internal fluid, one might expect the influence of membrane viscosity on the flow behaviour of the RBC across such small cross-sections to be large compared to the internal fluid viscosity [71].

The second set of experiments was performed in the J. Han Lab at MIT. The microfluidic device consists of two channels, 4.2 μm in height. Rows of 3 by 10 μm triangular obstacles are placed into the channels as shown in Fig. 10.13(a). The distance between the obstacles is 3 μm, while the distance between rows of obstacles is 10 μm. The only difference between the two channels in the device is the orientation of the obstacles; one channel is the other flipped by 180°.

For low-Reynolds number flows, the resistance and average fluid velocities in the absence of cells must be the same for both channels. When the RBC concentration is low, the cells move with different average velocities in the two channels. This indicates that for openings of the same minimal cross-section area, the geometry (rate) of constriction affects the amount of force required for cell traversal. Also, the channels appear to be sensitive to some specific properties of RBCs, therefore the device can be used to estimate these properties for a given cell from its velocity at known applied pressure gradient.

In simulations, the solid walls are assembled from randomly distributed DPD particles whose positions are fixed. In addition, bounce-back reflections are used to achieve no-slip conditions and prevent fluid particles from penetrating the walls [68]. A portion of the microfluidic device with dimensions 200 by 120 by 4.2 microns containing 5 rows of pillars (10 pillars in each row) is modelled. The fluid region is bounded by four walls while periodic boundary conditions are used in the flow direction. Here, the RBC is simulated using 5,000 DPD particles to obtain accurate results unlike most of the other simulations, including the previous example, where

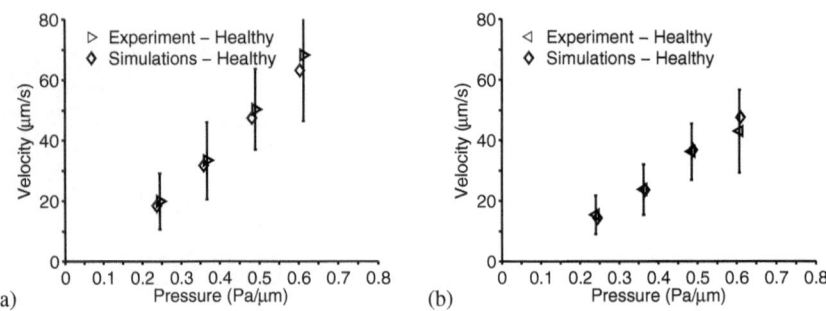

Fig. 10.14. Average velocity of healthy RBCs as a function of pressure gradient and comparison of simulation and experimental results. Results for converging (a) and diverging geometries (b) (from [67])

500 DPD points per RBC were sufficient. This is due to the fast dynamic changes of the RBC membrane as the RBC travels through the narrow constrictions. Parameters of the healthy cell model are derived from RBC spectrin network properties as described in previous sections. In addition, membrane fluctuation measurements and optical tweezers experiments are used to define simulation parameters. Specifically, we required that the amplitude of thermal fluctuations of the membrane at rest to be within the range of experimental observations [12]. We also required that the characteristic relaxation time of the RBC model in simulations be equal to the experimentally measured value of 0.18 seconds. The RBC model is immersed into the DPD fluid. The membrane particles interact with internal and external fluid particles through the DPD forces. By changing the direction of the body force, the motion of the cell through channels with converging and diverging pores is simulated using the same channel geometry.

The DPD model is able to capture the effect of obstacle orientation quite accurately. Quantitative comparison of simulation results with experimental data for healthy cell velocity as a function of applied pressure gradient is shown in Fig. 10.14.

In order to evaluate contributions of individual mechanical properties of the cell to overall dynamic behaviour, we run additional simulations. The DPD model provides a unique opportunity to perform this analysis, since experimental evaluation of these contributions is laborious or impossible. Larger cells are found to travel with lower velocities; however, the velocity variation due to cell size is not significant. Additional simulations were performed in which the membrane shear modulus and membrane viscosity were varied independently of each other. The results showed that the RBC velocity in the device is sensitive to shear modulus, while (in contrast to the device described above) variation of membrane viscosity did not affect the RBC traversal significantly. This finding may seem to be counter intuitive; when the membrane viscosity is increased one would expect higher energy dissipation and therefore lower RBC velocity. Indeed, increased membrane viscosity increases the time it takes for a RBC to traverse an individual opening between pair of obstacles. However, it also slows down the recovery of RBC shape when the cell is traveling between rows of obstacles, making it easier to enter the next opening. As a result, the particular design of this device lessens the dependence of the cell velocity on membrane viscosity.

10.4.2 Whole healthy blood

Next, we present simulation results for whole blood modelled as a suspension of healthy RBCs using the two RBC models without changing the parameters that we have established from single-cell measurements. We first consider flow in a tube in order to assess flow resistance in microvessels, and subsequently, we focus on Couette flow in order to compare the predicted blood viscosity from rheometric measurements.

10.4.2.1 Flow resistance

Here, we simulate blood flow in tubes of diameters ranging from 10 μm to 40 μm. In case of the MS-RBC model, it is important to model carefully the excluded volume (EV) interactions among cells, which are often implemented through a repulsive force between membrane vertices of different cells. A certain range (force cutoff radius) of the repulsive interactions may impose a non-zero minimum allowed distance between neighboring RBC membranes, which will be called "screening distance" between membranes. The choice of a smaller cutoff radius may result in overlapping of cells, while a larger one would increase the screening distance between cells, which may be unphysical and may strongly affect the results at high volume fractions of RBCs. A better approach is to enforce EV interactions among cells by employing reflections of RBC vertices on the membrane surfaces of other cells yielding essentially a zero screening distance between two RBC surfaces. In addition, we employ a *net repulsion* of RBCs from the wall by properly setting the repulsive force coefficient between the wall particles and the cell vertices.

Fig. 10.15 shows plots of the apparent blood viscosity with respect to the plasma viscosity. The apparent viscosity is defined as follows $\eta_{app} = \frac{\pi \Delta P D^4}{128 Q L}$, where ΔP is the pressure difference, Q is the flow rate, and L is the length of the tube. It increases for higher H_t values since higher cell crowding yields larger flow resistance. It is more convenient to consider the relative apparent viscosity defined as $\eta_{rel} = \frac{\eta_{app}}{\eta_s}$, where η_s is the plasma viscosity. Fig. 10.15(a) shows the simulated η_{rel} values in comparison with the empirical fit to experiments [72] for the tube diameter range 10–40 μm and H_t values in the range 0.15–0.45. Excellent agreement between simulations and experiments is obtained for the proper EV interactions for all cases tested. The pressure gradients employed here are 2.633×10^5, 1.316×10^5, and 6.582×10^4 Pa/m for tubes of diameters 10, 20, and 40 μm, respectively. In the case of low hematocrit H_t (e.g., 0.15) the velocity profiles closely follow parabolic curves in the near-wall region. In the central region of the tube a substantial reduction in velocity is found for all volume fractions in comparison with the parabolic profiles indicating a decrease in the flow rate [73]. Fig.10.15(b) shows results from both the MS-RBC and LD-RBC models for a wider range of tube diameters. The agreement is good between

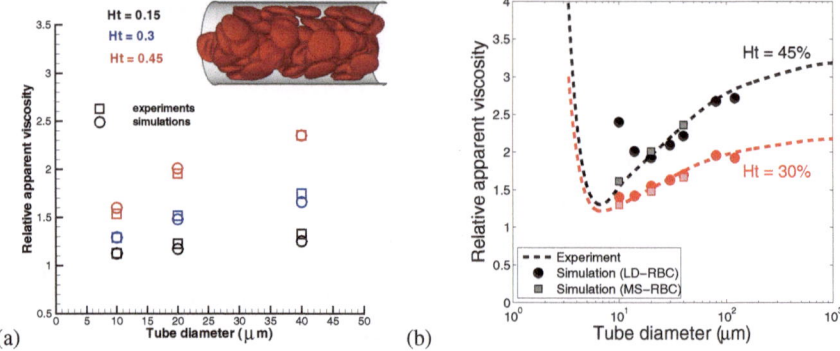

Fig. 10.15. Flow resistance in healthy blood: (a) Relative apparent viscosity compared with experimental data [72] for various hematocrit values and tube diameters. The inset plot is a snapshot of RBCs in Poiseuille flow in a tube of a diameter $D = 20\,\mu$m at $H_t = 0.45$. (b) Comparison of MS-RBC and LD-RBC models; the lines are the empirical correlation by Pries et al. [72] (from [51, 61])

the models and the experimental data represented by an empirical fit; however, it is clear that for vessels with diameter below $15 - 20$ microns the LD-RBC model fails as the membrane rheology becomes important, which the low-dimensional model does not account for.

RBCs in Poiseuille flow migrate to the tube center forming a core in the flow. The inset of Fig. 10.15 shows a sample snapshot of RBCs flowing in a tube of diameter $D = 20\,\mu$m. A RBC core formation is established with a thin plasma layer next to the tube walls called the *cell-free* layer (CFL) [73]. The thickness of the CFL is directly related to the Fahraeus and the Fahraeus-Lindqvist effects, both of which were accurately captured by the DPD model, see [73]. To determine the CFL thickness we computed the outer edge of the RBC core, which is similar to CFL measurements in experiments [74, 75]. Fig. 10.16 shows a sample CFL edge from simulations for $H_t = 0.45$ and $D = 20\,\mu$m and local CFL thickness distribution, which is constructed from a set of discrete local measurements of CFL thickness taken every $0.5\,\mu$m along the x (flow) direction. The fluid viscosity of the CFL region is much smaller than that of the tube core populated with RBCs providing an effective lubrication for the core to flow.

10.4.2.2 Aggregation and Rouleaux formation

Here, we present simulations in a wide range of shear rate values including the low shear rate regime with and without the aggregation models described in Sects. 10.2.3.4 and 10.2.4.2. The viscosity was derived from simulations of plane Couette flow using the Lees-Edwards periodic boundary conditions in which the shear rate and the density of cells were verified to be spatially uniform. The experimental viscosities of well-prepared erythrocytes without rouleaux and of whole blood were measured at hematocrit 45 % and at temperature $37°$ C by [76, 77, 78] using rotational Couette viscometers. At the same conditions for both the MS-RBC and

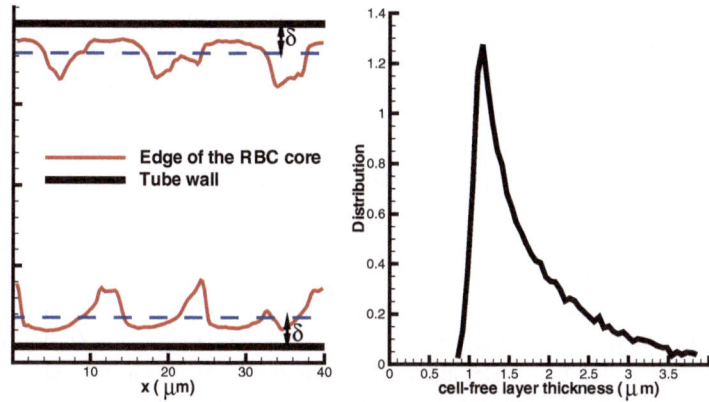

Fig. 10.16. An example of a CFL edge (left) and local CFL thickness distribution (right) for $H_t = 0.45$ and $D = 20\,\mu m$ (from [51])

the LD-RBC suspensions the viscosities were computed, with and without rouleaux, as functions of the shear rate over the range $0.005s^{-1}$ to $1000.0s^{-1}$. RBC suspension viscosities were normalized by the viscosity values of their suspending media. These data are compared in Fig. 10.17(a) as relative viscosity against shear rate at constant hematocrit. The MS-RBC model viscosity curves lie very close to the viscosities measured in three different laboratories. The model, consisting only of RBCs in suspension, clearly captures the effect of aggregation on the viscosity at low shear

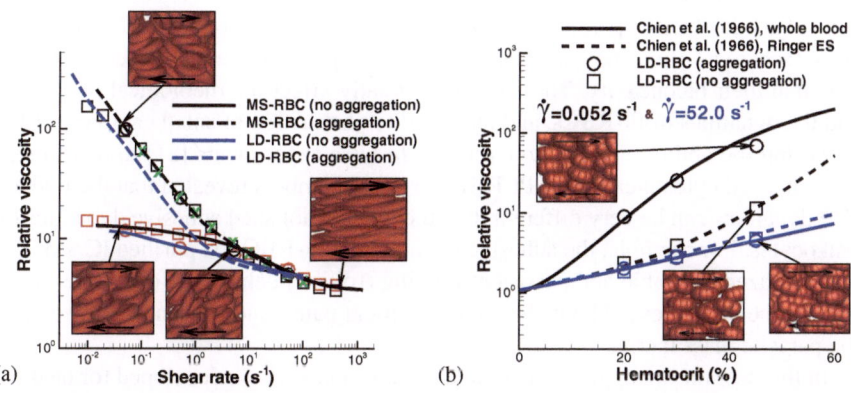

Fig. 10.17. Validation of simulation results for whole blood and Ringer erythrocyte solutions (ESs). (a) Plot of non-Newtonian viscosity relative to solvent viscosity as a function of shear rate at $H_t = 45\,\%$ and $37^{\circ}C$: *simulated* curves of this work, as indicated, and *experimental* points: Whole blood: green crosses - Merrill et al. [76]; black circles - Chien et al. [77], black squares - Skalak et al. [78]. Ringer ES: red circles – Chien et al. [77]; red squares - Skalak et al. [78]. (b) Plot of relative viscosity as a function of hematocrit (H_t) at shear rates 0.052 (black) and 52.0 (blue) s^{-1}: *simulated* (LD-RBC points), and Chien et al. [77] *experimental* fits for whole blood (solid lines) and Ringer ES (dashed lines) (from [61])

rates, and suggests that particles other than RBCs have little effect on the viscosity. The measured values for whole blood are more consistent than those for erythrocyte solutions, which may reflect differences in the preparation of the latter. The LD-RBC model underestimates somewhat the experimental data, but is generally in good agreement over the whole range of shear rates, and again demonstrates the effect of aggregation. This is remarkable in view of the simplicity and economy of that model.

The dependence of whole blood and erythrocyte solution viscosity on hematocrit (H_t) is demonstrated in Fig. 10.17(b). The curves are measured viscosities correlated with H_t at constant shear rate by Chien et al. [77], and the points are calculated with the LD-RBC model. This clearly shows how the latter captures the H_t dependence of viscosity, and that the model again demonstrates aggregation to be crucial for a quantitative account of the difference between the viscosity of whole blood and that of washed erythrocyte suspensions.

10.5 Application to malaria modelling

Plasmodium falciparum (Pf) causes one of the most serious forms of malaria resulting in several million deaths per year. Pf-parasitized cells (Pf-RBCs) experience progressing changes in their mechanical and rheological properties as well as in their morphology [79, 80] during intra-erythrocytic parasite development, which includes three stages from the earliest to the latest: ring → trophozoite → schizont. Progression through these stages leads to considerable stiffening of Pf-RBCs as found in optical tweezers stretching experiments [9] and in diffraction phase microscopy by monitoring the membrane fluctuations [12]. Pf development also results in vacuoles formed inside of RBCs possibly changing the cell volume. Thus, Pf-RBCs at the final stage (schizont) often show a "near spherical" shape, while in the preceding stages maintain their biconcavity. These changes greatly affect the rheological properties and the dynamics of Pf-RBCs, and may lead to obstruction of small capillaries [80] impairing the ability of RBCs to circulate. *In vitro* experiments [81] to investigate the enhanced cytoadherence of Pf-RBCs in flow chambers revealed that their adhesive dynamics can be very different than the well-established adhesive dynamics of leukocytes. For example, the adhesive dynamics of Pf-RBCs on purified ICAM-1 is characterized by stable and persistent flipping (rolling) behaviour for a wide range of wall shear stresses [81] but also by intermittent pause and sudden flipping due to the parasite mass inertia.

In this section, we apply the computational framework we developed for healthy RBCs to Pf-RBCs. In particular, we first consider single RBCs for validation purposes and subsequently we simulate whole infected blood as suspension of a mixture of healthy and Pf-RBCs. We examine the mechanical, dynamic and rheologic responses as well as the adhesive dynamics of infected RBCs.

10.5.1 Single cell

We include in this section comparison with optical tweezers experiments and with microfluidics to assess the fidelity of the RBC models to reproduce the mechanics and dynamics of Pf-RBCs.

10.5.1.1 Mechanics

In malaria disease, progression through the parasite development stages leads to a considerable stiffening of Pf-RBCs compared to healthy ones [9, 12]. Furthermore, in the schizont stage the RBC shape becomes near-spherical whereas in the preceding stages RBCs maintain their biconcavity. Fig. 10.18 shows simulation results for Pf-RBCs at different stages of parasite development. The simulation results were obtained with the MS-RBC model using 500 points. Table 10.1 presents the shear moduli of healthy and Pf-RBCs at different stages; these values are consistent with the experiments of [9, 12]. The bending rigidity for all cases is set to $2.4 \times 10^{-19} J$, which is the value of bending rigidity for healthy RBCs, as the membrane bending stiffness for different stages is not known. The curve for the schizont stage marked as "near-spherical" corresponds to stretching an ellipsoidal shape with axes $a_x = a_y = 1.2a_z$. Here, the membrane shear modulus of $40\,\mu N/m$ matches the stress-strain response

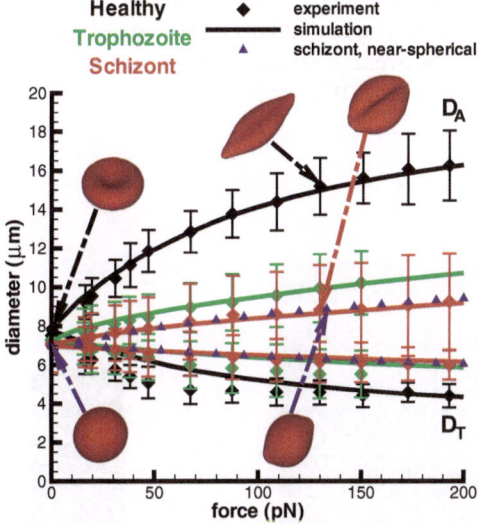

Fig. 10.18. Stretching response of Pf-RBCs using the MS-RBC for different stages compared with the experiments [9]. D_A and D_T refer to the axial and transverse diameters (from [51])

Table 10.1. Shear moduli of healthy and Pf-RBCs in $\mu N/m$ at $T = 23\,°C$. The "*" denotes a "near-spherical" RBC at the schizont stage

Healthy	Ring	Trophozoite	Schizont
6.3	14.5	29	60 & 40*

with the experiment, i.e., it is smaller than that for the biconcave-shape simulation. For the near-spherical cell the membrane is subject to stronger local stretching for the same uniaxial deformation compared to the biconcave shape. For the deflated biconcave shape, the inner fluid volume can be deformed in response to stretching, while in the near-spherical shape the fluid volume applies additional resistance onto the stretched membrane. Hence, the cell geometry plays an important role, and it has to be closely modelled for accurate extraction of parameters from the optical tweezers experiments.

10.5.1.2 Microfluidics

The microfluidic device with triangular obstacles described in Sect. 10.4.1 is used also here to perform experiments for the late ring-stage *P.falciparum*-infected RBCs that are infected with a gene encoding green fluorescent protein (GFP). For both the converging and diverging geometries infected RBCs exhibit lower average veloci-ties that healthy RBCs (see Fig. 10.19(a)). In the DPD simulations, the infected cells are modelled with increased shear modulus and membrane viscosity values obtained from optical tweezers as explained in the previous section. We model the parasite as a rigid sphere of two microns in diameter [82] placed inside the cell (see Fig. 10.1(b)). Snapshots from simulations showing passage of an infected RBC through channels with converging and diverging pore geometries are shown in Fig. 10.19(b). The DPD model is able to capture the effect of changes of RBC properties arising from para-sitization quite accurately. A quantitative comparison of the simulation results with experimental data for the average velocity of Pf-RBCs as a function of applied pres-sure gradient is shown in Fig. 10.20.

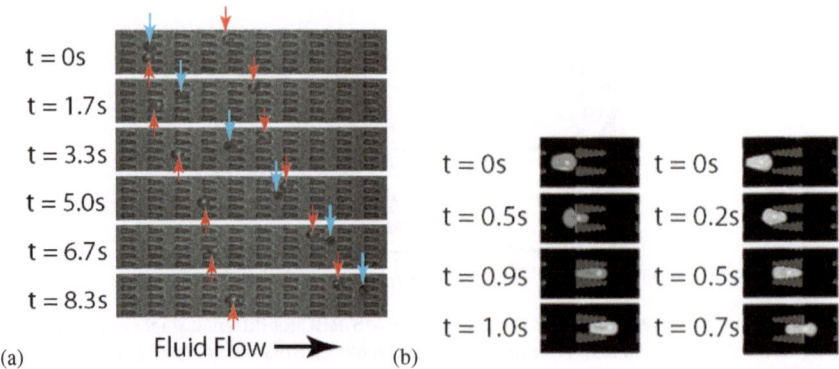

(a)   Fluid Flow ⟶   (b)

Fig. 10.19. (a) Experimental images of ring-stage P. falciparum-infected (red arrows) and unin-fected (blue arrows) RBCs in the channels at a pressure gradient of 0.24 Pa/μm. The small flu-orescent dot inside the infected cell is the GFP-transfected parasite. At 8.3 *s*, it is clear that the uninfected cell moved about twice as far as each infected cell. (b) DPD simulation images of P. falciparum-infected RBCs traveling in channels of converging (left) and diverging (right) opening geometry at 0.48 Pa/μm (from [67])

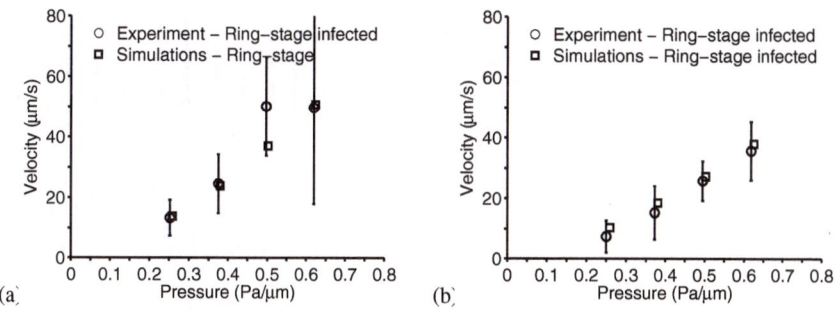

Fig. 10.20. Average velocity of ring-stage malaria infected RBCs as a function of pressure gradient and comparison of simulation and experimental results. Results for converging (a) and diverging (b) geometries (from [67])

10.5.1.3 Pf-RBC adhesive dynamics

Here, we present typical results for the adhesive dynamics of Pf-RBCs in shear flow for various values of the wall shear stress (WSS). The models employed are described in Sect. 10.2.3.5 with some modifications in order for the simulated RBC dynamics to be comparable with that found in experiments [81] using purified ICAM-1 as a wall coating. Fig. 10.21 shows several successive snapshots of a cell rolling on the wall for the schizont stage of a Pf-RBC. The dynamics of the Pf-RBC is characterized by a "flipping" behaviour initiated first by the cell peeling off the wall due to the force of the hydrodynamic flow after flat RBC adhesion (the first snapshot in Fig. 10.21). After the majority of the initial cell contact area with the wall is peeled off, a RBC flips over on its other side which is facilitated by the remaining

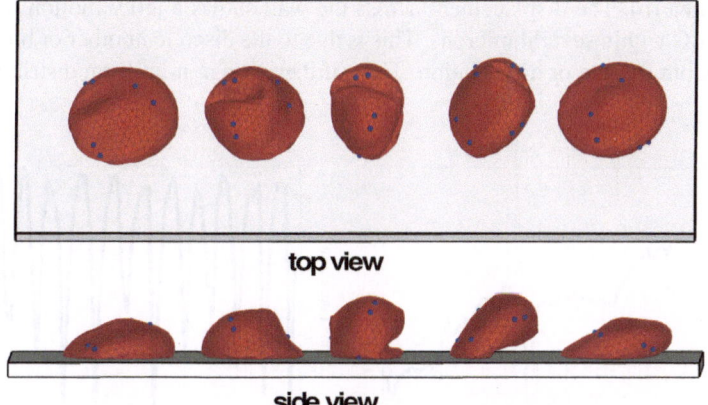

Fig. 10.21. Top and side views of successive snapshots of a single flipping of an infected RBC at the schizont stage. Coordinates along the wall for different snapshots are shifted in order to separate them for visual clarity. Blue particles are added as tracers during post-processing to illustrate the membrane dynamics (from [51])

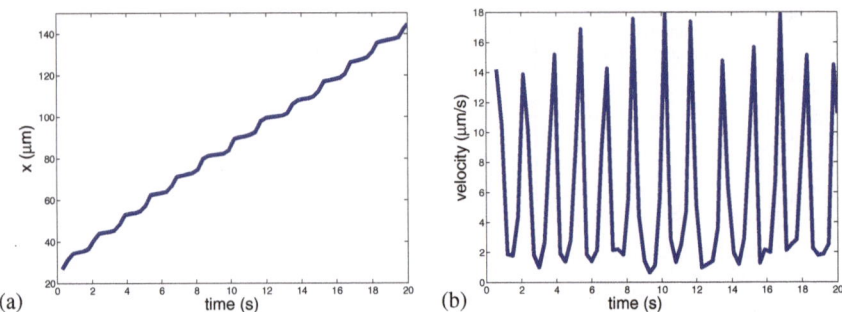

Fig. 10.22. Pf-RBC displacement (a) and velocity (b) along the wall for the schizont stage (from [51])

small contact area with the wall. During these steps Pf-RBCs undergo strong membrane deformations as illustrated in Fig. 10.21. A similar behaviour was found in experiments [81] of Pf-RBCs which showed flipping (rolling) along a wall coated with purified ICAM-1. In agreement with the simulations, RBCs in experiments also showed strong membrane deformations characterized by local membrane buckling.

Fig. 10.22 presents the corresponding displacement along the x coordinate (a) and instantaneous RBC velocity (b). An infected RBC rolls in a relatively stable motion which resembles a staircase. The segments of smaller displacements correspond to the stage of a flat RBC adhesion followed by its slow peeling off the wall (see Fig. 10.21), while the fragments of larger displacements represent the stage of RBC fast flipping described above. The RBC velocity is in agreement with its displacement showing high peaks or fast cell motion during the time segments with larger displacements. The average cell velocity is approximately 5.8 μm/s. Fig. 10.23 shows RBC displacement along the z cross-flow coordinate (a) and instantaneous contact area (b). The displacement across the wall shows a jerky motion of an infected RBC within several microns. This is due to the discrete number of bonds and their random rupture or dissociation. Thus, if there is a non-uniform distribution of

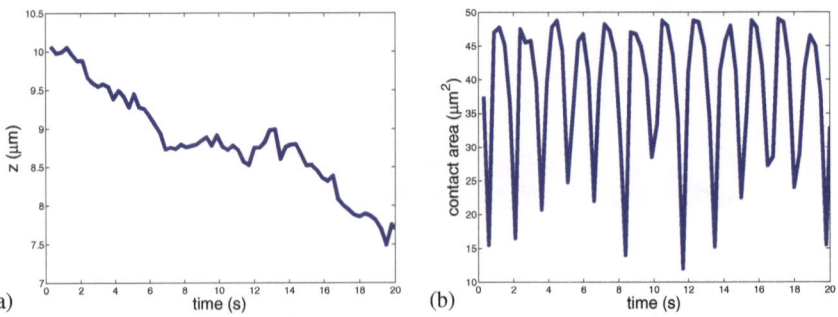

Fig. 10.23. RBC displacement across the wall (a) and the cell contact area (b) for the schizont stage (from [51])

bonds over the contact area at some instance of time, a Pf-RBC may be pulled to one side. In addition, the hydrodynamic force on the RBC may be non-zero in z direction, since the cell is not symmetric due to the local deformations shown in Fig. 10.21. The RBC contact area in Fig. 10.23(b) is correlated with its displacement and velocity in Fig. 10.22. Minima in the contact area coincide with maxima in the RBC velocity corresponding to the stage of fast cell flipping from its one side to the other. The cell contact area remains within the range of $10–50\,\mu m^2$, while the average value is equal to $38.6\,\mu m^2$.

To investigate the dependence of RBC adhesive dynamics on WSS, the velocity of the upper plate is changed. Note that the shear rate is altered at the same time. However, the WSS appears to be a key parameter which governs RBC adhesive dynamics, since adhered RBCs are driven by fluid stresses and roll along the wall with a much smaller velocity than that of the shear flow.

Several initial simulations with a varying WSS and other fixed parameters revealed that a Pf-RBC may exhibit firm adhesion at a WSS lower than $0.317\,Pa$ for the case described above and can completely detach from the wall at higher WSS. At low WSS, adhesion forces are strong enough to counteract the stress exerted on the cell by the flow resulting in its firm sticking to the lower wall. On the contrary, at high WSS existing bonds do not provide sufficiently strong adhesive interactions which yields RBC detachment from the wall. RBC visualizations showed that its detachment at high WSS occurs during the relatively fast motion of RBC flipping, since the contact area at that step corresponds to its minimum. However, in experiments [81] Pf-RBCs which moved on a surface coated with the purified ICAM-1 showed persistent and stable rolling over long observation times and for a wide range of WSS between $0.2\,Pa$ and $2\,Pa$. This suggests that there must be a mechanism which stabilizes rolling of infected RBCs at high WSS. This fact is not surprising since, for example, leukocyte adhesion can be actively regulated depending on flow conditions and biochemical constituents present [83, 84].

To stabilize RBC binding at high WSS we introduce adaptivity of the bond spring constant (k_s) see Eq. (10.29). As the first approximation we assume a linear dependence of k_s on the WSS, such that k_s is increased or decreased proportionally to an increase or decrease in the WSS. Fig. 10.24 presents the average rolling velocity of a Pf-RBC in comparison with experiments of cell rolling on a surface coated with purified ICAM-1 [81]. The simulated average velocities for the "linear" case show a near-linear dependence on the WSS and are in good agreement with experiments up to some WSS value; the simulated value remains between the 10th and the 90th percentiles found in experiments. However, the observed discrepancy at the highest simulated WSS suggests that a further strengthening of cell-wall bond interactions may be required. The dependence of the RBC rolling velocity on WSS found in experiments is clearly non-linear. Therefore, the assumption of linear dependence of k_s on the WSS is likely to be an oversimplification. The simulation results marked "non-linear" in Fig. 10.24 adopt a non-linear dependence of k_s on the WSS, and yield excellent agreement with experiments.

In addition, there may be a change in bond association and dissociation kinetics with WSS which would be able to aid in rolling stabilization of infected RBCs at

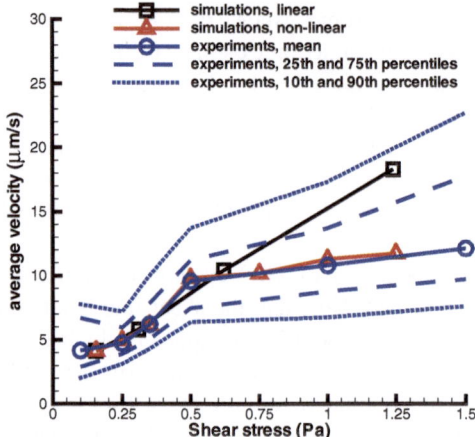

Fig. 10.24. Average rolling velocity of infected RBCs depending on the WSS in comparison with the experiments of cell rolling on purified ICAM-1 [81]. Experimental data include mean values and curves that correspond to the 10th, 25th, 75th, and 90th percentiles (from [51])

high shear rates. The DPD simulations suggest that the adhesive dynamics of Pf-RBCs is not very sensitive to a moderate change (below 30 %–40 %) in k_{on}^0 and k_{off}^0; however, cell dynamics may be strongly affected if these parameters are changed considerably. Moreover, experimental data show a much larger scatter in the average RBC velocity for different cells observed than that in simulations (not shown). This is likely to be related to non-uniform distributions of receptors on the RBC membrane and ligands on the wall. In the simulations, distributions of both receptors and ligands are fixed and are nearly homogeneous with approximately the same area occupied by each receptor or each ligand. A scatter in behaviour among distinct RBCs in the simulations is solely related to the stochastic nature of the adhesive model. However, in experiments irregular distributions of receptors and ligands are likely to significantly contribute to a scatter in RBC adhesive dynamics.

10.5.2 Whole infected blood

Finally, we simulate blood flow in malaria as a suspension of healthy and Pf-RBCs at the trophozoite stage and hematocrit $H_t = 0.45$. Several parasitemia levels (percentage of Pf-RBCs with respect to the total number of cells in a unit volume) from 5 % to 100 % are considered in vessels with diameters 10 and 20 μm. The inset of Fig. 10.25 shows a snapshot of RBCs flowing in a tube of diameter 20 μm at a parasitemia level of 25 %. The main result in Fig. 10.25(a) is given by the plot of the relative apparent viscosity in malaria – a measure of flow resistance as in Sect. 10.4.2 – obtained at different parasitemia levels. The effect of parasitemia level appears to be more prominent for small diameters and high H_t values. Thus, at $H_t = 0.45$ blood flow resistance in malaria may increase up to 50 % in vessels of diameters around 10 μm and up to 43 % for vessel diameters around 20 μm. These increases do not include any

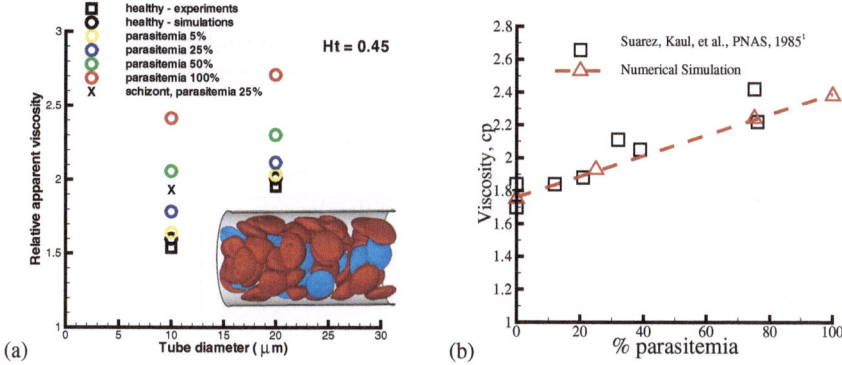

Fig. 10.25. Flow resistance in malaria: (a) Healthy (red) and Pf-RBCs (blue) in Poiseuille flow in a tube of diameter $D = 20\mu m$. $H_t = 0.45$, parasitemia level 25 %. Plotted is the relative apparent viscosity of blood in malaria for various parasitemia levels and tube diameters. Symbol "x" corresponds to the schizont stage with a near-spherical shape. Experimental data from the empirical fit by Pries et al. [72]. (From [51]). (b) Bulk viscosity versus parasitemia level for 30 % hematocrit using a Couette device setup at shear rate 230 s^{-1}. The square symbols are measurements from [85] and the triangles are simulations of Huan Lei (Brown University)

contributions from the interaction of Pf-RBCs with the glycocalyx [69, 86]; such important interactions are complex as they may include cytoadhesion. In Fig. 10.25(b) we also present the bulk viscosity of infected blood (schizont stage) simulated in a Couette type device at shear rate $\gamma = 230s^{-1}$. The DPD simulations compare favorably with the experimental data obtained with a corresponding rheometer in [85]. These validated predictions were obtained without an explicit adhesion model between Pf-RBCs. It seems that such cell-cell interactions are not important at this high shear rate value.

10.6 Summary

In this chapter we have presented a comprehensive simulation methodology based on dissipative particle dynamics (DPD), which is effective in predicting the blood flow behaviour (mechanics, dynamics and rheology) in health and disease. We emphasized, in particular, how single-RBC experiments – using optical tweezers and novel microfluidic devices – can provide data from which we can extract the macroscopic parameters of the model, which can then be related to the microscopic parameters required by the two RBC models we presented. In addition, these single-RBC data can serve as a validation test bed over a wide range of operating conditions. The success of the DPD models is then to predict *whole blood* behaviour in health or disease without any further "tuning" of the models' parameters. We demonstrated that this is indeed the case for healthy and malaria-infected whole blood in two different set ups, i.e., blood flow in a tube as well as in Couette flow. In particular, we

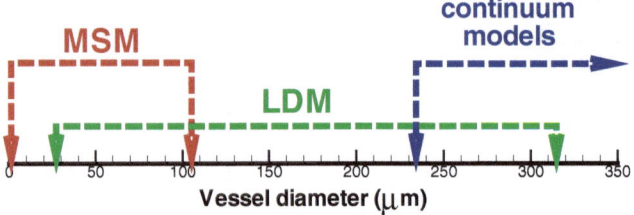

Fig. 10.26. Applicability of different models with respect to a characteristic vessel size (diameter)

presented accurate predictions of the bulk viscosity both for healthy blood as well as for infected blood with parasitemia levels up to 100 %.

The two RBC models we presented can be used in a complementary fashion in simulations of hematologic disorders. The multiscale model (MS-RBC) can resolve structures down to protein level on the lipid bilayer or the spectrin level of the cytoskeleton and can be used efficiently for whole blood simulations (up to 45 % hematocrit) for capillaries and arterioles of approximately up to 100 microns in diameter. On the other hand, the more economical low-dimensional model (LD-RBC) does not account for the membrane explicitly and it is only accurate, as we demonstrated in Sect. 10.4, for arterioles, i.e. vessels with diameter above 15–20 microns. That too, however, can become computationally expensive for high hematocrit values and for large arteries. To this end, a continuum model can be employed above a certain vessel size with the Newtonian constitutive law valid for blood for that size of arterial vessels. This multiscale approach, which is appropriate for the entire human arterial tree, is demonstrated diagrammatically in the sketch of Fig. 10.26.

Next, we comment further on the RBC model parameters used as input from experimental data, which can be roughly divided into three groups: (1) RBC properties, e.g., geometry, shear and bending moduli, membrane viscosity, parasite in malaria, polymerized hemoglobin in sickle cell anemia, etc.; (2) adhesion properties, e.g., receptor and ligand densities, on and off rates, and bond stiffness, and (3) flow properties, e.g., shear rate, fluid viscosity. The flow properties can be effectively reduced to a single parameter, e.g., the wall shear stress (WSS), which is the governing parameter and can influence greatly the adhesive dynamics as we demonstrated in the case of malaria. Often the dominant RBC parameter is the shear modulus not only in the static but also in the dynamic response. For example, the increased shear modulus of Pf-RBCs in comparison with healthy RBCs is the main reason for the flipping motion of Pf-RBCs. Cell geometry could be also very important, for example more spherical Pf-RBCs will likely roll on a surface than flip; shape changes are particularly influential in sickle cell anemia, spherocytosis and elliptocytosis. On the other hand, the RBC bending rigidity seems to play a secondary role in the mechanics or dynamics, and in most cases the membrane viscosity also plays a secondary role, but not for very small capillaries, as we discussed in Sect. 10.4.1. For diseased RBCs, the adhesion parameters govern the states of adhesion, e.g., firm adhesion, flipping, slipping, detachment. For leukocytes, these different states have been studied thoroughly, e.g. [87] and agree well with the plethora of experimental data. In summary, for fixed flow conditions and cell parameters, the space of adhesion parameters can be divided into sub-spaces with different adhesion states. A similar adhesion state diagram can be constructed for diseased RBCs with adhesive properties, but experimental data are currently lacking.

Finally, we want to clarify that we modelled here whole blood as suspension of RBCs in plasma, hence ignoring the effect of white cells (about 0.7 %) and platelets (less than 0.5 %) or the effect of other proteins in the plasma, although we modelled fibrinogen implicitly in Sect. 10.4 on rouleaux formation. From the numerical modelling standpoint, there is no particular difficulty in also modelling these other cells, which are significant in specific biomedical studies, e.g. in thrombosis, immune response. From the biophysical view point, however, as we demonstrated in Sects. 10.4 and 10.5 their presence is not important for the whole blood rheological properties.

Acknowledgements. This work was supported by NIH and NSF. Simulations were performed at the NSF supercomputing center NICS, at the BG/P at ANL via an INCITE DOE award, and at the Jülich supercomputing center.

References

[1] Fung Y.C.: Biomechanics: Mechanical properties of living tissues. Springer-Verlag, New York, second edition, 1993.

[2] Park Y., Best C., Auth T., Gov N., Safran G., Popescu G., Suresh S., Feld M.: Metabolic remodeling of the human red blood cell membrane. Proceedings of the National Academy of Sciences USA **107**: 1289, 2010.

[3] Diez-Silva M., Dao M., Han J., Lim C.-T., Suresh S.: Shape and biomechanical characteristics of human red blood cells in health and disease. MRS Bulletin **35**: 382–388, 2010.

[4] Higgins J., Eddington D., Bhata S., Mahadevan L.: Sickle cell vasoocclusion and rescue in a microfluidic deveice. Proceedings of the National Academy of Sciences USA **104**: 20496, 2007.

[5] Beck W.S. (ed.): Hematology. MIT Press, fifth edition, 1991.

[6] Discher D.E., Mohandas N., Evans E.A.: Molecular maps of red cell deformation: hidden elasticity and in situ connectivity. Science **266**: 1032–1035, 1994.

[7] Waugh R., Evans E.A.: Thermoelasticity of red blood cell membrane. Biophysical Journal **26**: 115–131, 1979.

[8] Henon S., Lenormand G., Richert A., Gallet F.: A new determination of the shear modulus of the human erythrocyte membrane using optical tweezers. Biophysical Journal **76**: 1145–1151, 1999.

[9] Suresh S., Spatz J., Mills J.P., Micoulet A., Dao M., Lim C.T., Beil M., Seufferlein T.: Connections between single-cell biomechanics and human disease states: gastrointestinal cancer and malaria. Acta Biomaterialia **1**: 15–30, 2005.

[10] Chaudhuri O., Parekh S., Lam W., Fletcher D.: Combined atomic force microscopy and side-view optical imaging for mechanical studies of cells. Nature Methods **6**(5): 383–387, 2009.

[11] Puig-de Morales-Marinkovic M., Turner K.T., Butler J.P., Fredberg J.J., Suresh S.: Viscoelasticity of the human red blood cell. American Journal of Physiology: Cell Physiology **293**: 597–605, 2007.

[12] Park Y.-K., Diez-Silva M., Popescu G., Lykotrafitis G., Choi W., Feld M.S., Suresh S.: Refractive index maps and membrane dynamics of human red blood cells parasitized by Plas-

modium falciparum. Proceedings of the National Academy of Sciences USA **105**: 13730–13735, 2008.

[13]   Popescu G., Park Y.-K., Dasari R.R., Badizadegan K., Feld M.S.: Coherence properties of red blood cell membrane motions. Physical Review E **76**: 031902, 2007.

[14]   Abkarian M., Faivre M., Viallat A.: Swinging of red blood cells under shear flow. Physical Review Letters **98**: 188302, 2007.

[15]   Fischer T.M.: Shape memory of human red blood cells. Biophysical Journal **86**: 3304–3313, 2004.

[16]   Suzuki Y., Tateishi N., Soutani M., Maeda N.: Deformation of erythrocytes in microvessels and glass capillaries: effect of erythrocyte deformability. Microcirculation **3**: 49–57, 1996.

[17]   Tomaiuolo G., Preziosi V., Simeone M., Guido S., Ciancia R., Martinelli V., Rinaldi C., Rotoli B.: A methodology to study the deformability of red blood cells flowing in microcapillaries in vitro. Ann Ist Super Sanita **43**: 186–192, 2007.

[18]   Rosenbluth M., Lam W., Fletcher D.: Analyzing cell mechanics in hematologic diseases with microfluidic biophysical flow cytometry. Lab on a Chip **8**: 1062–1070, 2008.

[19]   Li J., Lykotrafitis G., Dao M., Suresh S.: Cytoskeletal dynamics of human erythrocyte. Proceedings of the National Academy of Sciences USA **104**: 4937–4942, 2007.

[20]   Gov N.S.: Active elastic network: cytoskeleton of the red blood cell. Physical Review E **75**: 011921, 2007.

[21]   Gov N.S., Safran S.A.: Red blood cell membrane fluctuations and shape controlled by ATP-induced cytoskeletal defects. Biophysical Journal **88**: 1859–1874, 2005.

[22]   Magowan C., Brown J.T., Liang J., Heck J., Coppel R.L., Mohandas N., MeyerIlse W.: Intracellular structures of normal and aberrant Plasmodium falciparum malaria parasites imaged by soft x-ray microscopy. Proceedings of the National Academy of Sciences of the United States of America **94**: 6222–6227, 1997.

[23]   Eggleton C.D., Popel A.S.: Large deformation of red blood cell ghosts in a simple shear flow. Physics of Fluids **10**: 1834, 1998.

[24]   Evans E.A., Skalak R.: Mechanics and thermodynamics of biomembranes. CRC Press, Inc., Boca Raton, Florida, 1980.

[25]   Pozrikidis C.: Numerical Simulation of Cell Motion in Tube Flow. Annals of Biomedical Engineering **33**: 165–178, 2005.

[26]   Discher D.E., Boal D.H., Boey S.K.: Simulations of the erythrocyte cytoskeleton at large deformation. II. Micropipette aspiration. Biophysical Journal **75**: 1584–1597, 1998.

[27]   Li J., Dao M., Lim C.T., Suresh S.: Spectrin-level modeling of the cytoskeleton and optical tweezers stretching of the erythrocyte. Biophysical Journal **88**: 3707–3719, 2005.

[28]   Dupin M.M., Halliday I., Care C.M., Alboul L., Munn L.L.: Modeling the flow of dense suspensions of deformable particles in three dimensions. Physical Review E **75**: 066707, 2007.

[29]   Dzwinel W., Boryczko K., Yuen D.A.: A discrete-particle model of blood dynamics in capillary vessels. Journal of Colloid and Interface Science **258**: 163–173, 2003.

[30]   Noguchi H., Gompper G.: Shape transitions of fluid vesicles and red blood cells in capillary flows. Proceedings of the National Academy of Sciences USA **102**: 14159–14164, 2005.

[31]   Pivkin I.V., Karniadakis G.E.: Accurate coarse-grained modeling of red blood cells. Physical Review Letters **101**: 118105, 2008.

[32]   Noguchi H., Gompper G.: Dynamics of fluid vesicles in shear flow: Effect of the membrane viscosity and thermal fluctuations. Physical Review E **72**: 011901, 2005.

[33]   Skotheim J.M., Secomb T.W.: Red blood cells and other nonspherical capsules in shear flow: Oscillatory dynamics and the tank-treading-to-tumbling transition. Physical Review Letters **98**: 078301, 2007.

[34]   Succi S.: The Lattice Boltzmann equation for fluid dynamics and beyond. Oxford University Press, Oxford, 2001.

[35]   Malevanets A., Kapral R.: Mesoscopic model for solvent dynamics. Journal of Chemical Physics **110**: 8605–8613, 1999.

[36]   Hoogerbrugge P.J., Koelman J.M.V.A.: Simulating microscopic hydrodynamic phenomena with dissipative particle dynamics. Europhysics Letters **19**: 155–160, 1992.

[37] Fedosov D.A., Caswell B., Karniadakis G.E.: A multiscale red blood cell model with accurate mechanics, rheology, and dynamics. Biophysical Journal **98**: 2215–2225, 2010.

[38] Pan W., Caswell B., Karniadakis G.E.: A low-dimensional model for the red blood cell. Soft Matter **6**: 4366–4376, 2010.

[39] Groot R.D., Warren P.B.: Dissipative particle dynamics: Bridging the gap between atomistic and mesoscopic simulation. Journal of Chemical Physics **107**: 4423–4435, 1997.

[40] Lei H., Caswell B., Karniadakis G.: Direct comparison of mesoscopic models from microscopic simulations. Physical Review. E **81**: 026704, 2010.

[41] Espanol P., Warren P.: Statistical mechanics of dissipative particle dynamics. Europhysics Letters 30: 191–196, 1995.

[42] Fan X., Phan-Thien N., Chen S., Wu X.H., Ng T.Y.: Simulating flow of DNA suspension using dissipative particle dynamics. Physics of Fluids **18**: 063102, 2006.

[43] Fedosov D.A., Pivkin I.V., Karniadakis G.E.: Velocity Limit in DPD Simulations of Wall-Bounded Flows. Journal of Computational Physics **227**: 2540–2559, 2008.

[44] Symeonidis V., Karniadakis G.: A family of time-staggered schemes for integrating hybrid DPD models for polymers: Algorithms and applications. Journal of Computational Physics **218**: 82–101, 2006.

[45] Pan W., Pivkin I.V., Karniadakis G.E.: Single-particle hydrodynamics in DPD: A new formulation. Europhysics Letters **84**: 10012, 2008.

[46] Pan W., Caswell B., Karniadakis G.E.: Rheology, Microstructure and Migration in Brownian Colloidal Suspensions. Langmuir **26**: 133–142, 2010.

[47] Pryamitsyn V., Ganesan V.: A coarse-grained explicit solvent simulation of rheology of colloidal suspensions. Journal of Chemical Physics **122**: 104906, 2005.

[48] Fan X.J., Phan-Thien N., Chen S., Wu X.H., Ng T.Y.: Simulating flow of DNA suspension using dissipative particle dynamics. Physics of Fluids **18**: 063102, 2006.

[49] Pan W., Fedosov D.A., Caswell B., Karniadakis G.E.: Hydrodynamic interactions for single dissipative-particle-dynamics particles and their clusters and filaments. Physical Review E **78**: 046706, 2008.

[50] Fedosov D.A., Caswell B., Karniadakis G.E.: Systematic coarse-graining of spectrin-level red blood cell models. Computer Methods in Applied Mechanics and Engineering **199**: 1937–1948, 2010.

[51] Fedosov D.A.: Multiscale modeling of blood flow and soft matter. PhD thesis, Brown University, USA, 2010.

[52] Helfrich W.: Elastic properties of lipid bilayers: theory and possible experiments. Z. Naturforschung C **28**: 693–703, 1973.

[53] Espanol P.: Fluid particle model. Physical Review E **57**: 2930, 1998.

[54] Hammer D.A., Apte S.M.: Simulation of cell rolling and adhesion on surfaces in shear flow: general results and analysis of selectin-mediated neutrophil adhesion. Biophysical Journal **63**: 35–57, 1992.

[55] Howard R.J.: Malarial proteins at the membrane of Plasmodium falciparum-infected erythrocytes and their involvement in cytoadherence to endothelial cells. Progress in Allergy **41**: 98–147, 1988.

[56] Ho M., White N.J.: Molecular mechanisms of cytoadherence in malaria. American Journal of Physiology **276**: C1231–C1242, 1999.

[57] Nagao E., Kaneko O., Dvorak J.A.: Plasmodium falciparum-infected erythrocytes: qualitative and quantitative analyses of parasite-induced knobs by atomic force microscopy. Journal of Structural Biology **130**: 34–44, 2000.

[58] Chien S., Jan K.-M.: Ultrastructural basis of the mechanism of rouleaux formation. Microvascular Research **5**: 155–166, 1973.

[59] Neu B., Meiselman H.J.: Depletion-mediated red blood cell aggregation in polymer solutions. Biophysical Journal **83**: 2482–2490, 2002.

[60] Liu Y., Liu W.K.: Rheology of red blood cell aggregation by computer simulation. Journal of Computational Physics **220**: 139–154, 2006.

[61] Pan W.: Single particle DPD: Algorithms and applications. PhD Thesis, Brown University, 2010.

[62] Marinkovic M., Diez-Silva M., Pantic I., Fredberg J.J., Suresh S., Butler J.P.: Febrile temperature leads to significant stiffening of Plasmodium falciparum parasitized erythrocytes. American Journal of Physiology: Cell Physiology **296**: C59–C64, 2009.

[63] Tran-Son-Tay R., Sutera S.P., Rao P.R.: Determination of RBC membrane viscosity from rheoscopic observations of tank-treading motion. Biophysical Journal **46**: 65–72, 1984.

[64] Fischer T.M.: Tank-Tread Frequency of the Red Cell Membrane: Dependence on the Viscosity of the Suspending Medium. Biophysical Journal **93**: 2553–2561, 2007.

[65] Kessler S., Finken R., Seifert U.: Swinging and tumbling of elastic capsules in shear flow. Journal of Fluid Mechanics **605**: 207–226, 2008.

[66] Quinn D.J., Pivkin I.V., Wong S.Y., Chiam K.H., Dao M., Karniadakis G.E., Suresh S.: Combined simulation and experimental study of large deformation of red blood cells in microfluidic systems. Annals of Biomedical Engineering **39**: 1041–1050, 2011.

[67] Bow H., Pivkin I.V., Diez-Silva M., Goldfless S.J., Dao M., Niles J.C., Suresh S., Han J.: A microfabricated deformability-based flow cytometer with application to malaria. Lab on a Chip **11**: 1065–1073, 2011.

[68] Pivkin I.V., Karniadakis G.E.: A new method to impose no-slip boundary conditions in dissipative particle dynamics. Journal of Computational Physics **207**: 114–128, 2005.

[69] Popel A.S., Johnson P.C.: Microcirculation and hemorheology. Annual Review of Fluid Mechanics **37**: 43–69, 2005.

[70] Canham P.B., Burton A.C.: Distribution of size and shape in populations of normal human red cells. Circulation Research **22**, 1968.

[71] Evans E.A., Hochmuth R.M.: Membrane Viscoelasticity. Biophysical Journal **16**: 1–11, 1976.

[72] Pries A.R., Neuhaus D., Gaehtgens P.: Blood viscosity in tube flow: dependence on diameter and hematocrit. American Journal of Physiology **263**: H1770–H1778, 1992.

[73] Fedosov D.A., Caswell B., Popel A.S., Karniadakis G.E.: Blood Flow and Cell-Free Layer in Microvessels. Microcirculation **17**: 615–628, 2010.

[74] Maeda N., Suzuki Y., Tanaka J., Tateishi N.: Erythrocyte flow and elasticity of microvessels evaluated by marginal cell-free layer and flow resistance. American Journal of Physiology **271**: H2454–H2461, 1996.

[75] Kim S., Long L.R., Popel A.S., Intaglietta M., Johnson P.C.: Temporal and spatial variations of cell-free layer width in arterioles. American Journal of Physiology **293**: H1526–H1535, 2007.

[76] Merrill E.W., Gilliland E.R., Cokelet G., Shin H., Britten A., Wells R.E.: Rheology of human blood near and at zero flow. Biophysical Journal **3**: 199–213, 1963.

[77] Chien S., Usami S., Taylor H.M., Lundberg J.L., Gregersen M.I.: Effects of hematocrit and plasma proteins on human blood rheology at low shear rates. Journal of Applied Physiology **21**: 817, 1966.

[78] Skalak R., Keller S.R., Secomb T.W.: Mechanics of blood flow. Journal of Biomechanical Engineering **103**: 102–115, 1981.

[79] Cranston H.A., Boylan C.W., Carroll G.L., Sutera S.P., Williamson J.R., Gluzman I.Y., Krogstad D.J.: Plasmodium falciparum maturation abolishes physiologic red cell deformability. Science **223**: 400–403, 1984.

[80] Shelby J.P., White J., Ganesan K., Rathod P.K., Chiu D.T.: A microfluidic model for single-cell capillary obstruction by Plasmodium falciparum-infected erythrocytes. Proceedings of the National Academy of Sciences USA **100**: 14618–14622, 2003.

[81] Antia M., Herricks T., Rathod P.K.: Microfluidic modeling of cell-cell interactions in malaria pathogenesis. PLoS Pathogens **3**: 939–945, 2007.

[82] Enderle T., Ha T., Ogletree D.F., Chemla D.S., Magowan C., Weiss S.: Membrane specific mapping and colocalization of malarial and host skeletal proteins in the Plasmodium falciparum infected erythrocyte by dual-color near-field scanning optical microscopy. Proceedings of the National Academy of Sciences of the United States of America **94**: 520–525, 1997.

[83] Springer T.A.: Traffic signals on endothelium for lymphocyte recirculation and leukocyte emigration. Annual Review of Physiology **57**: 827–872, 1995.

[84] Finger E.B., Puri K.D., Alon R., Lawrence M.B., von Andrian U.H., Springer T.A.: Adhesion through L-selectin requires a threshold hydrodynamic shear. Nature (London) **379**: 266–269, 1996.

[85] Raventos-Suarez C., Kaul D., Nagel R.: Membrane knobs are required for the microcirculatory obstruction induced by Plasmodium falciparum-infected erythrocytes. Proceedings of the National Academy of Sciences USA **82**: 3829–3833, 1985.

[86] Weinbaum S., Tarbell J.M., Damiano E.R.: The structure and function of the endothelial glycocalyx layer. Annual Review of Biomedical Engineering **9**: 121–167, 2007.

[87] Korn C.B., Schwarz U.S.: Dynamic states of cells adhering in shear flow: From slipping to rolling. Physical Review E **77**: 041904, 2008.

[88] Seung H.S., Nelson D.R.: Defects in flexible membranes with crystalline order. Physical Review A **38**: 1005–1018, 1988

11

Multiscale computational analysis of degradable polymers

Paolo Zunino, Simone Vesentini, Azzurra Porpora, Joao S. Soares,
Alfonso Gautieri, and Alberto Redaelli

Abstract. Degradable materials have found a wide variety of applications in the biomedical field ranging from sutures, pins and screws for orthopedic surgery, local drug delivery, tissue engineering scaffolds, and endovascular stents. Polymer degradation is the irreversible chain scission process that breaks polymer chains down to oligomers and, finally, to monomers. These changes, which take place at the molecular scale, propagate through the space/time scales and not only affect the capacity of the polymer to release drugs, bu also hamper the overall mechanical behaviour of the device, whose spatial scale is denoted as macroscale. A bottom-up multiscale analysis is applied to model the degradation mechanism which takes place in PLA matrices. The macroscale model is based on diffusion-reaction equations for hydrolytic polymer degradation and erosion while the microscale model is based on atomistic

Paolo Zunino (✉)
MOX, Department of Mathematics, Politecnico di Milano, Italy
e-mail: paolo.zunino@polimi.it

Simone Vesentini
Department of Bioengineering, Politecnico di Milano, Italy
e-mail: simone.vesentini@polimi.it

Azzurra Porpora
MOX, Department of Mathematics, Politecnico di Milano, Italy
e-mail: azzurra.porpora@mail.polimi.it

Joao S. Soares
CEMAT – Center for Mathematics and its Applications, Deparment of Mathematics, Instituto Superior Técnico/UTL, Portugal
e-mail: joao.soares@math.ist.utl.pt

Alfonso Gautieri
Department of Bioengineering, Politecnico di Milano, Italy
e-mail: alfonso.gautieri@polimi.it

Alberto Redaelli
Department of Bioengineering, Politecnico di Milano, Italy
e-mail: alberto.redaelli@polimi.it

Ambrosi D., Quarteroni A., Rozza G. (Eds.): Modeling of Physiological Flows.
DOI 10.1007/978-88-470-1935-5_11, © Springer-Verlag Italia 2012

simulations to predict the water diffusion as a function of the swelling degree of the PLA matrix. The diffusion coefficients are then passed to the macroscale model. In conclusion, the proposed multiscale analysis is capable to predict the evolution with time of several properties of water/PLA mixtures, according to the change of relevant indicators such as the extent of degradation and erosion of the PLA matrix.

11.1 Introduction

Biodegradable materials offer tremendous potential for the development of implantable devices and systems for treating disease. Currently, biodegradable polymers are used in diverse applications ranging from absorbable sutures [24], orthopedic implants [37], drug delivery devices [23], scaffolds for tissue engineered constructs [1], medicated and biodegradable stents [42]. When applications involve either negligible or well-known design requirements, the design of these classes of implants are greatly facilitated. However, in situation where the requirements are more complex, either in implants featuring complex geometries or in implants under conditions that influence the course of degradation and erosion, the design process is usually inhibited by the lack of rational models of biodegradable material behaviour [32, 42]. In order to advance from prototype status to a reliable human-implant devices, device designers must therefore rely on a combination of intuition and trial-and-error approaches that often fail due to two major reasons: (i) the lack of models able to describe the evolution of the material as it degrades and erodes, and (ii) the difficulty to collect reliable experimental data quantifying and characterizing this behaviour. Theoretical models to predict polymer degradation and erosion would seem to be important tools for a number of different applications. If drug elution is to be part of the therapy, drug delivery profiles should be programmable at the design stage. For load bearing implants, mechanical properties and structural integrity of the implant as well as their evolution should be accounted for. Because the implant is ultimately absorbed, structural breakdown and loss of function must be predicted and carefully designed for.

Polymer degradation is the deleterious change in properties of the material due to irreversible changes in its chemical structure. A biodegradable polymer is a polymer in which the degradation is mediated at least partially by a biological system [35]. More precisely, polymer degradation is the chain scission process that breaks polymer chains down to oligomers and finally to monomers, ultimately resulting in a decrease of molecular weight. Polymers degrade by several different mechanisms, depending on their inherent chemical structure and on the environment conditions to which they are subjected. The prevailing mechanism of biological degradation for synthetic biodegradable aliphatic polyesters (the most commonly employed biodegradable polymers in the medical such as polyglycolic acid and polylactic acid [16]) is scission of the hydrolytically unstable backbone chain by passive hydrolysis. By tailoring the polymer backbone with hydrolysable functional groups, the polymer chains become labile to an aqueous environment and their ester linkages are cleaved by absorbed water. There are two key factors that influence: (i) co-polymer composition and (ii) water uptake. Water adsorption, the first step of degradation is dependent on polymer hydrophilicity.

The diffusion of water into the polymer bulk and the chain scission reaction compete against each other in the process of polymer erosion. Erosion is caused by degradation and is the process of dissolution or wearing away of degradation byproducts, resulting in mass loss from the polymer bulk. Erosion is by far much more complex than degradation inasmuch as the number of parameters that potentially might influence the process is considerably larger. Two main modes of erosion can be systematized from widely established empirical evidence [7]. If degradation is fast, the diffusing water is absorbed quickly by hydrolysis and is hindered from penetrating deep into the polymer bulk. In this case, degradation and consequently erosion are restricted to the surface of the polymer, a phenomenon referred to as heterogeneous or surface erosion [45]. This type of erosion changes if degradation is slower than the rate of diffusion of water through the polymer. In this case, water cannot be absorbed quickly enough to be hindered from reaching deep into the polymer and the reaction takes place through its entire swollen bulk, a behaviour which has been termed homogeneous or bulk erosion [45]. Nevertheless, surface or bulk erosion modes are two extremes and the erosion of a polymer usually shows characteristics of both.

Hydrolysis is a very intricate process that occurs at the molecular level, as a variety of different scission pathways can occur simultaneously and concurrently [48]. Although the reactivity of each bond might be equal when considered individually, the large number of repeating units and their inherent steric environment, weak links, and branches may influence locally the rate of reaction. Ultimately, experiments with gel permeation chromatography provide data to model the mechanism of degradation [33, 34], and kinetic parameters are obtained from the evolution of experimentally obtained molecular weight distributions. An approach pioneered by Kuhn [22] and Montroll and Simha [31] employs combinatorial statistic to derive analytical solutions of the evolution of molecular weight distribution assuming that bond scission can be described with a known probability density function (e.g. equiprobable random scission, central Gaussian, or parabolic) and only for some limited simple initial conditions. Unfortunately, the applicability of such elegant exact solutions to real systems is limited essentially due to simplifying assumptions necessary for the analytical treatment of the problem. A second technique to model polymer degradation relies on the system of differential equations which describe the depolymerization rates of individual bonds that upon integration yield the time evolution of the molecular weight distribution [4]. However, the complete kinetic scheme that includes all the individual rate constents for each reacting bond could represent an enormous number of coupled differential equations even for modest size macromolecules. A third common method employs Monte Carlo simulations applied to populations of polymer chains [6, 9, 21], a versatile approach that can technically overcome the simplifying assumptions needed on the others, but realistic simulations may require an excessive amount of computational resources, results are usually subjected to large statistical errors, and may be in fact unnecessary when compared with simpler approaches.

Erosion is a much more complex phenomenon to model, not only because of the interplay between different physical mechanisms as well as due to the dramatic changes that occur in the polymer as it erodes. The choice of effective modelling tools is, however, not straightforward, and two main approaches can be currently identified: models based on differential equations that consider the erodible material as a continuum and stochastic models that describe degradation and erosion as a probabilistic event (cf. [40] for comprehensive review). In the scope of the deterministic approach, Heller and Baker [17] pioneered with a simple model for degradation from bulk eroding polymers consisting of steady state water diffusion coupled with a reaction equation describing the kinetics of the degradation mechanism. Lee [25] proposed a simplified model for surface erosion and drug release from polymer films based on the movements of two fronts, a diffusion front and an erosion front. Thombre, Joshi, and Himmelstein [20, 46, 47] proposed a comprehensive theory for drug release, water penetration, and erosion and corroborated the theoretical findings with experimental results. Similar methods based on diffusion equations that account degradation and erosion in more complex systems have been developed since [5, 26, 36]. On the other hand, stochastic models complemented with Monte Carlo simulations to simulate surface or bulk eroding polymers have been developed (cf. Zygourakis [50] and Gopferich and co-workers [12, 13, 14]). Erosion is described as being a probabilistic event and the polymer bulk as a grid of pixels. By removing eroded pixels from the grid, the stochastic evolution of a polymeric matrix was obtained and experimentally measurable parameters, such as porosity and weight loss, were calculated. Erosion fronts and a distinction between erosion modes were inferred from the results and their fit to experimental data allowed the determination of erosion rate constents. Although such models have shown good performance because of their versatility to account a multitude of phenomena occurring due to degradation and erosion (e.g. the formation of voids inside the polymer bulk as well as in the treatment of moving erosion fronts), their associated computational cost is generally much larger than with the solution of partial differential equations. Nonetheless, the common difficulties associated with erosion modelling are still present: (i) the necessity of choosing a priori the mode of erosion to model for, (ii) the difficulty arising from modelling preferential degradation of the amorphous phase, and (iii), the most difficult aspect, the incorporation of changes in the microstructure caused by erosion , which are usually specified within phenomenological reasoning. While the first two aspects have been tackled to some extent, the latter is still an open problem that will need a huge amount of insightful theoretical modelling and careful experimental characterization.

The authors have introduced a general class of mixture models to study water uptake, degradation, erosion, and drug release from degradable polydisperse polymeric matrices [43]. The model is comprehensive starting from individual polymer scission reaction all the way up to the macroscale diffusion, allows for the systematic characterization of the mass loss during the erosion process, and unifies both bulk and surface extremes of the erosion mode spectrum. The mixture is characterized by a finite number of constituents describing the polydisperse polymeric system, i.e. each representing collection of chains whose size belongs to a finite interval of degree of polymerization. In order to account for water uptake and drug release, two additional constituents (water and drug) constitute the mixture. Our approach is based on basic and widely established physical laws: (i) constituents diffuse individually accordingly to Fick's law of diffusion, (ii) hydrolysis is accounted as chemical reactions that result in the production/destruction of polymeric chains and water consumption, and (iii) balances of mass of constituents yield coupled partial differential equations that govern the reaction-diffusion system. Constitutive relationships characterizing the diffusivity of each constituent and the scission reaction rates must be specified; once known, the problem is closed and can be solved given initial and boundary conditions. Previously, we have proposed several broad assumptions in the derivation of phenomenologically reasoned constitutive relationships; nonetheless, their sufficient generality resulted in a general class of mathematical models that describe the polymer degradation and erosion. Our approach truly unified the behaviours of surface and bulk erosion and the nondimensionalization of the governing equations allowed the identification of the Thiele modulus, the ratio between characteristic timescales of diffusion and reaction, which is a key parameter in conferring the continuous shift between bulk or surface erosion behaviour to the solution of the governing equations.

Our model would have immediate direct impact in the design of biodegradable implants if these phenomenologically derived general constitutive behaviours were better characterized in regard to particular polymeric systems of interest. One possible strategy would be to perform an unprecedented series of experiments with the goal of characterizing the diffusivity of each constituent (water, drug, monomers, oligomers, etc) in a changing media (as the network degrades and erodes). To overcome this unfeasible plan, we have developed a coupling between the macroscale of the biodegradable polymer bulk, which is governed by the reaction-diffusion system, and the microscale of chemical reactions and molecular diffusion, which with the aid of atomistic simulations characterizes locally the diffusion of constituents in the polymer bulk accordingly to its changing microstructure. Atomistic or molecular dynamics simulations cannot yet provide further insight into the rates of reactions taking place, but on the other hand, have been able to characterize the diffusion coefficient of molecules in a polymeric network with quite success (cf. [18] and references therein). With these new tools, we are able to provide unique information into the macroscale model in the specification of the local diffusivities of mixture constituents. More precisely, the macroscale model provides local molecular configurations (i.e. set of polymer chains and water) as input for the microscale model, which then outputs the local diffusivities of the constituents at the macroscale level. Unfortunately, molecular dynamics simulations are extremely expensive computationally, and hence this dynamic coupling strategy would not result in a practical solution. To this end, we developed lumping strategies to parametrize the range of local molecular configurations and statically couple both scales.

11.2 A mixture model for water uptake, degradation and erosion from polymeric matrices

Following the lines of [43], we develop a general tridimensional model for the degradation of PLA upon water absorption and subsequent hydrolysis. First, we describe the nature of the model and its governing equations and then complement it with specific initial and boundary conditions. The main difficulty consists in the definition of suitable constitutive laws for the model coefficients. The novelty of the present work consists in the application of atomistic simulation to estimate the most significant properties of PLA as it degrades and erodes, giving rise to a multiscale description of the material.

11.2.1 Macroscale governing equations for a polymer mixture

We describe a polydisperse polymeric network as a collection of different linear chains of repeating units. Each chain is characterized by its degree of polymerization, x, defined as the number of repeating units. The system is discretized into N number of constituents by defining N mutually exclusive equidistent partitions $P_i = [\bar{x}_i - \bar{x}_1/2, \bar{x}_i + \bar{x}_1/2[$, for $i = 1, \ldots, N$, of length $\bar{x}_1$ of the degree of polymerization spectrum. Each partition P_i represents the class of chains whose average degree of polymerization is $\bar{x}_i$.

Diffusion driven by negative density gradients is the driving force for mass transport. To account for diffusion, we introduce a spatial coordinate $\mathbf{x}$ characterizing the location of a particle of the mixture with volume dV. At each particle, water, drug, and N polymeric constituents coexist. Since the mascroscale spans over the entire length of the polymer matrix, external boundaries have to be taken into account. In this respect, an open system is considered as water penetrates into the polymer matrix from the outside aqueous environment and polymeric mass is lost to the exterior. The mass balances for each individual constituent yield the system of reaction-diffusion equations constituting the mathematical model.

Let $\rho_i = \rho_i(\mathbf{x},t)$ be the partial density of chains of average degree of polymerization $\bar{x}_i$ in a representative control volume, dV, that corresponds to point $\mathbf{x}$ at time t and let $w_i = \rho_i / \sum_i \rho_i$ be the weight fraction of chains of length $\bar{x}_i$. Let $\rho_w(\mathbf{x},t)$ be the partial density of water. We also denote with $\tilde{\rho} = \sum_i \rho_i$ the partial density of all polymer sub-fractions and with $\rho = \rho_w + \sum_i \rho_i$ the total density of the mixture. Finally, for the forthcoming description, we denote with $\Upsilon = (\rho_w, \rho_1, \ldots, \rho_N)$ the vector of partial densities of each component that completely identifies the state of the mixture, i.e. this is the *state space* of the macroscale model. For notation cleanliness, we shall omit the dependence on space and time ($\mathbf{x}$ and t) of partial densities.

Mass can be neither created nor destroyed, but in the present case, polymer chains and water can diffuse in/out of dV through its boundary. Moreover, polymeric constituents interconvert from one to another due to scission reactions. The mass balance of each constituent in dV states that the time rate of change of mass existing in dV is equal to the divergence of the diffusive flux plus time rates of production and/or destruction in chemical reactions. As $dV \rightarrow 0$, mass balances of each polymeric constituent yield the system of N reaction-diffusion equations

$$\partial_t \rho_i = \nabla \cdot (D_i \nabla \rho_i) + \sum_{j=1}^{N} A_{ij} \rho_j, \text{ for } i = 1, \ldots, N \qquad (11.1)$$

where D_i is the diffusivity of chains of average degree of polymerization $\bar{x}_i$ and coefficients A_{ij} are the reaction coefficients corresponding to chain cleavage due to hydrolysis. More precisely, a chain of average degree of polymerization $\bar{x}_i$ can be cleaved at $j = 1, \ldots, i-1$ different scission locations (each composed of $\bar{x}_1$ individual polymeric bonds) to yield chains of smaller average degree of polymerization, $\bar{x}_j$ and $\bar{x}_{i-j}$. All $i-1$ possible outcomes of scission of a chain of degree of polymerization $\bar{x}_i$ are

$$\bar{x}_i \xrightarrow{k_{i,j}} \bar{x}_j + \bar{x}_{i-j}, \text{ for } i = 1, \ldots, N, \; j = 1, \ldots, i-1.$$

The depolymerization kinetics of populations of individual molecules can be described by means of a system of ordinary differential equations (cf. [33, 34]). Let $n_i = n_i(t)$ denote the number of chains of average degree of polymerization $\bar{x}_i$ existing at time t. As $dV \rightarrow 0$, the relationship between partial density ρ_i and number of molecules n_i is $\rho_i(\mathbf{x}, t) = n_i(\mathbf{x}, t) \bar{x}_i M_0 / dV$, where M_0 is the molecular mass of one monomeric unit and $\bar{x}_i M_0$ is the molecular weight of polymer subfractions of length $\bar{x}_i$. In a closed system, the rate of change of n_i is given by,

$$\dot{n}_i = -\sum_{j=1}^{i-1} k_{i,j} n_i + \sum_{j=i+1}^{N} (k_{j,i} + k_{j,j-i}) n_j, \text{ for } i = 1, \ldots, N.$$

Then, Eq. (11.1) is complemented by the following expressions:

$$A_{ij} = \begin{cases} 0 & \text{if } j < i \\ -\sum_{m=1}^{i-1} k_{i,m} & \text{if } j = i \\ (k_{j,i} + k_{j,j-i}) \frac{\bar{x}_i}{\bar{x}_j} & \text{if } j > i. \end{cases}$$

The combination of Eqs. (11.1) and (11.2) results in a multiscale description of polymer degradation and erosion, as it combines a molecular description of chain scission at the molecular level (second term on the right hand side of (11.1)) with macroscopic Fick's law of diffusion (first term of the right hand side of (11.1)).

Water diffuses in the polymeric matrix and hydrolysis is accounted as a sink of water, i.e.

$$\partial_t \rho_w = \nabla \cdot (D_w \nabla \rho_w) - f_w \qquad (11.2)$$

where $\rho_w = \rho_w(\mathbf{x}, t)$ and D_w are the partial density and the diffusivity of water. Reaction term $f_w > 0$ accounts for water consumption: one water molecule is consumed with each scission reaction, i.e.

$$\bar{x}_1 \cdot \left[\begin{array}{c} \text{bond of scission} \\ \text{location } j \text{ of chain } \bar{x}_i \end{array} \right] + H_2O \xrightarrow{k_{i,j}} \text{scission event.}$$

The rate $k_{i,j}$ is the rate of reaction of scission location j composed of $\bar{x}_1$ bonds, hence the rate of water consumption in the scission event is $k_{i,j}/\bar{x}_1$. The rate of change of the number of water molecules $n_w = n_w(\mathbf{x},t)$ in a representative volume dV is given by

$$\dot{n}_w = -\sum_{i=1}^{N} \sum_{j=1}^{i-1} \frac{k_{i,j}}{\bar{x}_1} n_i$$

which, with the relationship between n_w and ρ_w as $dV \to 0$, $\rho_w = n_w M_w/dV$ (where M_w is the molecular mass of water), yields the water consumption due to the chemical reaction in Eq.(6)

$$f_w = \sum_{i=1}^{N} \sum_{j=1}^{i-1} k_{i,j} \frac{M_w}{M_0 \bar{x}_1} \frac{\rho_i}{\bar{x}_i}.$$

11.2.2 Initial and boundary value problem

Boundary and initial conditions depend on the application that one has in mind. In our case, we aim to model the coating of a medicated stent, which is a thin polymer layer covering the surface of a cardiovascular stent. For this reason, the geometry of the polymer matrix can be thought of as a thin slab. Then, the governing equations can be reduced to one spatial dimension, z, the coordinate across the thickness ($z \in [0,L]$ where L is the coating thickness). The polymer network starts out dry, i.e. $\rho_w(z,0) = 0$ and the initial state of the polydisperse polymeric system is homogeneous, i.e. independent of z, and is characterized by an initial degree of polymerization distribution $w^0 = w^0(x)$ and an initial total (dry) polymer density ρ^0. The initial weight fraction of chains of average degree of polymerization $\bar{x}_i, w_i$ are obtained with Eq. (1) and the initial conditions of the polymeric constituents densities are set as $\rho_i(z,0) = w_i^0 \rho^0$ for $i = 1,\ldots,N$. In what follows, indexes 0 and ∞ will refer to the state of the mixture at $t \to 0$ and $t \to \infty$, respectively.

At $z = 0$ we apply impermeable boundary conditions

$$\partial_z \rho_w|_{z=0} = 0, \; \partial_z \rho_i|_{z=0} = 0$$

with the meaning that any constituent is not able to leave from the polymeric matrix and penetrate the stainless steel bulk of the stent. At $z = L$, the polymer contacts surrounding water or biological tissue. Water permeates through the interface according to the following law,

$$-D_w \partial_z \rho_w|_{z=L} = \pi_w (\rho_w|_{z=L} - A)$$

where π_w is the permeability of the interface to water molecules and A is a partition coefficient (with $A \in [0,1]$), which accounts that water uptake increases as the polymer erodes, leading to an eventual total replacement of the swollen network with a mixture solely composed of pure water.

Proceeding similarly, boundary conditions for polymeric constituents are

$$-D_i \partial_z \rho_i|_{z=L} = \pi_i \rho_i|_{z=L}.$$

The permeability π_i of the polymer matrix to different polymer sub-fractions at the interface with aqueous outside will be defined later on in order to make sure that the boundary conditions for polymeric constituents shift from an approximation of a perfect sink condition for the smaller chains towards a no-flux condition for the larger molecules. More precisely, smaller chains are readily dissolved in the surrounding water or absorbed by the tissue when they reach the boundary, i.e. $\rho_1(L,t) \simeq 0$, $t > 0$, whereas the longer chains are unable to diffuse due to their size and entanglements in the polymeric matrix; hence $\partial_z \rho_N(L,t) \simeq 0$, $t > 0$. In between, a shift occurs gradually with both terms of $-D \partial_z \rho_i = \pi_i \rho_i$ having the same relevance for species of medium degree of polymerization.

11.2.3 Constitutive laws

Constitutive relationships for the diffusivity of each constituent and for the reaction rates must be specified. The mechanisms of diffusion and reaction are the only physical mechanisms that need constitutive specification and once known, the model is closed and can be solved.

Diffusion depends on the nature of the constituent in question and on the local characteristics of the mixture on which is diffusing. In the most general case this is taken into account by the following expressions

$$D_w = D_w(\rho_w, \rho_1, \ldots, \rho_N), \quad D_i = D_i(\rho_w, \rho_1, \ldots, \rho_N).$$

In the forthcoming sections, we will apply atomistic simulations of molecular diffusion into a polymer mixture in order to provide a simplified characterization of such models.

To describe the permeability of the mixture at the interface with the external medium, we assume that permeability and diffusivity are proportional. Then, the permeability π_w, π_i are given by

$$\frac{\pi_w}{\pi_w^0} = \frac{D_w}{D_w^0}, \quad \frac{\pi_i}{\pi_i^0} = \frac{D_i}{D_i^0}$$

where π_w^0 and and D_w^0 represent the permeability and the diffusivity of water in a non-degraded mixture, which will be provided by atomistic simulations. To determine the reference permeability π_i^0 we apply again the proportionality between permeability and diffusivity,

$$\frac{\pi_i^0}{\pi_w^0} = \frac{D_i^0}{D_w^0}$$

that combined with the previous expression gives,

$$\pi_i = D_i \frac{\pi_w^0}{D_w^0}.$$

The water saturation of the mixture is modelled with the partition factor A. The non-degraded mixture is characterized by a partition factor $0 < A^0 < 1$ (that will be later determined according to data taken by [18]). As erosion of the polymer at the interface leads to a decrease in the total polymer density, denoted with $\tilde{\rho} = \sum_i \rho_i$, we propose the linear relationship

$$A = 1 - (1 - A^0) \frac{\tilde{\rho}|_{z=L}}{\rho_w^\infty}$$

such that $A \to 1$ when $\tilde{\rho}|_{z=L} \to 0$, which at saturation equilibrium results in a mixture characterized by $\rho_w \to \rho_w^\infty$ and $\rho_i \to 0$ for all $i = 1, \ldots, N$, i.e. the network was replaced by pure surrounding fluid.

We finally take into account hydrolysis. One common tool to describe the localization of the scission event along large chains is a scission probability density function, which distinguishes the likelihood of scission among scission locations in a chain of average degree of polymerization $\bar{x}_i$, i.e. the relationships among $k_{i,j}$ with i fixed and $j \in 1, 2, \ldots, i-1$. Some frequently encountered scission probability density functions are: random, parabolic, and central scission (cf. [20, 21] for details). Aliphatic polyesters degrade by passive hydrolysis and are usually characterized by random scission events. Random scission is defined with a constent probability density function along the length of the chain, i.e. $k_{i,1} = k_{i,2} = \cdots = k_{i,i-1}$, for all $i = 2, \ldots, N$. Considering k as the rate of hydrolysis of the particular type of polymeric bond, all $k_{i,j}$ are given by

$$k_{i,j} = k\bar{x}_1, \text{ for } i = 2, \ldots, N \text{ and } j = 1, \ldots, i-1.$$

Hydrolysis happens due to the presence of water, hence its rate depends on the partial density ρ_w. Because water might not be homogeneously distributed over the network, the rate of reaction depends implicitly on space and leads to inhomogeneous degradation. Following similar studies (cf. [40]), random hydrolysis is a 1st order reaction with water, i.e. the polymeric bond has reaction rate that follows a linear relationship with partial density of water such that

$$k = k(\rho_w) = \bar{k}\rho_w$$

where $\bar{k}$ is a constent reaction rate. In such way, bilinear terms (each featuring ρ_w and one ρ_i) appear in the reaction terms of Eqs. (11.1) and (11.2). Autocatalysis, i.e. when the rate of scission depends on the presence of residual monomer, is not accounted with this constitutive specification.

11.3 Molecular modelling of polymers

In molecular modelling of amorphous polymers a volume element is filled with polymer chain segments and small permeant molecules (e.g, water, gases). In the molecular modelling paradigm all atoms are considered to be spheres of given diameter di and mass m_i. The bonded interactions between atoms resulting in bonds, bond angles, and conformation angles are then described by springs with spring constants related to the bond strength known experimentally or by quantum mechanics calculations. Non-bond interactions between atoms are considered via e. g. Lennard-Jones and Coulomb potentials. The sum of all interatomic interactions written as the potential energy of the atomistic model is then called a forcefield. Below is shown the typical forcefield structure for a system of N atoms with the cartesian atomic position vectors r_i,

$$V(\mathbf{r}_1,\ldots,\mathbf{r}_N) = \frac{1}{2} \sum_{\text{bounds}} k_l (l - l_0)^2 + \frac{1}{2} \sum_{\text{angles}} k_\theta (\theta - \theta_0)^2$$

$$+ \sum_{\text{dihedrals}} k_\phi (1 + \cos(n\phi - \delta)) + \sum_{i=1}^{N} \sum_{j=i+1}^{N} \left(\frac{q_i q_j}{4\pi\varepsilon r_{ij}} + \frac{A_{ij}}{\mathbf{r}_{ij}^{12}} + \frac{C_{ij}}{\mathbf{r}_{ij}^{6}} \right)$$

where l is the bond length between two atoms, k_l is the bond spring constant and l_0 the reference bond length; θ is the angle between three atoms, k_θ is the angle spring constant and θ the reference angle; ϕ and δ represent the cis and trans dihedral angles, n stands for the periodicity of the dihedral term and k_ϕ is the dihedral spring constant. The parameters in the last term regard non-bonded interactions where A_{ij} and C_{ij} are the van der Waals parameters for the interacting pair of atoms, r_{ij} is the non-bonded distance between the atoms, q_i and q_j are their charges and ε is the dielectric constant.

The evaluation of the nonbonded energy terms is the numerically most expensive part in molecular modelling calculations, because these terms include contributions from each pair of atoms in a model. This leads to restrictions on the maximum possible size of a simulated molecular system. With current computational resources, the number of atoms N can not be much higher than 10'000–20'000 for simulations on modern workstations, whereas N may be up to about a factor of ten higher for selected simulations on computer clusters. Therefore, only polymeric materials being homogeneous enough that a volume element of several thousand atoms is representative for the whole polymer structure can be modelled with atomistic simulations.

Forcefields may be used in two ways. On one hand, model systems can be subjected to a static structure optimization, i.e. the geometry of the simulated system is changed until the potential energy reaches the closest minimum value. This is performed for the reduction of unrealistic local tensions in a model structure. On the other hand, from the potential energy of a model system it is possible to calculate the forces $\mathbf{F}_i$ acting on each atom of the model via the gradient operator,

$$\mathbf{F}_i = \frac{\partial V(\mathbf{r}_1,\ldots,\mathbf{r}_N)}{\partial \mathbf{r}_i}.$$

Then Newton's equations of motion can be solved for every atom of the investigated system,

$$\mathbf{F}_i = m_i \frac{\partial^2 \mathbf{r}_i(t)}{\partial t^2}. \tag{11.3}$$

The starting velocities $\mathbf{v}_i(0)$ of all atoms are assigned through the well known relation between the average kinetic E_{kin} energy of an atomistic system and its temperature T,

$$E_{kin} = \sum_{i=1}^{N} \frac{1}{2} m_i \mathbf{v}_i^2 = \frac{3N-6}{2} k_B T$$

where k_B is the Boltzmann constant, $(3N-6)$ is the number of degrees of freedom of a N-atom model considering the fact that in the given case the centre of mass of the whole model with its 6 translation and rotation degrees of freedom does not move during the molecular dynamics simulation. Using this approach it is then possible to follow the motions of the atoms of a polymer matrix and the diffusive movement of penetrant molecules at a given temperature over a certain interval of time.

Since Eq. (11.3) represents a system of several thousand coupled differential equations of second order, it can be solved only numerically in small time steps Δt. This time step must not be greater than about a tenth of the oscillation time of the fastest vibrations in the investigated system, otherwise serious numerical errors occur. Since the fastest oscillation time (covalent stretching of a C-H bond) in a polymer is approximately 10 fs, the integration time step Δt is usually assumed as 1 fs. This imposes a further restriction in the processes that can be treated by molecular dynamics simulations. With currently available coputational power, it is possible to solve Eq. (11.3) for a system composed of several thousands of atoms for a few million time steps Δt, i. e. for a few nanoseconds. That means, any simulated process in a polymer bulk has to be fast enough that a few nanoseconds are sufficient to get the relevant information.

The restriction of the possible number of atoms N leads to a characteristic side length of the atomistic model in the order of a few nanometers. Thus, to avoid severe surface effects it is necessary to introduce periodic boundary conditions, i. e. the basic volume element is surrounded by virtual volume elements of identical contents. The atoms of the surrounding cells only contribute to the nonbond interactions of the atoms in the basic cell, which basically establishes conditions of an infinite solid. To avoid the consideration of an infinite number of pair interactions and artificial translation symmetry, for nonbond interactions it is indrocuced a cut-off distance, r_{cut}, smaller than half the shortest side length of the characteristic volume element. In this way, van der Waals and electrostatic interactions are considered to be zero for interatomic distances greater than r_{cut}.

A promising strategy to overcome the computational limitations of molecular dynamics simulations is the use of Graphical Processing Units (GPUs). In recent years, commercial GPUs have acquired non-graphical, general-purpose programmability. The most mature programming environment is the so called Compute Unified Device Architecture (CUDA) and have been the focus of the majority of investigation in the computational science field. Several groups have lately shown results for molecular dynamics codes which utilize CUDA-capable GPUs, reporting speed-up factors in the order of 100 with respect to conventional CPUs, [51, 52, 53].

11.3.1 Generation and equilibration of the atomistic models of PLA

Our specific objective is to predict the water and polymer diffusion as a function of the composition of the water/polymer mixture, characterized by the coordinates in the state space $\Upsilon = (\rho_w, \rho_1, \ldots, \rho_N)$ previously introduced. The diffusion coefficients are then passed to the macroscale model.

To generate atomistic models of PLA we have applied Materials Studio 4.4 (Accelrys, Inc.) and the COMPASS force field [44]. The repeating unit of PLA is available as standard model in Materials Studio. The polymeric chains with the desired length are generated starting from the repeating units using the "Build Polymers" tool of Materials Studio. Finally, we generate solvated amorphous bulk models containing PLA and water using the "Amorphous Cell" tool of Materials Studio. The polymeric chains are solvated in a periodic box with a total atom number varying from 6,000 atoms (for quasi-dry systems) to 35,000 atoms (for highly solvated systems), as depicted in Fig. 11.1

The initial geometries of the bulk models are refined following the procedure which consists in a series of MD calculations at different temperature and ensembles in order to obtain chain redistribution within the periodic cell [10, 11, 18, 19]. Specifically, we perform a preliminary minimization followed by a sequence of nine MD simulations (see Table 11.1). MD molecular dynamics simulations are carried using the Discover module and the COMPASS force field implemented in Materials Studio. Nonbonding interactions are computed using a cutoff for neighbor list at

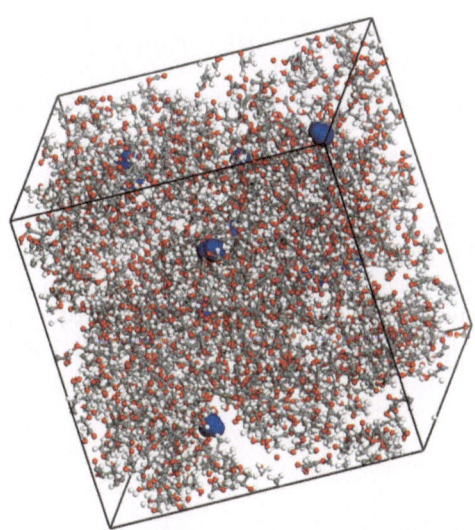

Fig. 11.1. Atomistic bulk model of quasi-dry PLA matrix (water molecules in blue color)

Table 11.1. Summary of the simulation protocol applied to obtain equilibrated bulk model of PLA matrices

Simulation	Ensemble	Time	Time step	Temperature	Pressure
#1	NVT	0.2 ps	0.1 fs	300 K	–
#2	NVT	2 ps	1 fs	600 K	–
#3	NVT	100 ps	1 fs	300 K	–
#4	NPT	60 ps	1 fs	300 K	0 GPa
#5	NVT	20 ps	1 fs	750 K	–
#6	NVT	20 ps	1 fs	600 K	–
#7	NVT	20 ps	1 fs	450 K	–
#8	NVT	100 ps	1 fs	300 K	–
#9	NPT	100 ps	1 fs	300 K	0 GPa

1 nm, with a switching function between 0.85 and 0.95 nm for Van der Waals and Coulomb interactions.

We carefully checked that the potential energy, temperature, pressure and density reached a stable value after each step of the equilibration procedure. The latest (#9) simulation has been employed to validate the models by comparing the average density during the NPT dynamics with the experimental one.

Finally, in order to obtain water and polymer diffusivity by means of an atomistic model of the polymer mixture, we select an ensemble of M (water or polymer) molecules in the model and we compute their mean square displacement $MSD(t)$. Denoting with

$$\langle \mathbf{r}(t)^2 \rangle = \frac{1}{M} \sum_{m=1}^{M} \mathbf{r}_m(t)^2$$

the averaging over all the particles, the mean square displacement is defined as

$$MSD(t) = \langle |\mathbf{r}(t) - \mathbf{r}(0)|^2 \rangle.$$

The diffusion coefficient D of water/polymer molecule is then calculated using Einstein's relation

$$D = \lim_{t \to \infty} MSD(t)/(6t).$$

Manipulating Einstein's formula one easily obtain that for sufficiently large times $\log(6D) + \log(t) = \log(MSD(t))$. Then, the realm of normal diffusion (also known as Fickian diffusion) is reached when $\log(MSD(t))$ is a linear function of time with unit slope. The validity of this fundamental property is equivalent to say that the application of Fick's law to derive Eqs. (11.1) and (11.2) is correct.

11.4 Multiscale analysis

We formulate here different strategies for the interaction of the micro and the macroscale models. We start from the algorithms that most tightly couple the two models and we progressively simplify them, in order to balance the accuracy of the methodology with a reasonable computational cost.

11.4.1 Input/Output description of the models

According to the previous derivation, the macroscale model for polymer mixtures is not computable, because it needs to determine water and polymer diffisivities in the mixture to close governing equations and boundary conditions. More precisely, the input/output structure of the macroscale model can be described as follows:

Input: given $D_w(\rho_w, \rho_1, \dots, \rho_N)$ and $D_i(\rho_w, \rho_1, \dots, \rho_N)$.
Compute: for any (z,t) and for $i = 1, \dots, N$ solve the following problem

$$
\begin{cases}
\partial_t \rho_w = \nabla \cdot (D_w \nabla \rho_w) - \bar{k} \frac{M_w}{M_0 \bar{x}_1} \sum_{i=1}^N \frac{i-1}{i} \rho_w \rho_i, \\
\partial_t \rho_i = \nabla (D_i \nabla \rho_i) - (i-1)\bar{k}\bar{x}_1 \rho_w \rho_i + 2\bar{k}\bar{x}_1 \rho_w \sum_{j=i+1}^N \frac{i}{j} \rho_j, \\
\partial_z \rho_w|_{z=0} = 0, \quad -D_w \partial_z \rho_w|_{z=L} = \pi_w (\rho_w|_{z=L} - A) \\
\partial_z \rho_i|_{z=0} = 0, \quad -D_i \partial_z \rho_i|_{z=L} = \pi_i \rho_i|_{z=L} \\
\rho_w(z,0) = 0, \quad \rho_i(z,0) = w_i^0 \tilde{\rho}^0.
\end{cases}
\tag{11.4}
$$

Output: determine the mixture partial densities: $\rho_w(t,z), \rho_1(t,z), \dots, \rho_N(t,z)$ that can be further post-processed to compute other indicators that characterize the mixture composition.

Reminding that $\Upsilon = (\rho_w, \rho_1, \dots, \rho_N)$ denotes the state variables of the model, the *macroscale* model can be described by means of the following input/output scheme:

$$
D_w(\Upsilon), D_i(\Upsilon) \overset{\text{macro}}{\Longrightarrow} \Upsilon(t,z).
$$

The solution of problem (11.4) is provided by numerical approximation schemes. In particular, we have exploited Lagrangian finite elements for the space discretization, by resorting to a semi-discrete problem that has been fully discretized with backward finite difference schemes to advance in time. For further details, we refer to [15, 39]. The main difficulty in the approximation of the problem at hand consists in the efficient solution of the nonlinear system of equations arising from the fully discrete scheme. To this aim, we have applied the damped Newton method proposed in [8]. For one or two space dimensions, the computational cost of the macroscale model is almost negligible with respect to the microscale one. Indeed, its characteristic computational cost does not exceeds the scale of minutes of CPU time. A similar cost can be estimated for three-dimensional applications, for reasonably simple geometrical configurations.

Concerning the microscale model, we observe that, since the dimension of the state space $\Upsilon(t,z) = (\rho_w(t,z),\rho_1(t,z),\ldots,\rho_N(t,z))$ is considerably large, namely $N+1$, it would be a challenging task to design an ensemble of atomistic simulations sufficient to characterize constitutive laws of type $D_w = D_w(\rho_w,\rho_1,\ldots,\rho_N)$, $D_i = D_i(\rho_w,\rho_1,\ldots,\rho_N)$. By consequence, we first introduce a *lumped state space*, identified by a vector valued function $\mathscr{F} : \mathbb{R}^{N+1} \to \mathbb{R}^M$ with $M \leq N+1$, such that it is possible to initialize an atomistic PLA model on the basis of $\mathscr{F}(\Upsilon)$ solely. This assumption will remarkably simplify the interacton between the micro and the macro scale models. According to the results reported in the forthcoming sections, we select a the follwing lumped state space with $M = 2$:

1. the degree of swelling (also related to the extent of erosion),

$$\mathscr{F}_1(\Upsilon(t,z)) = \phi_w(\Upsilon(t,z)) = \frac{\rho_w(t,z)}{\rho_w(t,z) + \sum_{i=1}^{N} \rho_i(t,z)}$$

   which quantifies the amount of water in the entire mixture;
2. the (weight) average degree of polymerization (the extent of degradation),

$$\mathscr{F}_2(\Upsilon(t,z)) = \bar{x}(\Upsilon(t,z)) = \sum_{i=1}^{N} w_i(t,z) \cdot \bar{x}_i$$

   which quantifies the average size of polymeric chains in the mixture and always satisfies $\bar{x}_1 \leq \mathscr{F}_2 \leq \bar{x}_N$.

We notice that this simplification involves some loss of information. On one hand, according to degradation of polymeric chains, a polymer mixture is generally *polydisperse*, i.e. it is composed by a collection of several sub-fractions. On the other hand, the selected lumped state space is not able to represent polydispersity and it replaces a given distribution of sub-fractions with *monodisperse* mixture of equivalent average degree of polymerization. Mathematically speaking, this corresponds to say that function $\mathscr{F}$ is surjective, for any $M < N+1$. In this framework, we sketch below the input/output description of the microscale model:

Input: given $(\phi_w,\bar{x})$ generate the corresponding atomistic model of PLA.
Compute: select some tracers molecules (either water or polymer) to evaluate molecular diffusion. Then, perform MD simulations to compute their trajectories $r(t)$ over a sufficiently large time span. Finally perform the mean square displacement analysis combined with Einstein's relation in order to estimate the molecular diffusivity $D_w(\phi_w,\bar{x})$, $D_i(\phi_w,\bar{x})$ of the water or polymer tracers in a given PLA mixture.
Output: the model provides the molecular diffusivity of water and polymer molecules, i.e. independent values $D_w(\phi_w,\bar{x})$, $D_i(\phi_w,\bar{x})$, into a given PLA/water mixture.

A synthetic input/output relation for the microscale model reads as follows:

$$\mathscr{F}(\Upsilon) \overset{\text{micro}}{\Longrightarrow} D_w(\mathscr{F}(\Upsilon)), D_i(\mathscr{F}(\Upsilon))$$

that will be complemented with the application of the macroscale model in the lumped state space,

$$D_w(\mathscr{F}(\Upsilon)), D_i(\mathscr{F}(\Upsilon)) \stackrel{\text{macro}}{\Longrightarrow} \Upsilon \Longrightarrow \mathscr{F}(\Upsilon).$$

We finally notice that the set up of the microscale model is pointwise or local, that is it applies to any single point (t, z) of the macroscale domain.

11.4.2 Multiscale coupling strategies

First, we address a *fully coupled* algorithm, that best exploits the interaction among micro and macro-sacles. Let t^n the time levels corresponding to time discretization of the macroscale model, and let $z_j \in [0, L]$ the nodes associated to space discretization. For any (t^n, z_j), the algorithm consists in the following iterative steps for $k = 1, \dots$:

1. Set a guess for the state of the system i.e. $\Upsilon^{(0)}(t^n, z_j)$.
2. Solve the microscale model:

$$\mathscr{F}(\Upsilon^{(k-1)}(t^n, z_j)) \stackrel{\text{micro}}{\Longrightarrow} D_w^{(k)}(t^n, z_j), D_i^{(k)}(t^n, z_j).$$

3. Solve the macroscale model:

$$D^{(k)}(t^n, z_j), D_i^{(k)}(t^n, z_j) \stackrel{\text{macro}}{\Longrightarrow} \Upsilon^{(k)}(t^n, z_j).$$

Then, compute $\mathscr{F}(\Upsilon^{(k)}(t^n, z_j))$.

Reminding that the computational cost of a single solution of the microscale model is considerable, see Sect. 11.3, this algorithm is not yet affordable with standard computational devices. Furthermore, we notice that the cost of this algorithm exponentially increases with the number of space dimensions accounted by the macroscale model. Indeed, realistic applications in three space dimensions would be practically unachievable. For this reason, we consider the following simplification, which can be classified as a *time staggered and space averaged* coupled algorithm. As it will be discussed later on, such simplification is acceptable for bulk eroding polymers, where spatial gradients of densities are almost negligible, i.e. $\nabla \Upsilon(t, \mathbf{x}) \simeq 0$ for any $t > 0$ and for any $\mathbf{x} \in \Omega$. Let us define the following spatial average for $\mathscr{F}$:

$$\overline{\mathscr{F}}(\Upsilon(t, z)) = \int_0^L \mathscr{F}(\Upsilon(t, z)) dz.$$

Then, given the initial state of the system i.e. $\Upsilon(t^0, z)$, for any $n > 0$ we perform the following steps:

1. Solve the microscale model at time t^{n-1}:

$$\overline{\mathscr{F}}(\Upsilon(t^{n-1}, z)) \stackrel{\text{micro}}{\Longrightarrow} \overline{D}_w(t^{n-1}), \overline{D}_i(t^{n-1}).$$

2. Advance in time and solve the macroscale model at time t^n:

$$\overline{D}_w(t^{n-1}), \overline{D}_i(t^{n-1}) \overset{\text{macro}}{\Longrightarrow} \Upsilon(t^n, z_j).$$

Then, compute $\overline{\mathscr{F}}(\Upsilon(t^n, z))$.

We notice that this algorithm considerably reduces the number of calls to the microscale model and makes it independent on the number of space dimensions of the macroscale equations. Since the coupling of the models is sequential in time, the main limitation of this approach consists in the different characteristic CPU efforts needed for the micro and macroscale models. Considering the aforementioned promising performance of GPU computing devices for molecular dynamics simulations, future works could be devoted to the development of an hybrid CPU-GPU coupled computational environment for an efficient implementation of a dynamic interaction between macroscale and atomistic models.

In alternative to the previous algorithms, which turn out to be computationally expensive in any case, we propose the following *static coupling strategy* to feed the macroscale model with data provided by microscale sumulations. For the sake of simplicity, we only refer to a generic diffusivity $D(\Upsilon)$ that stands for either $D_w(\Upsilon)$ or $D_i(\Upsilon)$. Let $\{\Upsilon_p\}_{p=1}^P$ be a collection of different mixture configurations (see Fig. 11.2 for an illustration) and let $\mathbf{D} = \{D(\overline{\mathscr{F}}(\Upsilon_p))\}_{p=1}^P$ be the vector defined as follows,

$$\overline{\mathscr{F}}(\Upsilon_p) \overset{\text{micro}}{\Longrightarrow} D(\overline{\mathscr{F}}(\Upsilon_p)).$$

We aim to determine a function $\mathscr{D}(\overline{\mathscr{F}}(\Upsilon))$ that suitably approximates its discrete analog $\mathbf{D} \in \mathbb{R}^P$. First, a parametrization for $\mathscr{D}$ with polynomials is obtained by defining as $\mathbf{q} \in \mathbb{N}_0^{M=2}$ a multi-index and

$$\mathscr{D}(\overline{\mathscr{F}}) = \mathscr{D}^{\mathbf{a}} \left(\sum_{q_1=0}^{M} \sum_{q_2=0}^{M} a_{\mathbf{q}} \overline{\mathscr{F}}_1^{q_1} \overline{\mathscr{F}}_2^{q_2} \right)$$

where $\mathbf{a}$ is the vector of coefficients $\{a_{\mathbf{q}}\}_{q=1}^Q$ with $Q = (M+1)^2 = 9$. For the well posedness of the forthcoming algorithm we have to make sure that $P \geq Q$. Then we introduce a vector function $\mathbf{D}^{\mathbf{a}} : \mathbb{R}^Q \to \mathbb{R}^P$ defined as $\mathbf{D}^{\mathbf{a}} = \{D^{\mathbf{a}}(\overline{\mathscr{F}}(\Upsilon_p))\}_{p=1}^P$ such that,

$$D^{\mathbf{a}}(\overline{\mathscr{F}}(\Upsilon_p)) = \mathscr{D}^{\mathbf{a}} \left(\sum_{q_1=0}^{M} \sum_{q_2=0}^{M} a_{\mathbf{q}} \overline{\mathscr{F}}_1^{q_1}(\Upsilon_p) \overline{\mathscr{F}}_2^{q_2}(\Upsilon_p) \right).$$

Then, the optimal parameters $\mathbf{a}$ such that $\mathscr{D}^{\mathbf{a}}$ best fits the data $\mathbf{D}$ are determined as follows,

$$\mathbf{a}^* = \operatorname{argmin}_{\mathbf{a}} \|\mathbf{D} - \mathbf{D}^{\mathbf{a}}\|_2.$$

In such way we determine an explicit representation of functions $\mathscr{D}_w(\Upsilon)$ and $\mathscr{D}_i(\Upsilon)$ representing a closure of the macroscale model (11.4) that is now completely solvable.

One advantage of the static coupling strategy consists in the fact that different calls to the microscale model, i.e. $\overline{\mathscr{F}}(\Upsilon_p) \overset{\text{micro}}{\Longrightarrow} D(\overline{\mathscr{F}}(\Upsilon_p))$ for $p = 1, \ldots, P$, are independent and thus can be performed in parallel on CPU or GPU clusters. By this way, the computational time needed for the miscroscale simulations is considerably reduced. Furthermore, the cost of the miscroscale model does not affect the macroscale computations, that could be extended to a more realistic three-dimensional setting without compromising the applicability of the entire model.

11.5 Computational analysis of polymer degradation

11.5.1 Computational results of the microscale model

The diffusion of water in PLA matrices is modelled considering monodisperse systems with different level of PLA degradation and swelling (i.e., water content). We generated 30 different molecular models of PLA matrices, characterized by different degree of polymerization (monodisperse systems with 600, 300, 150, 75, 30 and 1 monomers per chain, respectively) (see Fig. 11.2) and different degree of swelling (with 2 %, 20 %, 40 %, 60 % and 80 % of water). Additionally, we studied a system containing pure water for validating reasons.

As a preliminary validation of the atomistic models we analyze the density of the quasi-dry PLA matrices, see Table 11.2, showing that the predicted density ($\simeq 1.18 g/cm^3$) is close to experimental density of PLA ($1.24 g/cm^3$). On the other hand, the final density of the pure water systems ($0.98 g/cm^3$) well matches the water density. For the systems with intermediate water content the density decrease linearly from the density of quasi-dry PLA to the density of pure water.

We then proceed with the calculation of the water diffusivity in PLA matrices by means of in silico experiments run for a simulated time of 7 ns. As a validation of the approach we calculated the self diffusion coefficient of water, obtaining a value of $42.0 \cdot 10^{-6} cm^2/s$, close to the experimental value ($22.7 \cdot 10^{-6} cm^2/s$). The analysis of the water diffusivity values in PLA (see Table 11.3) shows that the diffusivity coefficient spans two orders of magnitude (from 10^{-7} to $10^{-5} cm^2/s$) and is mainly affected by the water content. Indeed, the diffusion coefficient increases

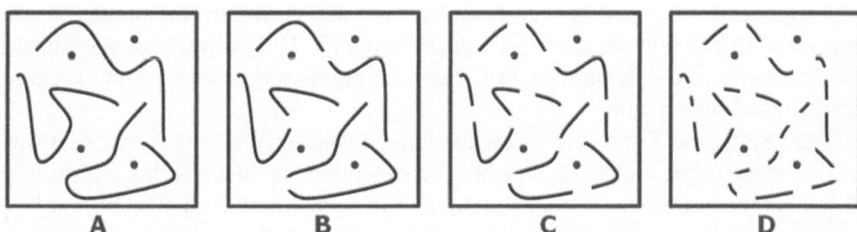

Fig. 11.2. Sketch of the quasi-dry PLA matrices (2 % water) studied in this work, from non degraded matrix (A) to partially degraded (B and C) and highly degraded (D)

Table 11.2. Density of the water/polymer mixture in g/cm^3 for different average degree of polymerization, $\bar{x}$, (colums) and % of water content, ϕ_w (rows)

| | ϕ_w | | | | | |
	0%	20%	40%	60%	80%	100%
600	1.1703	1.1424	1.0836	1.0566	1.0177	0.9871
300	1.1763	1.1415	1.0984	1.0623	1.0161	0.9871
150	1.1634	1.1419	1.0976	1.0582	1.0168	0.9871
$\bar{x}_i$ 75	1.1677	1.1482	1.1033	1.0536	1.0170	0.9871
30	1.1815	1.1422	1.0953	1.0570	1.0175	0.9871
1	1.0497	1.0196	1.0071	0.9953	0.9866	0.9871

Table 11.3. Diffusion coefficient of water $(10^{-6}cm^2/s)$ with respect to monomers per chain (colums) and % of water content ϕ_w (rows)

| | ϕ_w | | | | | |
	2%	20%	40%	60%	80%	100%
600	0.41	3.39	13.5	24.2	36.8	42.0
300	0.42	4.97	13.2	24.5	33.9	42.0
150	0.40	4.56	13.8	22.2	35.6	42.0
$\bar{x}_i$ 75	0.47	2.79	12.0	23.9	34.7	42.0
30	0.39	3.56	12.3	24.2	35.8	42.0
1	0.51	4.38	16.4	25.9	36.8	42.0

almost linearly with the degree of swelling, while it is little or no affected by the degree of polymerization. It is important to observe that the Einstein relation holds only if the regime of normal diffusion is reached. Reminding that normal diffusion is reached when $\log(MSD(t))$ is an affine linear function with respect to $\log(t)$ with unit slope, we observe that the normal diffusion of water is reliably reached during the data production runs of 7 ns, since the aforementioned slope ranges from 0.87 to 1.00. Consequently, the diffusivity coefficients obtained for water transport in the polymer matrix can be assumed to be reliable.

As similar procedure may be applied to calculate $D_i(\phi_w, \bar{x})$, resorting to a table similar to Table 11.3 for any $i = 1, \ldots, N$. Although all aforementioned simplifications, this task requires a large number of microscale simulations. To fulfill this task with a moderate computational cost, we have initially performed the simpler investigation of calculating the matrix of values $D_i(\phi_w, \bar{x}_i)$. This corresponds to estimate the diffusivity of a polymer chain of length $\bar{x}_i$ into a mixture of the same average degree of polymerization.

The results (see Table 11.4) show that the polymer diffuses much less than water due to its larger molecular weight. The exceptions are the systems with single PLA monomers (highly degraded matrices) for which a much higher PLA diffusion constant is obtained. This is likely due to the low molecular weight of PLA monomers. The analysis of the results for PLA diffusivity shows that the polymer diffusion coefficient depends on the swelling, while it is not affected by the poly-

Table 11.4. Diffusion coefficient $(10^{-8}cm^2/s)$ of PLA chains of length $\bar{x}_i$ into themselves (columns) with respect to % of water content ϕ_w (rows)

		2 %	20 %	40 %	ϕ_w 60 %	80 %	100 %
	600	0.33	1.3	3.5	3.7	42.5	–
	300	0.50	1.5	5.2	4.8	35.2	–
$\bar{x}_i$	150	0.45	0.8	8.4	6.8	38.6	–
	75	0.37	1.2	5.0	6.8	24.0	–
	30	0.51	1.6	3.8	6.8	27.6	–
	1	480	530	540	590	650	–

mer degradation, showing a behaviour similar to water transport. This consideration allows us to assume that the diffusivity of polymer sub-fractions of length $\bar{x}_i$ into a mixture of average degree of polymerization $\bar{x} \neq \bar{x}_i$ may be very similar to the sub-fraction of length $\bar{x}_i$ in itself. In practice, we conclude that $D_i(\phi_w, \bar{x}_i) \simeq D_i(\phi_w, \bar{x})$ for any $\bar{x}_1 \leq \bar{x} \leq \bar{x}_N$. This is equivalent to say that Table 11.4 represents the diffusivity of polymer subfraction $\bar{x}_i$ in all possible water/polymer mixtures.

Finally, we must note that normal diffusion is not reached for the polymer diffusivity, since the slope of $\log(MSD(t))$ presents values below 1 (with the exception of PLA monomers, for which the normal diffusion is instead reached). In this regard it is important to remind that normal diffusion behaviour holds only in the case that the observation time (i.e. the simulation time) is large enough to allow the particles to show uncorrelated motion. There are cases (i.e. simulation time) however, in which the mean squared displacement is not linear in time, as in the case of Einstein's relation, but displays a different power law, i.e. $MSD(t) \sim t^n$. If the exponent n has a value close to 1 it denotes the normal or Fickian diffusion, whereas it denotes linear or ballistic motion if is close to 2. When $1 < n < 2$ the regime is called superdiffusive motion whereas if $n < 1$ the regime is called subdiffusion or anomalous diffusion. An example of this latter behaviour the model of the *ant in the labyrinth*, see [54], where a particle (the ant) performs a random walk on a grid on which sites are randomly blocked for diffusion (the labyrinth). In this case the motion of a particle in a labyrinth is determined by the shape of the labyrinth itself and therefore it is not a random walk. Similarly, for the diffusion of water into a polymer network, if the simulation time is too short, the only motion captured is the very fast movement of a given molecule in the void space between polymer chains, namely an individual hole. In this case the trajectories are affected by the hole dimension or more in general by the microstructure of the material, contrarily to what the Fickian regime would request. The usual effect of anomalous diffusion is to create a smaller slope of the mean square displacement curve and the influence of this problem may extend up to several nanoseconds and increases with the size of the permeant molecule. This is the case of the diffusive behaviour observed in the polymer fragments in our systems.

11.5.2 Results and validation for the macroscale model

As proposed in [43] to perform numerical simulations it is convenient to rewrite problem (11.4) in non dimensional form. First, the longest PLA chains that we consider feature $\bar{x}_N = 600$ monomers,the shortest $\bar{x}_1 = 15$ units and the entire spectrum is subdivided into $N = 40$ classes. To this purpose, we select the polymer thickness $L = 10\mu m$ as the characteristic length, $\tau = \bar{k}\rho_w^\infty s$ as the characteristic time, $D_w^0 = 4.1 \times 10^{-7} cm^2/s$ as reference diffusivity and finally $\rho_w^\infty = 0.98$, $\rho^0 = 1.18 g/cm^3$ as the density of pure water and non degraded PLA respectively. Then, without change of notation, the non-dimensional form of Eq. (11.4) reads as follows,

$$
\begin{cases}
\partial_t \rho_w = \Lambda \nabla \cdot (\mathscr{D}\nabla\rho_w) - K \sum_{i=1}^N \frac{i-1}{i}\rho_w\rho_i, \\
\partial_t \rho_i = \Lambda \nabla (\mathscr{D}_i \nabla \rho_i) - (i-1)\bar{x}_1\rho_w\rho_i + 2\bar{x}_1\rho_w \sum_{j=i+1}^N \frac{i}{j}\rho_j, \\
\partial_z \rho_w|_{z=0} = 0, \quad -\mathscr{D}_w\partial_z\rho_w|_{z=L} = \Gamma\pi_w(\rho_w|_{z=L} - A) \\
\partial_z \rho_i|_{z=0} = 0, \quad -\mathscr{D}_i\partial_z\rho_i|_{z=L} = \Gamma\pi_i\rho_i|_{z=L} \\
\rho_w(z,0) = 0, \quad \rho_i(z,0) = w_i^0\tilde{\rho}^0
\end{cases}
\tag{11.5}
$$

where the diffusivity of water and polymer are determined by atomistic simulations, as previously described, while the non-dimensional numbers

$$
\Lambda = \frac{D_w^0}{L^2\bar{k}\rho_w^\infty}, \quad K = \frac{M_w\rho^0}{M_0\bar{x}_1\rho_w^\infty}, \quad \Gamma = \frac{L\pi_w^0}{D_w^0}.
\tag{11.6}
$$

We estimate the Thiele number as $\Lambda = 7 \times 10^4$, where $\bar{k} = 5 \times 10^{-2} day^{-1}$ is taken from [43] and references therein, which confirms that PLA is a bulk eroding polymer, $\Gamma = 10^{-3}$ with the assumption $\pi_w^0/D_w^0 = 1$, $K = 0.0042$ with $\bar{x}_1 = 15$, $M_0 = 90, M_w = 18 g/mol$ are the molecular weights of PLA and water respectively. Finally, the saturation of water into dry polymer is estimated by [41] as $1 g\ water/g\ PLA$, i.e. $A^0 = 0.5\%$.

For the estimation of the diffusivity of water and polymer with respect to the degree of swelling ϕ_w and the average degree of polymerization $\bar{x}$, we have applied the static multiscale coupling strategy previously described. In particular, for water we obtain,

$$
\mathscr{D}_w(\phi_w,\bar{x}) = -3.524 + 92.974\phi_w + 0.0137\bar{x} - 0.115\phi_w\bar{x} + 17.540\phi_w^2
$$
$$
- 0.0000213\bar{x}^2 + 0.0916\phi_w^2\bar{x} + 0.000183\phi_w\bar{x}^2 - 0.000142\phi_w^2\bar{x}^2.
$$

For the diffusivity D_i we have considered a slightly different approach. Denoting with $\mathbf{D}_i$ the data-set relative to Table 11.3 and with $\mathbf{D}_i^a$ the parametrized function to be estimated, we have solved the following least square problem,

$$
\mathbf{a}^* = \arg\min_{\mathbf{a}} \| \log(\mathbf{D}_i^a - \mathbf{D}_i)\|.
$$

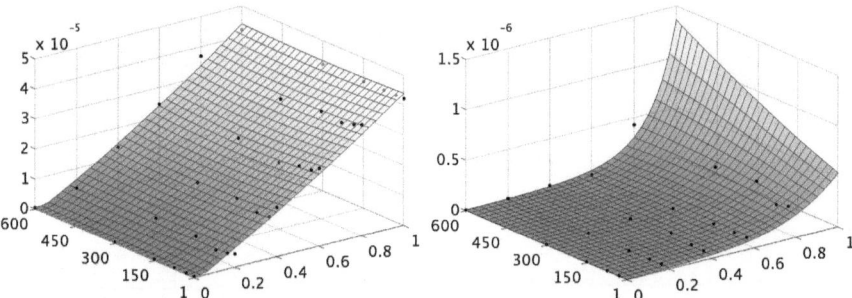

Fig. 11.3. A visual comparison of $\mathcal{D}_w(\phi_w,\bar{x}), \mathcal{D}_i(\phi_w,\bar{x})$ with the entries of Tables 11.3 and 11.4 respectively. The control points are visualized with dots

As a result of that we obtain,

$$\mathcal{D}_i(\phi_w,\bar{x}) = \exp(-4.531 + 5.483\phi_w + 0.000642\bar{x} - 0.00074\phi_w\bar{x} - 0.693\phi_w^2$$
$$- 0.00000115\bar{x}^2 + 0.002\phi_w^2\bar{x} - 0.000004\phi_w\bar{x}^2 + 0.000005\phi_w^2\bar{x}^2)$$

that ensures that $\mathcal{D}_i(\phi_w,\bar{x})$ is always positive. A visual comparison of the interpolants $\mathcal{D}_w(\phi_w,\bar{x}), \mathcal{D}_i(\phi_w,\bar{x})$ with the corresponding control points, that are entries of Tables 11.3 and 11.4 respectively, is reported in Fig. 11.3.

The results of numerical simulations for $\rho_w(t,z)$, $\tilde{\rho}(t,z) = \Sigma_i \rho_i(t,z)$, $\rho_{31}^{40} = \Sigma_{i=31}^{40} \rho_i(t,z)$, $\rho_1^{10} = \Sigma_{i=1}^{10} \rho_i(t,z)$ are reported in Fig. 11.4. Depending on the value of the Thiele modulus, Λ, different modes of degradation and erosion occurred. For PLA $\Lambda = 17000$ diffusion occurs at a much faster rate than the chemical reaction and water have saturated the polymer across the entire thickness before significant scission takes place (Fig. 11.4 top-left). Polymeric byproducts are produced almost homogeneously across the thickness of the coating and their consequent diffusion is responsible for conferring bulk erosion characteristics to the behaviour of the reaction-diffusion system (Fig. 11.4 bottom-left). Polymeric density $\tilde{\rho}$ decreases in a homogeneous fashion across the coating as smaller chains diffuse away (Fig. 11.4 top and bottom-right). Such qualitative interpretation of polymer degradation can be profitably complemented with the analysis of the evolution of the system in the lumped state space $(\phi_w,\bar{x})$, reported in Fig. 11.5. It shows that, because of fast water absorption and subsequent hydrolysis, the average degree of polymerization $\bar{x}$ quickly decreases and the water content of the mixture ϕ_w progressively increases. At the end of the process, most of the polymer in the mixture is in the range of small sub-fractions as confirmed by Fig. 11.4 bottom-left.

Further information is obtained by analyzing how the mean value of the partial density of water and polymer, respectively defined as follows,

$$\bar{\rho}_w(t) = \frac{1}{L}\int_0^L \rho_w(t,z)dz, \quad \bar{\tilde{\rho}} = \frac{1}{L}\int_0^L \tilde{\rho}(t,z)dz.$$

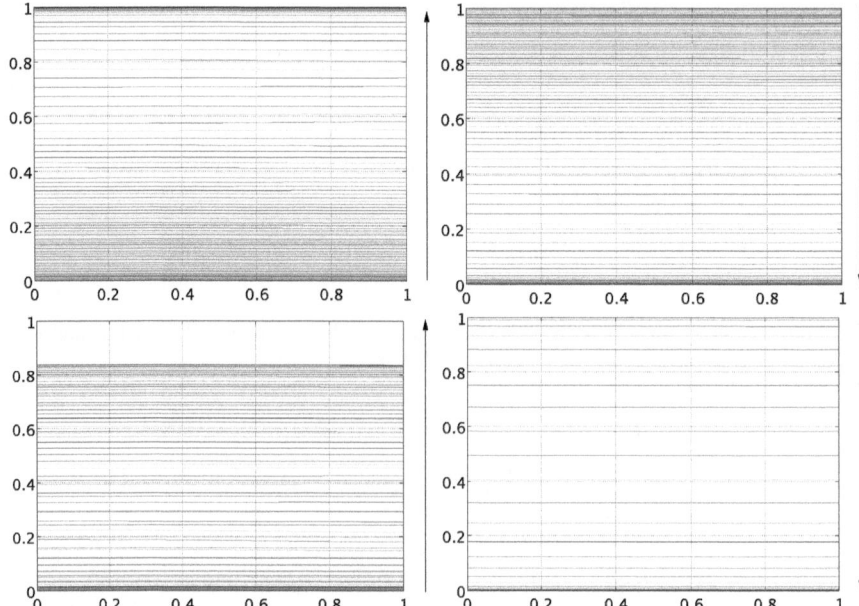

Fig. 11.4. The non-dimensional partial densities $\rho_w(t,z)$, $\tilde{\rho}(t,z) = \sum_i \rho_i(t,z)$ are reported on top, while $\rho_1^{10} = \sum_{i=1}^{10} \rho_i(t,z)$, $\rho_{31}^{40} = \sum_{i=31}^{40} \rho_i(t,z)$ are depicted at the bottom. The abscissa represents the non-dimensional coordinate along the coating thickness, z, and the time evolution is indicated with the arrows on the right of each picture

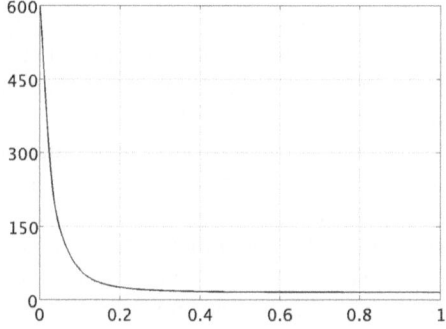

Fig. 11.5. The trajectory of the system in the $(\phi_w, \bar{x})$ phase space

For a bulk eroding polymer such diagrams are expected to feature a sigmoid shape as confirmed by the results, reported in Fig. 11.6.

Finally, we observe that the combination of microscale with macroscale models allows us to perform a quantitative validation, which is based on the comparison of the total density of the water/polymer mixture along the degradation process. On one hand, the microscale model provides, beside diffusivity, the total density of the

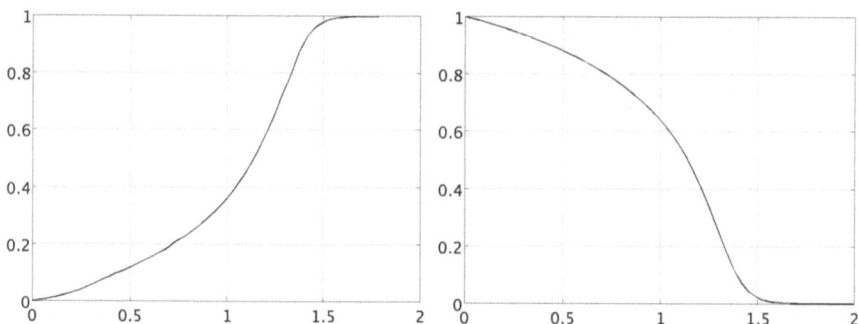

Fig. 11.6. The mean value over $(0,L)$ of the non-dimensional partial density of water (left) and polymer (right)

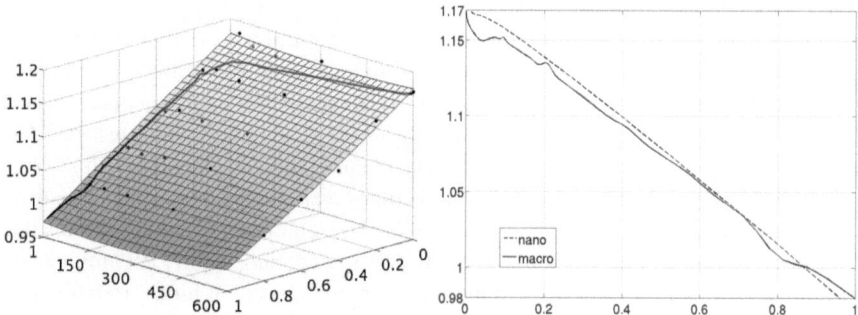

Fig. 11.7. A comparison of the predicted water/polymer mixture densities provided by micro and macroscale models (left). On the right we show a transverse section at $\bar{x}_{300}$

mixture for different configurations of the system in the lumped state space, i.e. for different combinations of $(\phi_w, \bar{x})$. The results are reported in Table 11.2. On the other hand, the macroscale model is capable to represent how the mean value of total mixture density evolves in the lumped state space, that is

$$\bar{\rho}(\phi_w, \bar{x}) = \int_0^L \rho\big(\phi_w(t,z), \bar{x}(t,z)\big)dz.$$

Such micro and macroscale data are only indirectly correlated, because the macroscale model was fitted on the diffusivity provided by atomistic simulations. By consequence, their comparison represents a significant and quantitative indicator of the multiscale model capability to represent polymer degradation. In Fig. 11.7, we represent on the left a biquadratic interpolation of the density dataset of Table 11.2 where the control points are depicted with dots. On this surface, we superpose the evolution of $\bar{\rho}(\phi_w, \bar{x})$ computed with the macroscale model. The agreement is rather satisfactory, as also confirmed by the transverse section at $\bar{x}_{300}$ of Fig. 11.7 (left), which is reported on the right.

11.6 Conclusions and perspectives

Biodegradable implants show great potential in many areas of medicine, but this technology has been hampered by the lack of reliable models to predict the behaviour and evolution of biodegradable materials. The design of biodegradable implants (a complex and challenging process) is largely based on designer's intuition and guess-work with trial-and-error attempts. In order to advance from prototype status to reliable human-implantable devices, a very resource-expensive product development process of several in vivo and in vitro iterations that often fail must be taken. A prime example of this failure is the fact that the concept pof biodegradable vascular stents dates back to the 1980s, yet there are no FDA or CE approved designs on the market today [32].

The application of multiscale computational models has been proved to be an effective and promising tool for the study of biodegradable polymers and for the design of biodegradable implants. On the one hand, molecular dynamics simulations provide a self-consistent method to quantify the constitutive laws that characterize the behaviour of such complex materials, and in particular, the diffusion of certain species in polymeric mixtures. On the other hand, continuum based models with its associated established and efficient numerical methods are the natural test bed for optimized computational design of biodegradable implants. The incorporation of descriptions of material behaviour at the molecular level into account at the constitutive specification of the response of the material at the continuum level represents the connection between both scales of the multiscale model, and, if fully accomplished, certainly provides a superior understanding and predictability of biodegradable material behaviour.

Unequivocally, the present work puts into evidence interesting perspectives for the interaction of different disciplines such as numerical analysis, computer science and bioengineering. The final objective of efficiently integrating atomistic simulations with macroscale computational models is still far from being reached, but the experience gathered with this work suggests that the combination of ad hoc numerical algorithms with the emerging high performance computational facilities could be a powerful and promising tool to achieve an accurate and reliable investigation of not only polymer degradation, but also a multitude of other complex phenomena. More interestingly, as the development of molecular dynamics as a field ensues and better descriptions of the behaviour of the material at the molecular level are achieved, these could supply valuable information about materials with complex and evolving microstructures (a very large class indeed, and very common within the field of biomechanics) to macroscale modelling approaches based on the continuum framework.

Acknowledgements. We acknowledge the Italian Institute of Technology, with the Grant: *Models and methods for degradable materials*, and the European Research Council Advanced Grant: *Mathcard – Mathematical Modelling and Simulation of the Cardiovascular System*, Project ERC 2008 AdG 227058. JSS thanks the Portuguese Fundaçao para a Ciência e Tecnologia for its support, Grant SFRH/BPD/63119/2009.

References

[1] Agrawal C.M., Ray R.B.: Biodegradable polymeric scaffolds for musculoskeletal tissue engineering, J. Biomed. Mater. Res. **55**: 141–150, 2001.

[2] Ali S.A., Doherty P.J., Williams D.F.: Mechanisms of polymer degradation in implantable devices. 2. Poly(dl-lactic acid), J. Biomed. Mater. Res. **27**: 1409–1418, 1993.

[3] Ali S.A., Zhong S.P., Doherty P.J., Williams D.F.: Mechanisms of polymer degradation in implantable devices. 1. Poly(caprolactone), Biomaterials **14**: 648–656, 1993.

[4] Ballauff M., Wolf B.A.: Degradation of chain molecules .1. Exact solution of the kinetic-equations, Macromolecules **14**: 654–658, 1981.

[5] Batycky R.P., Hanes J., Langer R., Edwards D.A.: A theoretical model of erosion and macromolecular drug release from biodegrading microspheres, J. Pharm. Sci. **86**: 1464–1477, 1997.

[6] Bose S.M., Git Y.: Mathematical modelling and computer simulation of linear polymer degradation: Simple scissions, Macromol. Theor. Simul. **13**: 453–473, 2004.

[7] Burkersroda F.v., Schedl L., Gopferich A.: Why degradable polymers undergo surface erosion or bulk erosion, Biomaterials **23**: 4221–4231, 2002.

[8] Deuflhard S.: A modified Newton method for the solution of ill-conditioned systems of nonlinear equations with application to multiple shooting, Numer. Math. **22**: 289–315, 1974.

[9] Emsley A.M., Heywood R.J.: Computer modeling of the degradation of linear-polymers, Polym. Degrad. Stabil. **49**: 145–149, 1995.

[10] Entrialgo-Castano M., Lendlein A., et al.: Molecular modeling investigations of dry and two water-swollen states of biodegradable polymers. Advanced Engineering Materials **8**(5): 434–439, 2008.

[11] A. Gautieri,, Ionita M., et al.: Computer-Aided Molecular Modeling and experimental validation of water permeability properties biosynthetic materials. Journal of Computational and Theoretical Nanoscience **7**: 1–7, 2010.

[12] Gopferich A.: Polymer degradation and erosion: Mechanisms and applications, Eur. J. Pharm. Biopharm. **4**: 1–11, 1996.

[13] Gopferich A.: Mechanisms of polymer degradation and elimination, in: A.J. Domb, J. Kost, D.M. Wiseman (eds.), Handbook of biodegradable polymers, Drug targeting and delivery, Harwood Academic Publishers, Australia, 1997, pp. 451–471.

[14] Gopferich A., Langer R.: Modeling polymer erosion, Macromolecules **26**: 4105–4112, 1993.

[15] Hairer E., Wanner G.: Solving ordinary differential equations II, Springer Series in Computational Mathematics 14, Springer-Verlag, Berlin, 2nd ed., 1996.

[16] Hayashi T.: Biodegradable polymers for biomedical uses, Prog. Polym. Sci. **19**: 663–702, 1994.

[17] Heller J., Baker R.W.: Theory and practice of controlled drug delivery from bioerodible polymers, in: R.W. Baker (ed.), Controlled release of bioactive materials, Academic Press, New York, 1980, pp. 1–18.

[18] Hofmann, D., L. Fritz, et al.: Detailed-atomistic molecular modeling of small molecule diffusion and solution processes in polymeric membrane materials. Macromolecular Theory and Simulations **9**(6): 293–327, 2000.

[19] Ionita M., Silvestri D., et al.: Diffusion of small molecules in bioartificial membranes for clinical use: molecular modelling and laboratory investigation. Desalination **200**(1–3): 157–159, 2006.

[20] Joshi A., Himmelstein K.J.: Dynamics of controlled release from bioerodible matrices, J. Control. Release **15**: 95–104, 1991.

[21] Kotliar A.M., Podgor S.: Evaluation of molecular size distributions and molecular weight averages resulting from random crosslinking and chain-scission processes, J. Polym. Sci. **55**: 423–436, 1961.

[22] Kuhn W.: The kinetics of the decomposition of high molecular chains, Ber. Deut. Chem. Ges. **63**: 1503–1509, 1930.

[23] Langer R.: Drug delivery and targeting, Nature **392**: 5–10, 1998.

[24] Laufman H., Rubel T.: Synthetic absorable sutures, Surg. Gynecol. Obstet. **145**: 597–608, 1977.

[25] Lee P.I.: Diffusional release of a solute from a polymeric matrix – approximate analytical solutions, Journal of Membrane Science **7**: 255–275, 1980.

[26] Lemaire V., Belair J., Hildgen P.: Structural modeling of drug release from biodegradable porous matrices based on a combined diffusion/erosion process, Int. J. Pharm. **258**: 95-107, 2003.

[27] Li S.M., McCarthy S.: Further investigations on the hydrolytic degradation of poly(dl-lactide), Biomaterials **20**: 35–44, 1999.

[28] Li S., Garreau H., Vert M.: Structure-property relationships in the case of degradation of massive poly(-hydroxy acids) in aqueous media – part 3: Influence of the morphology of poly(l-lactic acid), Journal of Materials Science: Materials in Medicine S: 198–206, 1990.

[29] Li S.M., Vert M.: Morphological-changes resulting from the hydrolytic degradation of stereocopolymers derived from l-lactides and dl-lactides, Macromolecules **27**: 3107–3110, 1994.

[30] Miller R.A., Brady J.M., Cutright D.E.: Degradation rates of oral resorbable implants (poly-lactates and polyglycolates): Rate modification with changes in pla/pga copolymer ratios, J. Biomed. Mater. Res. **11**: 711–719, 1977.

[31] Montroll E.W., Simha R.: Theory of depolymerization of long chain molecules, J. Chem. Phys. **8**: 721–727, 1940.

[32] Moore J., Soares J., Rajagopal K.: Biodegradable stents: Biomechanical modeling challenges and opportunities, Cardiovascular Engineering and Technology **1**: 52–65, 2010.

[33] Nguyen T.Q.: Kinetics of mechanochemical degradation by gel permeation chromatography, Polym. Degrad. Stabil. **46**: 99–111, 1994.

[34] Nguyen T.Q., Kausch H.H.: Gpc data interpretation in mechanochemical polymer degradation, Int. J. Polym. Anal. Ch. **4**: 447–470, 1998.

[35] Ottenbrite R.M., Albertsson A.C., Scott G.: Discussion on degradation terminology, in: M. Vert, J. Feijen, A.C. Albertsson, G. Scott, E. Chiellini (eds.), Biodegradable polymers and plastics, The Royal Society of Chemisty, Cambridge, 1992, pp. 73–92.

[36] Prabhu S., Hossainy S.: Modeling of degradation and drug release from a biodegradable stent coating, J. Biomed. Mater. Res. A 80A (2007) 732–741.

[37] Pietrzak W.S., Sarver D.R., Verstynen M.L.: Bioabsorbable polymer science for the practicing surgeon, J. Craniofac. Surg. **8**: 87–91, 1997.

[38] Pistner H., Bendix D.R., Muhling J., Reuther J.F.: Poly(l-lactide) – a long-term degradation study in vivo .3. Analytical characterization, Biomaterials **14**: 291–298, 1993.

[39] Quarteroni A., Valli A.: Numerical approximation of partial differential equations, Springer Series in Computational Mathematics 23, Springer-Verlag, Berlin, 1994.

[40] Siepmann J., Gopferich A.: Mathematical modeling of bioerodible, polymeric drug delivery systems, Adv. Drug Deliver. Rev. 48 (2001): 229–247.

[41] Siparsky G.L., Voorhees K.J., Dorgan J.R., Schilling K.: Water transport in polylactic acid (PLA), PLA/polycaprolactone copolymers, and PLA polyethylene glycol blends. Journal of Environmental Polymer Degradation **5**(3): 125–36, 1997.

[42] Soares J.S.: Bioabsorbable polymeric drug-eluting endovascular stents: A clinical review, Minerva Biotecnologica **21**: 217–230, 2009.

[43] Soares J.S., Zunino P.: A mixture model for water uptake, degradation, erosion, and drug release from polydisperse polymeric networks, Biomaterials **31**: 3032—3042, 2010.

[44] H. Sun,: COMPASS: An ab initio force-field optimized for condensed-phase applications – Overview with details on alkane and benzene compounds. Journal Of Physical Chemistry B **102**(38): 7338–7364, 1998.

[45] Tamada J.A., Langer R.: Erosion kinetics of hydrolytically degradable polymers, Proc. Natl. Acad. Sci. USA **90**: 552–556, 1993.

[46] Thombre A.G.: Theoretical aspects of polymer biodegradation: Mathematical modeling of drug release and acid-catalyzed poly(otho-ester) biodegradation, in: M. Vert, J. Feijen, A.C. Albertsson, G. Scott, E. Chiellini (eds.), Biodegradable polymers and plastics, The Royal Society of Chemisty, Cambridge, 1992, pp. 214–225.

[47] Thombre A.G., Himmelstein K.J.: A simultaneous transport-reaction model for controlled drug delivery from catalyzed bioerodible polymer matrices, AIChE Journal **31**: 759–766, 1985.

[48] Vert M., Li S., Garreau H., Mauduit J., Boustta M., Schwach G., Engel R., Coudane J.: Complexity of the hydrolytic degradation of aliphatic polyesters, Angew. Markomol. Chemie **247**: 239–253, 1997.

[49] Wu X.S., Wang N.: Synthesis, characterization, biodegradation, and drug delivery application of biodegradable lactic/glycolic acid polymers. Part ii: Biodegradation, J. Biomater. Sci. Polym. Ed. **12**: 21–34, 2001.

[50] Zygourakis K.: Development and temporal evolution of erosion fronts in bioerodible controlled release devices, Chemical Engineering Science **45**: 2359–2366, 1990.

[51] Harvey M.J., Giupponi G., De Fabritiis G.: ACEMD: Accelerating biomolecular dynamics in the microsecond time scale, Journal of Chemical Theory and Computation **5**: 1632–1639, 2009.

[52] Stone J.E., Phillips J.C., Freddolino P.L., Hardy D.J., Trabuco L.G., Schulten K.: Accelerating molecular modeling applications with graphics processors, Journal of Computational Chemistry **28**: 2618–2640, 2007.

[53] Friedrichs M.S., Eastman P., Vaidyanathan V., Houston M., Legrand S., Beberg A.L., Ensign D.L., Bruns C.M., Pande V.S.: Accelerating molecular dynamic simulation on graphics processing units, Journal of Computational Chemistry **30**: 864–872, 2009.

[54] Yuval G., Amnon A., Shlomo A.: Anomalous Diffusion on Percolating Clusters. Phys Rev Lett. **50**: 77–80, 1983.

Applications of variational data assimilation in computational hemodynamics

Marta D'Elia, Lucia Mirabella, Tiziano Passerini, Mauro Perego, Marina Piccinelli, Christian Vergara, and Alessandro Veneziani

Abstract. The development of new technologies for acquiring measures and images in order to investigate cardiovascular diseases raises new challenges in scientific computing. These data can be in fact merged with the numerical simulations for improving the accuracy and reliability of the computational tools. Assimilation of measured data and numerical models is well established in meteorology, whilst it is relatively new in computational hemodynamics. Different approaches are possible for the mathematical setting of this problem. Among them, we follow here a variational formulation, based on the minimization of the mismatch between data and numerical results by acting on a suitable set of control variables. Several modelling and methodological problems related to this strategy are open, such as the analysis of the impact of the noise affecting the data, and the design of effective numerical solvers. In this chapter we present three examples where a mathematically sound (variational) assimilation of data can significantly improve the reliability of the numerical models. *Accuracy* and *reliability* of computational models are increasingly important features in view of the progressive adoption of numerical tools in the de-

Marta D'Elia, Tiziano Passerini, Marina Piccinelli, Alessandro Veneziani (✉)
Department of Mathematics and Computer Science, Emory University, 400 Dowman Dr, 30322, Atlanta GA, USA
e-mail: {marta,tiziano,marina,ale}@mathcs.emory.edu

Lucia Mirabella
W.H. Coulter Department of Biomedical Engineering, Georgia Institute of Technology, 315 Ferst Dr., 30332 Atlanta GA, USA
e-mail: lucia.mirabella@bme.gatech.edu

Mauro Perego
Department of Scientific Computing, Florida State University, Tallahassee FL, USA
e-mail: mperego@fsu.edu

Christian Vergara
Department of Information Engineering and Mathematical Methods, University of Bergamo, Italy
e-mail: christian.vergara@unibg.it

Ambrosi D., Quarteroni A., Rozza G. (Eds.): Modeling of Physiological Flows.
DOI 10.1007/978-88-470-1935-5_12, © Springer-Verlag Italia 2012

sign of new therapies and, more in general, in the decision making process of medical doctors.

12.1 Introduction

In the last 20 years mathematical and numerical models have been progressively used as a tool for supporting medical research in the cardiovascular science. *In silico* experiments can provide remarkable insights into a physio-pathological process completing more traditional *in vitro* and *in vivo* investigations. Numerical models have been playing the role of "individual based" simulators, able to furnish a dynamical representation of the biology of a specific patient as a support to the prognostic activity. At the same time, the need for quantitative responses for diagnostic purposes has strongly stimulated the design of new methods and instruments for measurements and imaging. On the one hand, we can simulate in 3D large portions of the cardiovascular system of a real patient properly including simplified models for the peripheral sites (see e.g. [16, 21, 31, 32, 33]). On the other hand, thanks to new instruments, images and measures nowadays provide doctors and bioengineers with a huge amount of data. These data offer obviously new possible benchmarks for the numerical simulations (see e.g. [34]). However, beyond the validation, it is possible to merge simulations and measures by means of more sophisticated numerical techniques. This procedure is called *Data Assimilation* (DA) (see e.g. [22]). With this name we mean the ensemble of methods for merging observed (generally sparse and noisy) information into a numerical model based on the approximation of physical and constitutive laws. The merging improves the quality of the information brought both by numerical results and by measurements:

- numerical simulations are improved by the merging of data that allow to include effects otherwise difficult to model (at the qualitative or quantitative level), such as the presence of tissues surrounding an artery or the motion of heart affecting the aortic dynamics;
- measures are in general affected by noise, so that assimilation of results based on physical and constitutive laws introduces a sophisticated filter, forcing the consistency with basic principles.

In some fields, these techniques are quite mature and tested, in particular in geophysics and meteorology (see the excellent review of methods in [22]). There are basically two classes of methods for performing DA, both with pros and cons.

Variational methods. DA is performed by minimizing a functional, estimating the discrepancy between numerical results and measures. The optimization problem is solved by using the mathematical model as a constraint, upon the identification of a proper set of control variables. In environmental studies this is often the initial state of the system of interest. In some cases (*Nudging* or *Dynamic Relaxation Methods*) the functional to be minimized is properly "altered" so to include the data to be assimilated directly in the equations of the model.

Stochastic methods. These are based on the extension to nonlinear problems of the *Kalman filter*, which is a statistical approach for prediction of linear systems affected by uncertainty [8, 18, 35], relying upon a Bayesian maximum likelihood argument.

On the contrary, there are relatively few studies devoted to a mathematically sound assimilation of data in hemodynamics, probably because the availability of more and more accurate measurements is the result of truly recent advancements. In particular, we mention [36, 37, 38, 39, 40], essentially based on Kalman filtering techniques, and [41] for the variational approach.

In this chapter we consider three possible applications of DA based on *variational methods*. In particular we present possible techniques for:

- Merging velocity data into the numerical solution of the incompressible Navier-Stokes equations, so to eventually retrieve non primitive variables like the *Wall Shear Stress* (WSS);
- Including images into the simulation of blood flow in a moving domain, so to perform the fluid dynamics simulations including the measured movement of the vessel;
- Estimating physiological parameters of clinical interest by matching numerical simulations and available data.

In all these examples we face a common structure that can be depicted as a classical feedback loop illustrated in the scheme below.

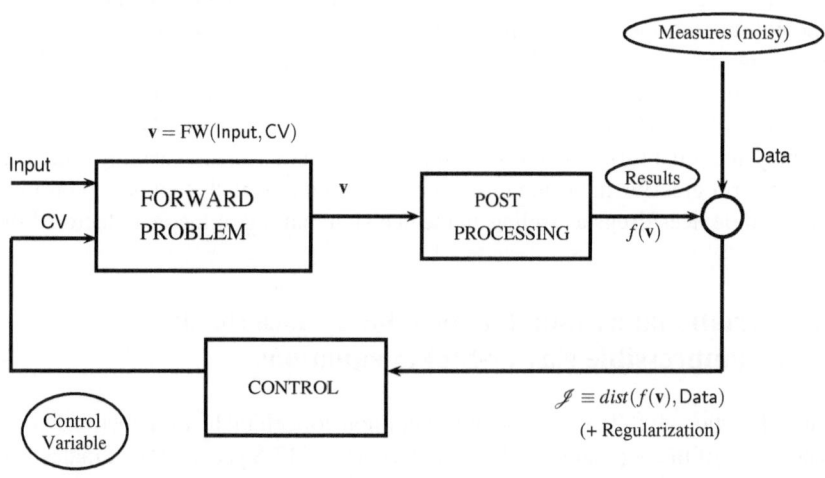

At an abstract level, all these applications actually lead to solve a problem in the form: Find the Control Variable CV (belonging to a suitable functional space) such that it minimizes the distance

$$\mathscr{J} \equiv dist(f(\mathbf{v}, \text{Data})) \quad (+ \text{Regularization}), \qquad (12.1)$$

where Data is the set of (noisy) measures, $\mathbf{v}$ the solution of the Forward Problem FW, which depends on some Input variables and CV. Finally, $f(\cdot)$ represents a post processing step for computing the quantity to be compared with the data. "Regularization" stands for some possible Tikhonov-like regularizing term with the role of making the mathematical and numerical problem more tractable (see e.g. [12, 15]). Control problems with constraints represented by partial differential equations have been studied since a long time ([3, 12, 17, 26, 29, 42, 43, 44]). In computational hemodynamics, these problems have been considered, for example, for the prescription of defective boundary conditions [45, 46, 47].

There are several issues when solving this kind of problems. Particularly relevant for our applications are:

- the *existence* of an admissible CV that attains the minimal distance between data and results. This can depend on the location (in space and time) of the available data and can be forced by a proper regularization term;
- the *noise* that invariably affects the data to be assimilated; this has a major impact on the reliability of the entire data assimilation process.

In the examples presented below we will partially address these issues, pointing out available results and open problems for each application. We will split each example in three sections after the presentation of the specific problem and its medical motivations, namely (i) the formalization of the problem in mathematical and numerical terms - with a specific link to the feedback loop above - (ii) the discussion of some preliminary numerical results and (iii) of the associated prospective research. Far from being a conclusive review of methods and applications, the present work pinpoints several open challenging problems in the adoption of variational methods for DA in computational hemodynamics. These are anticipated to become an important tool for pursuing more reliability of numerical simulations in the general perspective of *data driven simulations* [48] and *inverse cardiovascular mathematics*. Accuracy and reliability of scientific computing are in fact an increasingly critical issue for the progressive inclusion of numerical simulations in the validation protocol of medical devices/drugs as well as in the decision making of medical doctors [49].

12.2 Variational assimilation of velocity data for the incompressible Navier-Stokes equations

Bicuspid aortic valve (BAV) is the most common congenital heart defect, occurring in about 1 % of the population [14]. At a mean age of 17.8 years 52 % of males with normally functioning BAV already have aortic dilatation [9] which may eventually lead to aortic regurgitation or dissection or aortic aneurysms. Medical doctors are interested in developing a better understanding of the hemodynamics contributing to aortic dilatation not only in patients with BAV but also in other forms of congenital heart disease in which aortic dilatation is common [50]. Such an understanding may allow early risk stratification, possibly leading to guidelines for earlier intervention in high-risk groups, with an anticipated resultant reduction in morbidity and mortality for these patients. Some studies suggest that BAV morphology results in abnormal flow patterns in the ascending aorta, anticipating that valves with significant

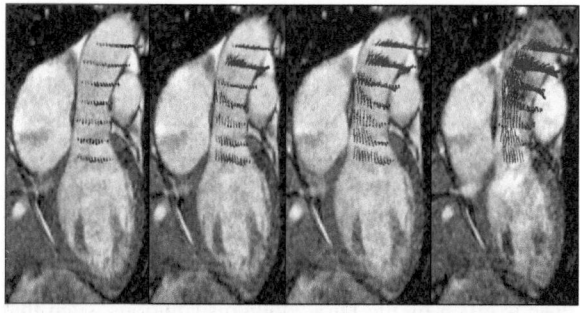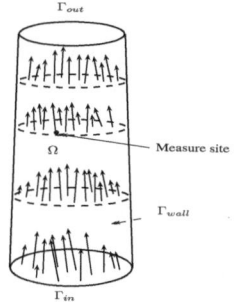

Fig. 12.1. Left: Blood velocity measured with Magnetic Resonance in the ascending aorta of a patient (courtesy of M. Brummer, Emory Children's Healthcare of Atlanta). Right: example of a possible region of interest Ω with velocity measures inside the domain

asymmetry would result in highly disturbed flow patterns. Consequently, the flow patterns, as detected by MRI flow-velocity encoding methodologies (see Fig. 12.1), have predictive value in determining which BAV morphology variants would be at greater risk of developing aortic dilatation.

In order to validate this hypothesis, clinical studies have been performed [51, 52] evaluating the WSS in BAV patients from MRI measurements. WSS is computed by a finite difference approximation based upon the velocity data and the blood viscosity measure. However, these estimates are clearly affected by both the discretization error and the noise of the data. Numerical simulations of blood flow can be carried out in the region of interest to improve this computation (see e.g. [53]). In this context, measures inside the domain of interest are not strictly needed for solving the incompressible fluid problem, that requires only initial and boundary conditions. However, they can be merged with the numerical results for obtaining a better estimate of the WSS. This leads to the following problem: How is it possible to incorporate velocity (noisy) data available in a domain of interest into the computation of the incompressible Navier-Stokes equations? A similar problem in the context of the fluid mechanics of the heart has been studied in [34] (and successively analyzed in [54]). In this work, available velocity data belong to a plane cutting the domain. As it has been observed in [34, 55], in principle, if the data belong to surfaces that split the region of interest into regular subdomains (as a plane), an immediate approach for the assimilation would be to solve the equations in each subdomain. In fact, the available data can be prescribed as standard boundary conditions. This naive approach, however, does not consider the presence of the noise. As a matter of fact, no filtering is introduced in this way and the noise is spread into each subdomain, resulting in significant inaccuracies (see [55]). For this reason, we resort here to a variational approach.

12.2.1 Mathematical formulation and numerical approximation

Let us denote by Ω a domain in $\mathbb{R}^d$ ($d = 2, 3$; in real applications $d = 3$, however here we limit numerical results to the 2D case). We assume that the domain of interest Ω (see Fig. 12.1, right) features an inflow boundary Γ_{in}, an outflow boundary Γ_{out} and the physical wall of the vessel Γ_{wall}. Γ_{in} and Γ_{out} can possibly consist of more parts (like in a vascular bifurcation). Variables of interest are the velocity $\mathbf{u}$ and the pressure $P = \rho_f p$ which are assumed to obey the incompressible Navier-Stokes equations in Ω. Here we have denoted by ρ_f the fluid density. At this stage, we consider the steady problem. We assume to have velocity measures (Data) $\mathbf{u}^m$ available at N_s sites[1] $\mathbf{x}_i \in \Omega$. Following the general description given in the Introduction, we assume that the CV is represented by the inflow normal stress $\mathbf{h}$. This is an arbitrary choice, an extensive comparison with other choices is still to be done. Post-processing $f(\cdot)$ in this case is given by the Dirac delta distributions, such that $f(\mathbf{u})$ is the vector of the values of the computed velocity at the measurement sites. Then, the distance $dist(f(\mathbf{u}), \mathbf{u}^m)$ is defined as $\sum_{i=1}^{N_s} (\mathbf{u}(\mathbf{x}_i) - \mathbf{u}^m(\mathbf{x}_i))^2$. The control problem reads: Find

$$\min_{\mathbf{h}} \mathscr{J}(\mathbf{u}, \mathbf{h}) = dist(f(\mathbf{u}), \mathbf{u}^m) + \text{Regularization}(\mathbf{h})$$

$$\text{s.t.} \begin{cases} -\mu \nabla \cdot (\nabla \mathbf{u} + \nabla^T \mathbf{u}) + (\mathbf{u} \cdot \nabla)\mathbf{u} + \nabla p = \mathbf{s} \text{ in } \Omega, \\ \nabla \cdot \mathbf{u} = 0 & \text{in } \Omega, \\ \mathbf{u} = \mathbf{0} & \text{on } \Gamma_{wall}, \\ -\mu (\nabla \mathbf{u} + \nabla^T \mathbf{u}) \cdot \mathbf{n} + p\mathbf{n} = \mathbf{h} & \text{on } \Gamma_{in}, \\ -\mu (\nabla \mathbf{u} + \nabla^T \mathbf{u}) \cdot \mathbf{n} + p\mathbf{n} = \mathbf{g} & \text{on } \Gamma_{out}, \end{cases} \quad (12.2)$$

where $\mathbf{n}$ denotes the outward unit vector normal to the boundary. A Newtonian rheology is supposed to hold, since it is a common assumption in large and medium vessels [16] and μ is the kinematic viscosity. Since we are considering fixed geometries, we assume homogeneous Dirichlet boundary conditions on Γ_{wall}. When solving problems in the form (12.2) there are in general two possibilities. In the first one, we first write the necessary conditions associated with the continuous constrained optimization problem, the so called *Karush Kuhn Tucker* (KKT) system [6, 12]. These are obtained by augmenting the original functional with the (variational formulation of) the constraint given by FW (in this case the steady Navier-Stokes problem), weighted by unknown multipliers and then by setting to zero the derivatives of the augmented functional with respect to the multipliers (so to obtain the *state problem*), to the variables (*adjoint problem*) and to CV (*optimality conditions*). Successively, the resulting problem is discretized (*Optimize then Discretize* – OD – approach). In the second approach, we first discretize the different components of the problem (the functional to be minimized and the constraints) and then perform the optimization of the discrete system (*Discretize then Optimize* – DO – approach). In [55] we compared the two strategies, and found that the DO is more efficient for the problem at hand. For this reason we proceed with the latter approach.

---

[1] Notice that we use the word "sites" for the location of measurements, as opposed to the word "nodes" for points where velocities are computed. We do not assume at this level particular positions for the sites, even though in the applications it is reasonable to assume that they are located on planes transverse to the blood stream.

12.2.1.1 The discrete DA Oseen problem

Let us consider preliminarily the linear Oseen problem. The nonlinear convection term $(\mathbf{u} \cdot \nabla)\mathbf{u}$ is replaced with $(\boldsymbol{\beta} \cdot \nabla)\mathbf{u}$, where $\boldsymbol{\beta}$ is a known advection field. The discretized optimization problem reads

$$
\min_{\mathbf{H}} \mathscr{J}(\mathbf{V},\mathbf{H}) = \frac{1}{2}\|D\mathbf{V} - \mathbf{U}^m\|_2^2 + \frac{a}{2}\|L\mathbf{H}\|_2^2 \quad \text{where} \quad
\begin{aligned}
\mathbf{V} &= \begin{bmatrix} \mathbf{U} \\ \mathbf{P} \end{bmatrix}, \\
S &= \begin{bmatrix} C + A^{\beta} & B^T \\ B & O \end{bmatrix}.
\end{aligned} \tag{12.3}
$$
$$
\text{s.t. } S\mathbf{V} = R_{in}^T M_{in}\mathbf{H} + \mathbf{F}.
$$

Here, $\mathbf{U}$ and $\mathbf{P}$ are the discretization of velocity and pressure. In particular, we resort to an inf-sup compatible finite element (FE) discretization (see e.g. [56], Chapters 7, 9). $\mathbf{H}$ is the discretization of the control variable $\mathbf{h}$. In formulating the minimization problem, we need to introduce some special matrices. Q is the discrete operator corresponding to f in (12.2), i.e. the matrix such that $[Q\mathbf{U}]_i$ is the numerical solution evaluated at the site $\mathbf{x}_i$ and corresponding to the data $[\mathbf{U}^m]_i$. Matrix D is defined as $D = [Q \ O]$. R_{in} is a *restriction matrix* which selects the degrees of freedom (DOF) of the velocity $\mathbf{U}$ on Γ_{in}; M_{in} is the mass matrix restricted to inlet boundary nodes; C, A^{β} and B are the discretization of the diffusion, advection and divergence operators respectively. For $a > 0$, $\frac{a}{2}\|L\mathbf{H}\|_2^2$ is a Tikhonov regularization term (see e.g. [57]). Matrix L is such that $L^T L$ is positive definite. The Lagrange functional associated with the problem (12.3) is

$$
\mathscr{L}(\mathbf{V},\mathbf{H},\boldsymbol{\Lambda}) = \frac{1}{2}\|D\mathbf{V} - \mathbf{U}^m\|_2^2 + \frac{a}{2}\|L\mathbf{H}\|_2^2 + \boldsymbol{\Lambda}^T(S\mathbf{V} - R_{in}^T M_{in}\mathbf{H} - \mathbf{F}), \tag{12.4}
$$

where $\boldsymbol{\Lambda} \in \mathbb{R}^{N_u + N_p}$ is the discrete Lagrange multiplier. The associated KKT system reads

$$
\begin{cases}
D^T(D\mathbf{V} - \mathbf{U}^m) + S^T\boldsymbol{\Lambda} = 0 \\
aL^T L\mathbf{H} - M_{in}^T R_{in}\boldsymbol{\Lambda} = 0 \\
S\mathbf{V} - R_{in}^T M_{in}\mathbf{H} - \mathbf{F} = 0.
\end{cases} \tag{12.5}
$$

In [58] we proved the following proposition.

Proposition 1. *Sufficient conditions for the well-posedness of the discrete optimization problem are:*

1. $a > 0$;
2. for $a = 0$, $Null(D) \cap Range(S^{-1}R_{in}^T M_{in}) = \{0\}$ $(\star)$.

This result basically states that, in absence of regularization, well-posedness is guaranteed if enough measurement sites are placed at the inflow boundary. This proposition stems from the analysis of the system obtained after the elimination of $\mathbf{V}$ and $\boldsymbol{\Lambda}$ from the system (12.5) (the so-called *reduced Hessian*). In Fig. 12.2 we report the singular values of the reduced Hessian when the sufficient condition $(\star)$ is fulfilled (left) and violated (right). In the latter case, it is evident that in general a violation of such condition may lead to a *discrete ill-posed problem* [57].

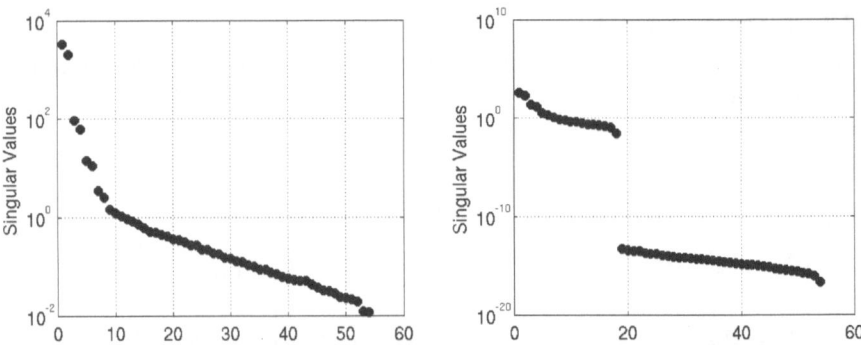

Fig. 12.2. Singular values of the reduced Hessian for a non regularized case ($a = 0$): on the left, the condition ($\star$) is fulfilled, on the right it is violated

On the contrary, no constraints on sites locations need to be fulfilled when the Tikhonov regularization is active ($a > 0$). However, in practice, the selection of a requires to find the proper trade-off between the requirement to solve a well conditioned problem (large a) and to keep the perturbation of the original problem as small as possible (small a). A possible approach (see [19, 57]) is to select the parameter according to the discrepancy principle (DP), i.e. to select a in such a way that the perturbation of the regularization term affects the solution with the same order of the discrepancy induced by the noise. The proper choice of the parameter following this approach may be however computationally expensive. There is another possible way for forcing the well-posedness exploiting the result of Proposition 1. Actually, let us assume that some data are available at the inflow, not necessarily fulfilling the well-posedness sufficient condition ($\star$). If we extend the given data to the entire set of DOF of Γ_{in} by interpolation of the available data (e.g. piecewise linear), the resulting problem satisfies condition ($\star$). This results in fact in an additional term to the functional $\mathcal{J}$ that plays the role of a regularizing term (see [58]). A more extensive analysis of this approach, and the interplay between the interpolation and the noise affecting the original data is currently under investigation.

12.2.1.2 The nonlinear Navier-Stokes problem

When we consider the nonlinear advection term $(\mathbf{u} \cdot \nabla)\mathbf{u}$ the problem becomes much more difficult since now we have a nonlinear constraint [6]. A possible approach is to combine the DA procedure for the linear case with classical fixed point linearization schemes (i.e. Picard and Newton). Thus, the DA assimilation problem is solved iteratively. We report the simple case of the Picard method. Given a guess for the velocity at step k, say $\mathbf{U}_k$, we solve

$$\min_{\mathbf{H}_{k+1}} \frac{1}{2}\|D\mathbf{V}_{k+1}(\mathbf{H}_{k+1}) - \mathbf{U}^m\|_2^2 + \frac{a}{2}\|L\mathbf{H}_{k+1}\|_2^2 \quad \text{where } S = \begin{bmatrix} C + A^{\mathbf{U}_k} & B^T \\ B & O \end{bmatrix} \quad (12.6)$$
$$\text{s.t.} \quad S_k \mathbf{V}_{k+1} = R_{in}^T M_{in} \mathbf{H}_{k+1} + \mathbf{F}$$

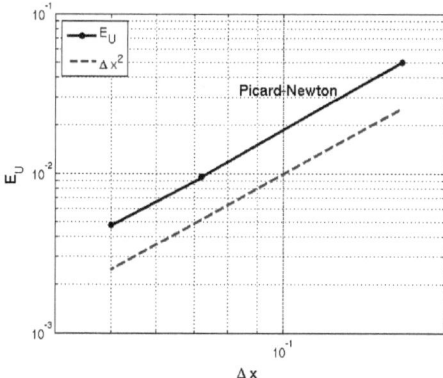

Fig. 12.3. Test of consistency of the DA procedure for noise-free velocity data. The accuracy of the computation, in terms of L^2 norm of the velocity error (E_U), is the same as for the solution of the FW problem

up to the fulfillment of a convergence criterion. When using the Newton method, the convergence strongly depends on the initial guess, so a common procedure is to perform a few Picard iterations (12.6) and use the resulting velocity as an initial guess for the Newton method. In our approach the loop for solving the nonlinear system is merged with the one for the optimization problem, thus reducing the computational cost. Numerical experience (next subsection) shows that convergence is not prevented by this further approximation. Other approaches can be however pursued, for an introduction to optimization with nonlinear constraints see [6].

12.2.2 Numerical results

We first present some simulations on an analytic test case, to investigate basic convergence properties of the DA procedure without and with the presence of the noise, in comparison with the FE convergence of the forward problem. Then we address a comparison between a classical Tikhonov regularization and the data interpolation method. Results have been obtained with the C++ finite element library LifeV [10].

12.2.2.1 A consistency test

Let Ω be the domain $\Omega = [-0.5, 1.5] \times [0, 2]$ with a flow described by the analytical solution $\mathbf{u}_1(x,y) = 1 - e^{\lambda x}\cos(2\pi y), \mathbf{u}_2(x,y) = \dfrac{\lambda}{2\pi}e^{\lambda x}\sin(2\pi y), p(x,y) = \dfrac{1}{2}e^{2\lambda x} + C$, with $\lambda = \frac{1}{2}(\mu^{-1} - \sqrt{\mu^{-2} + 16\pi^2})$, $\mu = 0.035$, and C is a constant chosen to give a zero mean pressure. Solution of the DA problem has been obtained by using inf-sup compatible FEs ($\mathbb{P}^1$bubble-$\mathbb{P}^1$). Regularization is obtained with L corresponding to the discrete gradient operator and a selected according to the DP. The nonlinear term has been solved by combining Picard and Newton methods.

As expected, in the noise-free case the assimilated velocity recovers the solution of the forward problem. In particular Fig. 12.3 shows the expected quadratic con-

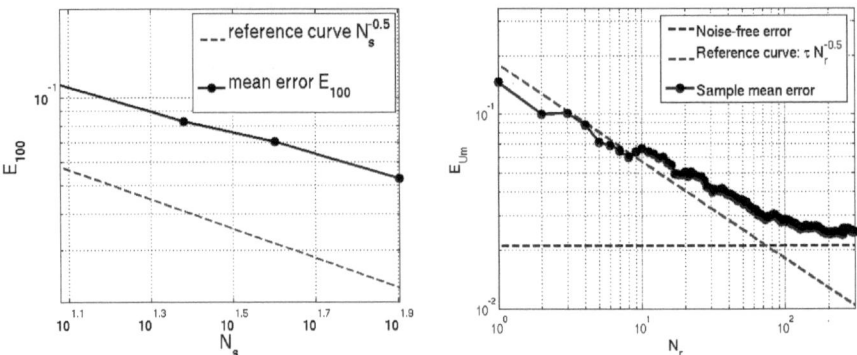

Fig. 12.4. Test of consistency for the DA procedure of velocity data: accuracy of the DA improves when the number of sites N_s (left) or of the noise realization N_r (right) increases

vergence rate for the L^2 norm of the velocity error, E_U, when the mesh size Δx tends to 0. In case of noisy data (Gaussian white noise), the error dynamics changes. In particular, the error decreases as more data are available (see Fig. 12.4, left). We observe that the convergence rate with respect to the number of sites N_s is of the order of $1/\sqrt{N_s}$. Moreover, when the number of sites and the mesh are fixed and we repeat the DA with different noise realizations, we observe a progressive convergence of the sample mean of the assimilated solution to the noise free solution, with a rate proportional to $1/\sqrt{N_r}$, being N_r the number or realizations. (see Fig. 12.4, right). Note that these results are consistent with the central limit theorem. More details can be found in [58].

12.2.2.2 Regularization & interpolation

We compare the results obtained with the regularized problem (parameter a tuned again with the DP) and the interpolation of inflow data when these do not fulfill condition $(\star)$ of Proposition 1. In Table 12.1 (left) in correspondence of different choices of locations and number N_{in} of sites on Γ_{in}, we report the relative error and the number of iterations (it) for solving the reduced Hessian. The case where neither Tikhonov regularization nor interpolation are performed is reported as a reference test. The addition of interpolated data on the DOF on Γ_{in} has the effect of forcing the well posedness of the problem, as can be inferred from the singular values of the reduced Hessian in the case reported in Fig. 12.2. Also, results reported in Table 12.1 show that, in terms of accuracy, the interpolation procedure is comparable with Tikhonov regularization. This fact, combined with the computational saving associated with the generation of the interpolating function, as opposed to applying the DP, makes interpolation an efficient regularization technique, competitive with common available methods.

Table 12.1. Left: Comparison of the results of a regularized DA vs a non-regularized interpolated DA. Right: **Relative** errors of the WSS computed with the DA procedure and a forward Navier-Stokes noisy simulation in a 2D carotid bifurcation for different values of the SNR

N_{in}	interpolation	a	E_U	it
14	no	0	0.068	14
14	no	0.021	0.061	15
14	yes	0	0.059	18
14	yes	0.021	0.056	16
8	no	0	0.199	11
8	no	0.038	0.137	18
8	yes	0	0.139	17
8	yes	0.038	0.129	17

SNR	$E_{WSS,DA}$	$E_{WSS,FW}$
100	0.2536	0.2667
20	0.2591	0.3030
10	0.2738	0.3861
5	0.3149	0.6114

12.2.2.3 Assimilated derived quantities in nontrivial geometries

In view of real hemodynamics applications, we present a demonstrative test case in non-trivial geometries (representing a 2D simplified model of the aortic arch and an arterial bifurcation). Since in these cases we do not have an analytical solution, we have computed a "reference" solution on an extremely fine mesh grid (using parabolic inflow conditions and homogeneous Neumann conditions at the outflow) in both cases. Successively, a noise generated with a uniform distribution has been added to the solution. We consider several values of Signal-to-Noise Ratio (SNR), defined as the ratio between the maximum magnitude of the velocity data and the noise standard deviation. This generates a set of noisy data to be assimilated represented by the black vector field in Fig. 12.5, left. Results of the assimilation are significantly close to the reference solution. As a matter of fact, we consider as an index of the accuracy for the solution the ratio $E_U^* = \frac{\|U - U_{ref}\|_2}{\|U_{ref}\|_2}$. To test the competitiveness of the DA procedure we compare the relative error of the assimilated velocity with the one of the velocity obtained from a forward simulation where noisy data on Γ_{in} are prescribed as a Dirichlet condition; in this case we obtain $E_U^* = 8.1e-2$ and $E_{U_{forward}}^* = 16.0e-2$. This pinpoints the role of DA as a process for de-noising the available data thanks to mathematical models. The DA procedure in fact corrects the measurements according to the physical principles underlying the mathematical model. This is evident not only for the primitive variable, but also checking non-primitive interesting quantities. In Fig. 12.5, right, we report the vorticity map recovered from an assimilated velocity field on a geometry approximating an arterial bifurcation. For the same simulation, we check the accuracy of the WSS. Accuracy results are reported in Table 12.1, right. The WSS is retrieved in two ways. In the first case, we perform the DA procedure and use the assimilated velocity field for extracting the WSS. In the second case, we use again the inflow noisy data as boundary conditions for a forward computation of the incompressible Navier-Stokes equations on the same mesh where DA is performed. In particular, in the table we report the relative errors, i.e. the difference of the WSS compared with the noise-free reference

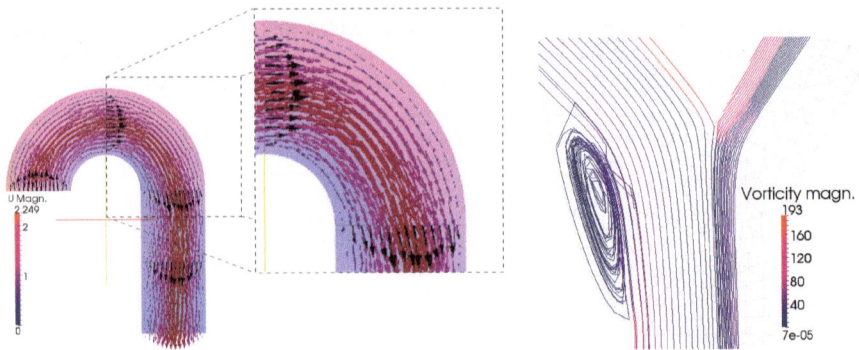

Fig. 12.5. Left: DA assimilation on 2D a curved domain. Black arrows are the data to be assimilated. The red arrows (colores refer to the pressure) are the results after the DA procedure. The results of DA are closer to the reference solution, highlighting the role of DA as a filtering procedure for the noise of the data. Right: vorticity in a 2D bifurcation computed by the DA procedure

solution on the fine mesh. It is evident that the DA leads to a more accurate estimate of the WSS, the improvement being more evident as the SNR gets smaller.

12.2.3 Perspectives

Assimilation of (velocity) data into the simulation of an incompressible fluid is a problem whose interest goes beyond the specific medical applications, and different methods are viable. In [34] Least Squares Finite Elements are used because of their versatility in managing different boundary conditions. In [41], merging of velocity data is carried out by using a "virtual" forcing term as CV. Here, we resorted to a control approach that in preliminary test cases provides promising results. Sufficient conditions for the well posedness of the linearized Oseen problem are given. Many challenges are however open by these preliminary results. Beyond (and before) an extensive use of DA in clinical practice, there are at least two main concerns that deserve an accurate consideration.

Analysis of the impact of the noise. An extensive analysis of this aspect is in order to identify the reliability of the results. Modelling the impact of uncertainty on the solution of partial differential equations is an up-to-date topic (see e.g. [59, 60] and the recent work [61], comparing Galerkin vs collocation methods). Different approaches can be pursued and different sources of noise should be considered, depending on the measurement devices (see e.g. [62] and Chapter 3 of [20]). A *sensitivity analysis* of quantities of interest such as the WSS on the noise affecting the data would clarify the robustness of the procedure to the perturbations. More advanced approaches are based on the moment method, the Bayesan approach, the polynomial chaos (see e.g. [27]). Extensive investigations on this aspects, with different approaches, will be carried out as a follow up of the present results.

Unsteady problems. When solving unsteady problems, following again a DO approach, we first discretize in time and at each instants solve the optimization problem. In this case, the extension of the method devised for the steady case is pretty immediate. However, possible computational concerns arise from the nesting of the time and the optimization loops. Selection of appropriate effective preconditioners is in order. Another issue refers to the initial conditions that in general are not known. In meteorological applications, these are included in the set of CV and used for driving the assimilation procedure. In cardiovascular applications an alternative approach consists of forcing periodicity of the solution. This approach will be investigated elsewhere.

12.3 Image assimilation in a moving domain simulation

Rigid-wall models for blood motion in arteries are often accurate enough for a quantitative analysis of hemodynamics (see e.g. [63]). However, there are situations in which the magnitude of the mechanical forces involved and the deformation experienced by the vessels cannot be neglected and their effects should be appropriately considered while modelling the coupled system.

The standard strategy to simulate the blood flow in a compliant vessel is to write the models for both the blood (the incompressible Navier-Stokes equations) and the wall (see e.g. [64]) together with appropriate matching conditions at the interface between the two domains (Fluid-Structure Interaction - FSI). At the numerical level, the coupled model is then solved either with a monolithic approach or by segregated solvers managing iteratively the sequence of fluid and solid problems (see e.g. [25]). This strategy allows the accurate computation of both fluid and solid mechanics and is challenging from both the modelling and the numerical point of view. In fact, the constitutive laws for modelling the arterial wall still deserve extensive investigations especially in the presence of vascular pathologies (see e.g. [65]), not to mention the difficulty to obtain *in vivo* measurements that can accurately estimate the model parameters for an individual patient (see Sect. 12.4). Moreover, vessels are subject to external loads due to the presence of the surrounding tissues, which are in general unknown or not easy to model. We mention for example the effects of cardiac motion on the aortic arch. From the numerical point of view, the strongly heterogeneous nature of the problem raises issues concerning numerical stability and efficiency of FSI algorithms (see e.g. [66, 67]).

Here we consider an alternative approach based on a DA procedure, that exploits the technological development experienced in the last decade by medical imaging techniques. The advent of high resolution imaging devices allows the fast acquisition of 4D (space + time) images. From those images it is possible to reconstruct anatomical structures not just in one specific instant, but in multiple ones over the cardiac cycle. Following this approach, the vessel motion, instead of being computed, is retrieved from images and plugged into the Navier-Stokes solver. The main advantage of this approach is the direct inclusion into the simulations of patient-specific data, i.e. the motion of the vessel (depending on its mechanical characteristics and those of the surrounding organs). This is done through the use of medical images at a limited additional computational cost with respect to the case in which the geometry is assumed to be fixed. We will denote this approach 4D Image Based (4DIB). A similar technique has been proposed in [63, 68] where the authors apply this image-based motion approach to intra-cranial aneurysms and coronary arteries respectively, even if implementing different strategies for some steps of the procedure.

12.3.1 Mathematical and numerical formulation

The workflow of the 4DIB approach consists in the following steps (for more details, we refer to [69]). We assume to have an image set that represents the vessel of interest at several time frames $\{t_k\}$ within a heart beat.

Segmentation. Depending on the nature of the source images, their dimensionality and the complexity of the geometry to be reconstructed, segmentation can be performed on single 2D planes or directly on 3D datasets. Different segmentation methodologies and different ways to represent the final models are available. For an introduction, see [24, 62]. In the applications presented here, a level set technique was used for the 3D segmentation of vessels, specifically the segmentation tool available within the Vascular Modelling Toolkit (VMTK) software package [3]. At the end of this step, a triangulated surface is available for each time frame.

Motion tracking. This consists in solving a *registration problem* (see e.g. [28]), i.e. finding the alignment of the geometries of two consecutive time frames, so to have a displacement field that maps the points on the surface of the lumen at a given time frame to the surface in the subsequent one. This is another example of inverse problems that can be cast in the form of the feedback loop in the Introduction. For this reason, we detail this step in the next subsection.

Simulation. From the sequence of maps describing the motion of the surface points from one time frame to the subsequent one, the displacement of the boundary of the moving domain is retrieved at the image acquisition times. Then, this is interpolated to define the displacement of the boundary at each time instant in the simulation. The velocity of the boundary is obtained as the time derivative of the displacement. To ensure also the continuity of the time derivative of the points velocity, a cubic spline time interpolation is chosen. The displacement and the grid velocity $\mathbf{w}$ of the whole domain, computed at each time step of the simulation, are obtained as the harmonic extension of the boundary fields. Once the domain motion is available, the incompressible Navier-Stokes equations for a Newtonian fluid in a moving domain can be written in the *Arbitrary Lagrangian Eulerian* (ALE) formulation (see, e.g., [70])

$$\frac{\partial \mathbf{u}}{\partial t} - \mu \nabla \cdot (\nabla \mathbf{u} + \nabla^T \mathbf{u}) + (\mathbf{u} - \mathbf{w}) \cdot \nabla \mathbf{u} + \nabla p = \mathbf{s}, \text{ in } \quad \Omega(t)$$

$$[1mm] \nabla \cdot \mathbf{u} = 0 \qquad\qquad\qquad\qquad \text{in} \quad \Omega(t),$$

$$\mathbf{u} = \mathbf{w} \qquad\qquad\qquad\qquad\qquad \text{on} \quad \Gamma_w(t),$$

$$+ \text{Boundary Conditions} \qquad\qquad \text{on} \quad \Gamma_{in}(t) \text{ and } \Gamma_{out}(t).$$

(12.7)

On the wall the fluid velocity is prescribed equal to the vessel velocity (Dirichlet condition), while inflow and outflow boundary data can be retrieved by measures or designed to reproduce a physiological or pathological behavior.

12.3.1.1 Assimilation of segmented vascular surfaces

Registration is a procedure to align images taken from different devices, from different viewpoints or at different time instants. Many different methodologies exist depending on the source of images, their dimensionality and the type of movement to be recovered, particularly whether we have small or large deformations. In particular, a wide number of different approaches has been detailed for surface registration (see e.g. [4, 11, 13, 71]).

Here we resort to an algorithm relying upon a minimization procedure [28]. The registration is performed over 3D surfaces representing the vessel at the different time frames. More precisely, given M+1 time frames corresponding to M+1 triangulated surfaces, the tracking process consists in M registration steps between each couple of consecutive time steps. Within each stage the points of one surface, the *source surface* $\mathscr{S}$, are mapped to the subsequent one, called the *target surface* $\mathscr{T}$. A displacement field for the whole surface mesh is computed so that at the end of this tracking procedure, M displacement fields are available describing the vessel wall motion at the instants of the image acquisitions.

The map between two consecutive frames is computed by minimizing a functional in the form (12.1). In particular, let us denote with $\varphi(\cdot)$ the (unknown) map from $\mathscr{S}$ to $\mathscr{T}$. Referring to the feedback loop in the Introduction, the forward problem FW is the actual application of φ to the source surface $\mathscr{S}$ (Input), so that[2] $\mathbf{v} = f(\mathbf{v}) = \varphi(\mathscr{S})$. The Data are represented by the target surface $\mathscr{T}$. The control variable set CV is given by the mathematical representation of the map $\varphi(\cdot)$. This can be parametrically described by assuming, e.g., that it belongs to a functional finite dimensional space spanned by a basis function set ψ_i so that $\varphi = \sum a_i \psi_i$, being a_i real coefficients. In this case a_i are the CV. However, since the map is supposed to be strongly space-dependent, in [69] we resorted to a *non-parametric map* implicitly defined with a collocation approach by the position of the nodes on the source image. This means that the coordinates of the vertices computed by the minimization process implicitly define the map point-wise. The map is then extended to the entire source surface by a piecewise linear interpolation of the values at the vertices.

Finally, to complete the picture, we need to specify the definition of the distance between $\varphi(\mathscr{S})$ and the data $\mathscr{T}$ and the regularizing term. Different choices are available, strictly problem dependent. Let us introduce the distance of the image of a point on $\mathscr{S}$ to the surface $\mathscr{T}$ as

$$\delta(\varphi(\mathbf{x}), \mathscr{T}) = \inf\{\|\varphi(\mathbf{x}) - \mathbf{y}\| : \mathbf{y} \in \mathscr{T}\}, \quad \mathbf{x} \in \mathscr{S}. \tag{12.8}$$

---

[2] The post-processing in this case is trivially the identity application.

The distance between $\mathscr{S}$ and $\mathscr{T}$ can be then defined as

$$dist(\varphi(\mathscr{S}),\mathscr{T}) \equiv \left(\frac{1}{|\mathscr{S}|}\int_{\mathscr{S}}(\delta(\varphi(\mathbf{x}),\mathscr{T}))^2\,d\mathscr{S}(\mathbf{x})\right)^{1/2} \tag{12.9}$$

where $|\mathscr{S}| := \int_{\mathscr{S}}d\mathscr{S}$ is a normalization factor. In practice the integral needs to be numerically approximated. For triangulated surfaces like $\mathscr{S}$ and $\mathscr{T}$ a reasonable and viable approximation is

$$dist(\varphi(\mathscr{S}),\mathscr{T}) = \sqrt{\frac{1}{n_S}\sum_j\min_i(d_{ji})^2} \tag{12.10}$$

where

$$d_{ji} = dist(\varphi(\mathbf{x}_j),\mathrm{tri}_i)$$

is the distance from vertex j of $\mathscr{S}$ to triangle i in $\mathscr{T}$, n_S (n_T) is the number of vertices (triangles) of $\mathscr{S}$ ($\mathscr{T}$). By using a tree search algorithm, it is possible to reduce the computational complexity to $\mathscr{O}(n_S\log(n_T))$ (see [23]).

This non-parametric registration by itself is in general ill-posed and multiple solutions are expected. Some of them are clearly unphysical and need to be filtered out. For this reason a regularizing term is introduced, forcing the solutions to be "physically acceptable" by adding some regularizing properties (see e.g. [5, 72]). In particular, we resort to a regularizing term stemming from a simplified physical model of the vascular wall as an elastic thin membrane [73] accounting for traction and bending internal forces. The membrane energy provides the regularizing term. In this way, displacements $\varphi(\cdot)$ that would cause a large increase to the membrane energy are heavily penalized (see [69]).

Additional constraints are required for preventing "flips" of triangles. Let

$$A_i = \mathrm{area}(\mathrm{tri}(\mathbf{x},\mathbf{y},\mathbf{z}))$$

be the area of the ith triangle before deformation and $\mathbf{x},\mathbf{y},\mathbf{z}$ its corresponding vertices. Correspondingly, let

$$\varphi(A_i) = \mathrm{area}(\mathrm{tri}(\varphi(\mathbf{x}),\varphi(\mathbf{y}),\varphi(\mathbf{z})))$$

be the area of the deformed i^{th} triangle.

Therefore, we add to the minimization of $\mathscr{J}$ the constraint of positive deformed area

$$C_i(\varphi) = \varphi(A_i) > 0. \tag{12.11}$$

The minimization problem has been solved by means of the L-BFGS procedure (Limited memory BFGS – see [6]), that requires only the computations of gradients and features (at least) a linear convergence even for non-smooth problems.

12.3.2 Numerical results

In the following application a 4D computed tomography (CT) dataset of a human aorta was employed as image source. The dataset was acquired at Ospedale Maggiore in Milan (Italy) using a Siemens SOMATOM Definition Flash Dual-Source CT scanner, which was able to capture 10 time frames per cardiac cycle. The 4D image refers to a 72-year-old man with a diagnosed abdominal aneurysm and covers the entire length of the ascending, thoracic and abdominal aorta. From this dataset the portion of the aorta including the aortic arch and the thoracic aorta was considered for a simulation in a moving domain. The aorta was then segmented with VMTK at all the 10 time frames available, and the tracking procedure was applied to extract the 10 displacement fields describing the vessel wall motion over the cardiac cycle. Fig. 12.6 represents some of the reconstructed surfaces at different time frames: they are simply superimposed prior to the registration procedure in order to highlight the misalignment due to their movement. Fig. 12.7 depicts the results of the registration procedure (performed with an *ad hoc* Matlab code) for two consecutive surfaces. In the rightmost panel frame 1 has been mapped to frame 2.

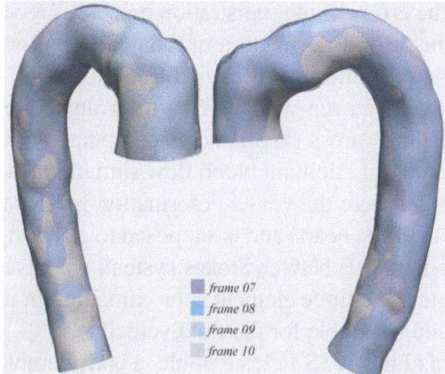

Fig. 12.6. Synopsis of the last 4 frames superimposed before tracking has been performed

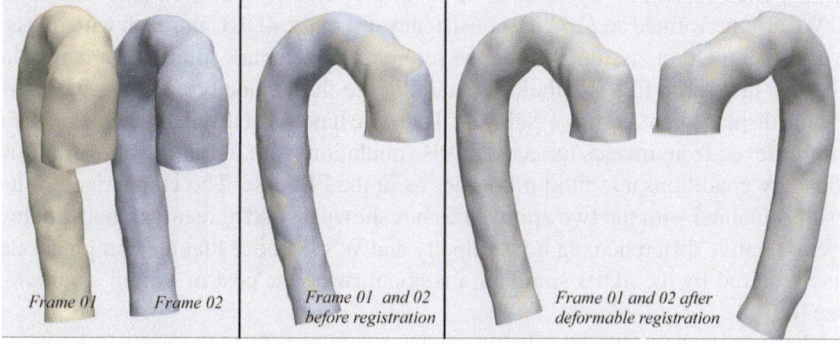

Fig. 12.7. Detail of two frames of the aorta before and after the registration

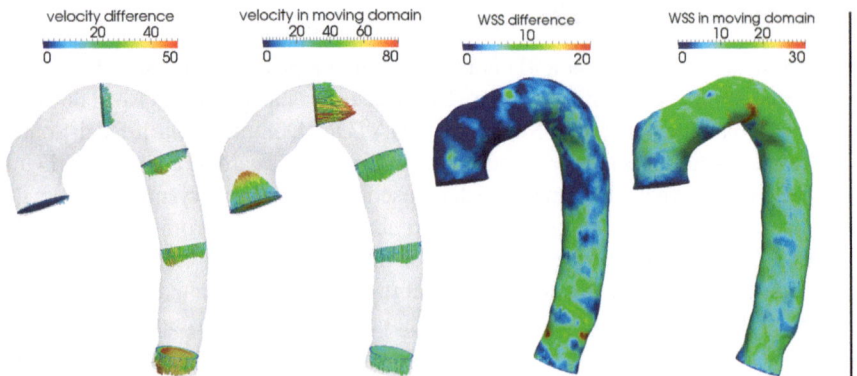

Fig. 12.8. Left panel: difference (in cm/s) between the fluid velocity computed in the rigid domain simulation and the one computed in the 4DIB simulation (left); velocity (in cm/s) computed with the 4DIB approach (right); both at peak systole. Right panel: difference (in dyne/cm^2) between the WSS computed in the rigid domain simulation and the one computed in the 4DIB simulation (left); WSS (in dyne/cm^2) computed with the 4DIB approach (right); both at peak systole

Quantification of the errors of the registration process is reported in [69]. A more detailed analysis of the error as a function of the number of nodes n_S and triangles n_T used in each couple of frames is however still missing.

Numerical tests have been run with LifeV to evaluate the difference between the velocity and wall shear stress (WSS) fields computed with the 4DIB approach and those computed on a rigid domain blood flow simulation. The choice of aorta is motivated by the fact that here the vessel deformation is relevant (mostly as a consequence of the motion of the heart) and is supposed to affect significantly the blood motion. To discretize the ALE Navier-Stokes system, we have chosen a first-order time advancing scheme and a finite element approximation for the space dependence ($\mathbb{P}^1$ for the pressure and $\mathbb{P}^1$ bubble for the fluid velocity).

Both the velocity and the WSS fields exhibits a considerable difference with respect to the rigid domain case, as shown in Fig. 12.8. In particular, the relative L^2 norm of the difference between the 4DIB fields and the rigid domain fields has an average over the cardiac cycle of 84.52 % for the velocity and 83.18 % for the WSS.

We also performed an *in-silico* consistency test of the 4DIB approach with respect to a FSI simulation, assumed to be the reference benchmark solution. In particular we have first run a FSI simulation, obtaining the fluid velocity and pressure fields and the displacement of the vessel wall. Then, we have used this displacement as if it was retrieved from images to feed a 4DIB simulation, with the same inflow/outflow boundary conditions and fluid properties as in the FSI case. The comparison of the results obtained with the two approaches has shown a good agreement, being below 1 % of relative difference, on both velocity and WSS. Notice that the computational time required by the 4DIB simulation is about twice the cost of a rigid simulation (see [69]).

These tests show that (i) when a relevant motion affects the vessel like in the aortic case, the 4DIB approach is a viable way for a more realistic description of the blood flow than a rigid simulation provided that available data can be properly assimilated; (ii) the results of the 4DIB method are consistent with the results of a traditional FSI simulation when the displacement field of the structure is the same.

12.3.2.1 A practical workaround for reduced data sets

The 4DIB approach presented here is based on the availability of 4D image data sets as it is made possible by recent devices. One of the limitations of the approach is that as for now only a few instruments are actually able to produce this kind of data set. This aspect will be naturally overcome in the future with a larger diffusion of those devices. However, a natural question arises now: is it possible to pursue a similar approach even for reduced data sets? The following example presents a possible workaround currently used in the analysis of the relations between WSS and atherogenesis, in collaboration with the group of Dr. W.R. Taylor at the Emory School of Medicine (Atlanta, GA, USA). In this case a mouse aorta was acquired with magnetic resonance imaging (MRI). The whole 3D geometry of the aorta and its main branches was reconstructed at a single time step, while the motion in time of the aorta was retrieved only at a number of locations along its centerline. At these points, in fact, the cross sections of the aorta were acquired in time by means of cine MRI sequences (Fig. 12.9). The lumen in the corresponding 2D slices was segmented and its area computed for each acquisition instant. The time evolution of each cross section area was then reconstructed by fitting these data with a cubic spline interpolation. The displacement of the whole aorta was suitably interpolated from the data available at each slice. More precisely, under the assumption that the longitudinal and circumferential motion of the vessel is negligible, and that each aortic section is circular, the time pattern of the area provided data on the wall displacement in correspondence of each slice. The displacement over the entire vessel at each instant was eventually retrieved by a cubic spline interpolation along the axial coordinate.

Since no information was available on the motion of the branching vessels, their presence was included in the simulation with the definition of proper stress boundary

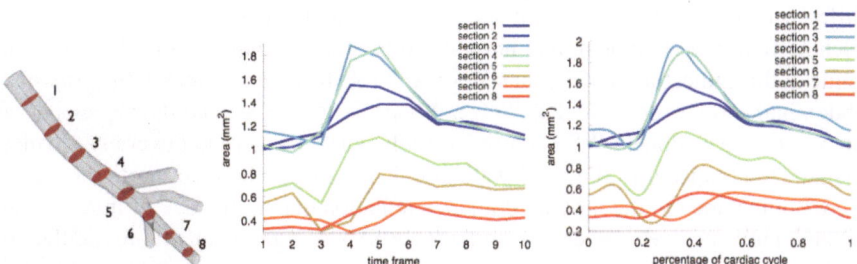

Fig. 12.9. Left: an example of a vascular structure of interest, the abdominal aorta in a mouse. In correspondence to the highlighted cross-sections, measurements of the arterial wall movement are available in the form of MRI cine sequences. Center: for each highlighted cross-section, the values of the area in time are plotted. The cross sectional area has been computed from MRI cine sequences, after segmentation of the images at each acquisition time. Right: The time pattern of the cross-section area has been reconstructed by fitting the values obtained from the images with cubic spline time interpolation

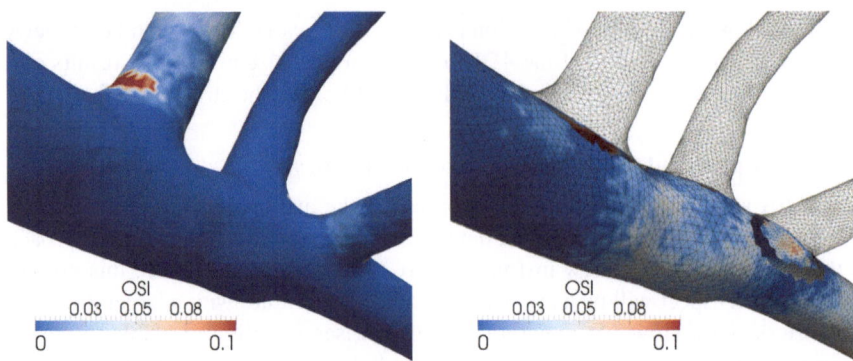

Fig. 12.10. Oscillatory Shear Index (OSI, [74]) on the arterial wall of the proximal abdominal aorta. Left: Results of the rigid wall simulation. Right: Results of the moving wall simulation

conditions for the fluid equations. More precisely, resistance conditions were prescribed at the outflow sections, yielding a normal stress proportional to the flow rate. The proportionality constant represents the downstream resistance and was tuned so that the desired distribution of abdominal flow among the branches was obtained. Should more data on the downstream network be available, more complex boundary conditions (including Windkessel-like conditions) could be considered as well.

Again, we compared the results obtained from a simulation of blood flow in a mouse aorta under the assumption that the vessel geometry is fixed, with the results of a simulation in moving domain with the "reduced" 4DIB approach.

The results of the rigid wall simulation (Fig. 12.10, left) showed that areas of disturbed flow characterize the branching points of the proximal abdominal aorta. High values of the *oscillatory shear index* (OSI – see e.g. [74]) were computed in very localized regions at the ostia of the main aortic branches. The hemodynamic environment was characterized overall by relatively low shear load. The results of the moving domain simulation (Fig. 12.10, right) provided an insight into the effects of the vessel dilatation in the region of interest. As a measure of the dilatation, the difference between the maximum and minimum radius (over the cardiac cycle) of each section was computed, and normalized by the minimum radius. The average value of this indicator on the eight slices was 35 %, being maximum in the proximal abdominal aorta (even more than 40 %). When taking into account the movement of the vessel, the computed WSS showed a similar spatial pattern but overall a smaller magnitude compared to the rigid wall case. The ostia of the main aortic branches were not included in the moving domain simulation due to the lack of information on their movement. However, the computed WSS was significantly more oscillating with respect to the rigid wall simulation in the entire proximal abdominal aorta, and in particular in the region surrounding the branching points. This was indeed experimentally found to be a typical site for atherosclerosis development.

Despite being only in a preliminary stage, these results suggest that neglecting the movement of the arterial wall may have a significant impact on the estimation of clinically relevant features, such as the presence of oscillatory flow. Validation of these results is ongoing (see [75]).

12.3.3 Perspectives

The 4DIB approach has some important drawbacks and limitations. It requires a large data set of images, which is not always available, even if some problem-specific workarounds can be devised to overcome this problem, as presented in Sect. 12.3.2.1. Moreover, this approach does not provide information on solid mechanics of the walls and it is therefore suitable when the focus of the study is on the flow features alone. However, this DA methodology splits the pipeline into a phase dedicated to the "offline" retrieval of the wall motion from images and a phase for the computation of the dynamics of the fluid alone, which has important computational advantages with respect to full FSI simulations. Furthermore, this approach could guarantee a reasonable reliability to patient-specific simulations of blood flow when the vascular motion is determined by external components that could not readily be included in a wall model, or more in general, when individual mechanical parameters for a single patient are not available.

Many open problems deserve to be addressed. As we have mentioned, a complete analysis of the accuracy of the registration process and how the registration errors affect the computation of fluid dynamics still needs to be carried out. In this context, it is particularly relevant the correlation of the numerical procedure with the noise affecting the image acquisition and segmentation. The effects of noise/errors in the recovery of the vessel wall motion affect indeed the boundary data for the flow simulation and eventually the estimation of WSS (or other post-processing quantities). A detailed error analysis is therefore in order.

Using the terminology in [76] this DA procedure can be considered as a *frame-to-frame pseudo-observational approach*. More mathematically advanced methods advocated in [76] which entail an integrated variational assimilation of data and images similar to the ones introduced in Sect. 12.2 could be considered as a future development.

12.4 Variational parameter estimation

Since mathematical and numerical models are earning more relevance in medical applications and are used as patient-specific tools, a precise estimation of individual physical parameters featured by the equations is needed. Moreover, by themselves, some parameters can play the role of landmarks of pathologies. This is for instance the case of the stiffness of soft tissues in detecting breast cancer. Significant changes of the stiffness of the tissue can identify the presence of tumors. On the other hand, a small value of the compliance of the tissue could be an indicator of atherosclerosis or hypertension, while an increase of the stiffness of the left ventricle wall is a clear marker of *diastolic dysfunction*, which can lead to an increase of the end diastolic left ventricle pressure and, possibly, to heart failure (see e.g. [1, 77]). This has motivated sophisticated image-based diagnostic approaches, such as the *elastography* (see e.g. [78, 79, 80, 81]).

Either for a direct diagnostic purpose or for an individual-based evaluation to be used in numerical simulations, a precise estimation of biological parameters *in vivo* is still a challenging problem demanding appropriate mathematical tools. In this section we suggest a DA procedure. The starting point is that the parameters of interest are non trivial functions of measurable quantities. For instance, the compliance of a tissue affects in a non trivial way its displacement, the latter being retrievable from images. In some cases (as in elastography) we can prescribe the forces inducing a measured displacement and formulate an inverse problem in the form: given the force and the consequent displacement, find the stiffness (or more precisely the Young's modulus, in the case of a linear elastic material) that fits at best the experimental stress-strain data. In other cases, practical reasons prevent the knowledge of some of the ingredients of this inverse problem. For instance, the natural periodic motion of a vessel is the result of the interaction with the blood (and the other tissues), in turn forced by the heart action. The forces exerted on the vascular wall by the blood are not explicitly known but can be computed by solving the complete FSI problem, as a function of (available) velocity/pressure values on the boundary of the region of interest. The basic idea of DA approach is then to use numerical simulations for bridging available data to the ingredients needed for solving the inverse parameter-estimation problem.

In the case of the vascular stiffness, usable data are the images of the vessel displacement (as in Sect. 12.3) and velocity and pressure (Sect. 12.2) on the boundary. Numerical simulations allow to compute the forces on the wall and eventually to solve an inverse problem. We present here a first step in this direction. However, it is worth stressing that this DA approach has potentially a more general use than for the evaluation of the vascular compliance (that can be currently achieved in several ways), for different (and more numerous) sets of parameters.

12.4.1 Mathematical and numerical formulation

We formulate the problem of estimating the compliance of a linearly elastic membrane filled by an incompressible fluid as follows. Let Ω be the volume of interest of the fluid, where we assume the incompressible Navier-Stokes equations (12.7) to hold. The membrane Γ_w is a portion of $\partial\Omega$, i.e. a 2D surface for a 3D fluid, which is assumed to obey the equation for an elastic membrane

$$\rho_w \frac{\partial^2 \eta}{\partial t^2} + E\theta\eta = s_w, \tag{12.12}$$

where η is the membrane displacement assumed to be normal to Γ_w, ρ_w is the density of the solid, s_w is the stress exerted by the fluid and by surrounding tissues (the latter will be neglected in the following), and θ is a function of the mean and Gaussian curvatures ρ_1 and ρ_2 of the membrane and of the Poisson's ratio v, in particular

$$\theta := \frac{h_s}{1 - v^2}(4\rho_1 - 2(1 - v)\rho_2).$$

The parameter θ accounts for the transversal membrane effects (see [82]). Young's modulus E is the parameter we want to estimate. The fluid subproblem (12.7) and the membrane one (12.12) are coupled at the FS interface Γ_w by the continuity of the normal component of the normal stress and of the velocity

$$-\mu\left(\left(\nabla\mathbf{u}+\nabla\mathbf{u}^T\right)\cdot\mathbf{n}\right)\cdot\mathbf{n}+p=s_w,\qquad \frac{\partial\eta}{\partial t}\mathbf{n}=\mathbf{u}\qquad \text{on }\Gamma_w. \qquad (12.13)$$

The grid velocity $\mathbf{w}$ is then computed as the harmonic extension of $\dfrac{\partial\eta}{\partial t}\mathbf{n}$ in Ω.

Now, we assume that the displacement of the vessel can be measured by a set of time resolved images and the sequence of steps *segmentation + registration*, as we have done in the previous section. After an appropriate cubic spline interpolation (see Sect. 12.3), we have the time dependent displacement field $\boldsymbol{\eta}^m(t,\mathbf{x})$ defined on Γ_w, that represents the **Data**. Displacement $\boldsymbol{\eta}^m$ is assimilated with the numerical model as indicated by the feedback loop in the Introduction. The FW problem is given by the system of Eqs. (12.7$_{1-2}$, 12.12, 12.13), the unknown being $\mathbf{v}=[\mathbf{u},p,\boldsymbol{\eta}]$. The post processing function selects the displacement, i.e. $f(\mathbf{v})=\boldsymbol{\eta}$. The CV is represented by the Young's modulus E. The functional $\mathscr{J}$ reads

$$\mathscr{J}=\int_0^T\int_{\Gamma_w}(\boldsymbol{\eta}-\boldsymbol{\eta}^m)^2\,d\mathbf{x}\,dt+\text{Regularization},$$

where T is the heart beat duration. Again, the regularizing term enhances the mathematical and numerical properties of the problem. A possible form is

$$\alpha\int_0^T\int_{\Gamma_w}\left(E-E_{ref}\right)^2 d\mathbf{x}\,dt,$$

where α is the usual parameter weighting the effect of the regularizing term on the minimization process and E_{ref} is a reference value of the Young's modulus available for instance from the literature. If we assume *a priori* that the CV is positive, we can also consider the term

$$\alpha\max_{\mathbf{x}\in\Gamma_w,t>0}\left(\log\left(\frac{E}{E_{ref}}\right)\right)^2.$$

In both cases the regularizing term penalizes the distance between the control variable E and the reference value for the Young's modulus E_{ref}.

The solution of this minimization problem is not trivial in many respects. Hereafter we present a first possible approach, under some simplifying assumptions. Even though in the more general case, the Young's modulus can be function of time and space, in the sequel we assume

- E constant in time in the interval $[0,T]$, significant changes of the compliance in an artery being expected over a longer time scale;
- E piecewise constant in space, as we distinguish basically healthy and pathological tissues featuring different values of compliance, each value being reasonably constant in each subregion.

Computational and algorithmic aspects of the numerical solution of the minimization problem are challenging. Here we resort to the workflow *Time Discretize, then Optimize, then Space Discretize*. This means that we first discretize in time the problem by collocating the minimization process at selected time instants t^k. Then we perform the minimization, by computing the KKT system for the space-continuous problem. Finally, we discretize the KKT system. In this way, the variational procedure for the minimization does not involve adjoint backward-in-time problems (see [12]) and the differentiation of the Lagrangian functional does not require to perform differentiation of the domain Ω (shape derivatives), since at each instant the domain Ω is frozen. The anticipated drawback of this approach is that the effect of noise over the time interval is not damped by a least square minimization, being the problem collocated pointwise at t^k. Post-processing for the estimates of E obtained at each step is required for filtering out the error. An extensive analysis of different solution methods for this problem is however an important follow up in this context.

After time discretization, at each instant t^k the problem reads (hereafter we omit to specify the time index k for the sake of readability): Find the piecewise constant function E defined on Γ_w that minimizes the functional

$$\mathscr{J} = \int_{\Gamma_w} (\boldsymbol{\eta} - \boldsymbol{\eta}^m)^2 \, d\mathbf{x} + \alpha \max_{\mathbf{x} \in \Gamma_w} \left(\log \left(\frac{E}{E_{ref}} \right) \right)^2$$

$$\text{s.t.} \quad \begin{cases} \gamma_1 \mathbf{u} + (\mathbf{u} - \mathbf{w}) \nabla \cdot \mathbf{u} - \mu \nabla \cdot (\nabla \mathbf{u} + \nabla \mathbf{u}^T) + \nabla p = \mathbf{g}_1 & \text{in } \Omega, \\ \nabla \cdot \mathbf{u} = 0 & \text{in } \Omega, \\ (\gamma_2 + \gamma_3 E \theta) \mathbf{u} \cdot \mathbf{n} - p + \mu \left((\nabla \mathbf{u} + \nabla \mathbf{u}^T) \cdot \mathbf{n} \right) \cdot \mathbf{n} = g_2, & \text{on } \Gamma_w, \\ \mathbf{n} \times \mathbf{u} \times \mathbf{n} = 0 & \\ + \text{Boundary Conditions} & \text{on } \Gamma_{in} \text{ and } \Gamma_{out}. \end{cases}$$

$$(12.14)$$

The functions $\mathbf{g}_1$, g_2 and γ_i depend on the time discretization. Notice that for the fluid at the interface with the wall, here we prescribe null tangential velocity conditions. Moreover, in the previous system, the FSI problem has been simplified by eliminating the displacement, leading to a fluid problem with Robin boundary conditions, as proposed in [82].

The explicit computation of the KKT system for this problem and its generalization to the case of a 3D thick elastic structure are reported in [83]. We have analyzed different choices for the space of the admissible CV (E). In particular, for the piecewise constant and piecewise linear cases, we can prove the following Proposition.

Proposition 2. *For $\alpha > 0$ the KKT system associated with the minimization problem has at least one solution.*

After the space discretization, the KKT system yields a non-linear algebraic minimization problem. In particular we can use again the gradient-based BFGS method (see e.g. [6]). For more details, see [83].

12.4.2 Numerical results

We present two test cases on simplified geometries, solved again with the library LifeV. These test cases have the role of assessing the overall performances of the method on synthetic data, in view of a more extensive analysis using real medical images. The "synthetically measured" displacement field $\boldsymbol{\eta}_{fwd}$ is therefore generated by a preliminary numerical simulations with a prescribed Young's modulus. Successively, the data are perturbed in order to mimic the presence of noise. The noise is generated with a uniform distribution $U\left(-\frac{\eta_M}{2\xi}, \frac{\eta_M}{2\xi}\right)$, where $\eta_M = \max_{\mathbf{x},t}|\eta_{fwd}(\mathbf{x}, t)|$. Smaller values of the parameter ξ represent a greater incidence of the noise.

In the first set of simulations (already reported in [83]), we solve the problem in a cylinder of radius $R = 0.5\,cm$ and height $H = 6\,cm$. The computation is performed in 2D under the assumption of axial symmetry of the problem. We impose the pressure drop $\Delta p = 10^4\,dyne/cm^2$ for the first 5 ms between the inlet and the outlet of the vessel. We set $\rho_f = 1\,g/cm^2$, $\rho_w = 1.1\,g/cm^2$, $\mu = 0.035\,Poise$, $h_s = 0.02\,cm$, $E = 1.3 \cdot 10^6 dyne/cm^2$, $v = 0.3$ and $\Delta t = 0.001\,s$.

Fig. 12.11 shows the geometry and the pressure along a longitudinal section of the cylinder, for different time instants.

The optimization problem has been solved by using the BFGS algorithm over 10 time steps, corresponding to the first 10 ms of the simulation. We run the optimization problem for 10 realizations of the noise. In Table 12.2, we report the average over the 10 realizations of the estimated values of E and the relative error. Different initial guesses E_0 and different ξ are considered. These results show that for large values of ξ, the estimate obtained by the method is accurate. A reasonable

Fig. 12.11. 2D axisymmetric case, Forward simulation. Geometry at time $t = 5\,ms$, $t = 7\,ms$ and $t = 9\,ms$. Colored with blood pressure, in $dyne/cm^2$

Table 12.2. Standard deviation of the ten estimates (to be multiplied by 10^6, top) and mean percentage error (bottom) for different values of the initial guess E_0 for the Young's modulus and of the percentage P. Exact E is $1.3 \cdot 10^6\,dyne/cm^2$

$\downarrow E_0 \setminus \xi \rightarrow$	10	5	3.3	2.5
$10^7\,dyne/cm^2$	1.302 ± 0.027	1.314 ± 0.054	1.330 ± 0.085	1.357 ± 0.103
	0.2%	1.1%	2.3%	4.4%
$10^5\,dyne/cm^2$	1.303 ± 0.027	1.315 ± 0.056	1.330 ± 0.087	1.348 ± 0.115
	0.2%	1.1%	2.3%	3.7%

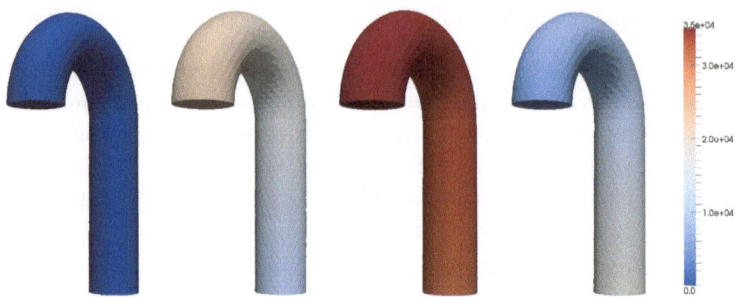

Fig. 12.12. 3D simplified aorta, Forward simulation. Geometry at time $t = 0\ ms$, $t = 75\ ms$, $t = 150\ ms$ and $t = 300\ ms$. Colored with blood pressure $[dyne/cm^2]$. The displacement has been scaled by a factor 2 for the sake of visualization

Table 12.3. Estimates (to be multiplied by 10^6, top) and relative error for different values of ξ. Exact E is $4 \cdot 10^6\ dyne/cm^2$

$\downarrow E_0 \setminus \xi \rightarrow$	40	20	10
$10^6\ dyne/cm^2$	3.955	3.915	3.679
	1.1%	2.1%	8.0%

accuracy is maintained also for smaller values of ξ. Moreover, the BFGS applied to the membrane case is pretty robust with respect to the noise, both in terms of accuracy and convergence.

Other 2D cases on non-trivial geometries (like a bifurcation) have been reported in [83].

We now consider the three dimensional geometry shown in Fig. 12.12, representing a simplified aorta. This geometry consists of a cylinder of radius $1.5cm$ and height $10cm$ and half a torus with curvature radius 4 cm. The physical parameters are $E = 4 \cdot 10^6$ dyne/cm^2, $\mu = 0.5$, $h_s = 0.2\ cm$. We prescribe a parabolic velocity profile at the inlet, with the peak velocity $u_M(t)$ given by $u_M = 100 \max_t(\sin(\frac{10}{3}\pi t))\ cm/s$, and absorbing boundary conditions at the outlet [82]. The time-step used is $\Delta t = 2.5\ ms$. In this case η_M is defined as $\eta_M(t) = \max_{\mathbf{x}} |\eta_{fwd}(\mathbf{x}, t)|$. The pressure field and the geometry displacement are shown in Fig. 12.12. As before, we estimate the compliance using the displacement of the forward simulation perturbed by an artificial noise. The estimation has been performed over the first 150 ms. In Table 12.3 we report the results related to a single realization of the noise.

In the 3D case, the method seems to be more sensitive to the noise. However even with a noise with $\xi = 10$, the parameter estimation is accurate enough for most of clinical applications.

12.4.3 Perspectives

As the preliminary results indicate, the assimilation of data and numerical models is a worthwhile approach for estimating patient-specific parameters to be used either for detecting possible anomalies or performing individual-based numerical simulations. Here, numerical differential models play the role of a bridge between the measurable data and the unknown parameters. There are many critical issues at the computational level to be addressed. Even if practical applications in general demand for less accuracy than the one usually considered acceptable from the numerical viewpoint, the noise is supposed to play a relevant role on the reliability of the entire approach. Moreover, the frequency of sampling of images, currently driven by technological limits, has probably a major impact on the accuracy of the results.

From the computational viewpoint, in these preliminary applications we resorted to standard numerical tools like the BFGS method. Extension of this approach to real 3D cases rises new issues on the computational effectiveness of the methods. An extensive comparison among different possible options (in particular for the sequence of optimization and discretization steps) and different possible algorithms is required for a massive use of these methods in practice.

A long term follow up of the present research is the extension of this optimization procedure to more complex sets of CV, such as the configuration and geometrical features of the cardiac fibres. In fact, we mention here that one of the open challenging problems in heart imaging and modelling is the estimation of the orientation of the fibres driving the mechanical contraction and the electrical potential propagation in the cardiac tissue.

12.5 Conclusions

A mathematically sound adoption of numerical models for investigating the vascular blood dynamics originates from pioneering works in the late 80s (among the others, see e.g. [30, 84, 85]). At that time, numerical simulations were carried out in idealized domains, moving from basic geometrical primitives to realistic shapes of regions of interest. Simulations were intended to provide an insight to physiological and pathological dynamics for a better understanding of the most relevant diseases. The impact of these simulations was mostly at a qualitative level, since data and geometries were realistic but not patient-specific. Successively, in the '90s, the advent of new imaging technologies and corresponding numerical methods allowed the introduction of "patient-specific" simulations. The geometry of the single patient at a given instant was reconstructed from digital subtraction angiographies or computed tomographies and used as the computational domain, possibly together with individual measures of data for the boundary conditions. This "sequential" merging of data and simulations (i.e., *first* the data, *then* the simulations fed by the data) led to a more quantitative relevance of numerical models, closer to the clinical activity. Reliability of numerical models have been progressively increased by removing many of the simplifying assumptions postulated in the first simulations, e.g. rigid geometries or Newtonian rheology (see e.g. [16]).

The development of more sophisticated mathematical and numerical models has been corresponded by the development of more sophisticated measurements and imaging tools. Nowadays, these instruments provide more data and more images, so that it is reasonable to think to a further step, moving from a "sequential" to an "integrated" use of data and simulations. The DA approach entails measures and images to be used not just for providing initial and boundary conditions, but to drive the results by a sophisticated integration with the mathematical models. The outcome of this process is an assimilated result where not only numerical computation is strictly consistent with the individual data, but the noise affecting the data has been filtered out by the mathematical modelling.

The "integrated" paradigm, which is well developed in other contexts such as the weather forecasting, opens many challenging problems at the methodological and practical level. The quality of the data in terms of their size, location in space and frequency in time plays obviously a major role in the mathematical properties (well posedness) of the assimilation problem (see [7]). Moreover, the noise that invariably affects the data has an impact on the reliability of the entire process. A precise evaluation of this aspect is strictly related to both the type of data and the methods used for the assimilation procedure. This leads to analyze and solve partial differential equations with stochastic terms (see e.g. [61, 86, 87, 88]).

When the assimilation problem is solved with variational methods, the mathematical structure almost invariably can be represented as a feedback control loop. The effective numerical solution of inverse problems in this form presents many open concerns to be properly addressed (see [6]).

In this chapter we have presented three basic examples sharing this control-loop structure, motivated by ongoing collaborations with medical doctors. The first preliminary results enlighten the great potential of DA as a way for improving both the reliability of numerical results and the quality of measures. As we have pointed out, a certified reliability is crucial since bioengineering and medical communities are increasingly resorting to scientific computing for taking decisions (see [49]). The accomplishment of the new integrated paradigm - requiring new advanced and increasingly interdisciplinary research - represents an exciting challenge of cardiovascular mathematics for the years to come.

Acknowledgements. Marina Piccinelli and Alessandro Veneziani thank Emory University Research Committee for the support of the Project "Image based numerical fluid structure interactions simulations in computational hemodynamics". Tiziano Passerini is supported by the NIH Grant 5R01HL070531-08 "Biology, Biomechanics and Atherosclerosis". The research of C. Vergara has been (partially) supported by the ERC Advanced Grant N.227058 MATHCARD. The authors wish to thank Marijn Brummer (Emory Children's Healthcare of Atlanta), Eldad Haber (University of British Columbia, Canada), Robert Taylor (Emory School of Medicine), Michelle Consolini (Emory School of Medicine), Michele Benzi (Emory University), Max Gunzburger (Florida State University), George E. Karniadakis (Brown University).

References

[1] Topol E.J. (ed.): Textbook of Cardiovascular Medicine. Lippincott-Raven Publisher, Philadelphia-New York, 1998.

[2] Antiga L. et al.: Vascular modeling toolkit, website. www.vmtk.org.

[3] Lions J.L.: Remarks on approximate controllability. Journal d'Analyse Mathématique **59**(1): 103–116, 1992.

[4] Brown L.G.: A survey of image registration techniques. ACM Computing Surveys (CSUR) **24**(4): 325–376, 1992.

[5] Vaillant M., Glaunes J.: Surface matching via currents. In Recent advances in parallel virtual machine and message passing interface: 11th European PVM/MPI Users' Group Meeting, Budapest, Hungary, September 19–22, 2004: proceedings, page 381. Springer-Verlag New York Inc, 2004.

[6] Nocedal J., Wright S.: Numerical Optimization. Springer, apr 2000.

[7] Lions J.L.: On the controllability of distributed systems. Proc Natl Acad Science **94**: 4828–4835, 1997.

[8] Welch G., Bishop G.: An introduction to the Kalman filter. University of North Carolina at Chapel Hill, Chapel Hill, NC, 1995.

[9] Ward C.: Clinical significance of the bicuspid aortic valve. Heart **83**(1): 81, 2000.

[10] DeParis S. et al.: Lifev – library for finite elements, website. www.lifev.org.

[11] Zitova B., Flusser J.: Image registration methods: a survey. Image and Vision Computing **21**(11): 977–1000, 2003.

[12] Gunzburger M.D.: Perspectives in flow control and optimization. Society for Industrial Mathematics, 2003.

[13] Maintz J.B., Viergever M.A.: A survey of medical image registration. Medical Image Analysis **2**(1): 1–36, 1998.

[14] Hoffman J.I.E., Kaplan S.: The incidence of congenital heart disease. Journal of the American College of Cardiology **39**(12): 1890, 2002.

[15] Engl H.W., Hanke M., Neubauer A.: Regularization of inverse problems. Springer Netherlands, 1996.

[16] Formaggia L., Quarteroni A., Veneziani A. (eds.): Cardiovascular Mathematics, vol. 1 of MM&S. Springer, Italy, 2009.

[17] Lions J.L.: Are there connections between turbulence and controllability? In 9th INRIA International Conference, Antibes, 1990.

[18] Robinson A.R., Lermusiaux P.F.J.: Overview of data assimilation. Technical Report 62, Harvard University, Cambridge, Massachusetts, aug 2000.

[19] Bonesky T.: Morozov's discrepancy principle and Tikhonov-type functionals. Inverse Problems **25**:015015, 2009.

[20] Kaipio J., Somersalo E.: Statistical and Computational Inverse Problems. Springer, 2005.

[21] Quarteroni A., Formaggia L., Veneziani A. (eds.): Complex Systems in Biomedicine. Springer, Italy, 2006.

[22] Blum J., Le Dimet F.X., Navon I.M.: Data Assimilation for Geophysical Fluids, vol. XIV of Handbook of Numerical Analysis, chap. 9. Elsevier, 2005.

[23] Barber C., Dobkin D., Hudhanpaa H.: The quickhull program for convex hulls. ACM Transactions on Mathematical Software **22**: 469–483, 1996.

[24] Antiga L., Steinman D.A., Peiró J.: From image data to computational domain. In: Formaggia L., Quarteroni A., Veneziani A. (eds.), Cardiovascular Mathematics, MM&S, chap. 4. Springer, Italy, 2009.

[25] Gerbeau J.F., Fernandez M.: Algorithms for fluid-structure interaction problems. In: Formaggia L., Quarteroni A., Veneziani A. (eds.), Cardiovascular Mathematics, MM&S, chap. 9. Springer, Italy, 2009.

[26] Lions J.L.: Optimal control of systems governed by partial differential equations. Springer-Verlag, 1971.

[27] Walters R.W., Huyse L.: Uncertainty analysis for fluid mechanics with applications, 2002.

[28] Fischer B., Modersitzki J.: Ill-posed medicine: an introduction to image registration. Inverse Problems **24**: 034008, 2008.

[29] Lions J.L.: Exact Controllability for distributed systems. Some trends and some problems. Applied and Industrial Mathematics: Venice-1, 1989, p. 59, 1991.

[30] Perktold K.: On numerical simulation of three-dimensional physiological flow problems. Technical report, Ber. Math.-Stat. Sekt. Forschungsges. Joanneum **280**, 1–32, 1987.

[31] Taylor C.A., Draney M.T., Ku J.P., Parker D., Steele B.N., Wang K., Zarins C.K.: Predictive medicine: Computational techniques in therapeutic decision-making. Computer Aided Surgery **4**(5): 231–247, 1999.

[32] Taylor C.A., Draney M.T.: Experimental and Computational Methods in Cardiovascular Fluid Mechanics. Ann. Rev. Fluid. Mech. **36**: 197–231, 2004.

[33] Grinberg L., Anor T., Cheever E., Marsden. J.P., Karniadakis G.E.: Simulation of the human intracranial arterial tree. Phil. Trans. R. Soc. A **367**: 2371–2386, 2009.

[34] Heys J.J., Manteuffel T.A., McCormick S.F., Milano M., Westerdale J., Belohlavek M.: Weighted least-squares finite elements based on particle imaging velocimetry data. Journal of Computational Physics **229**(1): 107–118, 2010.

[35] Ide K., Courtier P., Ghil M., Lorenc A.C.: Unified notation for data assimilation: Operational, sequential and variational. Journal of Meteorological Society of Japan **75**(Special): 181–189, 1997.

[36] Chapelle D., Moireau P.: Robust filter for joint state parameters estimation in distributed mechanical system. Discrete and Continous Dynamical Systems **23**(1–2): 65–84, 2009.

[37] Moireau P., Chapelle D.: Reduced-order Unscented Kalman Filtering with application to parameter identification in large-dimensional systems. ESAIM: Control, Optimisation and Calculus of Variations, 2010.

[38] Moireau P., Chapelle D., Le Tallec P.: Joint state and parameter estimation for distributed mechanical systems. Computer Methods in Applied Mechanics and Engineering **197**(6–8): 659–677, 2008.

[39] Moireau P., Chapelle D., Le Tallec P.: Filtering for distributed mechanical systems using position measurements: perspectives in medical imaging. Inverse Problems **25**: 035010, 2009.

[40] Sermesant M., Moireau P., Camara O., Sainte-Marie J., Andriantsimiavona R., Cimrman R., Hill D.L.G., Chapelle D., Razavi R.: Cardiac function estimation from MRI using a heart model and data assimilation: advances and difficulties. Functional Imaging and Modeling of the Heart, pp. 325–337, 2005.

[41] Funamoto K., Suzuki Y., Hayase T., Kosugi T., Isoda H.: Numerical validation of mr-measurement-integrated simulation of blood flow in a cerebral aneurysm. Ann. Biomed. Eng. **37**(6): 1105–1116, 2009.

[42] Glowinski R., Li C.H., Lions J.L.: A numerical approach to the exact boundary controllability of the wave equation (I) Dirichlet controls: Description of the numerical methods. Japan Journal of Industrial and Applied Mathematics **7**(1):1–76, 1990.

[43] Glowinski R., Lions J.L.: Exact and approximate controllability for distributed parameter systems. Acta Numerica **3**: 269–378, 1994.

[44] Zuazua E.: Controllability of partial differential equations and its semi-discrete approximations. Dynamical Systems **8**(2): 469–513, 2002.

[45] Ervin V.J.; Lee H.: Numerical approximation of a quasi-Newtonian Stokes flow problem with defective boundary conditions. SIAM J. Numer. Anal. **45**(5): 2120–2140, 2007.

[46] Formaggia L., Veneziani A., Vergara C.: A new approach to numerical solution of defective boundary value problems in incompressible fluid dynamics. SIAM Journal on Numerical Analysis **46**(6): 2769–2794, 2008.

[47] Formaggia L., Veneziani A., Vergara C.: Flow rate boundary problems for an incompressible fluid in deformable domains: formulations and solution methods. Computer Methods in Applied Mechanics and Engineering **199**(9–12): 677–688, 2010.

[48] Darema F.: Dynamic data driven applications systems (dddas) – a transformative paradigm. In ICCS (3), p. 5, 2008.

[49] Erdemir A., Guess T., Halloran J., Tadepalli S.C., Morrison T.M.: Recommendations for reporting finite element analysis studies in biomechanics. http://www.imagwiki.nibib.nih.gov/mediawiki/index.php?title=Reporting_in_FEA, 2010.

[50] Robicsek F., Thubrikar M.J., Cook J.W., Fowler B.: The congenitally bicuspid aortic valve: how does it function? Why does it fail? The Annals of Thoracic Surgery **77**(1): 177–185, 2004.

[51] Gurvitz M., Chang R.K., Drant S., Allada V.: Frequency of aortic root dilation in children with a bicuspid aortic valve. The American Journal of Cardiology **94**(10): 1337–1340, 2004.

[52] den Reijer P.M., Sallee D., van der Velden P., Zaaijer E., Parks W.J., Ramamurthy S., Robbie T. , Donati G. Lamphier C., Beekman R., and Brummer M.: Hemodynamic predictors of aortic dilatation in bicuspid aortic valve by velocity-encoded cardiovascular magnetic resonance. Journal of Cardiovascular Magnetic Resonance **12**(1): 4, 2010.

[53] Viscardi F., Vergara C., Antiga L., Merelli S., Veneziani A., Puppini G., Faggian G., Mazzucco A., Luciani G.B.: Comparative Finite Element Model Analysis of Ascending Aortic Flow in Bicuspid and Tricuspid Aortic Valve. Artificial Organs **34**(12): 1114–1120, 2010.

[54] Dwight R.P.: Bayesian inference for data assimilation using Least-Squares Finite Element methods. In IOP Conference Series: Materials Science and Engineering, vol. 10, p. 012224. IOP Publishing, 2010.

[55] D'Elia M., Veneziani A.: Methods for assimilating blood velocity measures in hemodynamics simulations: Preliminary results. Procedia Computer Science **1**(1): 1231–1239, 2010. ICCS 2010.

[56] Quarteroni A., Valli A.: Numerical Approximation of Partial Differential Equations. Springer, 1994.

[57] Hansen P.C.: Rank-deficient and discrete ill-posed problems: numerical aspects of linear inversion. Society for Industrial Mathematics, 1998.

[58] D'Elia M., Perego M., Veneziani A.: A variational data assimilation procedure for the incompressible Navier-Stokes equations in hemodynamics. Technical Report TR-2010-19, Department of Mathematics & CS, Emory University, 2010.

[59] Nobile F., Tempone R.: Analysis and implementation issues for the numerical approximation of parabolic equations with random coefficients. International Journal for Numerical Methods in Engineering **80**: 979–1006, 2009.

[60] Oden J.T., Babuska I., Nobile F., Feng Y., Tempone R.: Theory and methodology for estimation and control of errors due to modeling, approximation, and uncertainty. Computer Methods in Applied Mechanics and Engineering **194**(2–5): 195–204, 2005. Selected papers from the 11th Conference on The Mathematics of Finite Elements and Applications.

[61] Elman H.C., Miller C.W., Phipps E.T., Tuminaro R.S.: Assessment Of Collocation And Galerkin Approaches To Linear Diffusion Equations With Random Data. International Journal for Uncertainty Quantification **1**(1), 2011.

[62] Bertero M., Piana M.: Inverse problems in biomedical imaging: modeling and methods of solution. In: Quarteroni A., Formaggia L., Veneziani A. (eds.), Complex Systems in Biomedicine, chap. 1, pp. 1–33. Springer, 2006.

[63] Sforza D.M., Lohner R., Putman C., Cebral J.R.: Hemodynamic analysis of intracranial aneurysms with moving parent arteries: Basilar tip aneurysms. International Journal for Numerical Methods in Biomedical Engineering, 2010.

[64] Holzapfel G.A., Gasser T.C., Ogden R.W.: A new constitutive framework for arterial wall mechanics and a comparative study of material models. Journal of Elasticity **61**(1): 1–48, 2000.

[65] Wulandana R., Robertson A.M.: An inelastic multi-mechanism constitutive equation for cerebral arterial tissue. Biomechanics and Modeling in Mechanobiology **4**(4): 235–248, 2005.

[66] Causin P., Gerbeau J.F., Nobile F.: Added-mass effect in the design of partitioned algorithms for fluid-structure problems. Computer Methods in Applied Mechanics and Engineering **194**(42–44): 4506–4527, 2005.

[67] Gerardo-Giorda L., Nobile F., Vergara C.: Analysis and optimization of robin-robin partitioned procedures in fluid-structure interaction problems. SIAM J. Num. Anal. **48**(6): 2091–2116, 2010.

[68] Torii R., Keegan J., Wood N.B., Dowsey A.W., Hughes A.D., Yang G.Z., Firmin D.N., Thom S.A.M.G., Xu X.Y.: MR Image-Based Geometric and Hemodynamic Investigation of the

Right Coronary Artery with Dynamic Vessel Motion. Annals of Biomedical Engineering **8**: 1–15, 2010.

[69] Piccinelli M., Mirabella L., Passerini T., Haber E., Veneziani A.: 4d Image-Based CFD Simulation of a Compliant Blood Vessel. Technical Report TR-2010-27, Department of Mathematics & CS, Emory University, www.mathcs.emory.edu, 2010.

[70] Hughes T.J.R., Liu W.K., Zimmermann T.K.: Lagrangian-Eulerian finite element formulation for incompressible viscous flows. Computer Methods in Applied Mechanics and Engineering **29**(3): 329–349, 1981.

[71] Audette M.A., Ferrie F.P., Peters T.M.: An algorithmic overview of surface registration techniques for medical imaging. Medical Image Analysis **4**(3): 201–217, 2000.

[72] Perperidis D., Mohiaddin R.H., Rueckert D.: Spatio-temporal free-form registration of cardiac MR image sequences. Medical Image Analysis **9**(5): 441–456, 2005.

[73] Mollemans W., Schutyser F., Van Cleynenbreugel J., Suetens P.: Tetrahedral mass spring model for fast soft tissue deformation. In: Surgery Simulation and Soft Tissue Modeling, pp. 1002–1003, 2003.

[74] Ku D.N., Giddens D.P., Zarins C.K., Glagov S.: Pulsatile flow and atherosclerosis in the human carotid bifurcation. positive correlation between plaque location and low oscillating shear stress. Arterioscler. Thromb. Vasc. Biol. **5**(3): 293–302, 1985.

[75] Consolini M., Passerini T., Veneziani A., Taylor R.W.: Angiotensin II and Shear Stress in the Development and Localization of Abdominal Aortic Aneurysms. in preparation, 2011.

[76] Titaud O., Vidard A., Souopgui I., Le Dimet F.X.: Assimilation of image sequences in numerical models. Tellus A **62**(1): 30–47, 2010.

[77] Giuliani E.R., Gersh B.J., McGoon M.D., Hayes D.L., Schaff H.V.: Mayor Clinic Practice of Cardiolgy. Mosby Publisher, St. Louis, 1996.

[78] Manduca A., Muthupillai R., Rossman P. J., Greenleaf J. F.: Visualization of tissue elasticity by magnetic resonance elastography. Lecture Notes in Computer Science 1131: 63, 1996.

[79] Barbone P.E., Oberai A.A.: Elastic modulus imaging: some exact solutions of the compressible elastography inverse problem. Physics in Medicine and Biology **52**(6): 1577, 2007.

[80] Oberai A.A., Gokhale N.H., Goenezen S., Barbone P.E., Hall T.J., Sommer A.M., Jiang J.: Linear and nonlinear elasticity imaging of soft tissue in vivo: demonstration of feasibility. Physics in Medicine and Biology **54**(5): 1191, 2009.

[81] Gokhale N.H., Barbone P.E., Oberai A.A.: Solution of the nonlinear elasticity imaging inverse problem: the compressible case. Inverse Problems **24**(4): 045010, 2008.

[82] Nobile F., Vergara C.: An effective fluid-structure interaction formulation for vascular dynamics by generalized Robin conditions. SIAM J. Sc. Comp. **30**(2): 731–763, 2008.

[83] Perego M., Veneziani A., Vergara C.: A variational approach for estimating the compliance of the cardiovascular tissue: An inverse fluid-structure interaction problem. Technical Report TR-2010-18, Department of Mathematics & CS, Emory University, www.mathcs.emory.edu, 2010. to appear in SIAM J. Sc. Comp.

[84] Perktold K., Hilbert D.: Numerical simulation of pulsatile flow in a carotid bifurcation model. Journal of Biomedical Engineering **8**(3): 193–199, 1986.

[85] Rindt C.C.M., Vosse F.N., Steenhoven A.A., Janssen J.D., Reneman R.S.: A numerical and experimental analysis of the flow field in a two-dimensional model of the human carotid artery bifurcation. Journal of Biomechanics **20**(5): 499–509, 1987.

[86] Xiu D., Karniadakis G.E.: Modeling uncertainty in flow simulations via generalized polynomial chaos. Journal of Computational Physics **187**(1): 137–167, 2003.

[87] Xiu D., Lucor D., Su C.H., Karniadakis G.E.: Stochastic modeling of flow-structure interactions using generalized polynomial chaos. Journal of Fluids Engineering **124**: 51, 2002.

[88] Xiu D., Karniadakis G.E.: Modeling uncertainty in steady state diffusion problems via generalized polynomial chaos. Computer Methods in Applied Mechanics and Engineering **191**(43): 4927–4948, 2002.

13

Quality open source mesh generation for cardiovascular flow simulations

Emilie Marchandise, Paolo Crosetto, Christophe Geuzaine,
Jean-François Remacle, and Emilie Sauvage

Abstract. We present efficient algorithms for generating quality tetrahedral meshes for cardiovascular blood flow simulations starting from low quality triangulations obtained from the segmentation of patient specific medical images. The suite of algorithms that are presented in this paper have been implemented in the open-source mesh generator Gmsh [19]. This includes a high quality remeshing algorithm based on a finite element conformal parametrization and a volume meshing algorithm with a boundary layer generation technique. In the result section, we show that the presence of a boundary layer mesh plays an important role to reduce the problem size in cardiovascular flow simulations.

13.1 Introduction

Blood flow dynamics and arterial wall mechanics are thought to be an important factor in the pathogenesis and treatment of cardiovascular diseases. A number of specific hemodynamic and vascular mechanic factors - notably wall shear stress (WSS), pressure and mural stress, flow rate, and residence time - have been implicated in aneurysm growth and rupture [8, 39] or in the pathogenesis of atherosclerosis [25]. Judicious control of these hemodynamic factors may also govern the outcomes of vascular therapies [21, 22]. Blood flow simulations with either rigid or compliant

Emilie Marchandise (✉), Emilie Sauvage, Jean-François Remacl
Université catholique de Louvain, Institute of Mechanics, Materials and Civil Engineering (iMMC), Belgium
e-mail: {emilie.marchandise,emilie.sauvage,jean-francois.remacle}@uclouvain.be

Paolo Crosetto
Ecole Polytechnique Fédérale de Lausanne (EPFL), Switzerland
e-mail: paolo.crosetto@epfl.ch

Christophe Geuzaine
Université de Liège, Department of Electrical Engineering and Computer Science, Liège, Belgium
e-mail: cgeuzaine@ulg.ac.be

Ambrosi D., Quarteroni A., Rozza G. (Eds.): Modeling of Physiological Flows.
DOI 10.1007/978-88-470-1935-5_13, © Springer-Verlag Italia 2012

walls provide a viable option for understanding the complex nature of blood flow and arterial wall mechanics and for obtaining those relevant quantities. These numerical computations require meshes describing the patient-specific three-dimensional cardiovascular geometry.

The quality of the meshes is of great importance since it impacts both on the accuracy and the efficiency of the numerical method [5, 40]. For example, it is well known that for finite element computations, the discretization error in the finite element solution increases when the angles of the mesh elements become too large [4], and that the condition number of the finite element matrix increases with small angles [16] which hinders the numerical convergence.

Two important elements have to be taken into account in order to generate a high quality tetrahedral mesh for cardiovascular simulations from medical images: (i) the quality of the triangular surface mesh, (ii) the capability to generate boundary layers meshes.

Most of the current imaging techniques allow to extract only the inner wall of the arteries, called also lumen surface. The outcome of the segmentation procedure is then a triangulation of the surface. These triangulations are however not suited for subsequent numerical simulations since they are generally oversampled and of very low quality (with poorly shaped and distorted triangles). It is then desirable to modify the initial surface mesh to generate a new one with nearly equilateral triangles of given triangle density (e.g. density based on the vessel radius). There exists mainly two approaches for surface remeshing: mesh adaptation strategies [6, 23, 41] and meshing techniques that rely on a suitable surface parametrization [7, 27]. The mesh adaptation strategies belong to the direct meshing methods and use local mesh modifications in order both to improve the quality of the input surface mesh and to adapt the mesh to a given mesh size criterion. The parametrization techniques belong to the indirect meshing approach. The initial 3D surface mesh is first parametrized onto a 2D planar surface mesh; the initial surface can then be remeshed using any 2D mesh generation procedure with any given mesh size field by subsequently mapping the new mesh back to the original surface. In this paper we first propose an efficient approach based on parametrization for recovering a high quality surface mesh from a low quality input triangulation. The parametrization technique is based on discrete finite element conformal maps [29, 34] and the density of the new mesh can be for example adapted to the vessel radius or the discrete mean curvature (for an example of a curvature adapted mesh of an aneurysm, see [35]).

The proposed quality surface remeshing algorithm is of high importance for subsequent three-dimensional blood flow simulations. Indeed, the lumen surface triangulation is most of the time taken as input for the tetrahedral mesh generator (e.g. Delaunay, Frontal), which retains the remeshed surface as the boundary of the resulting tetrahedral mesh. Hence if the surface mesh contains low quality triangles with small angles, the resulting tetrahedral mesh might contain some degenerate tetrahedra with small volumes and small dihedral angles. Those degenerate triangles may lead to large interpolation errors, and have a negative effect on the convergence rate of the solution procedure. The worst impact results in an unresolvable system of equations.

In the context of cardiovascular flow simulations, another important point concerns the generation of a mesh boundary layer that is able to capture at the vicinity of the wall derived quantities of clinical interest. While many authors still use fully unstructured isotropic tetrahedral meshes, these meshes are not efficient in terms of computational time. Indeed, they require a huge number of elements in order to have sufficiently small elements near the wall to resolve the boundary layer and to be able to capture accurately derived quantities such as WSS. Moreover, some authors have reported that these meshes can produce spurious fluctuations for the WSS [36, 38]. Boundary layer meshes permit to capture those derived quantities accurately while keeping for efficiency purposes a reasonable number of mesh elements. Those meshes can be built with an advancing layer method [10, 17, 24] that extrudes the lumen surface mesh in the inward direction. The extruded prisms are then subsequently split into tetrahedra and the remaining of the lumen volume filled with tetrahedra. For blood flow simulations with compliant walls, one has to build also the vascular wall. Using the presented mesh boundary layer technique, the vascular wall can be built by extruding the lumen surface mesh in the outward direction with a given wall thickness. For arteries, the ratio of this wall thickness h to the inside radius is typically between 0.1 and 0.15.

Sect. 13.2 presents two meshing algorithms for cardiovascular simulations. The first algorithm is a surface remeshing method based on a finite element conformal parametrization and the second is volume meshing algorithm with a mesh boundary layer generation technique. Sect. 13.3 shows mesh quality statistics for cardiovascular meshes generated with the presented algorithms. Finally simulation results are given in Sect. 13.4 that show the impact of tetrahedral mesh quality for cardiovascular flow simulations.

13.2 Methods

In this section, we present the algorithms for the generation of high quality meshes of cardiovascular districts. Our proposed meshing pipeline consists mainly in two steps, which are described in the following subsections: (i) From a given triangulation of the lumen surface, use finite element conformal parametrizations to create a new surface mesh with a higher quality and a computational mesh size field (ii) extrude the lumen surface mesh outward and inward in order to build the vascular wall and the viscous boundary layer mesh for the blood flow.

The last meshing step involves the use of a tetrahedral mesh generator to mesh the two extruded volumes as well as the remaining volume of the lumen. All the presented algorithms are implemented in the open-source mesh generator Gmsh [19] and examples can be found on the Gmsh wiki[1].

---

[1] Gmsh's wiki: https://geuz.org/trac/gmsh (username: gmsh, password: gmsh).

13.2.1 Surface remeshing with finite element conformal maps

The remeshing technique we present is based on a least square conformal parametrization of a given mesh patch S that is also called discrete surface. We first assume we have automatically split our initial triangulation into different mesh patches that satisfy the three following conditions: have zero genus, have a boundary that is made of at least one closed curve, have a moderate geometrical aspect ratio (see [29, 30] for more details). Next we compute for each patch a finite element conformal map, and then remesh the patch in the parametric space using standard 2D mesh generators with a prescribed mesh size field. Fig. 13.1 shows the surface remeshing procedure applied to part for the green mesh patch. For this example, the chosen mesh size field is a constant.

Let us define now the discrete parametrization of a mesh patch S with a conformal map. Parametrizing such a patch S is defining a map $\mathbf{u}(\mathbf{x})$ (see Fig. 13.2):

$$\mathbf{x} \in S \subset \mathbb{R}^3 \mapsto \mathbf{u}(\mathbf{x}) \in S' \subset \mathbb{R}^2 \qquad (13.1)$$

that transforms continuously a 3D patch S into a patch S' embedded in $\mathbb{R}^2$ that has a well known parametrization. The least square conformal map as introduced by Levy at al. [28] asks that the gradient of u and the gradient of v shall be as orthogonal

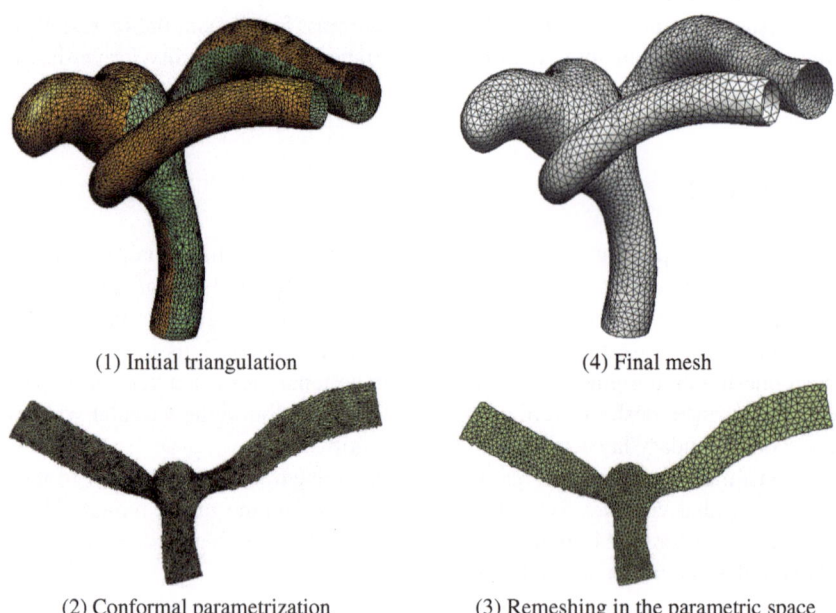

(1) Initial triangulation   (4) Final mesh

(2) Conformal parametrization   (3) Remeshing in the parametric space

Fig. 13.1. Illustration of the presented surface remeshing procedure. (1) We start from an initial triangulation that is automatically split into a minimal number of patches (green and orange patch in this case) [29, 30]. (2) Next we compute for each patch the conformal map, and (3) remesh the patch in the parametric space using standard 2D mesh generators. The final mesh is shown in (4)

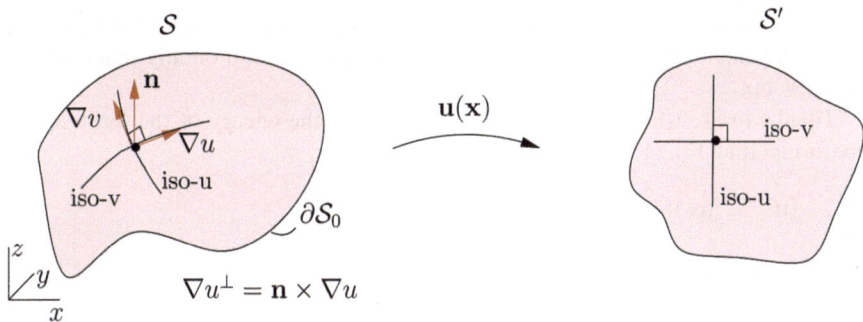

Fig. 13.2. Definitions for a conformal mapping $\mathbf{u}(\mathbf{x})$ of a 3D surface S. The boundary of S is a closed curve and is noted ∂S_0

as possible in the parametrization and have the same norm. This can bee seen as an approximation of the Cauchy-Riemann equations. For a piecewise linear mapping, the least square conformal map can be obtained by minimizing the conformal energy:

$$E_{\text{LSCM}}(\mathbf{u}) = \int_S \frac{1}{2} \left| \nabla u^\perp - \nabla v \right|^2 ds, \qquad (13.2)$$

where $\perp$ denotes a counterclockwise 90° rotation in S. For a 3D surface with normal vector $\mathbf{n}$, the counterclockwise rotation of the gradient can be written as: $\nabla u^\perp = \mathbf{n} \times \nabla u$ (see Fig. 13.2). Eq. (13.2) can be simplified and rewritten as follows:

$$\begin{aligned} E_{\text{LSCM}}(\mathbf{u}) &= \int_S \frac{1}{2} \left(\nabla u^\perp \cdot \nabla u^\perp + \nabla v \cdot \nabla v - 2\nabla u^\perp \cdot \nabla v \right) ds, \\ &= \int_S \frac{1}{2} \left(\nabla u \cdot \nabla u + \nabla v \cdot \nabla v - 2 \left(\mathbf{n} \times \nabla u \right) \cdot \nabla v \right) ds. \end{aligned} \qquad (13.3)$$

Recalling that a "dot" and a "cross" can be interchanged without changing the result, we have:

$$E_{\text{LSCM}}(\mathbf{u}) = \int_S \frac{1}{2} \left(\nabla u \cdot \nabla u + \nabla v \cdot \nabla v - 2\mathbf{n} \cdot (\nabla u \times \nabla v) \right) ds. \qquad (13.4)$$

We now derive the finite element formulation of the quadratic minimization problem (13.2):

$$\min_{\mathbf{u} \in U(S)} E_{\text{LSCM}}(\mathbf{u}), \quad \text{with} \quad U(S) = \{\mathbf{u} \in H^1(S), \ \mathbf{u} = \mathbf{u}_D(\mathbf{x}) \text{ on } \partial S_0\}, \qquad (13.5)$$

where ∂S_0 is a closed curve of S. We assume the following finite expansions for $\mathbf{u} = \{u, v\}$:

$$\mathbf{u}_h(\mathbf{x}) = \sum_{i \in I} \mathbf{u}_i \phi_i(\mathbf{x}) + \sum_{i \in J} \mathbf{u}_D(\mathbf{x}_i) \phi_i(\mathbf{x}) \qquad (13.6)$$

where I denotes the set of nodes of S that do not belong to the Dirichlet boundary, J denotes the set of nodes of S that belong to the Dirichlet boundary and where ϕ_i are the nodal shape functions associated with the nodes of the mesh. We assume here that nodal shape function ϕ_i is equal to 1 on vertex $\mathbf{x}_i$ and 0 on any other vertex: $\phi_i(\mathbf{x}_j) = \delta_{ij}$.

Thanks to (13.6) the functional E_{LSCM} defining the energy of the least square conformal map Eq. (13.3) can be rewritten as

$$E_{\text{LSCM}}(\mathbf{u}_1, \ldots, \mathbf{u}_N) =$$

$$\frac{1}{2} \sum_{i \in I} \sum_{j \in I} u_i u_j \int_S \nabla \phi_i \cdot \nabla \phi_j \, ds + \sum_{i \in I} \sum_{j \in J} u_i u_D(\mathbf{x}_j) \int_S \nabla \phi_i \cdot \nabla \phi_j \, ds +$$

$$\frac{1}{2} \sum_{i \in I} \sum_{j \in I} v_i v_j \int_S \nabla \phi_i \cdot \nabla \phi_j \, ds + \sum_{i \in I} \sum_{j \in J} v_i v_D(\mathbf{x}_j) \int_S \nabla \phi_i \cdot \nabla \phi_j \, ds +$$

$$\sum_{i \in I} \sum_{j \in J} u_D(\mathbf{x}_i) u_D(\mathbf{x}_j) \int_S \nabla \phi_i \cdot \nabla \phi_j \, ds + \sum_{i \in I} \sum_{j \in J} v_D(\mathbf{x}_i) v_D(\mathbf{x}_j) \int_S \nabla \phi_i \cdot \nabla \phi_j \, ds -$$

$$\sum_{i \in I} \sum_{j \in J} u_i v_j \int_S \mathbf{n} \cdot (\nabla \phi_i \times \nabla \phi_j) \, ds - \sum_{i \in I} \sum_{j \in J} u_D(\mathbf{x}_i) v_j \int_S \mathbf{n} \cdot (\nabla \phi_i \times \nabla \phi_j) \, ds -$$

$$\sum_{i \in I} \sum_{j \in J} u_i v_D(\mathbf{x}_i) \int_S \mathbf{n} \cdot (\nabla \phi_i \times \nabla \phi_j) \, ds - \sum_{i \in I} \sum_{j \in J} u_D(\mathbf{x}_i) v_D(\mathbf{x}_i) \int_S \mathbf{n} \cdot (\nabla \phi_i \times \nabla \phi_j) \, ds.$$

$$(13.7)$$

In order to minimize E_{LSCM}, we can simply compute the derivative of E_{LSCM} with respect to u_k for all $k \in I$.

$$\frac{\partial E_{\text{LSCM}}}{\partial u_k} = \sum_{j \in I} u_j \underbrace{\int_S \nabla \phi_k \cdot \nabla \phi_j \, ds}_{A_{kj}} + \sum_{j \in J} u_D(\mathbf{x}_j) \underbrace{\int_S \nabla \phi_k \cdot \nabla \phi_j \, ds}_{A_{kj}} -$$

$$\sum_{j \in I} v_j \underbrace{\int_S \mathbf{n} \cdot (\nabla \phi_k \times \nabla \phi_j) \, ds}_{C_{kj}} - \sum_{j \in I} v_D(\mathbf{x}_j) \underbrace{\int_S \mathbf{n} \cdot (\nabla \phi_k \times \nabla \phi_j) \, ds}_{C_{kj}} \ .$$

$$(13.8)$$

Then we solve for $(u_1, \ldots, u_N)$ such that all the derivatives are equal to zero (and similarly for $\partial/\partial v_k$)

The combination of the Eqs. (13.8) for the u_k and v_k derivatives gives us a linear system of $2I$ equations that can be written as follows:

$$\underbrace{\begin{pmatrix} \mathbf{A} & -\mathbf{C} \\ -\mathbf{C}^T & \mathbf{A} \end{pmatrix}}_{\mathbf{L_C}} \begin{pmatrix} U \\ V \end{pmatrix} = \begin{pmatrix} \mathbf{0} \\ \mathbf{0} \end{pmatrix}, \tag{13.9}$$

where $\mathbf{A}$ is a symmetric positive definite matrix and $\mathbf{C}$ is an antisymmetric matrix that are both built by assembling the elementary matrices A_{kj} and C_{kj}, and the vectors U and V denote respectively the vector of unknowns u_k and v_k. The resulting matrix $\mathbf{L_C}$ is then symmetric definite positive such that the linear system $\mathbf{L_C}U = \mathbf{0}$ can be efficiently solved using a direct sparse symmetric-positive-definite solver such as TAUCS [2].

It is necessary to impose appropriate boundary conditions to guarantee that the discrete minimization problem has a unique solution and that this unique solution defines a one-to-one mapping (and hence avoids the degenerate solution **u** =constant). For least square conformal maps, the mapping (13.9) has full rank only when the number of pinned vertices[3] is greater or equal to 2 [28]. Pinning down two vertices will set the translation, rotation and scale of the solution when solving the linear system $L_C U = 0$ and will lead to what is called a free-boundary parametrization. It was independently found by the authors of the LSCM [28] and the DCP [1] that picking two boundary vertices the farthest from each other seems to give good results in general.

13.2.2 Volume meshes with boundary layers

We have implemented an advancing layer method [10, 17, 24] for the generation of boundary layers. Those boundary layer meshes are attractive since they present high aspect ratio, orthogonal and possibly graded elements at the wall. The method starts from a surface mesh on which a boundary layer must be grown. From each surface node a direction is picked for placing the nodes of the boundary layer mesh. The direction is either computed using an estimate to the surface normal at the node using Gouraud shading, or specified directly as a three-dimensional vector field – obtained e.g. as the solution of a partial differential equation. The nodes are connected to form layers of prisms that are subsequently subdivided into tetrahedra. There technique is quite efficient in terms of computational time but cannot guarantee that there will not be any overlap at tight corners. Therefore, the user has to take care to produce elements of acceptable shape at sharp corners and to prevent element overlap in regions of tight corners.

As explained in the introduction, for cardiovascular simulations, there is a double necessity for boundary layer meshes: one for the viscous boundary layer mesh and one for the arterial wall of given thickness. These boundary layer meshes can be built by extruding outward and inward the lumen surface. Then a three-dimensional Delaunay mesh generator is called to fill the remaining of the lumen volume with isotropic tetrahedra. Fig. 13.3 shows an example of volume mesh with boundary layers that is well suited for blood flow simulations in compliant vessels.

It should be noted however that for realistic blood flow simulations, the thickness of the viscous boundary layer mesh and the mesh resolution for the inner tetrahedra are often unknown prior to the computation. An effective approach to overcome this difficulty is to start from the pre-defined boundary layer meshes as depicted in Fig. 13.3 and to apply an adaptive procedure [9, 36, 37] where the distribution of the spatial discretization errors are estimated and controlled by modifying the mesh resolution. For example, in the case of unsteady blood flow simulations, one could

---

[2] http://www.tau.ac.il/ stoledo/taucs/.

[3] A pinned vertex i is a vertex for which we have fixed the values of the mapping u_i and v_i.

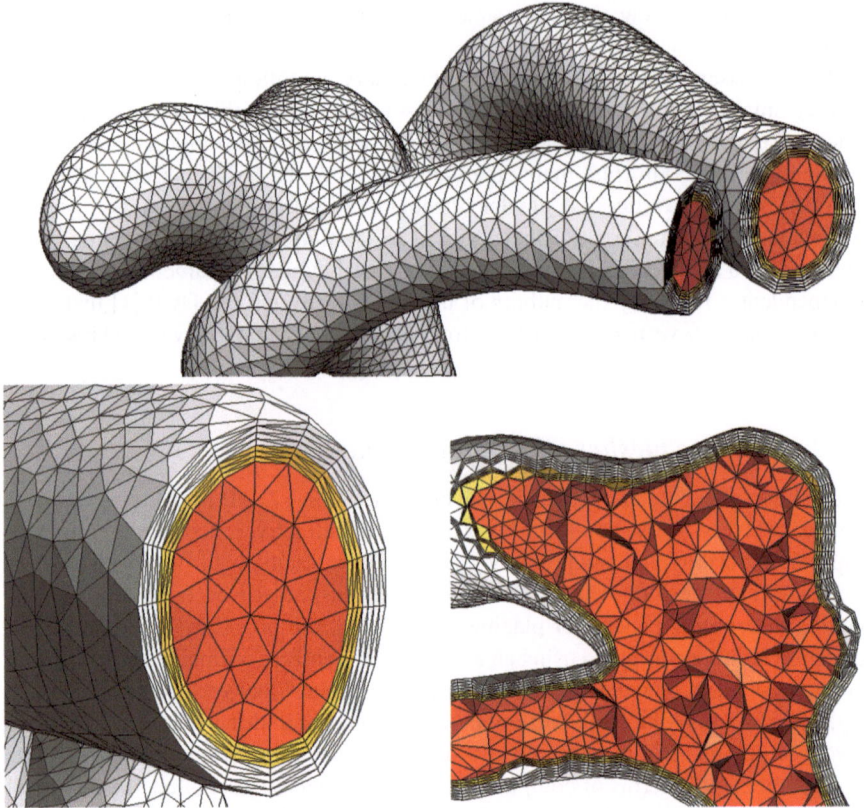

Fig. 13.3. Magnified views of the boundary layer volume mesh of an aneurysm. The white volume is a boundary layer mesh of the arterial wall, and the red and yellow volumes represent the arterial lumen. The yellow volume is the fluid boundary layer mesh that is built in order to capture accurately the wall shear stresses during the blood flow simulations and the red volume is the remaining of the lumen volume that is filled with isotropic tetrahedra

rely the adaptation of the mesh size field on an Hessian strategy [26] of the average flow speed over one cardiac cycles [36, 37].

13.3 High quality meshes

The aim of the mesh generation algorithms described in this paper is to build a mesh made of triangles and tetrahedra that have controlled element sizes and shapes.

As in this work the initial triangulations are remeshed with a given isotropic mesh size field $h(\vec{x})$, the quality of remeshed surfaces can be evaluated by computing the aspect ratio of every mesh triangle κ as follows [19]:

$$\gamma_\kappa = \alpha \frac{\text{inscribed radius}}{\text{circumscribed radius}} = 4 \frac{\sin \hat{a} \sin \hat{b} \sin \hat{c}}{\sin \hat{a} + \sin \hat{b} + \sin \hat{c}}, \qquad (13.10)$$

$\hat{a}, \hat{b}, \hat{c}$ being the three inner angles of the triangle. With this definition, the equilateral triangle has $\gamma_\kappa = 1$ and degenerated (zero surface) triangles have $\gamma_\kappa = 0$.

In order to measure the quality of the isotropic tetrahedral elements, we define another quality measure γ_τ based also on the radius ratio of the mesh element (tetrahedron) [18, 19]:

$$\gamma_\tau = \frac{6\sqrt{6}V_\tau}{S_F^{sum} L_E^{max}},$$

V_τ being the volume of the tetrahedron τ, S_F^{sum} being the sum of the areas of the 4 faces of the tetrahedron, and L_E^{max} being the maximum edge length of the 6 edges of the tetrahedron. This γ_τ quality measure lies in the interval $[0, 1]$, an element with $\gamma_\tau = 0$ being a sliver (zero volume).

We analyze the quality of the the lumen triangulations obtained with our remeshing algorithm based on finite element conformal maps. The quality of the volume meshes used for the simulations will be presented in the next section.

Fig. 13.4 shows two different steps in the parametrization-based remeshing algorithm of an initial triangulation of an iliac artery bifurcation[4] First the initial mesh is cut into different patches using the multiscale Laplacian partitioning method

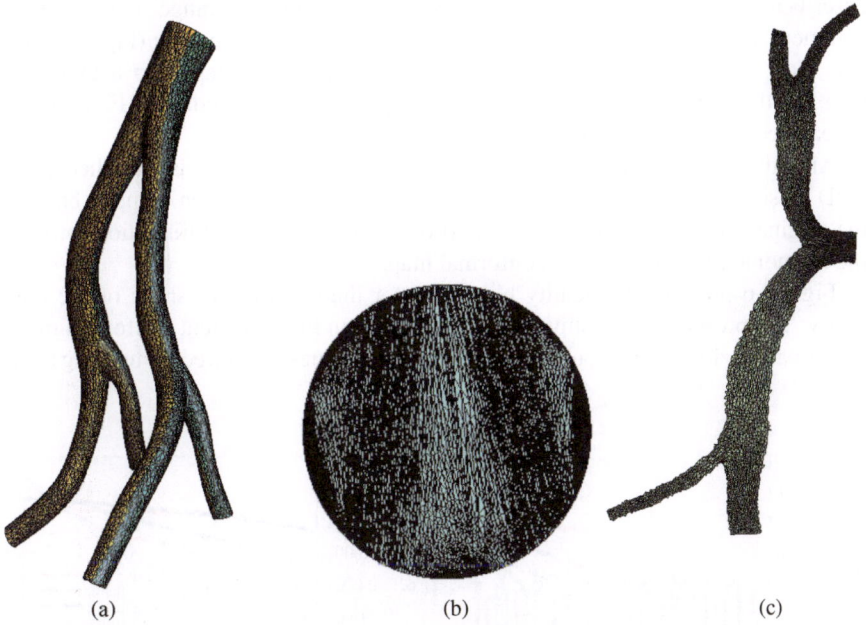

(a)   (b)   (c)

Fig. 13.4. Remeshing of an iliac bifurcation. The initial mesh is first split into two parts using the multiscale Laplacian partitioning method described in [30] (a). Each of those two parts is then mapped in the parametric space by computing a Laplacian harmonic map onto a unit disk (b) and the presented conformal map with open boundaries (c)

---

[4] It should be noted that the procedure is fully automatic as implemented in the open-source software Gmsh.

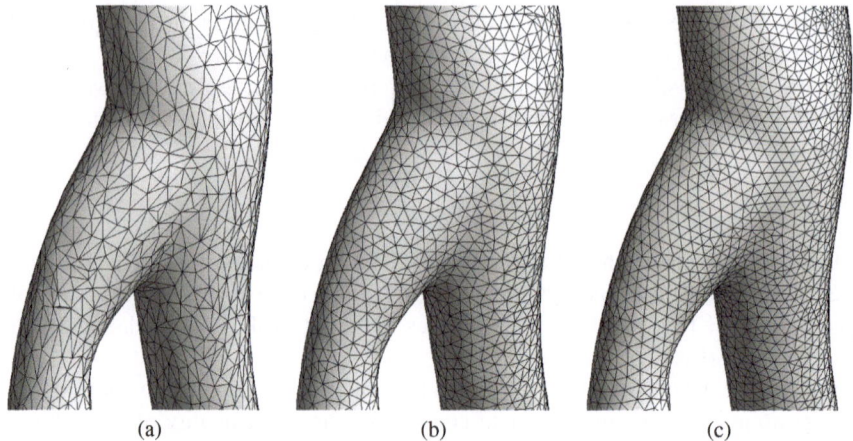

(a)   (b)   (c)

Fig. 13.5. Remeshing of an iliac bifurcation: a) part of the initial STL triangulation, b) remeshed geometry with the harmonic mapping using the MeshAdapt meshing algorithm, c) remeshed geometry with the presented conformal mapping using a frontal 2D meshing algorithm

described in [30] (Fig. 13.4a). Next, each mesh partition (orange and green) is parametrized onto a surface in $\mathbb{R}^2$ with a specific mapping algorithm (Fig. 13.4bc). We show two different mappings: a Laplacian harmonic map onto a unit disk (Fig. 13.4b) and the presented conformal map with open boundaries (Fig. 13.4c). As can be seen, the conformal mapping is much less distorted.

After the mapping has been computed, the parametrized surface is remeshed using a 2D mesh generation algorithm and the new triangulation is then mapped back to the original surface. Fig. 13.5 shows part of the remeshed iliac bifurcation for both the harmonic mapping and the conformal map.

Fig. 13.6 presents the quality histogram for the surface remeshing of the iliac artery. The presented remeshing algorithm based on finite element conformal maps is compared with different surface remeshing techniques: two direct remeshing tech-

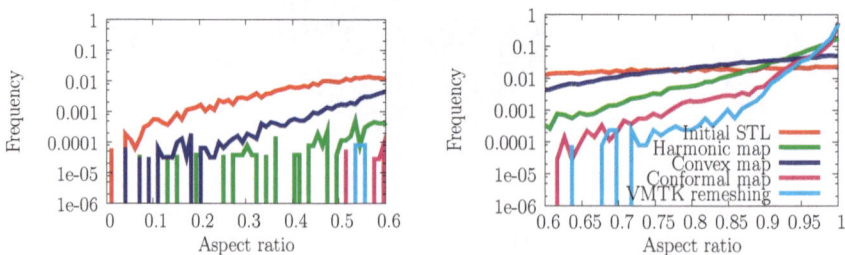

Fig. 13.6. Quality histogram for the remeshing of an iliac bifurcation and comparison of our remeshing method with other remeshing techniques based on parametrization. The left figure shows a zoom for range of small aspect ratio $\gamma_\kappa \in [0:0.6]$ and the right figure the range of high aspect ratio $\gamma_\kappa \in [0.6:1]$

Table 13.1. Quality of the surface mesh using different planar mesh generators for the remeshing of the parametric space computed with the conformal map. The qualities we look at are the the minimum aspect ratio γ_κ^{min} and the mean aspect ratio $\overline{\gamma_\kappa}$

Mesh generator	Surface quality	
	γ_κ^{min}	$\overline{\gamma_\kappa}$
Delaunay	0.18	0.966
MeshAdapt	0.36	0.972
Frontal	0.55	0.978

niques based respectively on the harmonic mapping [13, 34] and the convex combination map of Floater [14]) and one direct remeshing algorithm based on mesh adaptations (implemented in VMTK [2, 3]). As can be seen in Fig. 13.4, the harmonic map (and also the convex combination map) is a parametrization with fixed boundaries mapped on a unit circle in contrast with the conformal mapping which is a mapping with open boundaries. The quality histograms of Fig 13.6 show that our remeshing procedure based on conformal maps renders the highest quality mapping and has less small elements than the two other parametrization-based remeshing methods have. Furthermore, it performs as well as the direct remeshing algorithm of VMTK. Our surface remeshing algorithm is however more robust since it allows to remesh any kind of surface. For example, VMTK is not able to remesh the skull and pelvic surfaces but it is possible with our parametrization-based techniques [29].

An important element in the surface remeshing algorithm is the choice of the planar mesh generator to remesh the parametrized surface (see Figs. 13.1(2) and 13.1(3)). In Table 13.1, we compare the quality of the iliac surface meshes using three different planar mesh generators implemented in Gmsh: a Frontal-Delaunay algorithm [33], a planar Delaunay algorithm [18] and an algorithm based on local mesh adaptation (called MeshAdapt, see [19] for more details). Table 13.1 shows clearly that the best planar mesh generator for the conformal mapping is the Gmsh's Frontal-Delaunay algorithm. This is not a surprise: frontal techniques tend to produce meshes that are aligned with principal directions. If the planar domain that has to be meshed is equipped with a metric that conserves angles (i.e. when the mapping is conformal), then the angle between the principal directions is conserved. The use of conformal mapping helps therefore to obtain better results from the mesh generator, enabling us to produce high quality meshes.

13.4 Results

The numerical simulations aim at studying the effects of the mesh quality and the mesh algorithm on an important clinical indicator such as the Wall-Shear Stress (WSS). The comparison on different meshes shows how the mesh quality, and in particular wether accounting or not for the fluid boundary layer, affects the reliability of the simulations.

The blood flow simulations are performed with the open-source parallel finite element library LifeV[5] considering first an example with rigid walls, and then one with compliant walls. The fluid problem is discretized in space using P1-P1 finite elements stabilized with the interior penalty technique and discretized in time with an implicit Euler scheme. When considering compliant walls (FSI simulations), the wall structure is discretized in space with P1 finite elements and in time with a Newmark second order scheme. The FSI solver relies on a strongly coupled algorithm. More details on the FSI solver can be found in [12].

Blood is modelled as a Newtonian fluid and if compliant vascular walls are considered, the wall structure is modelled as a linear elastic material.

13.4.1 Unsteady blood flow in a rigid aortic arch

We study the unsteady blood flow in the same rigid aortic arch. We consider two different high quality tetrahedral meshes: one with a viscous boundary layer (such as depicted in yellow in Fig. 13.3) and one fully unstructured. Those two meshes have the same number of nodes (16k) and rely on a quality lumen surface triangulation that is obtained by remeshing the initial STL triangulation with a mesh size field h that is function of the vessel radius $R(\vec{x})$ (see Fig. 13.7b):

$$h(\vec{x}) = \frac{2\pi R(\vec{x})}{20}. \tag{13.11}$$

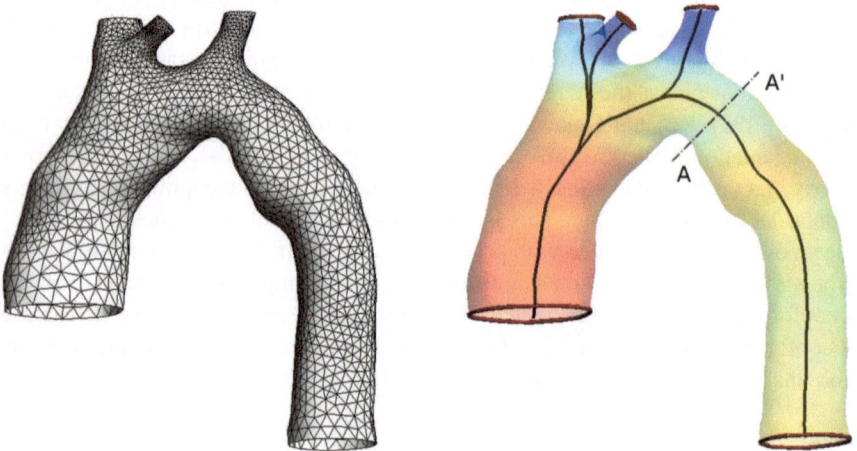

Fig. 13.7. Remeshed surface (Left) of the lumen of the aortic arch with a mesh size field h that is function of the vessel radius R. The right figure shows the computed centerlines (in black) and the distance of each surface point $\vec{x}$ to the centerlines, i.e the vessel radius $R(\vec{x})$ (the color corresponds to the magnitude of the vessel radius)

---

[5] http://www.lifev.org

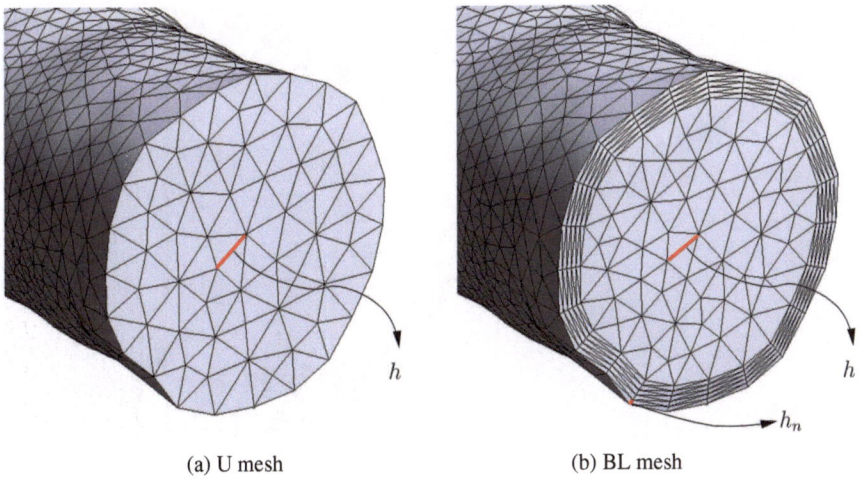

(a) U mesh   (b) BL mesh

Fig. 13.8. Definition of mesh size h and the mesh size normal to the wall h_n for (a) an unstructured tetrahedral mesh and (b) a boundary layer mesh. For this example, we have taken a value of $h_n = h/4$

The vessel radius is evaluated in Gmsh as the Euclidian distance to the centerlines (see Fig. 13.7a) which are extracted with the open source VMTK library [2, 3]. We consider two different meshes (see Fig. 13.8): (a) a fully unstructured mesh (called U mesh) of size h and (b) a boundary layer mesh (called BL mesh) with 4 layers having a mesh size normal to the wall $h_n = h/4$ and filled with tetrahedra of mesh size h. The U mesh has 6.10^4 tetrahedra while the BL mesh has 2.10^5 tetrahedra. We have also build a U mesh that has a similar mesh size than the BL mesh at the boundary (mesh size of $h/4$). This fine U mesh is made of $1.5.10^6$ tetrahedra.

The blood flow simulation is conducted over a time-span of four cardiac cycles with a realistic time-dependent inlet blood flow rate. Unlike in the steady case, flow rate boundary conditions are prescribed [31] at the different outlets with a magnitude of respectively $15\%, 10\%, 10\%$ and 65% of the inlet flow rate. We kept the same settings for the linear solver as in the steady case, except for the ILU preconditioner which was replaced by an algebraic multigrid preconditioner (implemented in LifeV as a wrapper to the ML package within the Trilinos library [6]). The coarse multigrid level was resolved with LU decomposition, and at the fine level Chebychev iterations were used as smoother.

Fig. 13.10 shows a mesh convergence analysis of the mean WSS at section AA' for those two different types of meshes We also compare the shear vector forces at peak systole along the circumference at section AA' (see Fig. 13.7) computed on both meshes. The results obtained on the boundary layer mesh (Fig. 13.9) show higher value of the wall shear stress than on unstructured mesh. This is due to the fact that for the fully unstructured meshes, the mesh adjacent to the wall is not fine enough to

---

[6] http://trilinos.sandia.gov

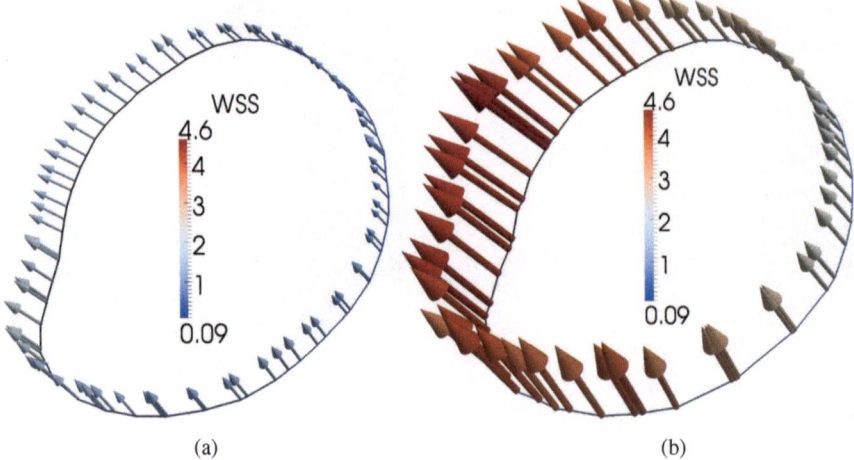

(a)   (b)

Fig. 13.9. Wall shear stress in $[dyn/cm^2]$ on the contour of the section $A - A'$: mesh without (a) and with (b) boundary layer

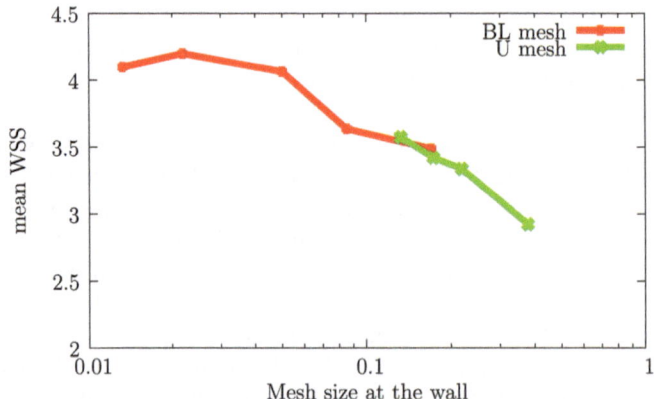

Fig. 13.10. Mesh convergence analysis for the mean WSS at peak systole along circumference AA' for boundary layer meshes and fully unstructured meshes of variable mesh size at the wall h_n

resolve the boundary layer flow. This in turn leads to a diffused boundary layer and hence a lower velocity gradient and WSS value.

It is interesting to note that unlike the simulations of Sahni et al. [36, 37, 38] we did not observe any oscillatory distribution of the WSS in the case of fully unstructured meshes (see the distribution of the WSS in Fig. 13.9a). This is probably due to higher quality of the generated tetrahedral meshes.

13.4.2 *Unsteady blood flow in a compliant femoral bypass*

For this test case, the geometric model of the distal anastomosis of a femoral bypass is obtained through a 3T MRI scanner of the left lower limb of a patient. The lumen geometry is subsequently reconstructed in 3D from the raw medical images using the open source software 3D slicer (see Fig. 13.11). The vascular wall is obtained by extruding the lumen surface in the outward normal direction with a given wall thickness of h of a tenth of the vessel radius. This extruded volume is then divided in 4 layers (see white volume in Figs. 13.11b and 13.11c).

For the meshing of the lumen volume (see red volume in Fig. 13.11b), we have considered different tetrahedral meshes: three fully unstructured meshes and three meshes with a fluid boundary layer (see table 13.2). We have considered three different mesh sizes h_{fine}, h_{medium} and h_{coarse} that are such that $h_{coarse}/h_{medium} = h_{medium}/h_{fine} = 2$.

The boundary conditions imposed at the inlet-outlet of the vessel are patient-specific measured flow rates (Fig. 13.11b), while an homogeneous Dirichlet condition is imposed on the occluded branch. The flow rate is imposed through a Lagrange multiplier as defective boundary conditions [15]. The Young modulus and Poisson coefficient characterizing the elastic material modelling the arterial wall are respectively $E = 4 \cdot 10^6 \, dyne/cm^2$ and $v = 0.45$. The fluid dynamic viscosity

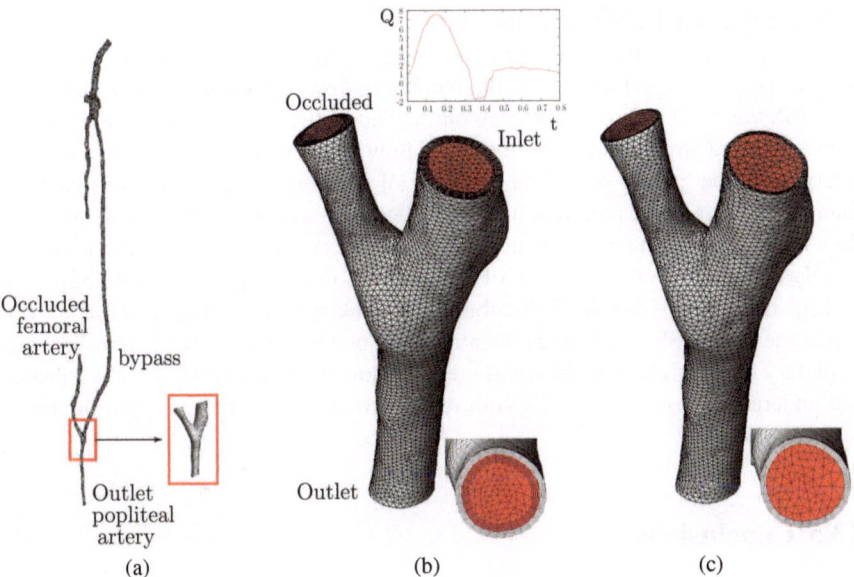

Fig. 13.11. Segmentation of the arteries of the left lower limb of a patient with a venous graft that bypasses the occluded femoral artery. We have focused on the geometry of the distal anastomosis. Two different type of meshes are considered for the lumen volume (in red): b) meshes with a viscous boundary layer (BL meshes) and c) fully unstructured tetrahedral meshes (U meshes). The mesh of the vascular wall is colored in white. The presented meshes are of middle mesh size (see Table 13.2)

Table 13.2. Different meshes considered for the numerical simulation: the boundary layer meshes (BL mesh) such as in Fig. 13.11b) and the fully unstructured meshes (U mesh) such as in Fig. 13.11c) with three different mesh sizes h: fine, middle and coarse. For the BL meshes, the ratio h/h_n is taken to be 10

BL Mesh	Fine	Middle	Coarse
# Nodes	212.633	57.318	23.697

U Mesh	Fine	Middle	Coarse
# Nodes	154.732	36.490	8.508

is $\mu = 0.035\,g/(cm\,s^2)$. The densities for blood and arterial wall are respectively $\rho_f = 1\,g/cm^3$ and $\rho_s = 1.2\,g/cm^3$. Timings and validation for FSI solver used were already discussed in [12], while in [11] the scalability issue was addressed. The simulations reported in this work were run on the Cray XT6 supercomputer in the UK National Supercomputing Service HECToR[7].

The FSI simulations are computationally expensive, so that they are run in parallel. As an example, the middle meshes are run on 48 cores, using 24 MPI processes per node; and the simulation takes about 8 hours to run for one heartbeat with the fine mesh. Due to the computational cost and since we are interested in the comparison of the WSS for the different meshes we did not run the simulation for several heartbeats, which would be necessary to reach periodicity and to obtain physiological results. We just ran for one heartbeat starting from a zero initial condition and we compared systolic hemodynamic values.

Fig. 13.12 shows the results obtained with the boundary layer mesh of middle size. The streamlines show clearly the secondary flows which are in agreement with the WSS values. The flow impinging on the bed of the junction creates a region of high wall shear stress. Moreover, the blood flow is accelerated in the outlet popliteal artery since the graft is sewed on an artery of smaller diameter. This mismatch in diameter creates also a region of high wall shear stress near the outlet. The observed flow behavior in such an end-to-side bypass do not occur naturally in arteries and is widely implicated in the initiation of the disease formation processes [20, 32].

Fig. 13.13 shows the WSS distribution at peak systole obtained for the six different meshes. WSS shows to be quantitatively better evaluated on the meshes with boundary layer, while even the finest mesh without boundary layer shows a substantial underestimation of the WSS with respect to all the the meshes with boundary layer.

13.5 Conclusions

We have presented a pipeline for generating spatially-adapted, high-quality tetrahedral meshes from a given STL triangulation and have shown its effectiveness for different cardiovascular simulations.

---

[7] http://www.hector.ac.uk

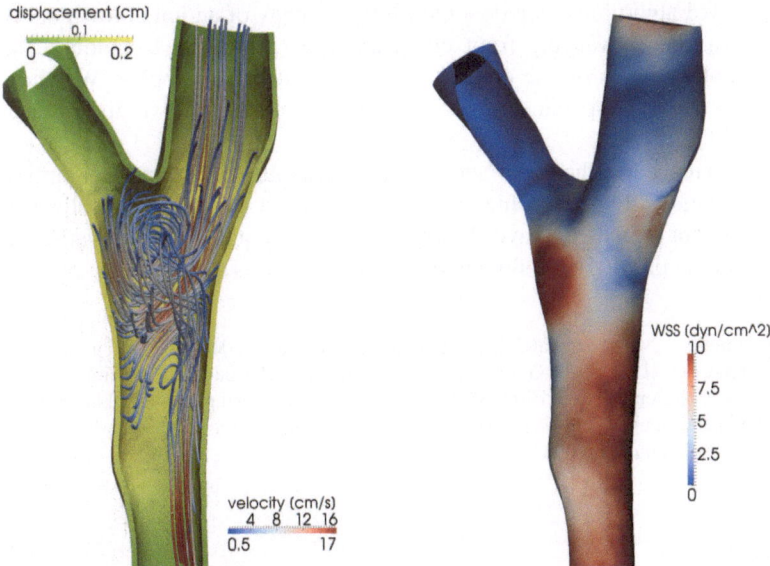

Fig. 13.12. Streamlines, wall displacement magnitude and distribution of WSS at end systole ($t = 0.3s$) obtained with the boundary layer mesh of middle mesh size

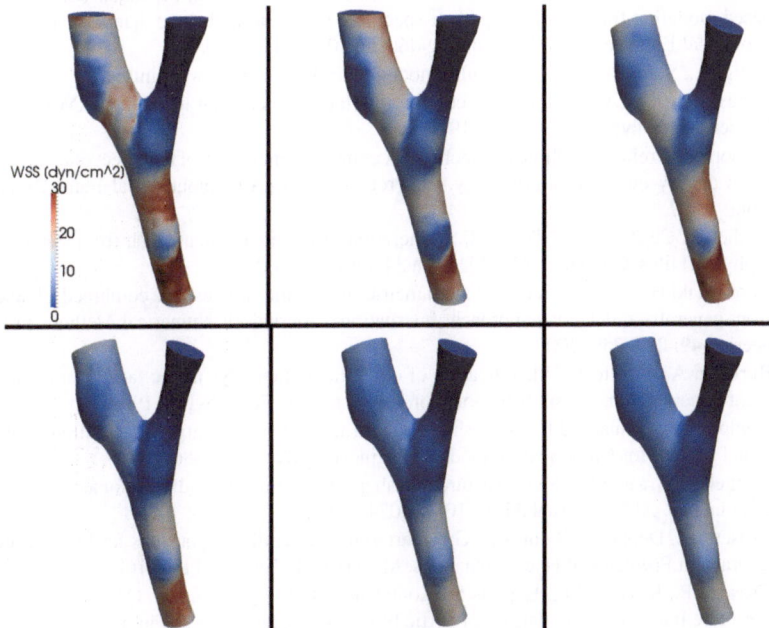

Fig. 13.13. Comparison of the WSS pattern for six meshes at peak systole ($t = 0.2s$). On the first row are the meshes with boundary layer, referring to table13.2, from left to right: FineBL, MiddleBL, CoarseBL. On the second row are the meshes without boundary layer. From left to right: Fine, Middle, Coarse

The presented algorithms include a remeshing strategy of an initial low-quality triangulation and an advancing front boundary layer generation technique. The remeshing strategy relies on the computation of a conformal mapping. We show that the remeshing in the parametric space is then optimal with a two-dimensional Frontal algorithm.

We have put to the fore the importance of the mesh quality and the mesh algorithm upon the simulations. By comparing an important clinical indicator, the wall shear stress, on different meshes,we have shown how the mesh quality, and in particular wether accounting or not for the fluid boundary layer, affects the reliability of these evaluations.

Acknowledgements. Swiss National Science Foundation under grant 200020-117587 and the European Research Council Advanced Grant "Mathcard, Mathematical Modelling and Simulation of the Cardiovascular System" Project ERC-2008-AdG 227058. The virtual pathological heart of the virtual physiological human (VPH2) project and the Swiss Platform for High-Performance and High-Productivity Computing (HP2C).

References

[1] Alliez P., Meyer M., Desbrun M.: Interactive geometry remeshing. Computer graphics (Proceedings of the SIGGRAPH 02) pp. 347–354, 2002.

[2] Antiga L., Piccinelli M., Botti L., Ene-Iordache B., Remuzzi A., Steinman D.: An image-based modeling framework for patient-specific computational hemodynamics. Medical and Biological Engineering and Computing **46**(11): 1097–1112, 2008.

[3] Antiga L., Steinman D.: The vascular modeling toolkit, http://www.vmtk.org

[4] Babuska I., Aziz A.: On the angle condition in the finite element method. SIAM Journal on Numerical Analysis **13**: 214–226, 1976.

[5] Batdorf M., Freitag L., Ollivier-Gooch, C.: A computational study of the effect of unstructured mesh quality on solution efficiency. In: Proc. 13th AIAA Computational Fluid Dynamics Conf., 1997.

[6] Bechet E., Cuilliere J.C., Trochu F.: Generation of a finite element mesh from stereolithography (stl) files. Computer-Aided Design **34**(1): 1–17, 2002.

[7] Borouchaki H., Laug P., George P.: Parametric surface meshing using a combined advancing-front generalized delaunay approach. International Journal for Numerical Methods in Engineering **49**: 223–259, 2000.

[8] Burleson A., Turitto V.: Identification of quantifiable hemody- namic factors in the assessment of cerebral aneurysm behavior. Thromb. Haemost. **76**: 118-123, 1996.

[9] Compère G., Remacle J.F.: A mesh adaptation framework for large deformations. International Journal for Numerical Methods in Engineering **82**(7): 843–867, 2009.

[10] Connell S., Braaten M.: Semistructured mesh generation for three-dimensional navier-stokes calculations. AIAA Journal **33**(6): 1017–1024, 1995.

[11] Crosetto P., Deparis S., Fourestey G., Quarteroni A.: Parallel Algorithms for Fluid-Structure Interaction Problems in Haemodynamics. MATHICSE Technical Report **8**, 2010.

[12] Crosetto P., Reymond P., Deparis S., Kontaxakis D., Stergiopulos N., Quarteroni A.: Fluid Structure Interaction simulations of aortic blood flow. Computers & Fluids, accepted, 2011.

[13] Eck M., DeRose T., Duchamp T., Hoppe H., Lounsbery M., Stuetzle W.: Multiresolution analysis of arbitrary meshes. In: SIGGRAPH '95: Proceedings of the 22nd annual conference on Computer graphics and interactive techniques, pp. 173–182, 1995.

[14] Floater M.S.: Parametrization and smooth approximation of surface triangulations. Computer aided geometric design **14**: 231–250, 1997.

[15] Formaggia L., Gerbeau J., Nobile, F, Quarteroni A.: Numerical treatment of defective boundary conditions for the navier-stokes equations. SIAM J. Numer. Anal. **40**: 376–401, 2002.

[16] Fried I.: Condition of finite element matrices generated from nonuniform meshes. AIAA Journal **10**: 219–221, 1972.

[17] Garimella R.V., Shephard M.S.: Boundary layer mesh generation for viscous flow simulations. International Journal for Numerical Methods in Engineering **49**(1): 193–218 (2000)

[18] George P.L., Frey P.: Mesh Generation. Hermes, 2000.

[19] Geuzaine C., Remacle J.F.: Gmsh: a three-dimensional finite element mesh generator with built-in pre- and post-processing facilities. International Journal for Numerical Methods in Engineering **79**(11): 1309–1331, 2009.

[20] Giddens D., Zarins C., Glagov S.: The role of fluid mechanics in localisation and detection of atherosclerosis. J Biomech Engng **115**: 588–594, 1993.

[21] Gobin Y., Counord J., Flaud P.: In vitro study of haemodynamics in a giant saccular aneurysm model: influence of flow dynamics in the parent vessel and effects of coil embolisation. J. Neuroradiology **36**: 530–536, 1994.

[22] Graves V.B., Strother C.M., Partington C.R., Rappe A.: Flow dynamics of lateral carotid artery aneurysms and their effects on coils and balloons: an experimental study in dogs. AJNR Am. J. Neuroradiol. **13**: 189–196, 1992.

[23] Ito Y., Nakahashi K.: Direct surface triangulation using stereolithography data. AIAA Journal **40**(3): 490–496, 2002.

[24] Kallinderis Y., Khawaja A., Mcmorris H.: Hybrid prismatic/tetrahedral grid generation for complex geometries. AIAA Paper **34**: 93–0669, 1996.

[25] Krams R., Wentzel J., Oomen J., Vinke R., Schuurbiers J., de Feyter P., Serruys P., Slager C.: Evaluation of endothelial shear stress and 3d geometry as factors determining the development of atherosclerosis and remodeling in human coronary arteries in vivo. Arteriosclerosis, Thrombosis, and Vascular Biology **17**: 2061–2065, 1997.

[26] Kunert G.: Anisotropic mesh construction and error estimation in the finite element method. Numer. Methods Partial Differential Equations **18**: 625–648, 2000.

[27] Laug P., Boruchaki H.: Interpolating and meshing 3d surface grids. International Journal for Numerical Methods in Engineering **58**: 209–225, 2003.

[28] Levy B., Petitjean S., Ray N., Maillot J.: Least squares conformal maps for automatic texture atlas generation. In: Computer Graphics (Proceedings of SIGGRAPH 02), pp. 362–371, 2002.

[29] Marchandise E., Compère G., Willemet M., Bricteux G., Geuzaine C., Remacle J.F.: Quality meshing based on stl triangulations for biomedical simulations. International Journal for Numerical Methods in Biomedical Engineering **83**: 876–889, 2010.

[30] Marchandise E., Carton de Wiart C., Vos W., Geuzaine C., Remacle J.F.: High quality surface remeshing using harmonic maps. Part II: Surfaces with high genus and of large aspect ratio. International Journal for Numerical Methods in Engineering, online 2011.

[31] Nicoud F., Schoenfeld T.: Integral boundary conditions for unsteady biomedical cfd applications. Int. J. Numer. Meth. Fluids **40**: 457–465, 2002.

[32] Noori N., Scherer R., Perktold K.: Blood flow in distal end-to-side anastomoses with ptfe and a venous patch: results of an in vitro flow visualisation study. Eur. J. Vasc. Endo. Vasc. Surg. **18**: 191–200, 1999.

[33] Rebay S.: Efficient unstructured mesh generation by means of Delaunay triangulation and bowyer-watson algorithm. Journal of Computational Physics **106**: 25–138, 1993.

[34] Remacle J.F., Geuzaine C., Compère G., Marchandise E.: High quality surface remeshing using harmonic maps. International Journal for Numerical Methods in Engineering **83**: 403–425, 2010.

[35] Remacle J.F., Lambrechts J., Seny B., Marchandise E.: Blossom-quad: a non-uniform quadrilateral mesh generator using a minimum cost perfect matching algorithm. International Journal for Numerical Methods in Engineering, submitted 2011.

[36] Sahni O., Jansen K., Shephard M.: Automated adaptive cardiovascular flow simulations. Engineering with Computers **25**(1): 25–36, 2009.

[37] Sahni O., Jansen K.E., Shephard M.S., Taylor C.A., Beall M.W.: Adaptive boundary layer meshing for viscous flow simulations. Eng. with Comput. **24**(3): 267–285, 2008.

[38] Sahni O., Mueller J., Jansen K., Shephard M., Taylor C.: Efficient anisotropic adaptive discretization of cardiovascular system. Comp. Meth. Appl. Mech. Eng. **195**: 5634–5655, 2006.

[39] Shaaban A.M., Duerinckx A.: Wall shear stress and early atherosclerosis. American Journal of Roentgenology **174**: 1657–1665, 2000.

[40] Szczerba D., McGregor R., Szekely G.: High quality surface mesh generation for multi-physics bio-medical simulations. In: Computational Science – ICCS 2007, vol. 4487, pp. 906–913. Springer Berlin, 2007.

[41] Wang D., Hassan O., Morgan K., Weatheril, N.: Enhanced remeshing from stl files with applications to surface grid generation. Commun. Numer. Meth. Engng. **23**: 227–239, 2007.

MS&A – Modeling, Simulation and Applications

Series Editors:

Alfio Quarteroni
MOX – Politecnico di Milano (Italy)
and
École Polytechnique Fédérale
de Lausanne (Switzerland)

Tom Hou
California Institute of Technology
Pasadena (USA)

Claude Le Bris
École des Ponts ParisTech
Paris (France)

Anthony T. Patera
Massachusetts Institute of Technology
Cambridge (USA)

Enrique Zuazua
Basque Center for Applied
Mathematics
Bilbao (Spain)

Editor at Springer:

Francesca Bonadei
francesca.bonadei@springer.com

THE ONLINE VERSION OF THE BOOKS PUBLISHED IN THE SERIES IS AVAILABLE ON SpringerLink

1   L. Formaggia, A. Quarteroni, A. Veneziani (eds.)
    Cardiovascular Mathematics
    2009, XIV+522 pp, ISBN 978-88-470-1151-9

2. A. Quarteroni (ed.)
    MATHKNOW
    2009, XII+264 pp, ISBN 978-88-470-1121-2

3. A. Quarteroni
    Numerical Models for Differential Problems
    2009, XVI+602 pp, ISBN 978-88-470-1070-3

4. A. Alonso Rodríguez, A. Valli
    Eddy Current Approximation of Maxwell Equations
    2010, XIV+348 pp, ISBN 978-88-470-1934-8

5. D. Ambrosi, A. Quarteroni, G. Rozza (eds.)
    Modeling of Physiological Flows
    2012, X+414 pp, ISBN 978-88-470-1934-8

For further information, please visit the following link:
http://www.springer.com/series/8377